Undergraduate Texts in Mathematics

Undergraduate Texts in Mathematics

Apostol: Introduction to Analytic Number Theory. Second edition.
Armstrong: Groups and Symmetry.
Armstrong: Basic Topology.
Bak/Newman: Complex Analysis.
Banchoff/Wermer: Linear Algebra Through Geometry. Second edition.
Brémaud: An Introduction to Probabilistic Modeling.
Bressoud: Factorization and Primality Testing.
Bressoud: Second Year Calculus.
Readings in Mathematics.
Brickman: Mathematical Introduction to Linear Programming and Game Theory.
Cederberg: A Course in Modern Geometries.
Childs: A Concrete Introduction to Higher Algebra.
Chung: Elementary Probability Theory with Stochastic Processes. Third edition.
Cox/Little/O'Shea: Ideals, Varieties, and Algorithms: An Introduction to Computational Algebraic Geometry and Commutative Algebra.
Curtis: Linear Algebra: An Introductory Approach. Fourth edition.
Dixmier: General Topology.
Driver: Why Math?
Ebbinghaus/Flum/Thomas: Mathematical Logic.
Edgar: Measure, Topology, and Fractal Geometry.
Fischer: Intermediate Real Analysis.
Flanigan/Kazdan: Calculus Two: Linear and Nonlinear Functions. Second edition.
Fleming: Functions of Several Variables. Second edition.
Foulds: Optimization Techniques: An Introduction.
Foulds: Combinatorial Optimization for Undergraduates.
Franklin: Methods of Mathematical Economics.
Halmos: Finite-Dimensional Vector Spaces. Second edition.
Halmos: Naive Set Theory.
Hämmerlin/Hoffmann: Numerical Mathematics.
Readings in Mathematics.
Iooss/Joseph: Elementary Stability and Bifurcation Theory. Second edition.
James: Topological and Uniform Spaces.
Jänich: Topology.
Kemeny/Snell: Finite Markov Chains.
Klambauer: Aspects of Calculus.
Lang: A First Course in Calculus. Fifth edition.
Lang: Calculus of Several Variables. Third edition.
Lang: Introduction to Linear Algebra. Second editon.
Lang: Linear Algebra. Third edition.
Lang: Undergraduate Algebra. Second edition.
Lang: Undergraduate Analysis.
Lax/Burstein/Lax: Calculus with Applications and Computing. Volume 1.
LeCuyer: College Mathematics with APL.
Lidl/Pilz: Applied Abstract Algebra.
Macki/Strauss: Introduction to Optimal Control Theory.
Malitz: Introduction to Mathematical Logic.
Marsden/Weinstein: Calculus I, II, III. Second edition.

(Continued after index.)

Larry Smith

Linear Algebra

Second Edition

With 21 Figures

Springer-Verlag
New York Berlin Heidelberg London Paris
Tokyo Hong Kong Barcelona Budapest

Larry Smith
Mathematisches Institut
Universität Göttingen
W3400 Göttingen
Germany

AMS Subject Classification: 15-01

Library of Congress Cataloging in Publication Data
Smith, Larry.
Linear algebra.
(Undergraduate texts in mathematics)
Includes index.
1. Algebras, Linear. I. Title. II. Series.
QA184.S63 1984 512'.5 84-5419

Typeset by Composition House Ltd., Salisbury, England.
Printed and bound by Edwards Brothers, Ann Arbor, MI.
Printed in the United States of America.

9 8 7 6 5 4 3 (Third corrected printing, 1992)

ISBN 0-387-96015-5 Springer-Verlag New York Berlin Heidelberg Tokyo
ISBN 3-540-96015-5 Springer-Verlag Berlin Heidelberg New York Tokyo

Preface

This text is written for a course in linear algebra at the (U.S.) sophomore undergraduate level, preferably directly following a one-variable calculus course, so that linear algebra can be used in a course on multidimensional calculus. Realizing that students at this level have had little contact with complex numbers or abstract mathematics, the book deals almost exclusively with real finite-dimensional vector spaces in a setting and formulation that permits easy generalization to abstract vector spaces. The parallel complex theory is developed in the exercises.

The book has as a goal the principal axis theorem for real symmetric transformations, and a more or less direct path is followed. As a consequence there are many subjects that are not developed, and this is intentional.

However, a wide selection of examples of vector spaces and linear transformations is developed, in the hope that they will serve as a testing ground for the theory. The book is meant as an *introduction* to linear algebra and the theory developed contains the essentials for this goal. Students with a need to learn more linear algebra can do so in a course in abstract algebra, which is the appropriate setting. Through this book they will be taken on an excursion to the algebraic/analytic zoo, and introduced to some of the animals for the first time. Further excursions can teach them more about the curious habits of some of these remarkable creatures.

For the second edition of the book I have added, amongst other things, a safari into the wilderness of canonical forms, where the hardy student can pursue the Jordan form with the tools developed in the preceding chapters.

Göttingen,
June 1984

LARRY SMITH

Contents

1 Vectors in the plane and space

In physics certain quantities such as *force*, *displacement*, *velocity*, and *acceleration* possess both a magnitude and a direction and they are most usually represented geometrically by drawing an arrow with the magnitude and direction of the quantity in question. Physicists refer to the arrow as a *vector*, and call the quantities so represented *vector quantities*. In the study of the calculus the student has no doubt encountered vectors, and their algebra, particularly in connection with the study of lines and planes and the differential geometry of space curves. Vectors can be described as *ordered pairs* of points (**P**, **Q**) which we call the *vector from P to Q* and often denote by $\overrightarrow{\mathbf{PQ}}$. This is substantially the same as the physics definition, since all it amounts to is a technical description of the word "arrow." **P** is called the *initial point* and **Q** the *terminal point*.

For our purposes it will be convenient to regard two vectors as being equal if they have the same length and the same magnitude. In other words we will regard $\overrightarrow{\mathbf{PQ}}$ and $\overrightarrow{\mathbf{RS}}$ as determining the *same vector* if $\overrightarrow{\mathbf{RS}}$ results by moving $\overrightarrow{\mathbf{PQ}}$ parallel to itself.

(*N.B.* Vectors that conform to this definition are called *free vectors*, since we are "free to pick" their initial point. Not all "vectors" that occur in nature conform to this convention. If the vector quantity depends not only on its direction and magnitude but its initial point it is called a *bound vector*. For example, torque is a bound vector. In the force-vector diagram represented by Figure 1.1 $\overrightarrow{\mathbf{PQ}}$ does not have the same effect as $\overrightarrow{\mathbf{RS}}$ in pivoting a bar. In this book we will consider only free vectors.)

With this convention of equality of vectors in mind it is clear that if we *fix a point* **O** *in space* called *the origin*, then we may regard all our vectors as having their initial point at **O**. The vector $\overrightarrow{\mathbf{OP}}$ will very often be abbreviated to $\vec{\mathbf{P}}$, if the point **O** which serves as the origin of all vectors is clear from

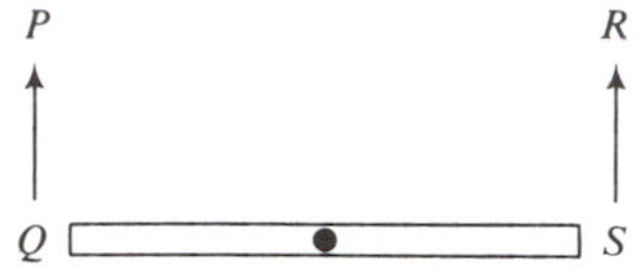

Figure 1.1

context. The vector $\vec{\mathbf{P}}$ is called the *position vector* of the point **P** relative to the origin **O**.

In physics vector quantities such as force vectors are often added together to obtain a resultant force vector. This process may be described as follows. **Suppose an origin O has been fixed**. Given vectors $\vec{\mathbf{P}}$ and $\vec{\mathbf{Q}}$ their sum is defined by the Figure 1.2. That is, draw the parallelogram determined by the three points **P**, **O** and **Q**. Let **R** be the fourth vertex and set $\vec{\mathbf{P}} + \vec{\mathbf{Q}} = \vec{\mathbf{R}}$.

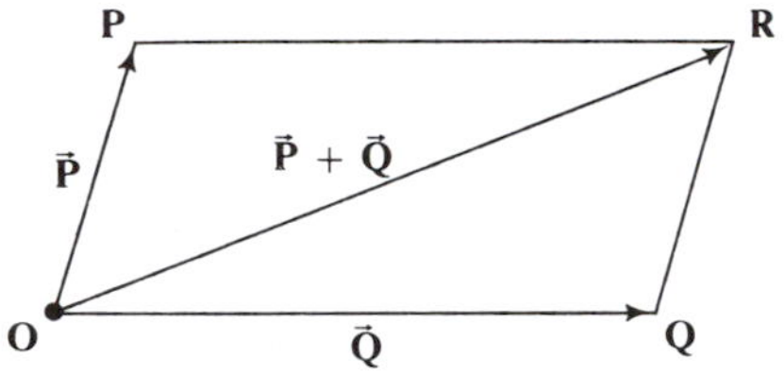

Figure 1.2

The following basic rules of vector algebra may be easily verified by elementary Euclidean geometry.

(1) $\vec{\mathbf{P}} + \vec{\mathbf{Q}} = \vec{\mathbf{Q}} + \vec{\mathbf{P}}$.

(2) $(\vec{\mathbf{P}} + \vec{\mathbf{Q}}) + \vec{\mathbf{R}} = \vec{\mathbf{P}} + (\vec{\mathbf{Q}} + \vec{\mathbf{R}})$.

(3) $\vec{\mathbf{P}} + \vec{\mathbf{O}} = \vec{\mathbf{P}} = \vec{\mathbf{O}} + \vec{\mathbf{P}}$.

It is also possible to define the operation of multiplying a vector by a number. Suppose we are given a vector $\vec{\mathbf{P}}$ and a number a. If $a > 0$ we let $a\vec{\mathbf{P}}$ be the vector with the same direction as $\vec{\mathbf{P}}$ only a times as long (see Figure 1.3). If $a < 0$ we set $a\vec{\mathbf{P}}$ equal to the vector of magnitude a times the magnitude of $\vec{\mathbf{P}}$ but having direction *opposite* of $\vec{\mathbf{P}}$ (see Figure 1.4). If $a = 0$ we set $a\vec{\mathbf{P}}$

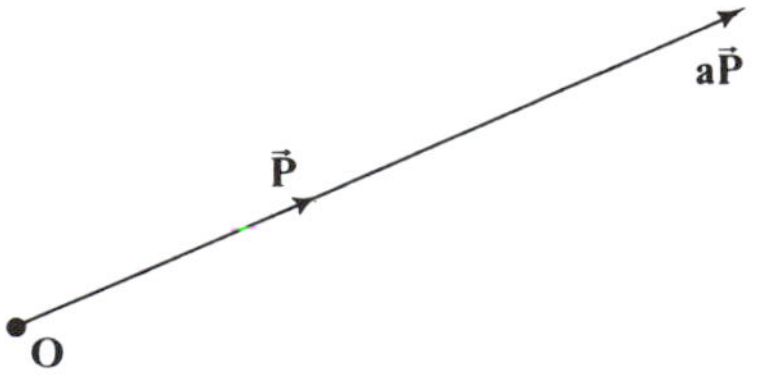

Figure 1.3

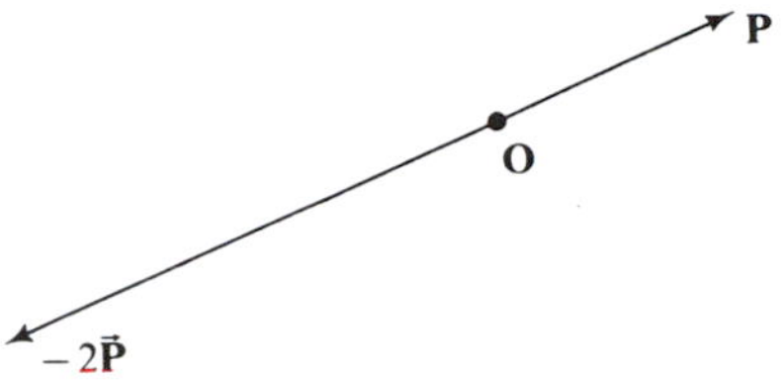

Figure 1.4

equal to $\vec{\mathbf{O}}$. It is then easy to show that vector algebra satisfies the following additional rules:

(4) $\vec{\mathbf{P}} + (-1\vec{\mathbf{P}}) = \vec{\mathbf{O}}$

(5) $a(\vec{\mathbf{P}} + \vec{\mathbf{Q}}) = a\vec{\mathbf{P}} + a\vec{\mathbf{Q}}$

(6) $(a + b)\vec{\mathbf{P}} = a\vec{\mathbf{P}} + b\vec{\mathbf{P}}$

(7) $(ab)\vec{\mathbf{P}} = a(b\vec{\mathbf{P}})$

(8) $0\vec{\mathbf{P}} = \vec{\mathbf{O}}$, $1\vec{\mathbf{P}} = \vec{\mathbf{P}}$

Note that Rule 6 involves two types of addition, namely addition of numbers and addition of vectors.

Vectors are particularly useful in studying lines and planes in space. Suppose that an origin **O** has been fixed and L is the line through the two points **P** and **Q** as in Figure 1.5. Suppose that **R** is any other point on L.

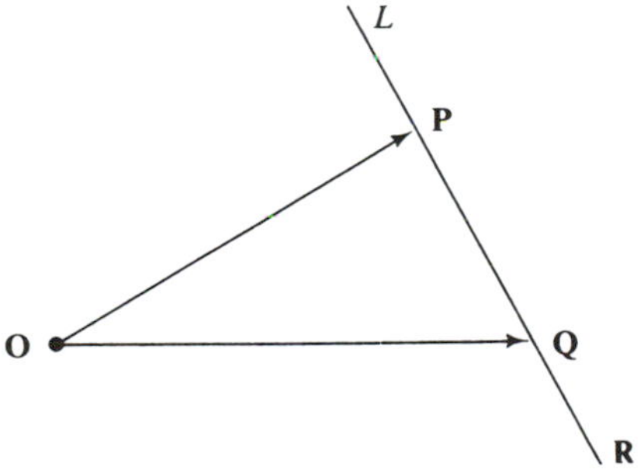

Figure 1.5

Consider the position vector $\vec{\mathbf{R}}$. Since the two points **P**, **Q** completely determine the line L, it is quite reasonable to look for some relation between the vectors $\vec{\mathbf{P}}$, $\vec{\mathbf{Q}}$, and $\vec{\mathbf{R}}$. One such relation is provided by Figure 1.6. Observe that

$$\vec{\mathbf{S}} + \vec{\mathbf{P}} = \vec{\mathbf{Q}}$$

Therefore if we write $-\vec{\mathbf{P}}$ for $(-1)\vec{\mathbf{P}}$ we see that

$$\vec{\mathbf{S}} = \vec{\mathbf{Q}} - \vec{\mathbf{P}}.$$

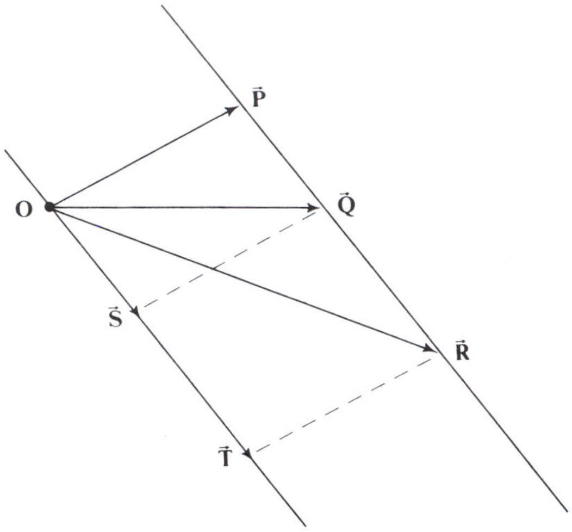

Figure 1.6

Notice that there is a number t such that

$$\vec{\mathbf{T}} = t\vec{\mathbf{S}}.$$

Moreover

$$\vec{\mathbf{R}} = \vec{\mathbf{P}} + \vec{\mathbf{T}}$$

and hence we find

$$(*) \qquad \vec{\mathbf{R}} = \vec{\mathbf{P}} + t(\vec{\mathbf{Q}} - \vec{\mathbf{P}}).$$

Equation $(*)$ is called the *vector equation* of the line L. To make practical computations with this equation it is convenient to introduce in addition to the origin **O** a cartesian coordinate system as in Figure 1.7. Every point **P** then has coordinates (x, y, z), and if we have two points **P** and **Q** with coordinates $(x_{\mathbf{P}}, y_{\mathbf{P}}, z_{\mathbf{P}})$ and $(x_{\mathbf{Q}}, y_{\mathbf{Q}}, z_{\mathbf{Q}})$ then it is quite easy to check that

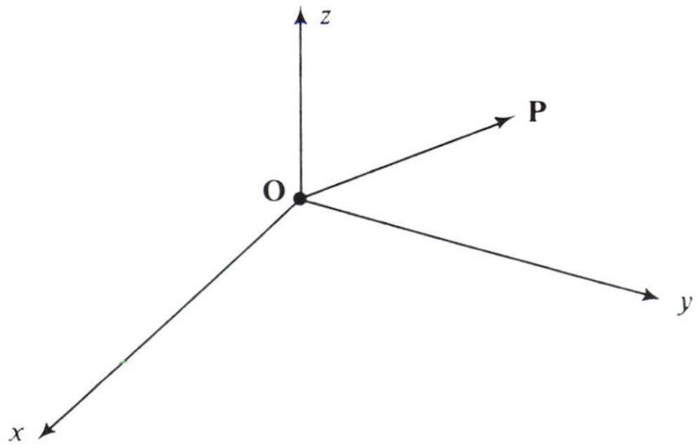

Figure 1.7

$\vec{\mathbf{P}} + \vec{\mathbf{Q}}$ is the position vector of the point with components $(x_{\mathbf{P}} + x_{\mathbf{Q}}, y_{\mathbf{P}} + y_{\mathbf{Q}}, z_{\mathbf{P}} + z_{\mathbf{Q}})$. Likewise for a number a the vector $a\vec{\mathbf{P}}$ is the position vector of the point with coordinates $(ax_{\mathbf{P}}, ay_{\mathbf{P}}, az_{\mathbf{P}})$. Thus we find by considering the coordinates of the points represented Equation (*) that (x, y, z) lies on the line L through **P**, **Q** iff

$$(**) \qquad \begin{aligned} x &= x_{\mathbf{P}} + t(x_{\mathbf{Q}} - x_{\mathbf{P}}), \\ y &= y_{\mathbf{P}} + t(y_{\mathbf{Q}} - y_{\mathbf{P}}), \\ z &= z_{\mathbf{P}} + t(z_{\mathbf{Q}} - z_{\mathbf{P}}). \end{aligned}$$

EXAMPLE 1. Does the point (1, 2, 3) lie on the line passing through the points (4, 4, 4) and (1, 0, 1)?

Solution. Let L be the line through $\mathbf{P} = (4, 4, 4)$ and $\mathbf{Q} = (1, 0, 1)$. Then the points of L must satisfy the equations

$$\begin{aligned} x &= 4 + t(1 - 4) = 4 - 3t, \\ y &= 4 + t(0 - 4) = 4 - 4t, \\ z &= 4 + t(1 - 4) = 4 - 3t, \end{aligned}$$

where t is a number. Let us check if this is possible:

$$\begin{aligned} 1 &= 4 - 3t, \\ 2 &= 4 - 4t, \\ 3 &= 4 - 3t. \end{aligned}$$

The first equation gives

$$-3 = -3t \qquad t = 1.$$

Putting this in the last equation gives

$$3 = 4 - 3 = 1$$

which is impossible. Therefore (1, 2, 3) does not lie on the line through (4, 4, 4) and (1, 0, 1).

EXAMPLE 2. Let L_1 be the line through the points (1, 0, 1) and (1, 1, 1). Let L_2 be the line through the points (0, 1, 0) and (1, 2, 1). Determine if the lines L_1 and L_2 intersect. If so find their point of intersection.

Solution. The equations of L_1 are

$$\begin{aligned} x &= 1 + t_1(1 - 1) = 1, \\ y &= 0 + t_1(1 - 0) = t_1, \\ z &= 1 + t_1(1 - 1) = 1. \end{aligned}$$

The equations of L_2 are

$$\begin{aligned} x &= 0 + (1 - 0)t_2 = t_2, \\ y &= 1 + (2 - 1)t_2 = 1 + t_2, \\ z &= 0 + (1 - 0)t_2 = t_2. \end{aligned}$$

If a point lies on both of these lines we must have

$$\begin{aligned} 1 &= t_2, \\ t_1 &= 1 + t_2, \\ 1 &= t_2. \end{aligned}$$

Therefore $t_2 = 1$ and $t_1 = 2$. Hence (1, 2, 1) is the only point these lines have in common.

EXAMPLE 3. Determine if the lines L_1 and L_2 with equations

$$L_1 \quad \begin{aligned} x &= 1 - 3t, \\ y &= 1 + 3t, \\ z &= t, \end{aligned}$$

$$L_2 \quad \begin{aligned} x &= -2 - 3t, \\ y &= 4 + 3t, \\ z &= 1 + t, \end{aligned}$$

have a point in common.

Solution. If a point (x, y, z) lies on both lines it must satisfy both sets of equations, so there is a number t_1 such that

$$\begin{aligned} x &= 1 - 3t_1, \\ y &= 1 + 3t_1, \\ z &= t_1, \end{aligned}$$

and a number t_2 with

$$\begin{aligned} x &= -2 - 3t_2, \\ y &= 4 + 3t_2, \\ z &= 1 + t_2, \end{aligned}$$

and the answer to the problem is reduced to determining if in fact two such numbers can be found, that is if the simultaneous equations

$$(*) \qquad \begin{aligned} 1 - 3t_1 &= -2 - 3t_2, \\ 1 + 3t_1 &= 4 + 3t_2, \\ t_1 &= 1 + t_2, \end{aligned}$$

have any solutions. Writing these equations in the more usual form they become

$$\begin{aligned} 3 &= 3t_1 - 3t_2, \\ -3 &= -3t_1 + 3t_2, \\ -1 &= -t_1 + t_2. \end{aligned}$$

By dividing the first equation by 3, the second by -3, and multiplying the third by -1 we get

$$\begin{aligned}1 &= t_1 - t_2,\\ 1 &= t_1 - t_2,\\ 1 &= t_1 - t_2,\end{aligned}$$

giving

$$t_1 = 1 + t_2.$$

What does this mean? It means that no matter what value of t_2 we choose there is a value of t_1, namely $t_1 = 1 + t_2$, which satisfies Equations (∗). By varying the values of t_2 we get all the points on the line L_2. For each such value of t_2 the fact that there is a (corresponding) value of t_1 solving Equations (∗) shows that every point of the line L_2 lies on the line L_1. Therefore these lines must be the same!

The lesson to be learned from this example is that the equations of a line are not unique. This should be geometrically clear since we only used two points of the line to determine the equations, and there are many such possible pairs of points.

EXAMPLE 4. Determine if the lines L_1 and L_2 with equations

$$L_1 \quad \begin{aligned}x &= 1 + t,\\ y &= 1 + t,\\ z &= 1 - t,\end{aligned}$$

$$L_2 \quad \begin{aligned}x &= 2 + t,\\ y &= 2 - t,\\ z &= 2 - t,\end{aligned}$$

have a point in common.

Solution. As in Example 3 our task is to determine if the simultaneous equations

$$(*) \qquad \begin{aligned}1 + t_1 &= 2 + t_2,\\ 1 + t_1 &= 2 - t_2,\\ 1 - t_1 &= 2 - t_2,\end{aligned}$$

has any solutions. In more usual form these equations become

$$\begin{aligned}-1 &= -t_1 + t_2,\\ -1 &= -t_1 - t_2,\\ -1 &= t_1 - t_2.\end{aligned}$$

Adding the first two equations gives

$$-2 = -2t_1,$$

so t_1 must equal 1. Putting this into the last equation we get

$$-1 = 1 - t_2,$$

so t_2 must equal 2. But substituting these values of t_1 and t_2 into either of the first two equations leads to a contradiction, namely

$$-1 = -1 + 2 = 1,$$
$$-1 = -1 - 2 = -3,$$

therefore no values of t_1 and t_2 can simultaneously satisfy Equations (*) so the lines have no point in common.

In Chapter 13 we will take up the study of solving simultaneous linear equations in detail. There we will explain various techniques and "tests" that will make the problems encountered in Examples 3 and 4 routine.

Suppose now that **P**, **Q**, and **R** are three noncolinear points. Then they determine a unique plane Π. If we introduce a fixed origin **O** then it is possible to deduce an equation that is satisfied by the position vectors of points of Π. Considering Figure 1.8 shows that

$$\vec{\mathbf{A}} - \vec{\mathbf{Q}} = s(\vec{\mathbf{P}} - \vec{\mathbf{Q}}) + t(\vec{\mathbf{R}} - \vec{\mathbf{Q}})$$

that is

$$(*) \qquad \vec{\mathbf{A}} = s(\vec{\mathbf{P}} - \vec{\mathbf{Q}}) + t(\vec{\mathbf{R}} - \vec{\mathbf{Q}}) + \vec{\mathbf{Q}}.$$

Equation (*) is called the *vector equation* of the plane Π. Compare it to the vector equation of a line. Note the presence of the two parameters s and t instead of the single parameter t

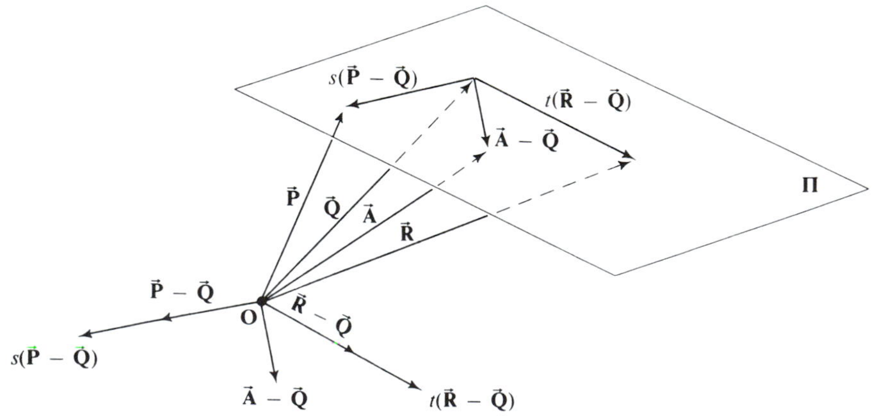

Figure 1.8

If we now introduce a coordinate system and pass to components in Equation (*) we obtain:

$$(**)\qquad \begin{aligned} x &= s(x_P - x_Q) + t(x_R - x_Q) + x_Q, \\ y &= s(y_P - y_Q) + t(y_R - y_Q) + y_Q, \\ z &= s(z_P - z_Q) + t(z_R - z_Q) + z_Q. \end{aligned}$$

We may regard Equation (**) as the equation of the plane Π or we may regard it as a system of three equations in the two unknowns s, t which we may formally eliminate and obtain the more familiar equation

$$(**)\qquad ax + by + cz + d = 0$$

where we *may* take (or twice these values, or -7 times, etc.)

$$\begin{aligned} a &= (y_R - y_Q)(z_P - z_Q) - (z_R - z_Q)(y_P - y_Q), \\ b &= (z_R - z_Q)(x_P - x_Q) - (x_R - x_Q)(z_P - z_Q), \\ c &= (x_R - x_Q)(y_P - y_Q) - (y_R - y_Q)(x_P - x_Q), \\ d &= -(ax_Q + by_Q + cz_Q). \end{aligned}$$

Equation (**) is also called the equation of the plane Π.

Example 5. Find the equation of the plane through the points

$$(1, 0, 1), \qquad (0, 1, 0), \qquad (1, 1, 1).$$

Determine if the point (0, 0, 0) lies in this plane.

Solution. We know that the equation has the form

$$ax + by + cz + d = 0$$

and all we must do is crank out values for a, b, c, d. (Remember they are not unique.) We must have

$$\begin{aligned} a + c + d &= 0, \\ b + d &= 0, \\ a + b + c + d &= 0, \end{aligned}$$

since the points (1, 0, 1), (0, 1, 0), and (1, 1, 1) lie in this plane. Thus

$$a + c = 0, \qquad d = 0, \qquad b = 0, \qquad a = -c.$$

So the plane has the equation

$$x - z = 0$$

and (0, 0, 0) lies in it.

Example 6. Determine the equation of the line of intersection of the planes

$$\begin{aligned} x - z &= 0, \\ x + y + z + 1 &= 0. \end{aligned}$$

Solution. The line in question has an equation of the form

$$\begin{aligned} x &= a + ut, \\ y &= b + vt, \\ z &= c + wt, \end{aligned}$$

for suitable numbers a, b, c, u, v, w. Since such points must lie in both planes we have

$$\begin{aligned} a + ut - (c + wt) &= 0, \\ a + ut + b + vt + c + wt + 1 &= 0, \end{aligned}$$

for all values of t. Put $t = 0$. Then

$$\begin{aligned} a - c &= 0, \\ a + b + c + 1 &= 0. \end{aligned}$$

The first equation yields $a = c$. Combining this with the second equation and setting $b = 1$ yields $2a + 2 = 0$. Hence $a = -1 = c$. Next put $t = 1$. Then

$$\begin{aligned} 0 &= a + ut - (c + wt) = -1 + u + 1 - w, \\ 0 &= a + ut + b + vt + c + wt + 1 \\ &= -1 + u + 1 + v - 1 + w + 1. \end{aligned}$$

The first equation yields $u = w$. Combining this with the second equation and setting $u = 1$ yields $w = u = 1$ and $v = -2$. Then

$$\begin{aligned} x &= -1 + t, \\ y &= 1 - 2t, \\ z &= -1 + t, \end{aligned}$$

are the equations of a line containing the two points $(-1, 1, -1)$ and $(0, -1, 0)$ which lie in both planes and hence must be the line of intersection.

Exercises

1. Suppose that an origin **O** and a coordinate system have been fixed. Let **P** be a point. Define vectors $\vec{\mathbf{E}}_1$, $\vec{\mathbf{E}}_2$, and $\vec{\mathbf{E}}_3$ by requiring that they be the position vectors of the points (1, 0, 0), (0, 1, 0), and (0, 0, 1), respectively. Let the coordinates of **P** be $(x_{\mathbf{P}}, y_{\mathbf{P}}, z_{\mathbf{P}})$. Show that

$$\vec{\mathbf{P}} = x_{\mathbf{P}}\vec{\mathbf{E}}_1 + y_{\mathbf{P}}\vec{\mathbf{E}}_2 + z_{\mathbf{P}}\vec{\mathbf{E}}_3.$$

The vectors

$$x_{\mathbf{P}}\vec{\mathbf{E}}_1, \qquad y_{\mathbf{P}}\vec{\mathbf{E}}_2, \qquad z_{\mathbf{P}}\vec{\mathbf{E}}_3$$

are called the *component vectors* of $\vec{\mathbf{P}}$ relative to the given coordinate system.

2. Find the equation of the line through the two points $(1, 0, -1)$, $(2, 3, -1)$. Does the point $(0, 1, -1)$ lie on this line?

3. Does the point $(1, 1, 1)$ lie in the plane through the points $(1, 1, 0)$, $(0, 1, 1)$, $(1, 0, 1)$?

4. Does the line through the points $(1, 1, 1)$, $(1, -1, 1)$ lie in the plane through the points $(1, -1, 0)$, $(1, 0, -1)$, $(-1, 1, 1)$?

5. Show that the point $(1, -2, 1)$ lies on the line through the two points $(0, 1, -1)$ and $(2, -5, 3)$.

6. Let $\mathbf{P} = (x_1, y_1, z_1)$, $\mathbf{Q} = (x_2, y_2, z_2)$ be two points. Show that the midpoint of the line segment $\overrightarrow{\mathbf{PQ}}$ is

$$\left(\frac{x_1 + x_2}{2}, \quad \frac{y_1 + y_2}{2}, \quad \frac{z_1 + z_2}{2}\right).$$

7. Find the equation of the line through the origin bisecting the angle formed by $\mathbf{A}\ \mathbf{O}\ \mathbf{B}$, where $\mathbf{A} = (1, 0, 0)$, $\mathbf{B} = (0, 0, 1)$.

8. Verify that vectors $\overrightarrow{\mathbf{PQ}}$ and $\overrightarrow{\mathbf{RS}}$ represent the same vector $\vec{\mathbf{T}}$ where $\mathbf{P} = (0, 1, 1)$, $\mathbf{Q} = (1, 3, 4)$, $\mathbf{R} = (1, 0, -1)$, $\mathbf{S} = (2, 2, 2)$. Find the coordinates of $\mathbf{T}$.

9. Find the sum of the vectors $\overrightarrow{\mathbf{PQ}}$ and $\overrightarrow{\mathbf{RS}}$ where $\mathbf{P} = (0, 1, 1)$, $\mathbf{Q} = (1, 0, 0)$, $\mathbf{R} = (1, 0, -1)$, $\mathbf{S} = (2, 2, 2)$.

10. Let $\mathbf{P} = (1, 1)$, $\mathbf{Q} = (2, 3)$, $\mathbf{R} = (-2, 3)$, $\mathbf{S} = (1, -1)$. Find $\overrightarrow{\mathbf{PQ}} - \overrightarrow{\mathbf{RS}}$, $\overrightarrow{\mathbf{PQ}} + \overrightarrow{\mathbf{RS}}$.

11. Show that the points **A**, **B**, **C**, **D** with the following coordinates form a parallelogram in a plane: $\mathbf{A} = (1, 1)$, $\mathbf{B} = (3, 2)$, $\mathbf{C} = (2, 3)$, $\mathbf{D} = (0, 2)$.

12. Let $\mathbf{P} = (1, 0, 1)$, $\mathbf{Q} = (1, 1, 1)$, and $\mathbf{R} = (-1, 1, -1)$. Find the coordinates of $\vec{\mathbf{T}}$ where

   (a) $\vec{\mathbf{T}} = 2\vec{\mathbf{P}} - \vec{\mathbf{Q}}$
   (b) $\vec{\mathbf{T}} = \overrightarrow{\mathbf{PQ}}$
   (c) $\vec{\mathbf{T}} = 2\vec{\mathbf{R}}$
   (d) $\vec{\mathbf{T}} = -\vec{\mathbf{R}}$
   (e) $\vec{\mathbf{T}} = \overrightarrow{\mathbf{PQ}} + \overrightarrow{\mathbf{PR}}$
   (f) $\vec{\mathbf{T}} = a\vec{\mathbf{P}} + b\vec{\mathbf{Q}} + c\vec{\mathbf{R}}$, where a, b, c are given constants.

13. In each of (a)–(g) find a vector equation of the line satisfying following conditions:

   (a) passing through the point $\mathbf{P} = (-2, 1)$ and having slope $\frac{1}{2}$
   (b) passing through the point $(0, 3)$ and parallel to the x-axis
   (c) the tangent line to $y = x^2$ at $(2, 4)$
   (d) the line parallel to the line of (c) passing through the origin
   (e) the line passing through points $(1, 0, 1)$ and $(1, 1, 1)$
   (f) the line passing through the origin and the midpoint of the line segment $\overrightarrow{\mathbf{PQ}}$ where $\mathbf{P} = (\frac{1}{2}, \frac{1}{2}, 0)$, $\mathbf{Q} = (0, 0, 1)$
   (g) the line on xy-plane passing through $(1, 1, 0)$ and $(0, 1, 0)$.

14. In each of (a)–(g) determine a vector equation of the plane satisfying the given conditions:

(a) the plane determined by (0, 0), (1, 0), and (1, 1)
(b) the plane determined by (0, 0, 1), (1, 0, 1), and (1, 1, 1)
(c) The plane determined by (1, 0, 0), (0, 1, 0), and (1, 1, 1) (Does the origin lie on this plane?)
(d) the plane parallel to the xy-plane and containing the point (1, 1, 1)
(e) the plane through the origin and containing the points $\mathbf{P} = (1, 0, 0)$, $\mathbf{Q} = (0, 1, 0)$
(f) the plane through three points **A**, **B**, **C**, where $\mathbf{A} = (1, 0, 1)$, $\mathbf{B} = (-1, 2, 3)$, and $\mathbf{C} = (2, 6, 1)$ (Does the origin lie on this plane?)
(g) the plane parallel to yz-plane passing through the point (1, 1, 1).

2 Vector spaces

In the previous chapter we reviewed the basic notions of vectors in space and their elementary application to the study of lines and planes. We derived elementary vector equations for lines and planes and saw how once a coordinate system was chosen these vector equations lead to the familiar equations of analytic geometry. However, particularly in application to physics, it is often very important to know the relation between the equations for the same plane (or line) in different coordinate systems. This leads us to the notion of a *coordinate transformation*. The appropriate domain in which to study such transformations are the abstract vector spaces to be introduced now.

Definition. A *vector space* is a set, whose elements are called vectors, together with two operations. The first operation, called *vector addition*, assigns to each pair of vectors $\mathbf{A}$ and $\mathbf{B}$ a vector denoted by $\mathbf{A} + \mathbf{B}$, called their sum. The second operation, called *scalar multiplication*, assigns to each vector $\mathbf{A}$ and each number[1] r a vector denoted by $r\mathbf{A}$. The two operations are required to have the following properties:

Axiom 1. $\mathbf{A} + \mathbf{B} = \mathbf{B} + \mathbf{A}$ for each pair of vectors $\mathbf{A}$ and $\mathbf{B}$ (Commutative law of vector addition).

Axiom 2. $(\mathbf{A} + \mathbf{B}) + \mathbf{C} = \mathbf{A} + (\mathbf{B} + \mathbf{C})$ for each triple of vectors, $\mathbf{A}$, $\mathbf{B}$ and $\mathbf{C}$.

Axiom 3. There is a unique vector $\mathbf{0}$, called the zero vector, such that $\mathbf{A} + \mathbf{0} = \mathbf{A}$ for every vector $\mathbf{A}$.

[1] For the moment we agree that the number r is a real number.

Axiom 4. For each vector $\mathbf{A}$ there corresponds a unique vector $-\mathbf{A}$ such that $\mathbf{A} + (-\mathbf{A}) = \mathbf{0}$.

Axiom 5. $r(\mathbf{A} + \mathbf{B}) = r\mathbf{A} + r\mathbf{B}$ for each real number r and each pair of vectors $\mathbf{A}$ and $\mathbf{B}$.

Axiom 6. $(r + s)\mathbf{A} = r\mathbf{A} + s\mathbf{A}$ for each pair of real numbers r and s and each vector $\mathbf{A}$.

Axiom 7. $(rs)\mathbf{A} = r(s\mathbf{A})$ for each pair r, s of real numbers and each vector $\mathbf{A}$.

Axiom 8. For each vector $\mathbf{A}$, $1\mathbf{A} = \mathbf{A}$.

In developing the mathematical theory of linear algebra we are going to follow the *axiomatic method*. That is a *vector*, *vector addition*, and *scalar multiplication* constitute the basic terms of the theory. They are not defined but rather our study of linear algebra will be based on the properties of these terms as specified by the preceding eight axioms. In the axiomatic treatment, what vectors, vector addition and scalar multiplication are is immaterial, rather what is important is the properties these quantities have as consequences of the axioms. Thus in our development of the theory we may not use properties of vectors that are not stated in or are consequences of the preceding axioms. We may use any properties of vectors, etc. that are stated in the axioms: for example, that the vector $\mathbf{O}$ is unique, or that $\mathbf{A} = 1\mathbf{A}$ for any vector $\mathbf{A}$. On the other hand we may *not* say that a vector is an arrow with a specified head and tail.

The advantage of the axiomatic approach is that results so obtained will apply to any special case or example that we wish to consider. The converse is definitely false. Presently we will see an enormous number of examples of vector spaces. Let us first begin with some elementary consequences of the axioms.

Proposition 2.1. $0\mathbf{A} = \mathbf{0}$.

PROOF. We have

$$\mathbf{A} = 1\mathbf{A} = (1 + 0)\mathbf{A} = 1\mathbf{A} + 0\mathbf{A} = \mathbf{A} + 0\mathbf{A}$$

by using Axioms 8, 6, and 8 again. To the equation

$$\mathbf{A} = \mathbf{A} + 0\mathbf{A}$$

we apply Axiom 1 getting

$$\mathbf{A} = 0\mathbf{A} + \mathbf{A}.$$

Now apply Axiom 4 and we obtain by Axiom 2:

$$\mathbf{0} = \mathbf{A} + (-\mathbf{A}) = (0\mathbf{A} + \mathbf{A}) + (-\mathbf{A}) = 0\mathbf{A} + (\mathbf{A} + (-\mathbf{A}))$$
$$= 0\mathbf{A} + \mathbf{0} = 0\mathbf{A}$$

by Axiom 3.

That is

$$\mathbf{0} = 0\mathbf{A}$$

which is the desired conclusion. □

Notational Convention: It should be clear by now that we will reserve capital letters for vectors and small letters for numbers.

The proof of (2.1) was given in considerable detail to illustrate how results are deduced by the axiomatic method. In the sequel we will not be so detailed in our proofs, leaving to the reader the task of providing as much detail as he feels needed.

Proposition 2.2. $(-1)\mathbf{A} = -\mathbf{A}$.

PROOF. We have by (2.1)

$$\mathbf{0} = 0\mathbf{A} = (1 - 1)\mathbf{A} = 1\mathbf{A} + (-1)\mathbf{A} = \mathbf{A} + (-1)\mathbf{A}.$$

Now add $-\mathbf{A}$ to both sides giving

$$-\mathbf{A} = -\mathbf{A} + (\mathbf{A} + (-1)\mathbf{A}) = (-\mathbf{A} + \mathbf{A}) + (-1)\mathbf{A}$$
$$= (\mathbf{A} - \mathbf{A}) + (-1)\mathbf{A} = \mathbf{0} + (-1)\mathbf{A} = (-1)\mathbf{A} + \mathbf{0}$$
$$= (-1)\mathbf{A}$$

as required. □

Proposition 2.3. $\mathbf{0} + \mathbf{A} = \mathbf{A}$.

PROOF. Exercise. □

These formal deductions may seem like a sterile intellectual exercise—an indication of the absurdity of too much reliance on abstraction and formalism. On the contrary, they help to point up the advantages of an abstract formulation of a mathematical theory. For if the basic terms are not defined, the possibility is opened of assigning to them content in new and unforeseen ways. If in this way the axioms become true statements when the meanings assigned to the basic terms vector, vector addition, and scalar multiplication are specified, we have constructed a **model** for the abstract theory. That is, if we can assign a meaning to the terms vector, vector addition and scalar multiplication such that Axioms 1–8 become true statements about this assignment then we say we have constructed a *model* or example of the axioms. Here then is one standard such model.

Cartesian or Euclidean spaces

Definition. Let k be a positive integer. The *Cartesian k-space* denoted by $\mathbb{R}^k$, is the set of all sequences $(a_1, a_2, \ldots, a_k)$ of k real numbers together with the two operations

$$(a_1, a_2, \ldots, a_k) + (b_1, b_2, \ldots, b_k) = (a_1 + b_1, a_2 + b_2, \ldots, a_k + b_k)$$

and

$$r(a_1, \ldots, a_k) = (ra_1, ra_2, \ldots, ra_k).$$

(In particular $\mathbb{R}^1 = \mathbb{R}$ is the set of real numbers with their usual addition and multiplication.) The number a_i is called the *i*th *component of* $(a_1, \ldots, a_k)$.

Theorem 2.4. *For each positive integer k, $\mathbb{R}^k$ is a vector space.*

Before beginning the proof of (2.4) let us consider exactly what it is that we are trying to prove. We are going to assign meanings to the three basic terms of the axioms for a vector space. Namely, by a vector we will mean a k-tuple $(a_1, \ldots, a_k)$. For vectors $\mathbf{A} = (a_1, \ldots, a_k)$ and $\mathbf{B} = (b_1, \ldots, b_k)$, the equality of the two vectors $\mathbf{A} = \mathbf{B}$ means that $a_1 = b_1, a_2 = b_2, \ldots, a_k = b_k$. By addition of $\mathbf{A}$ and $\mathbf{B}$, we shall mean the vector $(a_1 + b_1, \ldots, a_k + b_k)$, that is we define

$$\mathbf{A} + \mathbf{B} = (a_1 + b_1, \ldots, a_k + b_k).$$

Likewise we define

$$r\mathbf{A} = (ra_1, \ldots, ra_k).$$

Axioms 1–8 for a vector space then become statements about k-tuples and *we must verify that they are true statements.*

Proof of (2.4). We will verify the axioms in turn.

Axiom 1. Let $\mathbf{A} = (a_1, \ldots, a_k)$, $\mathbf{B} = (b_1, \ldots, b_k)$. Then

$$\begin{aligned}\mathbf{A} + \mathbf{B} &= (a_1 + b_1, \ldots, a_k + b_k)\\ &= (b_1 + a_1, \ldots, b_k + a_k) = \mathbf{B} + \mathbf{A}\end{aligned}$$

so Axiom 1 is true.

Axiom 2. Let $\mathbf{A} = (a_1, \ldots, a_k)$, $\mathbf{B} = (b_1, \ldots, b_k)$, and $\mathbf{C} = (c_1, \ldots, c_k)$. Then

$$\begin{aligned}(\mathbf{A} + \mathbf{B}) + \mathbf{C} &= (a_1 + b_1, \ldots, a_k + b_k) + (c_1, \ldots, c_k)\\ &= (a_1 + b_1 + c_1, \ldots, a_k + b_k + c_k)\\ &= (a_1, \ldots, a_k) + (b_1 + c_1, \ldots, b_k + c_k)\\ &= \mathbf{A} + (\mathbf{B} + \mathbf{C})\end{aligned}$$

so Axiom 2 holds.

Axiom 3. We let $\mathbf{0} = (0, \ldots, 0)$. Then for any $\mathbf{A} = (a_1, \ldots, a_k)$ we will have

$$\mathbf{A} + \mathbf{0} = (a_1 + 0, \ldots, a_k + 0) = (a_1, \ldots, a_k) = \mathbf{A}.$$

Moreover if $\mathbf{B} = (b_1, \ldots, b_k)$ is any vector such that

$$\mathbf{A} + \mathbf{B} = \mathbf{A}$$

then

$$(a_1 + b_1, \ldots, a_k + b_k) = (a_1, \ldots, a_k)$$

and therefore

$$\begin{aligned} a_1 + b_1 = a_1 \quad &\Rightarrow \quad b_1 = 0 \\ a_2 + b_2 = a_2 \quad &\Rightarrow \quad b_2 = 0 \\ &\vdots \\ a_k + b_k = a_k \quad &\Rightarrow \quad b_k = 0 \end{aligned}$$

i.e. $\mathbf{B} = \mathbf{0}$. Thus $\mathbf{0}$ is the unique vector with the property that $\mathbf{A} + \mathbf{0} = \mathbf{A}$, and Axiom 3 holds.

Axiom 4. Let $\mathbf{A} = (a_1, \ldots, a_k)$ and set $-\mathbf{A} = (-a_1, \ldots, -a_k)$. Then

$$\begin{aligned} \mathbf{A} + (-\mathbf{A}) &= (a_1, \ldots, a_k) + (-a_1, \ldots, -a_k) \\ &= (a_1 - a_1, \ldots, a_k - a_k) = (0, \ldots, 0) = \mathbf{0}. \end{aligned}$$

Moreover if $\mathbf{C} = (c_1, \ldots, c_k)$ is any vector such that

$$\mathbf{A} + \mathbf{C} = \mathbf{0}$$

then

$$(a_1 + c_1, \ldots, a_k + c_k) = (0, \ldots, 0)$$

and therefore

$$\begin{aligned} a_1 + c_1 = 0 \quad &\Rightarrow \quad c_1 = -a_1 \\ a_2 + c_2 = 0 \quad &\Rightarrow \quad c_2 = -a_2 \\ &\vdots \\ a_k + c_k = 0 \quad &\Rightarrow \quad c_k = -a_k \end{aligned}$$

i.e. $\mathbf{C} = -\mathbf{A}$. Thus $-\mathbf{A}$ is the unique vector with the property that $\mathbf{A} + (-\mathbf{A}) = \mathbf{0}$ and Axiom 4 holds.

Axiom 5. Let r be a real number and $\mathbf{A} = (a_1, \ldots, a_k)$ and $\mathbf{B} = (b_1, \ldots, b_k)$ be vectors. Then

$$\begin{aligned} r(\mathbf{A} + \mathbf{B}) &= r(a_1 + b_1, \ldots, a_k + b_k) = (r(a_1 + b_1), \ldots, r(a_k + b_k)) \\ &= (ra_1 + rb_1, \ldots, ra_k + rb_k) = (ra_1, \ldots, ra_k) + (rb_1, \ldots, rb_k) \\ &= r(a_1, \ldots, a_k) + r(b_1, \ldots, b_k) = r\mathbf{A} + r\mathbf{B} \end{aligned}$$

so that Axiom 5 is satisfied.

Axiom 6. Let r, s be numbers and $\mathbf{A} = (a_1, \ldots, a_k)$. Then

$$\begin{aligned} (r + s)\mathbf{A} &= ((r + s)a_1, \ldots, (r + s)a_k) \\ &= (ra_1 + sa_1, \ldots, ra_k + sa_k) \\ &= (ra_1, \ldots, ra_k) + (sa_1, \ldots, sa_k) \\ &= r(a_1, \ldots, a_k) + s(a_1, \ldots, a_k) \\ &= r\mathbf{A} + s\mathbf{A} \end{aligned}$$

so Axiom 6 holds.

Axiom 7. Let r, s be numbers and $\mathbf{A} = (a_1, \ldots, a_k)$. Then

$$\begin{aligned} (rs)\mathbf{A} &= (rsa_1, \ldots, rsa_k) = r(sa_1, \ldots, sa_k) \\ &= r(s(a_1, \ldots, a_k)) = r(s(\mathbf{A})) \end{aligned}$$

so Axiom 7 holds.

Axiom 8. Instant.

Therefore $\mathbb{R}^k$ is a vector space. □

Before turning to a few additional examples of vector spaces let us deduce some more elementary consequences of the axioms. One of these is the general associative law.

Proposition 2.5. *Let n be an integer $n \geq 3$. Then any two ways of associating a sum*

$$\mathbf{A}_1 + \cdots + \mathbf{A}_n$$

of n-vectors give the same vector. Consequently sums may be written without parentheses.

The proof of this proposition is elementary and may be carried out by induction on n. Similarly we have:

Proposition 2.6. *Let n be any integer ≥ 2. Then the sum of any n-vectors $\mathbf{A}_1, \ldots, \mathbf{A}_n$ is independent of the order in which the sum is taken.*

Notations. We will use the symbol $\in$ as an abbreviation for "is an element of." Thus $x \in S$ should read: x is an element of the set S.

The symbol $\subset$ is an abbreviation for "is contained in." Thus $S \subset T$ should be read: the set S is contained in the set T.

If S and T are sets then the collection of elements contained in either set is denoted by $S \cup T$. Thus $x \in S \cup T$ is equivalent to $x \in S$ *or* $x \in T$. The collection of all elements common to both sets is denoted $S \cap T$. Thus $x \in S \cap T$ is equivalent to $x \in S$ *and* $x \in T$. The set $S \cup T$ is called the **union** of S and T and $S \cap T$ the **intersection** of S and T. We denote by $\varnothing$ the empty set.

The axioms for a vector space that we have given are the axioms for a *real* vector space, that is a vector space whose scalars are the *real numbers*, which we denoted by $\mathbb{R}$. It is also possible and often important to study vector spaces whose scalars are the *complex numbers*, which we denote by $\mathbb{C}$. A vector space with complex scalars is called a *complex vector space*. The axioms for a complex vector space are exactly as for a real vector space except that the **numbers** (=scalars) are to be complex. The generic example of a complex vector space is the complex Cartesian space $\mathbb{C}^k$ of k-tuples $\mathbf{A} = (a_1, \ldots, a_k)$ of *complex* numbers where, for vectors $\mathbf{A} = (a_1, \ldots, a_k)$ and $\mathbf{B} = (b_1, \ldots, b_k)$ and for scalars $r \in \mathbb{C}$, vector addition and scalar multiplication are given by

$$\mathbf{A} + \mathbf{B} = (a_1 + b_1, \ldots, a_k + b_k)$$

and

$$r\mathbf{A} = (ra_1, \ldots, ra_k).$$

For a while at least we will study only real vectors spaces, indicating where necessary the modifications required in the complex case.

Exercises

1. Assume that the plane is equipped with a coordinate system. The set $\mathscr{V}$ of all vectors $\vec{\mathbf{P}}$ with initial point at the origin and terminal point at $\mathbf{P} = (x, y)$ of Chapter 1 is a vector space with the operation described in Chapter 1.
2. Let $A = \{(2a, a) | a \in \mathbb{R}\}$, $B = \{(b, b) | b \in \mathbb{R}\}$. Find $A \cup B$ and $A \cap B$.
3. $A = \{(2n, n) | n \in \text{integers}\}$, $B = \{(k + 1, k) | k \in \text{integers}\}$. Find $A \cup B$, $A \cap B$.
4. If $A \subset B$, $B \subset C$ then $A \subset C$.
5. If $A \subset B$, $A \subset C$ then $A \subset B \cap C$.
6. If $A \supset B$, $A \supset C$ then $A \supset B \cup C$.
7. Show $A \cap (B \cap C) = (A \cap B) \cap (A \cap C)$ and $A \cap (B \cup C) = (A \cup B) \cap (A \cap C)$, where A, B, C are sets.
8. Let $A = \{x \in \mathbb{R} | |x| > 1\}$, $B = \{x \in \mathbb{R} | -2 < x < 3\}$. Find $A \cup B$ and $A \cap B$.

9. Let $\mathscr{V}$ be a vector space. Prove each of the following statements:

(a) If $\mathbf{A} \in \mathscr{V}$ and a is a number then $a\mathbf{A} = \mathbf{0}$ iff $a = 0$ or $\mathbf{A} = \mathbf{0}$, or both.
(b) If $\mathbf{A} \in \mathscr{V}$ and a is a number then $a\mathbf{A} = \mathbf{A}$ iff $a = 1$ or $\mathbf{A} = \mathbf{0}$, or both.

10. Let V be the set of all ordered pairs of real numbers (a, b). Define an addition for the elements of V by the rule

$$(a, b) \oplus (c, d) = (a + d, b + c).$$

and a multiplication of elements of V by numbers by the rule

$$a \cdot (c, d) = (ac, d).$$

Is V, with these two operations, a vector space? Justify your answer.

11. A *translation* of the plane Π is a function $\mathsf{T}: \Pi \to \Pi$ with the following two properties:

(1) There is a constant k, called the **length** of the translation, such that for every point $p \in \Pi$ the distance from p to $\mathsf{T}(p)$ is equal to k.
(2) For any two points $p, q \in \Pi$ the distance from p to q is the same as the distance from $\mathsf{T}(p)$ to $\mathsf{T}(q)$.

Introduce cartesian coordinates in Π and show:

(a) If T is a translation, then for every $p = (x, y) \in \Pi$, $\mathsf{T}(p) = (x + l_1, x + l_2)$ where $\mathsf{T}(0, 0) = (l_1, l_2)$. T is said to be the translation by $l = (l_1, l_2)$.
(b) If $l = (l_1, l_2) \in \Pi$ show that

$$\mathsf{T}(p) = (x + l_1, y + l_2), \qquad p = (x, y) \in \Pi$$

defines a translation of Π.

If T, S are translations define their sum $\mathsf{T} \oplus \mathsf{S}$ by

$$(\mathsf{T} \oplus \mathsf{S})(p) = \mathsf{T}(\mathsf{S}(p)).$$

(c) Show that $\mathsf{T} \oplus \mathsf{S}$ is again a translation. If T is the translation by $l = (l_1, l_2)$ and a is a number define $a \cdot \mathsf{T}$ to be the translation by (al_1, al_2).
(d) Show that the set of translations with the addition $\oplus$ and scalar multiplication $\cdot$ form a vector space.

12. Show that if a vector space contains two elements then it contains infinitely many.

3 Subspaces

Definition. A nonempty subset $\mathscr{U}$ of a vector space $\mathscr{V}$ is called a *linear subspace* of $\mathscr{V}$ iff the following two conditions are satisfied:

(1) if $\mathbf{A} \in \mathscr{U}$ and $\mathbf{B} \in \mathscr{U}$ then $\mathbf{A} + \mathbf{B} \in \mathscr{U}$
(2) if $\mathbf{A} \in \mathscr{U}$ and $r \in \mathbb{R}$ then $r\mathbf{A} \in \mathscr{U}$.

These two conditions assert that applying the two basic vector operations to elements of the collection $\mathscr{U}$ give again elements of the collection $\mathscr{U}$. If the vector space $\mathscr{V}$ is complex then Condition (2) should be replaced by (2$\mathbb{C}$) if $\mathbf{A} \in \mathscr{U}$ and $c \in \mathbb{C}$ then $c\mathbf{A} \in \mathscr{U}$, and likewise in the sequel.

Proposition 3.1. If $\mathscr{U}$ is a linear subspace of the vector space $\mathscr{V}$ then $\mathscr{U}$ is itself a vector space if we define vector addition and scalar multiplication as in $\mathscr{V}$.

Proof. Notice that Conditions (1), (2) assure us that we have operations on $\mathscr{U}$, i.e., if $\mathbf{A}$ and $\mathbf{B}$ belong to $\mathscr{U}$ so do $\mathbf{A} + \mathbf{B}$ and if r belongs to $\mathbb{R}$, $r\mathbf{A}$ also belongs to $\mathscr{U}$. The properties expressed by Axioms 1, 2, 5, 6, 7, and 8 are valid for vectors in $\mathscr{V}$ and hence for vectors in the smaller set $\mathscr{U}$. To verify Axiom 3 we first show that $\mathbf{0} \in \mathscr{U}$. Since $\mathscr{U}$ is nonempty there exists at least one vector $\mathbf{A} \in \mathscr{U}$. By (2.1) and Condition (2) for a subspace $\mathbf{0} = 0\mathbf{A} \in \mathscr{U}$. Axiom 3 is now immediate since it holds in $\mathscr{V}$. To verify Axiom 4 suppose that $\mathbf{A} \in \mathscr{U}$. Then $(-1)\mathbf{A} \in \mathscr{U}$, but by (2.2), $(-1)\mathbf{A} = -\mathbf{A}$ and therefore $-\mathbf{A} \in \mathscr{U}$ and Axiom 4 holds. □

Examples

(1) $\mathscr{V}$ is always a subspace of $\mathscr{V}$.
(2) The set consisting of the zero vector alone $\{\mathbf{0}\}$ is always a subspace of $\mathscr{V}$. We often abuse notation and write $\mathbf{0}$ for this subspace.

Definition. If $\mathbf{A}_1, \ldots, \mathbf{A}_n$ are vectors of $\mathscr{V}$, then a *linear combination* of $\mathbf{A}_1, \ldots, \mathbf{A}_n$ is a vector of the form

$$\mathbf{A} = a_1\mathbf{A}_1 + \cdots + a_n\mathbf{A}_n$$

where $a_1, \ldots, a_n$ are numbers.

Definition. If the vectors $\mathbf{A}_1, \ldots, \mathbf{A}_n$ are fixed, the *linear span* of $\mathbf{A}_1, \ldots, \mathbf{A}_n$, denoted $\mathscr{L}(\mathbf{A}_1, \ldots, \mathbf{A}_n)$, is the set of all vectors of $\mathscr{V}$ which are linear combinations of $\mathbf{A}_1, \ldots, \mathbf{A}_n$.

Proposition 3.2. *Suppose that* $\mathbf{A}_1, \ldots, \mathbf{A}_n$ *are vectors of* $\mathscr{V}$, *then* $\mathscr{L}(\mathbf{A}_1, \ldots, \mathbf{A}_n)$ *is a linear subspace of* $\mathscr{V}$.

PROOF. We must verify that the two conditions of the definition of linear subspace are satisfied by the linear combinations of $\mathbf{A}_1, \ldots, \mathbf{A}_n$. So suppose

$$\mathbf{A}', \mathbf{A}'' \in \mathscr{L}(\mathbf{A}_1, \ldots, \mathbf{A}_n).$$

Then

$$\mathbf{A}' = a'_1\mathbf{A}_1 + \cdots + a'_n\mathbf{A}_n$$
$$\mathbf{A}'' = a''_1\mathbf{A}_1 + \cdots + a''_n\mathbf{A}_n$$

for suitable numbers $a'_1, \ldots, a'_n, a''_1, \ldots, a''_n$. Then using the generalized associative and commutative laws we find

$$\begin{aligned}\mathbf{A}' + \mathbf{A}'' &= a'_1\mathbf{A}_1 + \cdots + a'_n\mathbf{A}_n + a''_1\mathbf{A}_1 + \cdots + a''_n\mathbf{A}_n \\ &= a'_1\mathbf{A}_1 + a''_1\mathbf{A}_1 + \cdots + a'_n\mathbf{A}_n + a''_n\mathbf{A}_n \\ &= (a'_1 + a''_1)\mathbf{A}_1 + \cdots + (a'_n + a''_n)\mathbf{A}_n\end{aligned}$$

which shows that $\mathbf{A}' + \mathbf{A}''$ is again a linear combination of $\mathbf{A}_1, \ldots, \mathbf{A}_n$, that is, $\mathbf{A}' + \mathbf{A}'' \in \mathscr{L}(\mathbf{A}_1, \ldots, \mathbf{A}_n)$. Similarly if $r \in \mathbb{R}$ and $\mathbf{A} \in \mathscr{L}(\mathbf{A}_1, \ldots, \mathbf{A}_n)$ then

$$\mathbf{A} = a_1\mathbf{A}_1 + \cdots + a_n\mathbf{A}_n$$

for suitable numbers $a_1, \ldots, a_n$, so

$$\begin{aligned}r\mathbf{A} &= r(a_1\mathbf{A}_1 + \cdots + a_n\mathbf{A}_n) \\ &= ra_1\mathbf{A}_1 + \cdots + ra_n\mathbf{A}_n\end{aligned}$$

showing $r\mathbf{A} \in \mathbf{L}(\mathbf{A}_1, \ldots, \mathbf{A}_n)$. Therefore $\mathscr{L}(\mathbf{A}_1, \ldots, \mathbf{A}_n)$ is a subspace of $\mathscr{V}$. □

The idea of the linear span is not restricted to finite sets of vectors, but may be extended to arbitrary sets of vectors as follows.

Definition. Let $\mathscr{V}$ be a vector space and $E \subset \mathscr{V}$, that is, E is a collection of vectors in $\mathscr{V}$. A linear combination of vectors in E is a vector in $\mathscr{V}$ of the form

$$a_1\mathbf{A}_1 + a_2\mathbf{A}_2 + \cdots + a_n\mathbf{A}_n$$

where $\mathbf{A}_1, \ldots, \mathbf{A}_n \in E$. The *linear span of E*, denoted by $\mathscr{L}(E)$, is the set of all vectors that are linear combinations of vectors of E.

Proposition 3.3. *Let $\mathscr{V}$ be a vector space and $E \subset \mathscr{V}$. Then $\mathscr{L}(E)$ is a linear subspace of $\mathscr{V}$.*

The proof of (3.3) follows closely the proof of (3.2) and will be left to the diligent student. Note that the linear span allows us to assign to each sub*set* of $\mathscr{V}$ a sub*space* of $\mathscr{V}$. Note $E \subset \mathscr{L}(E)$.

Proposition 3.4. *Let $\mathscr{V}$ be a vector space and $E \subset \mathscr{V}$. Then $E = \mathscr{L}(E)$ iff E is a linear subspace of $\mathscr{V}$.*

PROOF. Suppose that E is a linear subspace of $\mathscr{V}$. Then if $\mathbf{A}_1, \ldots, \mathbf{A}_n \in E$ and $a_1, \ldots, a_n$ are numbers the vector $a_1\mathbf{A}_1 + \cdots + a_n\mathbf{A}_n$ belongs to E because E is closed under the operation of scalar multiplication and vector addition. Therefore $\mathscr{L}(E) \subset E$. Since $E \subset \mathscr{L}(E)$ we must conclude that $E = \mathscr{L}(E)$.

Conversely, suppose that $E = \mathscr{L}(E)$. If $\mathbf{A}, \mathbf{B} \in E$ then $\mathbf{A} + \mathbf{B}$ is certainly a linear combination of vectors in E and hence $\mathbf{A} + \mathbf{B}$ belongs to $\mathscr{L}(E)$, which since $\mathscr{L}(E) = E$ leads us to conclude $\mathbf{A} + \mathbf{B} \in E$. Likewise $a\mathbf{A}$ is a linear combination of vectors of E and hence belongs to $\mathscr{L}(E) = E$. Therefore E is closed under vector addition and scalar multiplication, and hence E is a linear subspace of $\mathscr{V}$. □

The preceding propositions show that in general a vector space has an abundance of subspaces.

EXAMPLE. In $\mathbb{R}^3$ consider the subspace spanned by the two vectors $\mathbf{A} = (1, 0, 1)$ and $\mathbf{B} = (0, 1, 0)$. (See Figure 3.1.) Note that this is just the plane through the origin, $x - z = 0$. That is the vectors in $\mathscr{L}(A, B)$ are those vectors $(x, y, z) \in \mathbb{R}^3$ whose coordinates satisfy the equation $x - z = 0$.

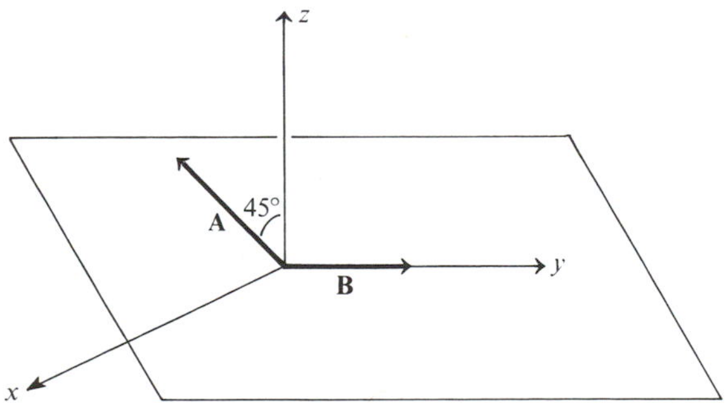

Figure 3.1

Proposition 3.5. *Let $\mathscr{S}$ and $\mathscr{T}$ be subspaces of $\mathscr{V}$. Then $\mathscr{S} \cap \mathscr{T}$ is also a subspace of $\mathscr{V}$.*

PROOF. Suppose that $\mathbf{A} \in \mathscr{S} \cap \mathscr{T}$ and $\mathbf{B} \in \mathscr{S} \cap \mathscr{T}$. Then $\mathbf{A} \in \mathscr{S}$ and $\mathbf{B} \in \mathscr{S}$. Since $\mathscr{S}$ is a subspace $\mathbf{A} + \mathbf{B} \in \mathscr{S}$. Likewise $\mathbf{A} \in \mathscr{T}$ and $\mathbf{B} \in \mathscr{T}$ and since $\mathscr{T}$ is a subspace $\mathbf{A} + \mathbf{B} \in \mathscr{T}$. Therefore $\mathbf{A} + \mathbf{B} \in \mathscr{S} \cap \mathscr{T}$. If r is a number, then since $\mathscr{S}$ and $\mathscr{T}$ are subspaces $r\mathbf{A} \in \mathscr{S}$ and $r\mathbf{A} \in \mathscr{T}$ so $r\mathbf{A} \in \mathscr{S} \cap \mathscr{T}$ showing that $\mathscr{S} \cap \mathscr{T}$ is again a subspace of $\mathscr{V}$. □

Definition. If $\mathscr{S}$ and $\mathscr{T}$ are subspaces of $\mathscr{V}$, their *sum*, denoted by $\mathscr{S} + \mathscr{T}$, is defined to be the set of all vectors $\mathbf{C}$ in $\mathscr{V}$ of the form

$$\mathbf{C} = \mathbf{A} + \mathbf{B}$$

where $\mathbf{A} \in \mathscr{S}$ and $\mathbf{B} \in \mathscr{T}$.

Proposition 3.6. *If $\mathscr{S}$ and $\mathscr{T}$ are subspaces of the vector space $\mathscr{V}$ then so is $\mathscr{S} + \mathscr{T}$.*

PROOF. Suppose that $\mathbf{C}_1, \mathbf{C}_2 \in \mathscr{S} + \mathscr{T}$. Write

$$\begin{aligned} \mathbf{C}_1 &= \mathbf{A}_1 + \mathbf{B}_1 \qquad \mathbf{A}_1 \in \mathscr{S}, \mathbf{B}_1 \in \mathscr{T} \\ \mathbf{C}_2 &= \mathbf{A}_2 + \mathbf{B}_2 \qquad \mathbf{A}_2 \in \mathscr{S}, \mathbf{B}_2 \in \mathscr{T}. \end{aligned}$$

Then

$$\begin{aligned} \mathbf{C}_1 + \mathbf{C}_2 &= \mathbf{A}_1 + \mathbf{B}_1 + \mathbf{A}_2 + \mathbf{B}_2 \\ &= (\mathbf{A}_1 + \mathbf{A}_2) + (\mathbf{B}_1 + \mathbf{B}_2) \end{aligned}$$

Let $\mathbf{A} = \mathbf{A}_1 + \mathbf{A}_2$, $\mathbf{B} = \mathbf{B}_1 + \mathbf{B}_2$. Since $\mathscr{S}$ and $\mathscr{T}$ are subspaces $\mathbf{A} \in \mathscr{S}$ and $\mathbf{B} \in \mathscr{T}$ while since

$$\mathbf{C}_1 + \mathbf{C}_2 = \mathbf{A} + \mathbf{B}$$

it follows that $\mathbf{C}_1 + \mathbf{C}_2 \in \mathscr{S} + \mathscr{T}$.

Next suppose that $\mathbf{C} \in \mathscr{S} + \mathscr{T}$ and $r \in \mathbb{R}$. Then we may write

$$\mathbf{C} = \mathbf{A} + \mathbf{B} \qquad \mathbf{A} \in \mathscr{S}, \mathbf{B} \in \mathscr{T}$$

and hence

$$r\mathbf{C} = r(\mathbf{A} + \mathbf{B}) = (r\mathbf{A}) + (r\mathbf{B}).$$

Since $\mathscr{S}$ and $\mathscr{T}$ are subspaces $r\mathbf{A} \in \mathscr{S}$, $r\mathbf{B} \in \mathscr{T}$ and hence $\mathbf{C} \in \mathscr{S} + \mathscr{T}$. □

PROBLEM. Suppose that $\mathscr{S}, \mathscr{T}$ are subspaces of $\mathscr{V}$. When is $\mathscr{S} \cup \mathscr{T}$ again a subspace of $\mathscr{V}$?

Answer. Iff $\mathscr{S} \subset \mathscr{T}$ or $\mathscr{T} \subset \mathscr{S}$.

Exercises

1. Which of the following collections of vectors in $\mathbb{R}$ are subspaces?

   (a) $\mathscr{U} = \{(x_1, x_2, x_3) \in \mathbb{R}^3 \mid x_1 = 0\}$
   (b) $\mathscr{U} = \{(x_1, x_2, x_3) \in \mathbb{R}^3 \mid x_2 = 0\}$
   (c) $\mathscr{U} = \{(x_1, x_2, x_3) \in \mathbb{R}^3 \mid x_1 + x_2 = 0\}$
   (d) $\mathscr{U} = \{(x_1, x_2, x_3) \in \mathbb{R}^3 \mid x_1 + x_2 = 1\}$
   (e) $\mathscr{U} = \{(x_1, x_2, x_3) \in \mathbb{R}^3 \mid x_1 + x_2 \geq 0\}$.

2. Determine the subspace of $\mathbb{R}^3$ which is the linear span of the three vectors $(1, 0, 1)$, $(0, 1, 0)$, $(0, 1, 1)$.

3. Repeat Exercise 2 for $(1, 0, 0)$, $(0, 1, 0)$, $(1, 1, 1)$.

4. Suppose $\mathscr{S}$, $\mathscr{T}$ are subspaces of $\mathscr{V}$ and $\mathscr{S} \cap \mathscr{T} = \mathbf{0}$. Show that every vector in $\mathscr{S} + \mathscr{T}$ can be written *uniquely* in the form $\mathbf{A} + \mathbf{B}$, $\mathbf{A} \in \mathscr{S}$, $\mathbf{B} \in \mathscr{T}$. Construct an example to show that this is false if $\mathscr{S} \cap \mathscr{T} \neq \mathbf{0}$.

5. Show that any nonzero vector spans $\mathbb{R}^1$.

6. Show that the two sets of vectors

$$\{\mathbf{A} = (1, 1, 0), \mathbf{B} = (0, 0, 1)\}$$

   and

$$\{\mathbf{C} = (1, 1, 1), \mathbf{D} = (-1, -1, 1)\}$$

   span the same subspace of $\mathbb{R}^3$.

7. Let $\mathscr{V}$ be the set of pairs of numbers $\mathbf{A} = (a_1, a_2)$. If $\mathbf{A}, \mathbf{B} \in \mathscr{V}$ define ($\mathbf{B} = (b_1, b_2)$)

$$\mathbf{A} + \mathbf{B} = (a_1 + b_1, a_2 + b_2).$$

   If a is a number define

$$a\mathbf{A} = (aa_1, 0)$$

   Is $\mathscr{V}$ a vector space? Why?

8. Suppose $\mathscr{V}$ is a vector space and E, F are subsets of $\mathscr{V}$. Show

   (a) $E \subset F \Rightarrow \mathscr{L}(E) \subset \mathscr{L}(F)$.
   (b) $\mathscr{L}(E \cup F) = \mathscr{L}(E) + \mathscr{L}(F)$.
   (c) $\mathscr{L}(E \cap F) \subset \mathscr{L}(E) \cap \mathscr{L}(F)$.

9. Let $\mathscr{V}$ be a vector space and $E, F \subset \mathscr{V}$. Suppose $\mathscr{L}(E) \subset \mathscr{L}(F)$. Is it true that $E \subset F$?

10. Suppose that $\mathscr{V}$ is a vector space and $E \subset \mathscr{V}$. If $\mathscr{U}$ is a subspace containing E then $\mathscr{U}$ contains $\mathscr{L}(E)$.

11. Suppose $\mathscr{V}$ is a vector space and $E \subset \mathscr{V}$. Show that $\mathscr{L}(E) = \cap \{\mathscr{U} \mid \mathscr{U}$ is a subspace of $\mathscr{V}$ and $\mathscr{U}$ contains $E\}$.

12. For any subspace $\mathscr{U}$ of $\mathscr{V}$ show that $\mathscr{U} + \mathscr{U} = \mathscr{U}$.

13. Find all the linear subspaces of $\mathbb{R}^2$.

14. Find all the linear subspaces of $\mathbb{R}^3$.

15. Show that a subset E of a vector space $\mathscr{V}$ which does not contain $\mathbf{0}$ is not a subspace of $\mathscr{V}$.

16. Let E be the subset of $\mathbb{R}^2$ defined by $E = \{(x, y) | x \geq 0, y \in \mathbb{R}\}$. Is E a subspace of $\mathbb{R}^2$?

17. Let $E = \{(x, 2x + 1) | x \in \mathbb{R}\}$. E is a subset of $\mathbb{R}^2$. Is E a subspace of $\mathbb{R}^2$?

18. (a) Let $E = \{(2a, a) | a \in \mathbb{R}\}$. Is E a subspace of $\mathbb{R}^2$?
(b) Let $B = \{(b, b) | b \in \mathbb{R}\}$. Is B a subspace of $\mathbb{R}^2$?
(c) What is $E \cap B$?
(d) Is $E \cup B$ a subspace of $\mathbb{R}^2$?
(e) What is $E + B$?

19. Let $\mathscr{S}$ and $\mathscr{T}$ be subspaces of $\mathscr{T}$. Prove:

(a) $\mathscr{S} + \mathscr{T} = \mathscr{L}(\mathscr{S} \cup \mathscr{T})$.
(b) $\mathscr{S} \cap (\mathscr{S} + \mathscr{T}) = \mathscr{S}$.
(c) $\mathscr{S} + \mathscr{T} = \mathscr{T} + \mathscr{S}$.
(d) If $\mathscr{S} \subset \mathscr{T}$, then $\mathscr{S} + \mathscr{T} = \mathscr{T}$.

20. Let $\mathscr{V}$ be a vector space, $\mathbf{A}, \mathbf{B}, \mathbf{C} \in \mathscr{V}$. Suppose $\mathbf{A} + \mathbf{B} + \mathbf{C} = 0$. Show that $\mathscr{L}(\mathbf{A}, \mathbf{B}) = \mathscr{L}(\mathbf{B}, \mathbf{C})$.

21. Let $\mathscr{V}$ be a vector space and $\mathscr{W}$ a subspace of $\mathscr{V}$. Show that

$$\{\mathbf{A} \in \mathscr{V} | \mathbf{A} \notin \mathscr{W}\}$$

is not a vector subspace of $\mathscr{V}$.

22. Let $\mathscr{V}$ be a vector space, $\mathscr{W}$ a subspace of $\mathscr{V}$, and $\mathbf{A}, \mathbf{B} \in \mathscr{V}$. Assume that

$$\mathbf{A} \notin \mathscr{W} \quad \text{but} \quad \mathbf{A} \in \mathscr{L}(\mathscr{W} \cup \{\mathbf{B}\}).$$

Show that $\mathbf{B} \in \mathscr{L}(\mathscr{W} \cup \{\mathbf{A}\})$.

23. Let $\mathscr{S}, \mathscr{T}, \mathscr{U}$ be subspaces of a vector space $\mathscr{V}$. Is it always true that

$$\mathscr{S} \cap (\mathscr{T} + \mathscr{U}) = (\mathscr{S} \cap \mathscr{T}) + (\mathscr{S} + \mathscr{W})?$$

24. Can it happen for two subspaces $\mathscr{S}$ and $\mathscr{T}$ of $\mathscr{V}$ that $\mathscr{S} \cap \mathscr{T} = \varnothing$?

25. Can you find two vectors $\mathbf{A}, \mathbf{B} \in \mathbb{R}^3$ such that $\mathscr{L}(A, B) = \mathbb{R}^3$?

4 Examples of vector spaces

Before continuing with our study of the elementary properties of vector spaces and their linear subspaces let us collect a list of examples of vector spaces. We have already encountered the cartesian k-space $\mathbb{R}^k$ and so for the sake of completeness let us begin by listing this example:

EXAMPLE 1. $\mathbb{R}^k$

The first new example that we have in this chapter is primarily designed to destroy the belief that a vector is a quantity with both direction and magnitude and to give meaning to the phrase in our comments on axiomatics in Chapter 2, that "the possibility is opened of assigning to them (the axioms of a vector space) content in new and unforeseen ways."

EXAMPLE 2. $\mathscr{P}_n(\mathbb{R})$

The vectors in $\mathscr{P}_n(\mathbb{R})$ are polynomials

$$p(x) = a_0 + a_1 x + \cdots + a_m x^m$$

of degree less than or equal to n, that is, $m \leq n$. Addition of vectors is to be ordinary addition of polynomials and multiplication of a polynomial by a number, the ordinary product of a polynomial by a number. With these interpretations of the basic terms:

$$\text{vector} \leftrightarrow \text{polynomial of degree} \leq n$$
$$\text{vector addition} \leftrightarrow \text{addition of polynomials}$$
$$\text{scalar multiplication} \leftrightarrow \text{multiplication of a polynomial by a number,}$$

we obtain an example of a vector space. To verify that $\mathscr{P}_n(\mathbb{R})$ is indeed a vector space we must check that the eight declarative sentences obtained from these interpretations of the basic terms vector, vector addition, scalar multiplication, are true sentences. This is a straightforward deduction from the (assumed) properties of real numbers following the pattern of (2.4) and will be left to the diligent reader.

Note that in this example it is very difficult to say what direction or length a vector has.

Example 3. $\mathbb{C}$

Let us denote by $\mathbb{C}$ the *complex numbers*. (We will not here be concerned with the technical details of constructing the complex numbers, but will take them as we learned them in grammar school.) Recall that a complex number looks like

$$a + bi$$

where a, b are real numbers, and i is a number with $i^2 = -1$.

The vectors in our vector space will be complex numbers. Addition of vectors is to be the ordinary addition of complex numbers, and scalar multiplication the familiar process of multiplying a complex number by a real number. With these interpretations of the basic terms

vector ↔ complex number
vector addition ↔ addition of complex numbers
scalar multiplication ↔ multiplication of a complex number by a real number,

we obtain an example of a vector space. The verifications are again routine.

Note that in Example 3 we are not using all of the structure that we have, for it is possible to multiply two complex numbers, that is, *in this example* we may multiply two vectors, something it is not always possible to do in a vector space. This is a possibility worthy of further study, and we will do just that when we study spaces of linear transformation and *linear algebras*.

Note. the product of two polynomials of degree at most n will have degree at most $2n$, so you cannot multiply elements of $\mathscr{P}_n(\mathbb{R})$ in any obvious way.

Example 2 is a very important example, and a prototype for many others of the same type. These examples are characterized by the fact that their "vectors" are actually functions of some type or other.

Example 4. $\mathscr{P}(\mathbb{R})$

The simplest way to obtain a space akin to but different from $\mathscr{P}_n(\mathbb{R})$ is simply to remove the restriction that the polynomials have degree at most n. In this way we obtain the vector space $\mathscr{P}(\mathbb{R})$ whose vectors are the polynomials

$$p(x) = a_0 + a_1x + \cdots + a_mx^m$$

with no restriction on m. The interpretation of the basic terms we propose in this example is:

$$\text{vector} \leftrightarrow \text{polynomial}$$
$$\text{vector addition} \leftrightarrow \text{addition of polynomials}$$
$$\text{scalar multiplication} \leftrightarrow \text{multiplication of a polynomial by a number.}$$

It is again a routine verification that the vector space axioms are satisfied.

EXAMPLE 5. $\mathscr{C}(a, b)$

This is a very fancy example, and included only to indicate the wealth of possible examples of vector spaces.

Let a and b be numbers with $a < b$. The vectors of $\mathscr{C}(a, b)$ are the continuous functions defined for $a \leq x \leq b$. Addition of vectors is to be addition of functions. That is if f and g are functions defined and continuous for $a \leq x \leq b$ then $f + g$ is the function defined by

$$(f + g)(x) = f(x) + g(x)$$

for all $a \leq x \leq b$. It is an important theorem of the calculus that $f + g$ is again a continuous function for $a \leq x \leq b$. Scalar multiplication is to be defined as the ordinary product of a function by a number. If the function is continuous then its product by a number is also continuous. The basic terms of a vector space are to be interpreted as follows in this example:

$$\text{vector} \leftrightarrow \text{continuous function on } a \leq x \leq b$$
$$\text{vector addition} \leftrightarrow \text{addition of functions}$$
$$\text{scalar multiplication} \leftrightarrow \text{multiplication of a function by a number.}$$

Again the vector space axioms are easily verified.

EXAMPLE 6. *Linear Homogenous equations.* A *linear homogenous equation* in the variables $x_1, \ldots, x_n$ is an equation of the form

$$(*) \qquad a_1x_1 + a_2x_2 + \cdots + a_nx_n = 0.$$

A solution to this equation is a sequence of n-numbers $(s_1, \ldots, s_n)$ such that

$$a_1s_1 + \cdots + a_ns_n = 0.$$

If $\mathbf{A} = (s_1, \ldots, s_n)$ and $\mathbf{B} = (t_1, \ldots, t_n)$ are solutions to $(*)$ define

$$\mathbf{A} + \mathbf{B} = (s_1 + t_1, \ldots, s_n + t_n).$$

We claim that $\mathbf{A} + \mathbf{B}$ is again a solution to $(*)$. For we have

$$\begin{aligned} &a_1(s_1 + t_1) + a_2(s_2 + t_2) + \cdots + a_n(s_n + t_n) \\ &\qquad = a_1s_1 + a_1t_1 + a_2s_2 + a_2t_2 + \cdots + a_ns_n + a_nt_n \\ &\qquad = a_1s_1 + a_2s_2 + \cdots + a_ns_n + a_1t_1 + a_2t_2 + \cdots + a_nt_n \\ &\qquad = 0 + 0 = 0 \end{aligned}$$

as we claimed. Next define $a\mathbf{A}$ for a number a to be

$$a\mathbf{A} = (as_1, \ldots, as_n).$$

Simple manipulation shows

$$\begin{aligned} a_1(as_1) + \cdots + a_n(as_n) &= aa_1s_1 + \cdots + aa_ns_n \\ &= a(a_1s_1 + \cdots + a_ns_n) \\ &= a(0) = 0 \end{aligned}$$

so that $a\mathbf{A}$ is again a solution to $(*)$. We now define a vector space $\mathscr{V}$ by the interpretation

vector $\leftrightarrow$ solution to $(*)$
vector addition $\leftrightarrow$ as defined above
scalar multiplication $\leftrightarrow$ as defined above.

To show that $\mathscr{V}$ is a vector space we will show that it is actually a linear subspace of $\mathbb{R}^n$. For by definition the vectors of $\mathscr{V}$ are sequences $(s_1, \ldots, s_n)$ of numbers and hence are vectors in $\mathbb{R}^n$. The process of adding solutions and multiplying solutions by scalars is exactly the process of adding vectors in $\mathbb{R}^n$ and multiplying a vector of $\mathbb{R}^n$ by a number. In our preceding discussion we checked

(1) If $\mathbf{A}, \mathbf{B} \in \mathscr{V}$ then $\mathbf{A} + \mathbf{B} \in \mathscr{V}$.
(2) If $\mathbf{A} \in \mathscr{V}$ then $a\mathbf{A} \in \mathscr{V}$ for any number a.

Thus we may apply (3.1) to conclude that $\mathscr{V}$ is a vector space. But wait! In order to apply (3.1) to $\mathscr{V}$ we must know that $\mathscr{V}$ is nonempty, that is, that $(*)$ has at least one solution. Happily this is a simple point, because $(0, \ldots, 0)$ is a solution to $(*)$ as one easily sees, since

$$a_10 + a_20 + \cdots + a_n0 = 0 + \cdots + 0 = 0.$$

Thus $\mathscr{V}$ is a linear subspace of $\mathbb{R}^n$.

The preceding example may be extended from one equation to many, but this is a topic for future study. (Await Chapter 13.)

Continuing our list of examples we introduce:

Example 7. Let S be a set and $\mathscr{F}(S)$ the set of all functions $\mathbf{f}: S \to \mathbb{R}$. If $\mathbf{f}, \mathbf{g} \in \mathscr{F}(S)$ define

$$\mathbf{f} + \mathbf{g}: S \to \mathbb{R}$$

by

$$(\mathbf{f} + \mathbf{g})(s) = \mathbf{f}(s) + \mathbf{g}(s)$$

and for a real number r define

$$r\mathbf{f}: S \to \mathbb{R}$$

by

$$(r\mathbf{f})(s) = r(\mathbf{f}(s)),$$

for all $s \in S$.

Equipped with this vector addition and scalar multiplication the set $\mathscr{F}(S)$ becomes a vector space. The zero vector of $\mathscr{F}(S)$ is the function

$$\mathbf{0} : S \to \mathbb{R}$$

defined by

$$\mathbf{0}(s) = 0$$

for all $s \in S$; that is, $\mathbf{0}$ is the constant function which takes the value 0 for all $s \in S$. The negative of $\mathbf{f} \in \mathscr{F}(S)$ is the function

$$-\mathbf{f} : S \to \mathbb{R}$$

defined by

$$(-\mathbf{f})(s) = -\mathbf{f}(s).$$

It is now routine to verify that the axioms of a real vector space are satisfied for $\mathscr{F}(S)$.

We can also make the set of all complex-valued functions $\mathscr{F}_{\mathbb{C}}(S) = \{\mathbf{f} : S \to \mathbb{C}\}$ into a complex vector space by setting

$$(\mathbf{f} + \mathbf{g})(s) = \mathbf{f}(s) + \mathbf{g}(s)$$
$$(c\mathbf{f})(s) = c(\mathbf{f}(s))$$

for all $\mathbf{f}, \mathbf{g} \in \mathscr{F}_{\mathbb{C}}(S)$, $s \in S$, and $c \in \mathbb{C}$.

If $T \subset S$ then we denote by $\mathscr{F}(S, T)$ the set of all functions

$$\mathbf{f} : S \to \mathbb{R}$$

such that

$$\mathbf{f}(s) = 0 \quad \text{for all } s \in T.$$

The set $\mathscr{F}(S, T) \subset \mathscr{F}(S)$ is in fact a subspace. For if $\mathbf{f}, \mathbf{g} \in \mathscr{F}(S, T)$ and $s \in T$ then

$$(\mathbf{f} + \mathbf{g})(s) = \mathbf{f}(s) + \mathbf{g}(s) = 0 + 0 = 0$$

and if $r \in \mathbb{R}$ then

$$(r\mathbf{f})(s) = r\mathbf{f}(s) = r \cdot 0 = 0,$$

so that $\mathbf{f} + \mathbf{g} \in \mathscr{F}(S, T)$, $r\mathbf{f} \in \mathscr{F}(S, T)$, and finally since $\mathbf{0} \in \mathscr{F}(S, T)$ we see that $\mathscr{F}(S, T)$ is a subspace of $\mathscr{F}(S)$.

Finally we will close the **introduction** to examples of vector spaces by describing a rather artificial example.

EXAMPLE 8. Let $\mathscr{V}$ be the set of all *positive* real numbers and define for $\mathbf{A}, \mathbf{B} \in \mathscr{V}$ a vector sum by

$$\mathbf{A} + \mathbf{B} = \mathbf{A} \cdot \mathbf{B}$$

where the product on the right is the usual product of numbers. If a is a number and $\mathbf{A} \in \mathscr{V}$ define

$$a \cdot \mathbf{A} = \mathbf{A}^a$$

that is the number $\mathbf{A}$ raised to the a power. Note that since $\mathbf{A} > 0$ the $(1/a)$th root of $\mathbf{A}$ will always exist. For example, with these definitions

$$2 + 3 = 6$$
$$2 \cdot 3 = 9.$$

We claim that with these definitions of vector, vector addition and scalar multiplication $\mathscr{V}$ becomes a vector space. The details of verification are left to you.

The preceding list only barely scratches the surface of the enormous variety of examples of vector spaces. More examples will appear as we progress through the book and will by no means exhaust the possibilities.

EXERCISES

1. Show that $\mathscr{P}_r(\mathbb{R})$ is a linear subspace of $\mathscr{P}_s(\mathbb{R})$ whenever $r \leq s$.
2. Show that $\mathscr{P}_t(\mathbb{R})$ is always a subspace of $\mathscr{P}(\mathbb{R})$.
3. What is the span of $\{1 + x, 1 - x\}$ in $\mathscr{P}(\mathbb{R})$?
4. What is the span of $\{1, x^2, x^4\}$ in $\mathscr{P}_4(\mathbb{R})$?
5. Find a vector that spans the subspace $2x - 3y = 0$ of $\mathbb{R}^2$.
6. Find a pair of vectors that span the subspace $x + y - 2z = 0$ of $\mathbb{R}^3$.
7. Verify that example 7 is indeed a vector space. What is the zero vector in this example?
8. Let $\mathscr{E}$ be the subset of $\mathscr{P}_r(\mathbb{R})$ defined by

$$\mathscr{E} = \{p(x) \mid p(x) \in \mathscr{P}_r(\mathbb{R}) \text{ and } p(-x) = p(x)\}$$

   Show that $\mathscr{E}$ is a linear subspace of $\mathscr{P}_r(\mathbb{R})$.
9. The set of all continuous functions $y = f(x)$, $-\infty < x < \infty$ satisfying the differential equation

$$y'' - y' - 2y = 0$$

   is a vector space. (In fact any solution of this differential equation is a linear combination of $y = e^{-x}$ and $y = e^{2x}$.)

10. The set of all continuous solutions of a linear differential equation

$$a_0(x)y^{(n)} + a_1(x)y^{(n-1)} + \cdots + a_{n-1}(x)y' + a_n(x)y = 0$$

where $a_0(x)$ is not zero on $[a, b]$, and $a_i(x)$ are continuous on $[a, b]$, $i = 1, 2, \ldots, n$, is a vector space.

11. Let $\mathscr{D} = \{f \in \mathscr{C}(a, b) \mid f \text{ is differentiable on } (a, b)\}$. Show that $\mathscr{D}$ is a subspace of $\mathscr{C}(a, b)$.

12. Let $\mathscr{P}(a, b) = \{\text{Polynomials of } x, a \leq x \leq b\}$. Then $\mathscr{P}(a, b) \subset \mathscr{D} \subset \mathscr{C}(a, b)$. Show $\mathscr{P}(a, b)$ is a subspace of $\mathscr{D}$ and also a subspace of $\mathscr{C}(a, b)$.

13. Consider the set $\mathscr{D}^{(n)}$ of all n-times differentiable functions on the interval $[a, b]$. $\mathscr{D}^{(n)}$ is also a vector space. Is $\mathscr{P}(a, b) \subset \mathscr{D}^{(n)}$?

14. Let S be a set and $\mathscr{F}(S)$ the vector space of real-valued functions on S. If $A, B \subset S$ show that

$$\mathscr{F}(S, A) \cap \mathscr{F}(S, B) = \mathscr{F}(S, A \cup B).$$

If $A \subset B$ show that $\mathscr{F}(S, A)$ contains $\mathscr{F}(S, B)$ as a subspace.

15. Let S be a set and $T \subset S$. Is it true that

$$\mathscr{F}(S) = \mathscr{F}(S, T) + \mathscr{F}(S, S - T)?$$

16. Let $\mathscr{V}$ be a vector space and S a set. Define $\text{Fun}(S, \mathscr{V}) = \{\mathbf{f}: S \to \mathscr{V}\}$, that is, the set of all functions from S into $\mathscr{V}$. (In this notation $\mathscr{F}(S) = \text{Fun}(S, \mathbb{R})$.) If $\mathbf{f}, \mathbf{g} \in \text{Fun}(S, \mathscr{V})$ define $\mathbf{f} + \mathbf{g}: S \to \mathscr{V}$ by

$$(\mathbf{f} + \mathbf{g})(s) = \mathbf{f}(s) + \mathbf{g}(s).$$

If $\mathbf{f} \in \text{Fun}(S, \mathscr{V})$ and $r \in \mathbb{R}$ define $r \cdot \mathbf{f}: S \to \mathscr{V}$ by

$$(r \cdot \mathbf{f})(s) = r \cdot \mathbf{f}(s).$$

(a) Show that $\text{Fun}(S, \mathscr{V})$ equipped with these two operations becomes a vector space.
(b) If $\mathscr{V}$ is a subspace of $\mathscr{W}$ show $\text{Fun}(S, \mathscr{V})$ is a subspace of $\text{Fun}(S, \mathscr{W})$.
(c) If $\mathscr{S}$ and $\mathscr{T}$ are subspaces of $\mathscr{V}$ show

$$\text{Fun}(S, \mathscr{S}) + \text{Fun}(S, \mathscr{T}) = \text{Fun}(S, \mathscr{S} + \mathscr{T}).$$

17. Let $\mathscr{V}$ be a vector space and

$$\varphi: \mathscr{V} \to \mathbb{R}$$

such that

(LF 1) $\qquad \varphi(\mathbf{A} + \mathbf{B}) = \varphi(\mathbf{A}) + \varphi(\mathbf{B}), \qquad \forall \mathbf{A}, \mathbf{B} \in \mathscr{V},$

(LF 2) $\qquad \varphi(r\mathbf{A}) = r\varphi(\mathbf{A}), \qquad \forall \mathbf{A} \in \mathscr{V}, \forall r \in \mathbb{R}.$

Show that $\mathscr{N} = \{\mathbf{A} \in \mathscr{V} \mid \varphi(\mathbf{A}) = 0\}$ is a vector subspace of $\mathscr{V}$.

18. Let $\mathscr{V}$ be a vector space and S be a set and $s \in S$ a fixed element. Define

$$e_S: \text{Fun}(S, \mathscr{V}) \to \mathscr{V}$$

by

$$e_s(\mathbf{f}) = \mathbf{f}(s).$$

Show that

$$\mathscr{K} = \{\mathbf{f} \in \text{Fun}(S, \mathscr{V}) \mid e_s(\mathbf{f}) = 0\}$$

is a subspace of Fun(S, $\mathscr{V}$).

19. Let $\mathscr{V}$ and $\mathscr{W}$ be vector spaces. Show that the cartesian product $\mathscr{V} \times \mathscr{W}$ consisting of all the ordered pairs $(\mathbf{A}, \mathbf{B})$, $\mathbf{A} \in \mathscr{V}$, $\mathbf{B} \in \mathscr{W}$ becomes a vector space if we define vector addition and scalar multiplication componentwise, that is

$$(\mathbf{A}', \mathbf{B}') + (\mathbf{A}'', \mathbf{B}'') = (\mathbf{A}' + \mathbf{A}'', \mathbf{B}' + \mathbf{B}''),$$

$$r(\mathbf{A}, \mathbf{B}) = (r\mathbf{A}, r\mathbf{B}).$$

20. Let $\mathscr{V}$ be a vector space, S a set and $s, t \in \mathscr{S}$. Define

$$\varphi\colon \text{Fun}(S, \mathscr{V}) \to \mathscr{V} \times \mathscr{V}$$

by

$$\varphi(\mathbf{f}) = (\mathbf{f}(s), \mathbf{f}(t)).$$

Show that

$$\mathscr{N} = \{\mathbf{f} \in \text{Fun}(S, \mathscr{V}) \mid \varphi(\mathbf{f}) = 0\}$$

is a vector subspace of Fun(S, $\mathscr{V}$).

21. Let $\mathscr{V}$ be a vector space, $\mathscr{W}$ a subspace of $\mathscr{V}$, S a set and $s \in S$. Show that

$$\mathscr{U} = \{\mathbf{f} \in \text{Fun}(S, \mathscr{V}) \mid \mathbf{f}(s) \in W\}$$

is a subspace of Fun(S, $\mathscr{V}$).

5 Linear independence and dependence

Definition. A set of vectors E is said to be *linearly dependent* if there exist distinct vectors $\mathbf{A}_1, \ldots, \mathbf{A}_k$ in E and numbers $a_1, \ldots, a_k$, not all zero, such that

$$(*) \qquad a_1\mathbf{A}_1 + a_2\mathbf{A}_2 + \cdots + a_k\mathbf{A}_k = \mathbf{0}.$$

Equation $(*)$ is called a *linear relation* between $\mathbf{A}_1, \ldots, \mathbf{A}_k$.

EXAMPLE 1. Let E be the set of vectors

$$E = \{(0, 1, 0), (0, 1, 1), (0, 0, 1)\}$$

in $\mathbb{R}^3$. Then E is a linearly dependent set of vectors because

$$1(0, 1, 0) + 1(0, 0, 1) + (-1)(0, 1, 1) = (0, 0, 0).$$

EXAMPLE 2. Let E be the set of vectors

$$E = \{p(x) \mid \text{such that degree } p(x) \text{ is at most } 1\}$$

in $\mathscr{P}_3(\mathbb{R})$. Then E is a linearly dependent set of vectors because the vectors $1, x, 1 + x$ belong to E and

$$1(1 + x) + (-1)x + (-1)(1) = \mathbf{0}.$$

Definition. A set of vectors E that is *not* linearly dependent is said to be *linearly independent.*

EXAMPLE 3. Let E be the vectors

$$\{(1, 1, 1), (0, 1, 1), (0, 0, 1)\}$$

in $\mathbb{R}^3$. Then E is a linearly independent set of vectors. In order to prove this suppose to the contrary that E is linearly dependent. Then there must exist numbers a_1, a_2, a_3, not all zero, such that

$$a_1(1, 1, 1) + a_2(0, 1, 1) + a_3(0, 0, 1) = (0, 0, 0).$$

But if these were so, then since

$$a_1(1, 1, 1) + a_2(0, 1, 1) + a_3(0, 0, 1) = (a_1, a_1 + a_2, a_1 + a_2 + a_3)$$

we would have

$$(0, 0, 0) = (a_1, a_1 + a_2, a_1 + a_2 + a_3)$$

and hence that

$$a_1 = 0 \qquad a_1 + a_2 = 0 \qquad a_1 + a_2 + a_3 = 0$$

from which we see that of necessity

$$a_1 = 0, \qquad a_2 = 0, \qquad a_3 = 0.$$

But this contradicts our original assumption that not all of a_1, a_2, and a_3 are zero. Therefore the set E cannot be linear dependent, and hence must be linearly independent.

EXAMPLE 4. Let E be the set of vectors $\{1, i\}$ in $\mathbb{C}$. Then E is linearly independent. For if we suppose that $\{1, i\}$ is dependent, then there are real numbers a_1, a_2, not both zero, such that

$$a_1(1) + a_2(i) = \mathbf{0}.$$

Now let $a_1 - a_2 i$ be the conjugate complex number. Then $\mathbf{0} = a_1 + a_2 i$ implies

$$\mathbf{0} = (a_1 + a_2 i)(a_1 - a_2 i) = a_1^2 + a_2^2$$

which is impossible. Therefore E cannot be linearly dependent and hence must be linearly independent.

Remark. The proof above uses the conjugate complex number. There is an alternative proof as follows. Since we consider only real vector spaces now, suppose there are two real numbers a_1, a_2 so that

$$a_1 1 + a_2 i = \mathbf{0}.$$

Then

$$a_1 = -a_2 i \qquad (*).$$

a_1 is real from assumption, and since a_2 is real $(*)$ says that a_1 is also purely imaginary. The only way to avoid a contradiction is for $a_1 = 0$ and $a_2 = 0$. Thus $\{1, i\}$ is a set of linearly independent vectors.

Example 5. Let E be the set of vectors $\{1 + x, 1 - x\}$ in $\mathscr{P}_1(\mathbb{R})$. Then E is linearly independent. For suppose to the contrary that $\{1 + x, 1 - x\}$ is linear dependent. Then there exist numbers a_1, a_2, not both zero, such that

$$a_1(1 + x) + a_2(1 - x) = \mathbf{0}.$$

Then we will have

$$\begin{aligned}\mathbf{0} &= a_1(1 + x) + a_2(1 - x) = a_1 + a_1x + a_2 - a_2x \\ &= (a_1 + a_2) + (a_1 - a_2)x.\end{aligned}$$

Remember that a polynomial is identically zero iff all its coefficients are zero. Therefore we have

$$a_1 + a_2 = 0, \qquad a_1 - a_2 = 0.$$

Solving these equations we find

$$a_1 = 0, \qquad a_2 = 0$$

which is a contradiction to the assumption that $\{1 + x, 1 - x\}$ is a linearly dependent set of vectors. Therefore it is linearly independent.

Example 6. Let S be a set. For each $s \in S$ the *characteristic function of* s is the function

$$\chi_s : S \to \mathbb{R}$$

defined by

$$\chi_s(t) = \begin{cases} 1 & \text{if } t = s, \\ 0 & \text{if } t \neq s. \end{cases}$$

If $s_1, \ldots, s_k \in S$ are distinct points then their characteristic functions $\chi_{s_1}, \ldots, \chi_{s_k} \in \mathscr{F}(S)$ are linearly independent. To see this suppose that

$$a_1 \chi_{s_1} + \cdots + a_k \chi_{s_k} = \mathbf{0}$$

is a linear relation between $\chi_{s_1}, \ldots, \chi_{s_k}$. Then

$$\begin{aligned}0 &= (a_1 \chi_{s_k} + \cdots + a_k \chi_{s_k})(s_i) \\ &= a_1 \chi_{s_1}(s_i) + \cdots + a_k \chi_{s_k}(s_i) \\ &= a_1 0 + \cdots + a_{i-1} 0 + a_i 1 + a_{i+1} 0 + \cdots + a_k 0 = a_i\end{aligned}$$

so $a_1 = 0, a_2 = 0, \ldots, a_k = 0$ and $\chi_{s_1}, \ldots, \chi_{s_k}$ are linearly independent.

A very quick test for a linearly dependent set is the following:

Proposition 5.1. *If a set of vectors E contains the vector $\mathbf{0}$, it is linear dependent.*

Proof. Clearly

$$1 \cdot \mathbf{0} = \mathbf{0}$$

so letting $\mathbf{A}_1 = \mathbf{0} \in E$ and $a_1 = 1$, $k = 1$ we satisfy the condition of linear dependence. □

Corollary 5.2. *If E is a linear subspace of $\mathscr{V}$ then E is a linearly dependent set of vectors.*

PROOF. A linear subspace always contains $\mathbf{0}$. Apply (5.1). □

Definition. A vector $\mathbf{A}$ is said to be *linearly dependent on a set of vectors E* iff $\mathbf{A} \in \mathscr{L}(E)$.

Proposition 5.3. *A set of vectors E is linearly dependent iff there is a vector $\mathbf{A}$ in E linearly dependent on the remaining vectors of E.*

PROOF. Suppose that E is linearly dependent. Then we may find distinct vectors $\mathbf{A}_1, \mathbf{A}_2, \ldots, \mathbf{A}_k$ in E, and numbers $a_1, \ldots, a_k$, not all zero, such that

$$a_1\mathbf{A}_1 + a_2\mathbf{A}_2 + \cdots + a_k\mathbf{A}_k = \mathbf{0}.$$

Since not all the numbers $a_1, \ldots, a_k$ are zero, we can by changing the order arrange so that $a_1 \neq 0$. Then we have

$$a_1\mathbf{A}_1 = -a_2\mathbf{A}_2 - a_3\mathbf{A}_3 - \cdots - a_k\mathbf{A}_k$$

and since $a_1 \neq 0$,

$$\mathbf{A}_1 = \frac{-a_2}{a_1}\mathbf{A}_2 + \frac{-a_3}{a_1}\mathbf{A}_3 + \cdots + \frac{-a_k}{a_1}\mathbf{A}_k$$

and hence $\mathbf{A}_1 \in \mathscr{L}(\mathbf{A}_2, \ldots, \mathbf{A}_k)$ which shows (since $\mathbf{A}_1, \ldots, \mathbf{A}_k$ are distinct) that $\mathbf{A}_1$ is linear dependent on the remaining vectors of E.

Conversely, if there is a vector $\mathbf{A}$ in E which is linearly dependent on the remaining vectors of E we may find distinct vectors $\mathbf{A}_2, \ldots, \mathbf{A}_k$, different from $\mathbf{A}$, such that

$$\mathbf{A} = a_2\mathbf{A}_2 + \cdots + a_k\mathbf{A}_k.$$

Then

$$\mathbf{0} = (-1)\mathbf{A}_1 + a_2\mathbf{A}_2 + \cdots + a_k\mathbf{A}_k$$

is a linear relation between $\mathbf{A}_1 = \mathbf{A}, \mathbf{A}_2, \ldots, \mathbf{A}_k$ showing that E is a linearly dependent set. □

Theorem 5.4. *If E is a finite set of vectors spanning the linear subspace $\mathscr{U}$ of $\mathscr{V}$, that is $\mathscr{L}(E) = \mathscr{U}$, then there exists a subset F of E such that F is a linearly independent set of vectors and $\mathscr{L}(F) = \mathscr{U} = \mathscr{L}(E)$.*

PROOF. If E is linearly independent there is nothing to prove. So suppose that E is a linearly dependent set of vectors. By (5.3) there exists a vector $\mathbf{A}$ that is linearly dependent on the remaining vectors of E. Denote this set of remaining vectors by E'. Thus $\mathbf{A} \in \mathscr{L}(E')$. Therefore $\mathscr{L}(E') = \mathscr{L}(E)$ because (see Chapter 3, Exercises 8 and 12) $\mathscr{L}(E') \subset \mathscr{L}(E)$ and

$$\begin{aligned}\mathscr{L}(E) = \mathscr{L}(\{\mathbf{A}\} \cup E') \subset \mathscr{L}(A) + \mathscr{L}(E') \subset \mathscr{L}(E') + \mathscr{L}(E') \\ = \mathscr{L}(E').\end{aligned}$$

Now we can repeat our argument on E'. That is, either E' is linearly independent, in which case we are done, or it is linearly dependent and we can use the preceding argument to reduce the size of E' by one vector. Since the set E of vectors that we began with is finite the theorem will follow by repeating the argument a finite number of times. □

Example 7. Let $E = \{(1, 0, 0), (0, 1, 0), (0, 0, 1), (1. 1, 1)\}$ be vectors in $\mathbb{R}^3$. Find a linearly independent set F which is a subset of E such that $\mathscr{L}(F) = \mathscr{L}(E)$.

Solution 1. The proof of Theorem (5.4) suggests that we look for a vector in E linearly dependent on the remaining vectors of E and throw it away. If that doesn't work do it again, etc. Now we observe that

$$(1, 1, 1) = 1(1, 0, 0) + 1(0, 1, 0) + 1(0, 0, 1)$$

so that setting $F = \{(1, 0, 0), (0, 1, 0), (0, 0, 1)\}$ we have $\mathscr{L}(F) = \mathscr{L}(E)$. The set F, however, is a linearly independent set of vectors and so we are done.

Solution 2. Proceeding as in Solution 1 we note that $(1, 0, 0) = 1(1, 1, 1) + (-1)(0, 1, 0) + (-1)(0, 0, 1)$ and setting $H = \{(1, 1, 1), (0, 1, 0), (0, 0, 1)\}$ we have $\mathscr{L}(H) = \mathscr{L}(E)$. The set H, however, is a linearly independent set of vectors and so we are done.

Moral. If E is a finite set of vectors in $\mathscr{V}$ then there does *not* exist a unique subset $F \subset E$ with F linearly independent and $\mathscr{L}(F) = \mathscr{L}(E)$.

Example 8. Let S be a *finite* set and $X = \{\chi_s | s \in S\} \subset \mathscr{F}(S)$ the set of characteristic functions of the elements of S. Then X spans $\mathscr{F}(S)$, because for any $\mathbf{f} \in \mathscr{F}(S)$ we have the equality

$$\mathbf{f} = \sum_{s \in S} \mathbf{f}(s)\chi_s$$

as is seen by evaluating the right-hand side at an arbitrary element $t \in S$ giving

$$\left[\sum_{s \in S} \mathbf{f}(s)\chi_s\right](t) = \sum_{s \in S} \mathbf{f}(s) \cdot \chi_s(t)$$
$$= \left[\sum_{s \neq t} \mathbf{f}(s) \cdot 0\right] + \mathbf{f}(t) = \mathbf{f}(t)$$

as required.

Remark. Example 8 cannot be extended to infinite sets. For if S is an infinite set then the function

$$\mathbf{f}: S \to \mathbb{R}$$

defined by $\mathbf{f}(s) = 1, s \in S$ is not a (*finite*) linear combination of characteristic functions.

Exercises

1. Which of the following sets of vectors in $\mathbb{R}^3$ are linearly dependent and which are linearly independent?

$$E = \{(1, 1, 1), (0, 1, 0), (1, 0, 1)\}$$
$$F = \{(1, 1, 1), (1, 1, 0), (1, 0, 0)\}$$
$$G = \{(1, 1, 1), (1, 1, 0), (1, 0, 1)\}$$
$$H = \{(1, 0, 0), (0, 1, 0), (1, 1, 1)\}$$
$$K = \{(1, 1, 1), (0, 1, 0), (0, 0, 1)\}$$

2. Which of the following sets of vectors in $\mathscr{P}_2(\mathbb{R})$ are linearly dependent and which are linearly independent?

$$E = \{1, x, x^2\}$$
$$F = \{1 + x, 1 - x, x^2, 1\}$$
$$G = \{x^2 - 1, x + 1, x^2 - x, x^2 + x\}$$
$$H = \{x - x^2, x^2 - x\}$$
$$K = \{1, 1 - x, 1 - x^2\}$$

3. Show that the set of vectors in $\mathbb{R}^3$

$$E = \{(1, 0, 0), (0, 1, 0), (0, 0, 1), (1, 1, 1)\}$$

is linearly dependent, but that any set of three of them is linearly independent.

4. Let $\mathscr{U}$ be the subspace of $\mathbb{R}^5$ spanned by the vectors

$$E = \{(1, 1, 0, 0, 1), (1, 1, 0, 1, 1), (0, 1, 1, 1, 1), (2, 1, -1, 0, 1)\}$$

Find a linearly independent subset F of E with $\mathscr{L}(F) = \mathscr{U}$.

5. Let $\mathscr{U}$ be the subspace of $\mathscr{P}_3(\mathbb{R})$ spanned by

$$E = \{x^3, x^3 - x^2, x^3 + x^2, x^3 - 1\}$$

find a linearly independent subset F of E spanning $\mathscr{U}$.

6. Under what conditions on the numbers a and b are the vectors $(1, a)$, $(1, b)$ linearly independent in $\mathbb{R}^2$.

7. Suppose that E and F are sets of vectors in $\mathscr{V}$ with $E \subset F$. Prove that if E is linearly dependent then so is F.

8. Suppose that E and F are sets of vectors in $\mathscr{V}$ with $E \subset F$. Prove that if F is linearly independent then so is E.

9. Show that functions e^x and e^{2x} form a set of linearly independent vectors in $\mathscr{C}(-\infty, \infty)$.

10. Show that

$$\left\{\cos x, \sin x, \sin\left(x + \frac{\pi}{2}\right)\right\}$$

is a set of linearly dependent vectors in $\mathscr{C}(-\infty, \infty)$.

11. Is the pair of complex numbers $\alpha + \beta i$, $\alpha - \beta i$, a set of linearly independent vectors in $\mathbb{C}$ where α, β are any nonzero real numbers?

12. Show that the set of polynomials $E = \{x^2, 1 + x^2\}$ is a set of linearly independent vectors in $\mathscr{P}_2(\mathbb{R})$. What is the space spanned by E? Is $\mathscr{L}(E) = \mathscr{P}_2(\mathbb{R})$? If not, find a vector in $\mathscr{P}_2(\mathbb{R})$ which does not belong to $\mathscr{L}(E)$, and show together with E they form a set of linearly independent vectors in $\mathscr{P}_2(\mathbb{R})$.

13. Let E be a set of vectors in the vector space $\mathscr{V}$. Show that E is linearly independent iff for every proper subset E' of E, $\mathscr{L}(E') \neq \mathscr{L}(E)$. Equivalently, show that E is linearly dependent iff there exists a proper subset E' of E such that $\mathscr{L}(E') = \mathscr{L}(E)$.

14. Let $S = \{x_1, x_2, x_3\}$ and consider the functions $\mathbf{f}, \mathbf{g}, \mathbf{h} \in \mathscr{F}(S)$ defined by

$$\mathbf{f}(x_1) = 0, \qquad \mathbf{f}(x_2) = 1, \qquad \mathbf{f}(x_3) = 1,$$
$$\mathbf{g}(x_1) = 1, \qquad \mathbf{g}(x_2) = 0, \qquad \mathbf{g}(x_3) = 1,$$
$$\mathbf{h}(x_1) = 1, \qquad \mathbf{h}(x_2) = 1, \qquad \mathbf{h}(x_3) = 0.$$

Do $\{\mathbf{f}, \mathbf{g}, \mathbf{h}\}$ form a linearly independent set?

15. Let E', E'' be linearly independent sets of vectors in $\mathscr{V}$. Show that:
(a) $E' \cap E''$ is linearly independent.
(b) $E' \cup E''$ is linearly independent iff $\mathscr{L}(E') \cap \mathscr{L}(E'') = \{0\}$.

16. Let $\mathbf{A}, \mathbf{B}, \mathbf{C} \in \mathscr{V}$ be linearly independent vectors. Show that $\{\mathbf{A} + \mathbf{B}, \mathbf{B} + \mathbf{C}, \mathbf{C} + \mathbf{A}\}$ are linearly independent.

17. Let $\mathscr{S}, \mathscr{T}$ be subspaces of a vector space $\mathscr{V}$, $\mathbf{S} \in \mathscr{S}$, $\mathbf{T} \in \mathscr{T}$, $\mathbf{S} \neq 0$, $\mathbf{T} \neq 0$. Assume $\mathscr{S} \cap \mathscr{T} = \{0\}$. Show that $\{\mathbf{S}, \mathbf{T}\}$ is a linearly independent set of vectors.

18. Show that any four vectors in $\mathbb{R}^3$ must be linearly dependent.

19. Let $\mathscr{S}, \mathscr{T}$ be subspaces of $\mathscr{V}$ and S a set. Suppose $\mathbf{f} \in \text{Fun}(S, \mathscr{S})$, $\mathbf{g} \in \text{Fun}(S, \mathscr{T})$, and $\mathscr{S} \cap \mathscr{T} = 0$. Show that, regarded as vectors in $\text{Fun}(S, \mathscr{V})$, the vectors $\mathbf{f}$ and $\mathbf{g}$ are linearly independent.

20. Let $\mathbf{A}, \mathbf{B}, \mathbf{C} \in \mathscr{V}$. If $\mathbf{C} \notin \mathscr{L}(\mathbf{A}, \mathbf{B})$ show that $\mathbf{A}$ and $\mathbf{B}$ are linearly independent iff $\mathbf{A} + \mathbf{C}$ and $\mathbf{B} + \mathbf{C}$ are linearly independent.

21. Let $\mathbf{f}_1, \ldots, \mathbf{f}_k \in \mathscr{P}(\mathbb{R})$ be polynomials of positive degree. Suppose $\mathbf{f}_1, \ldots, \mathbf{f}_k$ are linearly independent. Does it follow that $(d/dt)\mathbf{f}_1, \ldots, (d/dt)\mathbf{f}_k$ are linearly independent?

22. Let S be a set. A permutation σ of S is a bijective (= one-to-one and onto) mapping $\sigma\colon S \to S$. For a permutation σ of S and $\mathbf{f} \in \mathscr{F}(S)$ define

$$\mathbf{f}_\sigma\colon S \to \mathbb{R} \,|\, \mathbf{f}_\sigma(s) = \mathbf{f}(\sigma(s)).$$

Suppose $\mathbf{f}_1, \ldots, \mathbf{f}_n \in \mathscr{F}(S)$ are linearly independent. Show that $(\mathbf{f}_1)_\sigma, \ldots, (\mathbf{f}_n)_\sigma$ are also linearly independent.

23. If $\mathbf{f}_1, \ldots, \mathbf{f}_n \in \mathscr{P}_m(\mathbb{R})$ are linearly dependent then so are $(d/dt)\mathbf{f}_1, \ldots, (d/dt)\mathbf{f}_n$.

24. Let $\mathscr{V}$ be the set of all polynomials $p(x) \in \mathscr{P}_m(\mathbb{R})$, $m \geq 2$ such that

$$p(0) = 0 \quad \text{and} \quad \left.\frac{d}{dx}\right|_{x=0} p(x) = 0.$$

Show that $\mathscr{V}$ is a subspace of $\mathscr{P}_m(\mathbb{R})$ and find a basis for this subspace. (See page 43 for the definition of basis.)

6 Bases and finite-dimensional vector spaces

Definition. A vector space $\mathscr{V}$ is said to be *finite dimensional* iff there exists a *finite* set of vectors E with $\mathscr{L}(E) = \mathscr{V}$.

EXAMPLE 1. $\mathbb{R}^k$ is finite dimensional.

To see this we introduce the vectors

$$\mathbf{E}_i = (0, 0, \ldots, \underset{\substack{\uparrow \\ i\text{th place}}}{1}, 0, \ldots, 0), \qquad i = 1, \ldots, k.$$

For example, if $k = 4$ then

$$\begin{aligned} \mathbf{E}_1 &= (1, 0, 0, 0), \\ \mathbf{E}_2 &= (0, 1, 0, 0), \\ \mathbf{E}_3 &= (0, 0, 1, 0), \\ \mathbf{E}_4 &= (0, 0, 0, 1). \end{aligned}$$

If $\mathbf{A}$ is any vector in $\mathbb{R}^k$ then $\mathbf{A} = (a_1, \ldots, a_k)$ and hence we find

$$\mathbf{A} = a_1\mathbf{E}_1 + \cdots + a_k\mathbf{E}_k.$$

For example in $\mathbb{R}^4$ we have

$$(1, 2, 3, 4) = 1\mathbf{E}_1 + 2\mathbf{E}_2 + 3\mathbf{E}_3 + 4\mathbf{E}_4.$$

Thus $\mathbb{R}^k = \mathscr{L}(E_1, E_2, \ldots, E_k)$ and since $E = \{E_1, \ldots, E_k\}$ is a finite set, $\mathbb{R}^k$ is finite dimensional.

EXAMPLE 2. $\mathscr{P}_k(\mathbb{R})$ is finite dimensional.

To see this let $E = \{1, x, x^2, \ldots, x^k\}$. If $p(x)$ is any vector in $\mathscr{P}_k(\mathbb{R})$ then

$$p(x) = a_0 1 + a_1 x + \cdots + a_k x^k$$

and hence $\mathscr{P}_k(\mathbb{R}) = \mathscr{L}(1, x, x^2, \ldots, x^k)$. Since the set $E = \{1, x, \ldots, x^k\}$ is finite $\mathscr{P}_k(\mathbb{R})$ is finite dimensional.

EXAMPLE 3. Let S be a *finite* set then $\mathscr{F}(S)$ is finite dimensional because the vectors (characteristic functions) $X = \{\chi_s | s \in S\}$ span $\mathscr{F}(S)$ (Chapter 5, Example 8).

Theorem 6.1. *Let $\mathscr{V}$ be a finite-dimensional vector space. Then there exists a finite set of linearly independent vectors F that spans $\mathscr{V}$, that is $\mathscr{L}(F) = \mathscr{V}$.*

PROOF. Since $\mathscr{V}$ is finite dimensional there is a finite set of vectors E with $\mathscr{L}(E) = \mathscr{V}$. By (5.5) we may find a linearly independent set $F \subset E$ with $\mathscr{L}(F) = \mathscr{L}(E)$. Then F is the required set of vectors. □

The property that a set of vectors be both linearly independent and span $\mathscr{V}$ is most important and fundamental to further developments. We therefore introduce:

Definition. A set of vectors E in a vector space $\mathscr{V}$ is called a *basis* for $\mathscr{V}$ iff E is linearly independent and $\mathscr{L}(E) = \mathscr{V}$.

EXAMPLE 4. The vectors (1, 1), (1, 0) are a basis for $\mathbb{R}^2$.

To show this we must check that (1, 1), (1, 0) are linearly independent and span $\mathbb{R}^2$. To see they span, suppose $(x, y) \in \mathbb{R}^2$. We wish to find numbers a, b such that

$$(x, y) = a(1, 1) + b(1, 0) = (a + b, a)$$

therefore

$$y = a, \qquad x = a + b,$$

so solving for a and b we get

$$a = y, \qquad b = x - y,$$

and therefore

$$(x, y) = y(1, 1) + (x - y)(1, 0)$$

showing $(x, y) \in \mathscr{L}\{(1, 1), (1, 0)\}$, and therefore that $\{(1, 1), (1, 0)\}$ spans $\mathbb{R}^2$. To show (1, 1), (1, 0) linearly independent, we suppose that

$$a(1, 1) + b(1, 0) = (0, 0)$$

is a linear relation between them. Then, as above we get

$$0 = a, \qquad 0 = a + b$$

whence $a = 0$, $b = 0$ as required.

EXAMPLE 5. The vectors

$$1, x - 1, (x - 2)(x - 1)$$

form a basis for $\mathscr{P}_2(\mathbb{R})$.

We must again check linear independence and spanning. To check linear independence suppose

$$0 = a_1 \cdot 1 + a_2(x - 1) + a_3(x - 2)(x - 1)$$

is a linear relation. Multiplying out gives

$$\begin{aligned} 0 &= a_1 + a_2(x - 1) + a_3(x^2 - 3x + 2) \\ &= a_1 + a_2 x - a_2 + a_3 x^2 - 3a_3 x + 2a_3 \\ &= (a_1 - a_2 + 2a_3) + (a_2 - 3a_3)x + a_3 x^2 \end{aligned}$$

so we must have (a polynomial is zero iff all its coefficients are zero)

$$\begin{aligned} 0 &= a_1 - a_2 + 2a_3 \\ 0 &= a_2 - 3a_3 \\ 0 &= a_3 \end{aligned}$$

which, solving in reverse order, gives

$$a_3 = 0, \qquad a_2 = 0, \qquad a_1 = 0$$

as required.

To show $\{1, x - 1, (x - 2)(x - 1)\}$ is a spanning set let $p(x) = a_0 + a_1 x + a_2 x^2 \in \mathscr{P}_2(\mathbb{R})$. We are looking for numbers b_0, b_1, b_2 such that

$$a_0 + a_1 x + a_2 x^2 = b_0 + b_1(x - 1) + b_2(x - 2)(x - 1).$$

Multiplying out and equating coefficients we obtain the simultaneous equations

$$\begin{aligned} a_0 &= b_0 - b_1 + 2b_2 \\ a_1 &= b_1 - 3b_2 \\ a_2 &= b_2 \end{aligned}$$

which may be solved to yield

$$\begin{aligned} b_2 &= a_2 \\ b_1 &= a_1 + 3a_2 \\ b_0 &= a_0 + a_1 + a_2 \end{aligned}$$

so

$$p(x) = (a_0 + a_1 + a_2)1 + (a_1 + 3a_2)(x - 1) + a_2(x - 2)(x - 1)$$

and therefore $\{1, x - 1, (x - 2)(x - 1)\}$ spans $\mathscr{P}_2(\mathbb{R})$.

EXAMPLE 6. In $\mathbb{R}^3$ the vectors $(1, -1, 0)$, $(0, 1, -1)$ are a basis for the subspace

$$\mathscr{V} = \{(x, y, z) \in \mathbb{R}^3 \mid x + y + z = 0\}.$$

To see this we first show that $(1, -1, 0)$, $(0, 1, -1)$ span $\mathscr{V}$. So let $(x, y, z) \in \mathscr{V}$. Then

$$(*) \qquad x + y + z = 0.$$

We wish to write

$$\begin{aligned}(x, y, z) &= a(1, -1, 0) + b(0, 1, -1)\\ &= (a, b - a, -b)\end{aligned}$$

so we must have

$$x = a, \qquad y = b - a, \qquad z = -b.$$

To solve for a and b use the first and last equations giving

$$a = x, \qquad b = -z.$$

But is this consistent with the middle equation? Substituting gives

$$y = -z - x$$

which recalling $(*)$ we see is valid precisely for the vectors of $\mathscr{V}$. Thus if $(x, y, z) \in \mathscr{V}$ then

$$(x, y, z) = x(1, -1, 0) - z(0, 1, -1),$$

and therefore the vectors $(1, -1, 0)$, $(0, 1, -1)$ span $\mathscr{V}$. The check for linear independence of $\{(1, -1, 0), (0, 1, -1)\}$ is easy and we omit it.

Example 6 is illustrated in Figure 6.1.

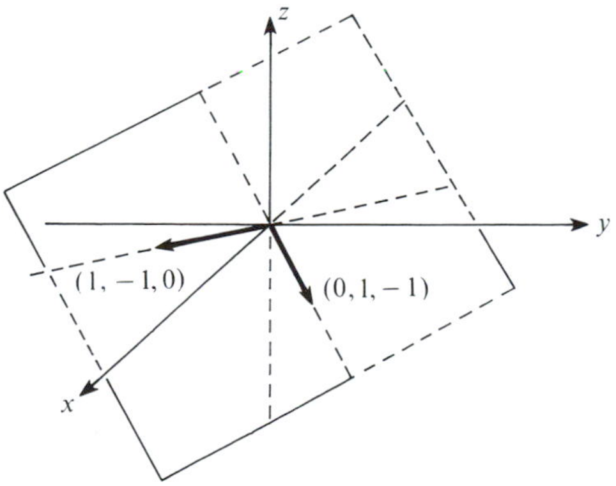

Figure 6.1

The preceding theorem shows that every finite-dimensional vector space has a basis. Actually more is true, namely any two bases for a finite-dimensional vector space contain the same number of vectors. To prove this important result we will need a few preliminary steps, the first of which is:

Proposition 6.2. *Let $E = \{\mathbf{A}_1, \ldots, \mathbf{A}_k\}$ be a finite set of vectors. Then E is linearly dependent iff*

$$\mathbf{A}_m \in \mathscr{L}(\mathbf{A}_1, \ldots, \mathbf{A}_{m-1})$$

for some $m \leq k$.

PROOF. If for some $m \leq k$ we have $\mathbf{A}_m \in \mathscr{L}(\mathbf{A}_1, \ldots, \mathbf{A}_{m-1})$ then $\mathbf{A}_m$ is linearly dependent on the remaining vectors of E and hence by (5.2) E is a linear dependent set of vectors.

On the other hand suppose that E is a linearly dependent set of vectors. Then there exists a linear relation

$$a_1\mathbf{A}_1 + a_2\mathbf{A}_2 + \cdots + a_k\mathbf{A}_k = 0.$$

Choose m to be the largest integer between 1 and k for which $a_m \neq 0$. Then since $a_{m+1} = \cdots = a_n = 0$ we may write our linear relation as

$$a_1\mathbf{A}_1 + a_2\mathbf{A}_2 + \cdots + a_m\mathbf{A}_m = 0$$

and since $a_m \neq 0$ we may solve for $\mathbf{A}_m$ obtaining

$$\mathbf{A}_m = \frac{-a_1}{a_m}\mathbf{A}_1 + \frac{-a_2}{a_m}\mathbf{A}_2 + \cdots + \frac{-a_{m-1}}{a_m}\mathbf{A}_m$$

which shows that $\mathbf{A}_m \in \mathscr{L}(\mathbf{A}_1, \ldots, \mathbf{A}_{m-1})$. □

Proposition 6.3. *Let $E = \{\mathbf{A}_1, \ldots, \mathbf{A}_k\}$ be a finite set of vectors in $\mathscr{V}$. If F is any linearly independent set of vectors in $\mathscr{L}(E)$ then F is finite and the number of elements in F is at most k.*

PROOF. Suppose that H is a finite subset of F. Let the vectors of H be $\mathbf{B}_1, \ldots, \mathbf{B}_s$. Note that $\mathbf{B}_1, \ldots, \mathbf{B}_s$ are linearly independent. We must show that $s \leq k$. To do this we consider the set of vectors

$$G_1 = \{\mathbf{B}_s, \mathbf{A}_1, \ldots, \mathbf{A}_k\}.$$

Note that this set is linearly dependent by (5.3) since

$$\mathbf{B}_s \in \mathscr{L}(E) = \mathscr{L}(\mathbf{A}_1, \ldots, \mathbf{A}_k).$$

Therefore we may apply (6.2) to conclude that G_1 is linearly dependent in another way, namely there exists a vector in the set G_1 linearly dependent on the *preceding* vectors. It cannot be $\mathbf{B}_s$ because $\mathbf{B}_s \neq 0$ (Why?). Therefore it must be an $\mathbf{A}_i$, and by rearranging terms we may assume that it is $\mathbf{A}_k$. That is

$$\mathbf{A}_k \in \mathscr{L}(\mathbf{B}_s, \mathbf{A}_1, \ldots, \mathbf{A}_{k-1}).$$

Now let $E_1 = \{\mathbf{B}_s, \mathbf{A}_1, \ldots, \mathbf{A}_{k-1}\}$ and note $\mathscr{L}(\mathbf{B}_s, \mathbf{A}_1, \ldots, \mathbf{A}_{k-1}) = \mathscr{L}(\mathbf{A}_1, \ldots, \mathbf{A}_k)$. Reasoning as before we see that some vector in the set

$$G_2 = \{\mathbf{B}_{s-1}, \mathbf{B}_s, \mathbf{A}_1, \ldots, \mathbf{A}_{k-1}\}$$

must be linearly dependent on the vectors that precede it. It cannot be $\mathbf{B}_{s-1}$ since $\mathbf{B}_{s-1} \neq 0$. Nor can it be $\mathbf{B}_s$ since $\{\mathbf{B}_{s-1}, \mathbf{B}_s\}$ are linearly independent (Why?). Therefore it must be an $\mathbf{A}_j$ and again by shuffling we may assume that it is $\mathbf{A}_{k-1}$. Thus

$$\mathbf{A}_{k-1} \in \mathscr{L}(\mathbf{B}_{s-1}, \mathbf{B}_s, \mathbf{A}_1, \ldots, \mathbf{A}_{k-2})$$

and hence that

$$\begin{aligned}\mathscr{L}(\mathbf{B}_{s-1}, \mathbf{B}_s, \mathbf{A}_1, \ldots, \mathbf{A}_{k-2}) &= \mathscr{L}(\mathbf{B}_s, \mathbf{A}_1, \ldots, \mathbf{A}_{k-1}) \\ &= \mathscr{L}(\mathbf{A}_1, \ldots, \mathbf{A}_k)\end{aligned}$$

Suppose now that $s > k$. Then we may continue the above process to show that

$$\mathscr{L}(\mathbf{B}_{s-k+1}, \mathbf{B}_{s-k+2}, \ldots, \mathbf{B}_s) = \mathscr{L}(\mathbf{A}_1, \ldots, \mathbf{A}_k)$$

by repeating the above argument k times. But then

$$\mathbf{B}_{s-k} \in \mathscr{L}(\mathbf{A}_1, \ldots, \mathbf{A}_k) = \mathscr{L}(\mathbf{B}_{s-k+1}, \ldots, \mathbf{B}_s)$$

and hence $\{\mathbf{B}_1, \ldots, \mathbf{B}_s\}$ is a linearly dependent set by (5.3) contrary to our hypothesis. Therefore $s \leq k$ as required. □

Theorem 6.4. *A finite-dimensional vector space $\mathscr{V}$ has a basis. Any two bases contain the same number of elements.*

Proof. We proved the first statement in (6.1). To prove the second statement suppose that

$$E = \{\mathbf{A}_1, \ldots, \mathbf{A}_n\}, \qquad F = \{\mathbf{B}_1, \ldots, \mathbf{B}_m\}$$

are bases for $\mathscr{V}$. Then F is a linearly independent set in $\mathscr{L}(E) = \mathscr{V}$ so that $m \leq n$ by (6.3). On the other hand E is a linearly independent set in $\mathscr{V} = \mathscr{L}(F)$ so $n \leq m$, also by (6.3). Therefore $m = n$. □

Example 7. $\mathscr{P}(\mathbb{R})$ is not finite dimensional.

To see this consider the set $F = \{1, x, x^2, \ldots, x^n, \ldots\}$ of vectors in $\mathscr{P}(\mathbb{R})$. This set is linearly independent. If $\mathscr{P}(\mathbb{R})$ were finite dimensional, say $\mathscr{P}(\mathbb{R}) = \mathscr{L}(A_1, \ldots, A_k)$ then applying (6.4) we would conclude that F is a finite set which it is not. Therefore $\mathscr{P}(\mathbb{R})$ is not finite dimensional.

If a vector space $\mathscr{V}$ is not finite dimensional, then we say it is *infinite dimensional.* It is possible to combine (6.4) with the ideas of the preceding example to characterize finite- and infinite-dimensional vector spaces in terms of the sets of linearly independent vectors that they contain.

Theorem 6.5. *A vector space $\mathscr{V}$ is finite dimensional iff every linearly independent set of vectors in $\mathscr{V}$ is finite. A vector space $\mathscr{W}$ is infinite dimensional iff there exists in $\mathscr{W}$ an infinite linearly independent set of vectors.*

PROOF. If $\mathscr{V}$ is finite dimensional then (6.3) says that every linear independent set of vectors in $\mathscr{V}$ is finite. On the other hand suppose that every linearly independent set in $\mathscr{V}$ is finite but the vector space $\mathscr{V}$ is not finite dimensional. (Remember that this means $\mathscr{V}$ is not spanned by any finite set of vectors in $\mathscr{V}$.) Let $\mathbf{A}_1 \neq 0 \in \mathscr{V}$. Then $\{\mathbf{A}_1\}$ is a linearly independent set of vectors. Since $\mathscr{V}$ is not finite dimensional $\mathscr{L}(\mathbf{A}_1) \neq \mathscr{V}$. Therefore we may select a vector $\mathbf{A}_2$ in $\mathscr{V}$ that is not in $\mathscr{L}(\mathbf{A}_1)$. From (6.2) it follows that $\{\mathbf{A}_1, \mathbf{A}_2\}$ is a linearly independent set. Let us repeat this process. In this way we obtain vectors $\mathbf{A}_1, \mathbf{A}_2, \mathbf{A}_3, \ldots$ such that $\mathbf{A}_{i+1} \notin \mathscr{L}(\mathbf{A}_1, \ldots, \mathbf{A}_i)$ and hence by (6.2) the infinite set $\{\mathbf{A}_1, \mathbf{A}_2, \ldots\}$ of vectors in $\mathscr{V}$ is linearly independent. This is a contradiction of the fact that every linearly independent set of vectors in $\mathscr{V}$ is finite. Therefore the assumption that $\mathscr{V}$ is not finite dimensional must be false, so $\mathscr{V}$ is finite dimensional. □

EXAMPLE 8. If S is an infinite set then $\mathscr{F}(S)$ is infinite dimensional.

To see this notice that in Example 8 of Chapter 5 it was shown that the set $X = \{\chi_s \mid s \in S\}$ of characteristic functions is a linearly independent set of vectors in $\mathscr{F}(S)$. If S is infinite, this set is infinite.

Notice that (6.5) says that the concept of finite dimensionality, which we defined in terms of spanning properties of vectors, is equivalent to certain linear independence properties of vectors. It is for this reason that bases are so important because they combine the spanning and linear independence concepts.

Definition. Let $\mathscr{V}$ be a finite-dimensional vector space. Then the number of vectors in any basis for $\mathscr{V}$ is called the *dimension* of $\mathscr{V}$, and is written $\dim \mathscr{V}$.

EXAMPLE 9. $\dim \mathbb{R}^k = k$.

We have already seen that the set $E = \{\mathbf{E}_1, \ldots, \mathbf{E}_k\}$ is a basis for $\mathbb{R}^k$.

EXAMPLE 10. $\dim \mathscr{P}_n(\mathbb{R}) = n + 1$.

We have already seen that the set $E = \{1, x, x^2, \ldots, x^n\}$ is a basis for $\mathscr{P}_n(\mathbb{R})$.

Notation. If S is any set we write $|S|$ for the number of elements of S when S is finite and let $|S| = \infty$ when S is not finite.

By combining Examples 3 and 8 we obtain:

EXAMPLE 11. $\dim \mathscr{F}(S) = |S|$.

Theorem 6.6. *Let $\mathscr{V}$ be a finite-dimensional vector space with basis $\mathbf{A}_1, \ldots, \mathbf{A}_n$. Then any vector $\mathbf{A} \in \mathscr{V}$ may be written uniquely as a linear combination*

$$\mathbf{A} = a_1\mathbf{A}_1 + a_2\mathbf{A}_2 + \cdots + a_n\mathbf{A}_n.$$

The numbers in the sequence $(a_1, \ldots, a_n)$ *are called* the coordinates *(or components) of* $\mathbf{A}$ *relative to the ordered basis* $\mathbf{A}_1, \ldots, \mathbf{A}_n$.

PROOF. Since $\mathbf{A}_1, \ldots, \mathbf{A}_n$ is a basis for $\mathscr{V}$ we have $\mathscr{V} = \mathscr{L}(\mathbf{A}_1, \ldots, \mathbf{A}_n)$ so since $\mathbf{A} \in \mathscr{V} = \mathscr{L}(\mathbf{A}_1, \ldots, \mathbf{A}_n)$ there is at least one way to write $\mathbf{A}$ as a linear combination

$$\mathbf{A} = a_1\mathbf{A}_1 + a_2\mathbf{A}_2 + \cdots + a_n\mathbf{A}_n.$$

Suppose that

$$\mathbf{A} = b_1\mathbf{A}_1 + b_2\mathbf{A}_2 + \cdots + b_n\mathbf{A}_n$$

were another way to write $\mathbf{A}$ as a linear combination of $\mathbf{A}_1, \ldots, \mathbf{A}_n$. Then

$$\mathbf{0} = \mathbf{A} - \mathbf{A} = (a_1 - b_1)\mathbf{A}_1 + (a_2 - b_2)\mathbf{A}_2 + \cdots + (a_n - b_n)\mathbf{A}_n.$$

Since $\{\mathbf{A}_1, \ldots, \mathbf{A}_n\}$ is a basis it is linearly independent, hence none of the coefficients in the preceding equation can be nonzero. That is

$$a_1 - b_1 = 0, \qquad a_2 - b_2 = 0, \qquad \ldots, \qquad a_n - b_n = 0$$

and hence

$$a_1 = b_1, \qquad a_2 = b_2, \qquad \ldots, \qquad a_n = b_n$$

which establishes uniqueness. □

EXAMPLE 12. Find the coordinates of $2 - x$ relative to the basis $\{(1 - x), (1 + x)\}$ for $\mathscr{P}_1(\mathbb{R})$.

Solution. We have by (6.6)

$$2 - x = a(1 - x) + b(1 + x)$$

for suitable numbers a and b. Multiplying out gives

$$2 - x = a + b + (b - a)x$$

and equating coefficients gives

$$2 = a + b, \qquad -1 = -a + b.$$

Therefore $b = \frac{1}{2}$ and $a = \frac{3}{2}$. So the answer is $(\frac{3}{2}, \frac{1}{2})$.

EXAMPLE 13. Find the coordinates of $(1, 1, 1)$ relative to the basis $\{(1, 1, 1), (1, 1, 0), (1, 0, 0)\}$ for $\mathbb{R}^3$.

Answer. $(1, 0, 0)$. Think about this one!

EXAMPLE 14. Find the coordinates of $1 + x + x^2$ relative to the basis $\{1, x - 1, (x - 2)(x - 1)\}$ for $\mathscr{P}_2(\mathbb{R})$.

Solution. We require numbers a, b, c such that

$$1 + x + x^2 = a + b(x - 1) + c(x - 2)(x - 1).$$

Multiplying out gives

$$\begin{aligned}1 + x + x^2 &= a + b(x - 1) + c(x^2 - 3x + 2)\\ &= a + bx - b + cx^2 - 3cx + 2c\\ &= (a - b + 2c) + (b - 3c)x + cx^2\end{aligned}$$

and so equating coefficients gives

$$\begin{aligned}1 &= a - b + 2c\\ 1 &= b - 3c\\ 1 &= c\end{aligned}$$

which solve to give

$$c = 1, \qquad b = 4, \qquad a = 3$$

whence

$$1 + x + x^2 = 3 \cdot 1 + 4(x - 1) + (x - 2)(x - 1),$$

so that the coordinates of $1 + x + x^2$ relative to the basis $\{1, x - 1, (x - 2)(x - 1)\}$ are $(3, 4, 1)$.

Theorem 6.7. *Let $\mathscr{V}$ be a finite-dimensional vector space and $\mathbf{A}_1, \ldots, \mathbf{A}_m$ linearly independent vectors in $\mathscr{V}$. Then there exist vectors $\mathbf{B}_1, \ldots, \mathbf{B}_n$ in $\mathscr{V}$ such that the set $\{\mathbf{A}_1, \ldots, \mathbf{A}_m, \mathbf{B}_1, \ldots, \mathbf{B}_n\}$ is a basis for $\mathscr{V}$.*

PROOF. Suppose that $\mathscr{L}(\mathbf{A}_1, \ldots, \mathbf{A}_m) = \mathscr{V}$. Then there is nothing to prove. So we may suppose that $\mathscr{L}(\mathbf{A}_1, \ldots, \mathbf{A}_m) \neq \mathscr{V}$. Let $\mathbf{B}_1$ be a vector in $\mathscr{V}$ that is not in $\mathscr{L}(\mathbf{A}_1, \ldots, \mathbf{A}_m)$. Then the set $\{\mathbf{A}_1, \ldots, \mathbf{A}_m, \mathbf{B}_1\}$ is linearly independent. For if it were linearly dependent then by (6.2) some vector would be linearly dependent on the preceding ones. The vector cannot be an $\mathbf{A}$ because $\mathbf{A}_1, \ldots, \mathbf{A}_m$ are linearly independent. It cannot be $\mathbf{B}_1$ because we chose $\mathbf{B}_1$ so that $\mathbf{B}_1 \notin \mathscr{L}(\mathbf{A}_1, \ldots, \mathbf{A}_m)$. Therefore

$$\{\mathbf{A}_1, \ldots, \mathbf{A}_m, \mathbf{B}_1\}$$

is a linearly independent set. If $\mathscr{L}(\mathbf{A}_1, \ldots, \mathbf{A}_m, \mathbf{B}_1) = \mathscr{V}$ we are done. If not we may repeat the argument starting with the vectors, $\mathbf{A}_1, \ldots, \mathbf{A}_m, \mathbf{B}_1$. In this way we obtain a set

$$\{\mathbf{A}_1, \ldots, \mathbf{A}_m, \mathbf{B}_1, \ldots, \mathbf{B}_n\}$$

of linearly independent vectors in $\mathscr{V}$. By (6.4) this process must stop when $m + n = \dim \mathscr{V}$, in which case

$$\mathscr{L}(\mathbf{A}_1, \ldots, \mathbf{A}_m, \mathbf{B}_1, \ldots, \mathbf{B}_n) = \mathscr{V}$$

as required. □

Theorem 6.8. *Let $\mathscr{V}$ be a finite-dimensional vector space and $\mathscr{U}$ a linear subspace of $\mathscr{V}$. Then $\mathscr{U}$ is finite dimensional and* $\dim \mathscr{U} \leq \dim \mathscr{V}$.

Proof. Suppose that $\mathscr{U}$ is not finite dimensional. Then by (6.5) there is an infinite linearly independent set E of vectors in $\mathscr{U}$. But then E is certainly an infinite linearly independent set of vectors in $\mathscr{V}$, and hence again by (6.5) $\mathscr{V}$ is infinite dimensional, contradiction! Therefore $\mathscr{U}$ is finite dimensional. Let $\{\mathbf{B}_1, \ldots, \mathbf{B}_m\}$ be a basis for $\mathscr{U}$ and $\{\mathbf{A}_1, \ldots, \mathbf{A}_n\}$ a basis for $\mathscr{V}$. Then by (6.3) $m \leq n$. □

Very often in dealing with problems of a concrete nature, we will be working in one of the finite-dimensional vector spaces $\mathbb{R}^n$, $\mathscr{P}_n(\mathbb{R})$ whose dimension we already know. If we wish to check that some set $\{\mathbf{A}_1, \ldots, \mathbf{A}_k\}$ of vectors is a basis in one of these spaces it turns out that we must only check that the set contains the correct number of vectors to be a basis and has *one* of the two characteristic properties; *linear independence*, *spanning*, of a basis. More precisely we have:

Theorem 6.9. *Let $\mathscr{V}$ be a finite-dimensional vector space of dimension n. If the vectors $\mathbf{A}_1, \ldots, \mathbf{A}_n$ in $\mathscr{V}$ are linearly independent then they are a basis.*

Proof. By (6.7) if $\mathbf{A}_1, \ldots, \mathbf{A}_n$ is not a basis we may find vectors $\mathbf{B}_1, \ldots, \mathbf{B}_m$ so that $\mathbf{A}_1, \ldots, \mathbf{A}_n, \mathbf{B}_1, \ldots, \mathbf{B}_m$ is a basis for $\mathscr{V}$. But then $\dim \mathscr{V} = m + n \neq n$ a contradiction. □

Theorem 6.10. *Let $\mathscr{V}$ be a finite-dimensional vector space of dimension n. If the vectors $\mathbf{A}_1, \ldots, \mathbf{A}_n$ span $\mathscr{V}$ they are a basis.*

Proof. Since the vectors $\mathbf{A}_1, \ldots, \mathbf{A}_n$ span $\mathscr{V}$ some subset of $\{\mathbf{A}, \ldots, \mathbf{A}_n\}$ is a basis for $\mathscr{V}$ by (5.4). By rearranging the ordering we may assume that $\{\mathbf{A}_1, \ldots, \mathbf{A}_m\}$ is a basis for $\mathscr{V}$ for some $m \leq n$. Then $m = \dim \mathscr{V} = n$ by (6.5) which shows $m = n$ and hence that $\mathbf{A}_1, \ldots, \mathbf{A}_n$ is a basis for $\mathscr{V}$. □

Corollary 6.11. *Let $\mathscr{V}$ be a finite-dimensional vector space of dimension n. Suppose that $\mathscr{U}$ is a linear subspace of $\mathscr{V}$ and* $\dim \mathscr{U} = n$. *Then $\mathscr{U} = \mathscr{V}$.*

Proof. Since $\dim \mathscr{U} = n$ there is a basis $\{\mathbf{A}_1, \ldots, \mathbf{A}_n\}$ for $\mathscr{U}$. By (6.9) $\{\mathbf{A}_1, \ldots, \mathbf{A}_n\}$ must be a basis for $\mathscr{V}$. Thus

$$\mathscr{U} = \mathscr{L}(\mathbf{A}_1, \ldots, \mathbf{A}_n) = \mathscr{V}$$

as required. □

Example 15. The vectors (1, 1, 1), (1, 1, 0), (1, 0, 0) are a basis for $\mathbb{R}^3$.

Proof. There are three vectors above and $\mathbb{R}^3$ has dimension 3, so by (6.9) we need only check that the three vectors are linearly independent. If

$$a_1(1, 1, 1) + a_2(1, 1, 0) + a_3(1, 0, 0) = (0, 0, 0)$$

then

$$(a_1 + a_2 + a_3, a_1 + a_2, a_1) = (0, 0, 0)$$

so

$$\begin{aligned} a_1 + a_2 + a_3 &= 0 \\ a_1 + a_2 \quad\; &= 0 \\ a_1 \qquad\qquad &= 0 \end{aligned}$$

and $a_1 = a_2 = a_3 = 0$. So they are linearly independent.

EXAMPLE 16. The vectors $(1, 1, -2)$, $(0, 3, -3)$ are a basis for the subspace

$$\mathscr{V} = \{(x, y, z) | x + y + z = 0\}.$$

PROOF. We already know, by Example 6, that $\dim \mathscr{V} = 2$. So it will suffice to check

(0) $(1, 1, -2)$ and $(0, 3, -3)$ belong to $\mathscr{V}$,

(1) $(1, 1, -2)$ and $(0, 3, -3)$ are linearly independent.

Condition (0) is easy since

$$1 + 1 - 2 = 0 \quad \text{and} \quad 0 + 3 - 3 = 0.$$

To check (1), suppose

$$a(1, 1, -2) + b(0, 3, -3) = \mathbf{0}$$

is a linear relation. Then

$$\begin{aligned} (0, 0, 0) &= (a, a, -2a) + (0, 3b, -3b) \\ &= (a, a + 3b, -2a - 3b) \end{aligned}$$

so equating coefficients gives

$$\begin{aligned} 0 &= a \\ 0 &= a + 3b \\ 0 &= -2a - 3b \end{aligned}$$

and $a = 0$, $b = 0$ as required.

EXAMPLE 17. For what values of r are the vectors $(r, 1, 1)$, $(1, r, 1)$, $(1, 1, r)$ a basis for $\mathbb{R}^3$?

Solution. Since $\dim \mathbb{R}^3 = 3$ it will suffice to find when the vectors in question are linearly independent. So suppose that

$$a(r, 1, 1) + b(1, r, 1) + c(1, 1, r) = \mathbf{0}$$

is a linear relation. Then

$$(ar + b + c, a + br + c, a + b + cr) = \mathbf{0} = (0, 0, 0),$$

so taking coordinates gives

$$\begin{aligned} ar + b + c &= 0 \\ a + br + c &= 0 \\ a + b + cr &= 0. \end{aligned}$$

From the last equation we get

$$(*) \qquad cr = -(a + b).$$

Multiplying the second equation by r and substituting gives

$$\begin{aligned} 0 &= ar + br^2 - (a + b) = a(r - 1) + b(r^2 - 1) \\ &= (r - 1)[a + b(r + 1)]. \end{aligned}$$

So if $r - 1 \neq 0$ we may divide by $r - 1$ and solve for a to get

$$(**) \qquad a = -(r + 1)b.$$

Multiplying the first equation by r and substituting in $(*)$ and $(**)$ we obtain

$$\begin{aligned} 0 &= [-(r + 1)b]r^2 + br - (a + b) \\ &= -r^2(r + 1)b + rb - [-(r + 1)b + b] \\ &= [-r^2(r + 1) + r + (r + 1) - 1]b \\ &= (-r^3 - r^2 + r + r)b = -(r^3 + r^2 - 2r)b \\ &= -r(r^2 + r - 2)b = -r(r + 2)(r - 1)b. \end{aligned}$$

So assuming $r \neq 0, -2, 1$ we may divide by $-r(r + 2)(r - 1)$ and conclude $b = 0$, where $a = 0$ by $(**)$ and $c = 0$ by $(*)$. Therefore the vectors are independent when $r \neq 0, 1, -2$. For $r = 0$, the original system simplifies to

$$\begin{aligned} 0 &= b + c, \\ 0 &= a + c, \\ 0 &= a + b. \end{aligned}$$

Substituting the first equation into the second gives

$$0 = a - b$$

and adding to the third gives

$$0 = 2a$$

giving $a = 0$ and $b = 0$, $c = 0$, so again the vectors are independent. For $r = 1$, the three vectors are the same vector three times, so cannot be a basis for $\mathbb{R}^3$. And finally for $r = -2$, there is the linear relation

$$(-2, 1, 1) + (1, -2, 1) + (1, 1, -2) = 0$$

so the vectors are not linearly independent. In conclusion, the vectors $(r, 1, 1)$, $(1, r, 1)$, $(1, 1, r)$ are a basis for $\mathbb{R}^3$ if and only if $r \neq 1, -2$.

Exercises

1. Which of the following sets of vectors are bases for $\mathbb{R}^3$?

$$\begin{aligned} E &= \{(1, 1, 1), (1, 1, 0), (1, 0, 0)\} \\ F &= \{(1, 0, 1), (0, 1, 0), (1, 1, 1)\} \\ G &= \{(-1, 1, 1), (1, -1, 1), (1, 1, -1)\} \\ H &= \{(1, -1, 0), (1, 0, -1), (-1, 0, 1)\} \end{aligned}$$

2. Which of the following sets of vectors are bases for $\mathscr{P}_2(\mathbb{R})$?

$$\begin{aligned} E &= \{1, 1 + x, 1 + x + x^2\} \\ F &= \{1, (x - 1), (x - 1)(x - 2)\} \\ G &= \{x^2 + x + 1, x^2 - x + 1, x^2 - x - 1\} \\ H &= \{x^2 + x, x^2 + 1, x + 1\}. \end{aligned}$$

3. Find a basis for each of the following subspaces of $\mathbb{R}^3$.

$$\begin{aligned} \mathscr{U} &= \{(x, y, z) | x - z = 0\} \\ \mathscr{S} &= \{(x, y, z) | x + y + z = 0\} \\ \mathscr{T} &= \{(x, y, z) | x - y + z = 0\} \\ \mathscr{W} &= \{(x, y, z) | x = 0 \text{ and } y + z = 0\} \end{aligned}$$

4. Find a basis for the subspaces of $\mathscr{P}_3(\mathbb{R})$ given by

$$\begin{aligned} \mathscr{V} &= \{p(x) | p(0) = 0\} \\ \mathscr{S} &= \left\{p(x) \,\middle|\, \frac{d}{dx} p(x) = 0\right\} \\ \mathscr{T} &= \{p(x) | p(x) = a_0 + a_1 x + a_3 x^3\} \\ \mathscr{W} &= \{p(x) | p(-x) = p(x)\}. \end{aligned}$$

5. Calculate the dimension of each of the subspaces in (3) and (4).

6. Show that if $\mathscr{V}$ is a vector space that is *not* finite dimensional then there exists in $\mathscr{V}$ an infinite sequence $\mathbf{A}_1, \mathbf{A}_2, \ldots,$ of linearly independent vectors.

7. Suppose that $\mathscr{V}$ is a finite-dimensional vector space and $\mathscr{S}$ is a linear subspace of $\mathscr{V}$. Show that there exists a linear subspace $\mathscr{T}$ of $\mathscr{V}$ such that $\mathscr{S} \cap \mathscr{T} = \{\mathbf{0}\}$ and $\mathscr{S} + \mathscr{T} = \mathscr{V}$. (Hint: Study the proof of (6.7).)

8. Suppose that $\mathscr{S}$ and $\mathscr{T}$ are subspaces of a finite-dimensional vector space $\mathscr{V}$, and $\mathscr{S} \cap \mathscr{T} = \{\mathbf{0}\}$. Show that

$$\dim(\mathscr{S} + \mathscr{T}) = \dim \mathscr{S} + \dim \mathscr{T}.$$

9. Suppose that $\mathscr{S}$ and $\mathscr{T}$ are subspaces of a finite-dimensional vector space $\mathscr{V}$. show that

$$\dim(\mathscr{S} + \mathscr{T}) = \dim \mathscr{S} + \dim \mathscr{T} - \dim(\mathscr{S} \cap \mathscr{T})$$

(Hint: Choose a basis for $\mathscr{S} \cap \mathscr{T}$, extend it to a basis for $\mathscr{S}$, extend it to a basis for $\mathscr{T}$, and count.)

10. Suppose that $\mathscr{U}$ is a subspace of a finite-dimensional vector space $\mathscr{V}$. Show $\mathscr{U} = \mathscr{V}$ iff $\dim \mathscr{U} = \dim \mathscr{V}$.

11. What is the dimension of $\mathbb{C}$ as a vector space over $\mathbb{R}$?

12. Let $\mathscr{S} = \{(a, 0, 0)\}$ and $\mathscr{T} = \{(0, b, b)\}$ be sets of vectors in $\mathbb{R}^3$. Show $\mathscr{S}$ and $\mathscr{T}$ are subspaces. Find a basis for $\mathscr{S} + \mathscr{T}$.

13. Let $\mathscr{S}$ be the subspace of $\mathbb{R}^3$ given by

$$\mathscr{S} = \{(x, y, z) | y - z = 0\}.$$

Find a subspace $\mathscr{T}$ of $\mathbb{R}^3$ such that $\mathscr{S} \cap \mathscr{T} = \{\mathbf{0}\}$ and $\mathscr{S} + \mathscr{T} = \mathbb{R}^3$.

14. Under what conditions on the number a will the vectors $(a, 1, 0), (1, a, 1), (0, 1, a)$ be a basis for $\mathbb{R}^3$?

15. Let $\mathbf{A}_1, \ldots, \mathbf{A}_n$ be vectors in $\mathscr{V}$. Suppose that $n = \dim \mathscr{V}$. Show that $\{\mathbf{A}_1, \ldots, \mathbf{A}_n\}$ is linearly independent iff $\dim \mathscr{L}(\mathbf{A}_1, \ldots, \mathbf{A}_n) = n$.

16. The equation $y = 3x$ defines a straight line in the xy-plane. Show that if A, B are on this line then $\vec{\mathbf{A}}, \vec{\mathbf{B}}$ are linearly dependent vectors.

17. Let A, B, C, D be four distinct points on a plane $\mathbf{\Pi}$. Show that the vectors $\overrightarrow{\mathbf{AB}}, \overrightarrow{\mathbf{AC}}, \overrightarrow{\mathbf{AD}}$ form a set of linearly dependent vectors.

18. Let $\mathbf{A}, \mathbf{B}, \mathbf{C}$ be three points in a space. Show that $\mathbf{A}, \mathbf{B}, \mathbf{C}$ are not colinear iff $\{\overrightarrow{\mathbf{AB}}, \overrightarrow{\mathbf{AC}}\}$ is a set of linearly independent vectors.

19. Let $S = \{u, v, w\}$. Show that the functions $\mathbf{f}, \mathbf{g}, \mathbf{h} \in \mathscr{F}(S)$ defined by

$$\begin{array}{lll} \mathbf{f}(u) = 1 & \mathbf{g}(u) = 1 & \mathbf{h}(u) = 1 \\ \mathbf{f}(v) = 1 & \mathbf{g}(v) = 1 & \mathbf{h}(v) = 0 \\ \mathbf{f}(w) = 1 & \mathbf{g}(w) = 0 & \mathbf{h}(w) = 0 \end{array}$$

are a basis for $\mathscr{F}(S)$. Find the coordinates of the characteristic functions χ_u, χ_v, χ_w relative to this basis.

20. Let S be a finite set and $\mathscr{V}$ a finite-dimensional vector space. Show that $\text{Fun}(S, \mathscr{V})$ is finite dimensional and find its dimension.

21. Let S be a finite set and $T \subset S$. Find the dimension of $\mathscr{F}(S, T)$.

22. Let $\mathscr{V}$ be a finite-dimensional vector space over $\mathbb{C}$. By forgetting a part of the structure, $\mathscr{V}$ becomes a real vector space. Show $\dim_{\mathbb{R}} \mathscr{V} = 2 \dim_{\mathbb{C}} \mathscr{V}$ where $\dim_{\mathbb{C}} \mathscr{V}$ is the dimension of $\mathscr{V}$ as a complex vector space, and $\dim_{\mathbb{R}} \mathscr{V}$ is its dimension regarded as a real vector space.

7 The elements of vector spaces: a summing up

Our objective in this section is to work out a number of numerical examples to illustrate and illuminate the theory of vector spaces we have developed so far.

EXAMPLE 1. Determine whether or not the vector

$$\mathbf{A} = (1, -2, 0, 3)$$

is a linear combination of the vectors

$$\mathbf{B}_1 = (3, 9, -4, -2), \qquad \mathbf{B}_2 = (2, 3, 0, -1), \qquad \mathbf{B}_3 = (2, -1, 2, 1).$$

That is, does $\mathbf{A}$ belong to the linear span of $\mathbf{B}_1$, $\mathbf{B}_2$, and $\mathbf{B}_3$, or in symbols is $\mathbf{A} \in \mathscr{L}(\mathbf{B}_1, \mathbf{B}_2, \mathbf{B}_3)$?

Solution. Suppose that $\mathbf{A} \in \mathscr{L}(\mathbf{B}_1, \mathbf{B}_2, \mathbf{B}_3)$. Then there are numbers b_1, b_2, b_3 such that

$$\mathbf{A} = b_1\mathbf{B}_1 + b_2\mathbf{B}_2 + b_3\mathbf{B}_3.$$

Therefore

$$\begin{aligned}(1, -2, 0, 3) &= b_1(3, 9, -4, -2) + b_2(2, 3, 0, -1) + b_3(2, -1, 2, 1)\\ &= (3b_1 + 2b_2 + 2b_3, 9b_1 + 3b_2 - b_3,\\ &\quad -4b_1 + 2b_3, -2b_1 - b_2 + b_3)\end{aligned}$$

and therefore

(1) $$1 = 3b_1 + 2b_2 + 2b_3,$$

(2) $$-2 = 9b_1 + 3b_2 - b_3,$$

(3) $$0 = -4b_1 + 0b_2 + 2b_3,$$

(4) $$3 = -2b_1 - b_2 + b_3.$$

Add (4) to (2) to get

(5) $$1 = 7b_1 + 2b_2.$$

(5) yields

(6) $$\begin{aligned} 2b_2 &= 1 - 7b_1 \\ b_2 &= \frac{1 - 7b_1}{2}. \end{aligned}$$

(3) gives

(7) $$\begin{aligned} 2b_3 &= 4b_1 \\ b_3 &= 2b_1. \end{aligned}$$

Putting (6), (7) into (4) gives

$$\begin{aligned} 3 &= -2b_1 + \frac{7b_1 - 1}{2} + 2b_1 \\ 6 &= -4b_1 + 7b_1 - 1 + 4b_1 \\ 7 &= 7b_1 \\ 1 &= b_1. \end{aligned}$$

So

$$b_1 = 1, \qquad b_2 = -3, \qquad b_3 = 2$$

and hence

$$\mathbf{A} = \mathbf{B}_1 - 3\mathbf{B}_2 + 2\mathbf{B}_3$$

so $\mathbf{A}$ is a linear combination of $\mathbf{B}_1, \mathbf{B}_2, \mathbf{B}_3$ and hence $\mathbf{A}$ belongs to $\mathscr{L}(\mathbf{B}_1, \mathbf{B}_2, \mathbf{B}_3)$.

Example 2. Determine whether or not the vector

$$\mathbf{A} = 1 + x - 2x^2 + 4x^3$$

belongs to the subspace $\mathscr{P}_3(\mathbb{R})$ spanned by the vectors

$$\mathbf{B}_1 = 1 - x, \qquad \mathbf{B}_2 = 1 - x^2, \qquad \mathbf{B}_3 = 1 - x^3.$$

Solution. Assume that $\mathbf{A}$ belongs to $\mathscr{L}(\mathbf{B}_1, \mathbf{B}_2, \mathbf{B}_3)$. Then there are numbers b_1, b_2, b_3 such that

$$\mathbf{A} = b_1\mathbf{B}_1 + b_2\mathbf{B}_2 + b_3\mathbf{B}_3$$

that is

$$\begin{aligned} 1 + x - 2x^2 + 4x^3 &= b_1(1 - x) + b_2(1 - x^2) + b_3(1 - x^3) \\ &= b_1 + b_2 + b_3 - b_1x - b_2x^2 - b_3x^3 \end{aligned}$$

and hence

$$\begin{aligned} 1 &= b_1 + b_2 + b_3, \\ 1 &= -b_1, \\ -2 &= -b_2, \\ 4 &= -b_3. \end{aligned}$$

Therefore the only possibility is:

$$b_1 = -1, \qquad b_2 = 2, \qquad b_3 = -4.$$

But then we receive the impossible equation

$$1 = -1 + 2 - 4 = -3.$$

Therefore there are no numbers b_1, b_2, b_3 such that

$$\mathbf{A} = b_1\mathbf{B}_1 + b_2\mathbf{B}_2 + b_3\mathbf{B}_3$$

and hence **A** does not belong to the subspace spanned by $\mathbf{B}_1$, $\mathbf{B}_2$, and $\mathbf{B}_3$.

EXAMPLE 3. In $\mathbb{R}^4$ let $\mathscr{S}$ and $\mathscr{T}$ be the subspaces defined by

$$\begin{aligned} \mathscr{S} &= \{\mathbf{A} = (a_1, a_2, a_3, a_4) | a_1 - a_2 + a_3 - a_4 = 0\} \\ \mathscr{T} &= \{\mathbf{A} = (a_1, a_2, a_3, a_4) | a_1 + a_2 + a_3 + a_4 = 0\}. \end{aligned}$$

Find a basis for $\mathscr{S} \cap \mathscr{T}$.

Solution. A vector A in $\mathbb{R}^4$ belongs to $\mathscr{S} \cap \mathscr{T}$ iff

$$\begin{aligned} a_1 - a_2 + a_3 - a_4 &= 0 \\ a_1 + a_2 + a_3 + a_4 &= 0 \end{aligned}$$

or

$$\begin{aligned} a_1 + a_3 &= 0 \\ a_2 + a_4 &= 0 \end{aligned}$$

or

$$\begin{aligned} a_1 &= -a_3 \\ a_2 &= -a_4. \end{aligned}$$

Let

$$\begin{aligned} \mathbf{B}_1 &= (1, 0, -1, 0) \\ \mathbf{B}_2 &= (0, -1, 0, 1). \end{aligned}$$

Note that $\mathbf{B}_1$ is a vector in $\mathscr{S} \cap \mathscr{T}$ because

$$\begin{aligned} 1 &= -(-1) \\ 0 &= -0, \end{aligned}$$

and $\mathbf{B}_2$ is a vector in $\mathscr{S} \cap \mathscr{T}$ because

$$\begin{aligned} 0 &= -0 \\ -1 &= -(1). \end{aligned}$$

We claim that $\{\mathbf{B}_1, \mathbf{B}_2\}$ is a basis for $\mathscr{S} \cap \mathscr{T}$. To prove this we must demonstrate two facts. First that $\{\mathbf{B}_1, \mathbf{B}_2\}$ is linearly independent and second that $\mathscr{L}(\mathbf{B}_1, \mathbf{B}_2) = \mathscr{S} \cap \mathscr{T}$.

(1) $\{\mathbf{B}_1, \mathbf{B}_2\}$ *is linearly independent*. Suppose to the contrary that $\{\mathbf{B}_1, \mathbf{B}_2\}$ is linearly dependent. Then there are numbers b_1, b_2, not both zero, such that

$$\begin{aligned} \mathbf{0} &= b_1\mathbf{B}_1 + b_2\mathbf{B}_2 \\ (0, 0, 0, 0) &= b_1(1, 0, -1, 0) + b_2(0, -1, 0, 1) \\ &= (b_1, -b_2, -b_1, b_2) \end{aligned}$$

and hence

$$0 = b_1, \qquad 0 = -b_2$$

so $b_1 = 0 = b_2$, a contradiction. Hence $\{\mathbf{B}_1, \mathbf{B}_2\}$ must be linearly independent.

(2) $\mathscr{L}(\mathbf{B}_1, \mathbf{B}_2) = \mathscr{S} \cap \mathscr{T}$. Suppose that $\mathbf{A} \in \mathscr{S} \cap \mathscr{T}$. Then we have seen

$$\mathbf{A} = (a_1, a_2, -a_1, -a_2).$$

Therefore

$$\mathbf{A} = a_1(1, 0, -1, 0) - a_2(0, -1, 0, 1)$$

and hence

$$\mathbf{A} = a_1\mathbf{B}_1 - a_2\mathbf{B}_2$$

so $\mathbf{A} \in \mathscr{L}(\mathbf{B}_1, \mathbf{B}_2)$. Therefore $\mathscr{S} \cap \mathscr{T}$ is contained in $\mathscr{L}(\mathbf{B}_1, \mathbf{B}_2)$. But since $\mathbf{B}_1, \mathbf{B}_2$ belong to $\mathscr{S} \cap \mathscr{T}$, $\mathscr{L}(\mathbf{B}_1, \mathbf{B}_2)$ is contained in $\mathscr{S} \cap \mathscr{T}$. The only conclusion possible is therefore that $\mathscr{L}(\mathbf{B}_1, \mathbf{B}_2) = \mathscr{S} \cap \mathscr{T}$.

Hence we have found that the vectors

$$\mathbf{B}_1 = (1, 0, -1, 0), \qquad \mathbf{B}_2 = (0, -1, 0, 1)$$

are a basis for $\mathscr{S} \cap \mathscr{T}$. Note that $\mathscr{S} \cap \mathscr{T}$ therefore has dimension 2.

EXAMPLE 4. Show that the vectors

$$\mathbf{A}_1 = (1, -1, -1, 1), \qquad \mathbf{A}_2 = (1, -2, -2, 1), \qquad \mathbf{A}_3 = (0, 1, 1, 0)$$

and

$$\mathbf{B}_1 = (1, 0, 0, 1), \qquad \mathbf{B}_2 = (0, -1, -1, 0)$$

span the same linear subspace of $\mathbb{R}^4$. Find a basis for this subspace.

Solution. If you think about it for a moment, you will see that we must prove the following facts:

$$\mathbf{A}_1 \in \mathscr{L}(\mathbf{B}_1, \mathbf{B}_2)$$
$$\mathbf{A}_2 \in \mathscr{L}(\mathbf{B}_1, \mathbf{B}_2)$$
$$\mathbf{A}_3 \in \mathscr{L}(\mathbf{B}_1, \mathbf{B}_2)$$
$$\mathbf{B}_1 \in \mathscr{L}(\mathbf{A}_1, \mathbf{A}_2, \mathbf{A}_3)$$
$$\mathbf{B}_2 \in \mathscr{L}(\mathbf{A}_1, \mathbf{A}_2, \mathbf{A}_3).$$

To do this we note the following

$$\mathbf{A}_1 = \mathbf{B}_1 + \mathbf{B}_2$$
$$\mathbf{A}_2 = \mathbf{B}_1 + 2\mathbf{B}_2$$
$$\mathbf{A}_3 = -\mathbf{B}_2$$
$$\mathbf{B}_1 = \mathbf{A}_1 + \mathbf{A}_3$$
$$\mathbf{B}_2 = -\mathbf{A}_3.$$

These equations are obtained as in Example 7.1.

If we next note that $\{\mathbf{B}_1, \mathbf{B}_2\}$ is linearly independent we see that $\{\mathbf{B}_1, \mathbf{B}_2\}$ is a basis for $\mathscr{L}(\mathbf{B}_1, \mathbf{B}_2) = \mathscr{L}(\mathbf{A}_1, \mathbf{A}_2, \mathbf{A}_3)$ and hence $\dim \mathscr{L}(\mathbf{B}_1, \mathbf{B}_2) = 2 = \dim \mathscr{L}(\mathbf{A}_1, \mathbf{A}_2, \mathbf{A}_3)$.

EXAMPLE 5. Let $\mathscr{S}$ and $\mathscr{T}$ be the subspaces of $\mathbb{R}^3$ defined by

$$\mathscr{S} = \{(x, y, z) \mid x = y = z\}$$
$$\mathscr{T} = \{(x, y, z) \mid x = 0\}.$$

Show that $\mathscr{S} + \mathscr{T} = \mathbb{R}^3$.

Solution. We must show that any vector $\mathbf{A}$ in $\mathbb{R}^3$ may be written in the form

$$\mathbf{A} = \mathbf{S} + \mathbf{T}$$

where $\mathbf{S}$ is a vector in $\mathscr{S}$ and $\mathbf{T}$ is a vector in $\mathscr{T}$. So let $\mathbf{A} = (a_1, a_2, a_3)$ and note

$$\mathbf{A} = (a_1, a_1, a_1) + (0, a_2 - a_1, a_3 - a_1).$$

The vector (a_1, a_1, a_1) belongs to $\mathscr{S}$ and the vector $(0, a_2 - a_1, a_3 - a_1)$ belongs to $\mathscr{T}$. Therefore

$$\mathbf{A} = \mathbf{S} + \mathbf{T}$$

as required.

EXAMPLE 6. Show the polynomials

$$\mathbf{A}_1 = 1, \qquad \mathbf{A}_2 = t - 1, \qquad \mathbf{A}_3 = (t-1)^2 = t^2 - 2t + 1$$

form a basis for $\mathscr{P}_2(\mathbb{R})$. Find the coordinates of the vector

$$\mathbf{B} = 2t^2 - 5t + 6$$

relative to this ordered basis.

Solution. To show that $\{\mathbf{A}_1, \mathbf{A}_2, \mathbf{A}_3\}$ is a basis we may apply Example 5 of Chapter 6 and (6.9) to conclude that it suffices to show $\mathbf{A}_1, \mathbf{A}_2, \mathbf{A}_3$ are linearly independent. This is easy. For if

$$\mathbf{0} = a_1\mathbf{A}_1 + a_2\mathbf{A}_2 + a_3\mathbf{A}_3 = (a_1 - a_2 + a_3) + (a_2 - 2a_3)t + a_3t^2$$

then

$$\begin{aligned} a_3 &= 0 \\ a_2 - 2a_3 &= 0 \\ a_1 - a_2 + a_3 &= 0 \end{aligned}$$

or

$$a_3 = 0, \qquad a_2 = 0, \qquad a_1 = 0$$

so $\mathbf{A}_1, \mathbf{A}_2, \mathbf{A}_3$ cannot be linearly dependent. Hence they form a basis for $\mathscr{P}_2(\mathbb{R})$ by (6.9) and Example 5 of Chapter 6.

To find the coordinates of B we write

$$\begin{aligned} \mathbf{B} &= b_1\mathbf{A}_1 + b_2\mathbf{A}_2 + b_3\mathbf{A}_3 \\ 2t^2 - 5t + 6 &= b_1 + b_2(t - 1) + b_3(t - 1)^2 \\ 6 - 5t + 2t^2 &= b_1 - b_2 + b_3 + (b_2 - 2b_3)t + b_3t^2 \end{aligned}$$

so

$$\begin{aligned} 6 &= b_1 - b_2 + b_3 \\ -5 &= \phantom{b_1 - {}} b_2 - 2b_3 \\ 2 &= \phantom{b_1 - b_2 + {}} b_3. \end{aligned}$$

Therefore

$$\begin{aligned} b_3 &= 2 \\ b_2 &= -5 + 2b_3 = -1 \\ b_1 &= 6 + b_2 - b_3 = 6 - 1 - 2 = 3. \end{aligned}$$

Therefore the coordinates of $6 - 5t + 2t$ relative to the ordered basis $\{1, (1 - t), (1 - t)^2\}$ are $(3, -1, 2)$.

EXAMPLE 7. Find the dimension of the subspace of $\mathbb{R}^3$ given by

$$\mathscr{V} = \{(x, y, z) | x + 2y + z = 0, x + y + 2z = 0, 2x + y + z = 0\}.$$

Solution. A vector $\mathbf{A} = (x, y, z)$ belongs to $\mathscr{V}$ iff

$$\begin{aligned} x + 2y + z &= 0 \\ x + y + 2z &= 0 \\ 2x + y + z &= 0 \end{aligned}$$

or

$$\begin{aligned} -x + y &= 0 \\ -x + z &= 0 \\ 2x + y - z &= 0 \end{aligned}$$

or

$$y = x, \qquad z = x, \qquad 2x + x - x = 0$$

or

$$y = x, \qquad z = x, \qquad 2x = 0$$

or

$$y = 0, \qquad z = 0, \qquad x = 0$$

so $\mathscr{V} = \{\mathbf{0}\}$ and $\dim \mathscr{V} = 0$.

EXAMPLE 8. Let $S = \{u, v, w\}$ and $\mathscr{V} = \{\mathbf{f} \in \mathscr{F}(S) \mid \mathbf{f}(u) = \mathbf{f}(w)\}$. Show that $\mathscr{V}$ is a subspace of $\mathscr{F}(S)$ and find its dimension.

Solution. To check that $\mathscr{V}$ is a subspace note first of all that it is nonempty since $\mathbf{0} \in \mathscr{V}$. If $\mathbf{f}, \mathbf{g} \in \mathscr{V}$ then

$$(\mathbf{f} + \mathbf{g})(u) = \mathbf{f}(u) + \mathbf{g}(u) = \mathbf{f}(w) + \mathbf{g}(w) = (\mathbf{f} + \mathbf{g})(w)$$

so $\mathbf{f} + \mathbf{g} \in \mathscr{V}$. Likewise for a number r

$$(r\mathbf{f})(u) = r\mathbf{f}(u) = r\mathbf{f}(w) = (r\mathbf{f})(w),$$

so $r\mathbf{f} \in \mathscr{V}$, and therefore $\mathscr{V}$ is a subspace of $\mathscr{F}(S)$.

To find the dimension of $\mathscr{V}$ recall that $\dim \mathscr{F}(S) = 3$. Moreover, $\mathscr{V} \neq \mathscr{F}(S)$ because $\boldsymbol{\chi}_u \notin \mathscr{V}$. Therefore $\dim \mathscr{V} < 3$, i.e. $\dim \mathscr{V} = 0, 1$, or 2. A little guessing shows that the two vectors $\boldsymbol{\chi}_u - \boldsymbol{\chi}_w$, $\boldsymbol{\chi}_v$ belong to $\mathscr{V}$. They are also linearly independent, for if

$$a(\boldsymbol{\chi}_u - \boldsymbol{\chi}_w) + b\boldsymbol{\chi}_v = \mathbf{0}$$

is a linear relation, then

$$a\boldsymbol{\chi}_u + b\boldsymbol{\chi}_v - a\boldsymbol{\chi}_w = \mathbf{0}$$

and evaluating at u and v gives

$$\begin{aligned} 0 &= a\boldsymbol{\chi}_u(u) + b\boldsymbol{\chi}_v(u) - a\boldsymbol{\chi}_w(u) = a \\ 0 &= a\boldsymbol{\chi}_u(v) + b\boldsymbol{\chi}_v(v) - a\boldsymbol{\chi}_w(v) = b \end{aligned}$$

so $a = 0$, $b = 0$ as required. Therefore $\dim \mathscr{V} \geq 2$, which since $\dim \mathscr{V} \leq 2$, shows $\dim \mathscr{V} = 2$.

Exercises

1. Let $\mathbf{A} = (1, 0, 1)$, $\mathbf{B} = (-1, 1, 0)$, $\mathbf{C} = (0, 1, 1)$ be three vectors of $\mathbb{R}^3$. Show $\mathscr{L}(\mathbf{A}, \mathbf{B}) = \mathscr{L}(\mathbf{B}, \mathbf{C})$.

2. $\{\mathbf{B}_1, \mathbf{B}_2\}$ is a basis for $\mathscr{L}(\mathbf{B}_1, \mathbf{B}_2)$ in Example 4 of this chapter. Find the coordinates (in components) of $\mathbf{A}_1$ relative to this basis $\{\mathbf{B}_1, \mathbf{B}_2\}$. Do the same problem for $\mathbf{A}_2, \mathbf{A}_3$.

3. Let $\mathbf{P} = (a, b)$ be a vector in $\mathbb{R}^2$.

   (a) Show that a, b are the coordinates of $\mathbf{P}$ relative to the basis $\mathbf{E}_1 = (1, 0)$, $\mathbf{E}_2 = (0, 1)$.
   (b) Find the coordinates of $\mathbf{P} = (a, b)$ relative to the basis $\mathbf{E}_1 = (1, 0)$ and $\mathbf{F}_2 = (0, 2)$.
   (c) Find the coordinates of $\mathbf{P}$ relative to the basis $\mathbf{G} = (1, 1)$ and $\mathbf{H} = (-1, 2)$.

4. Are the polynomials $\{x + x^3, 1 + x^2\}$ a set of linearly independent vectors of $\mathscr{P}_3(\mathbb{R})$? If so, is $\{x + x^3, 1 + x^2\}$ a basis for $\mathscr{P}_3(\mathbb{R})$? What is the dimension of $\mathscr{L}(x + x^3, 1 + x^2)$?

5. The solution space $\mathscr{S}$ of $x - 2y + 3z = 0$ is a subspace of $\mathbb{R}^3$. Show that $\dim \mathscr{S} = 2$. Find a basis for $\mathscr{S}$.

6. Let $\mathscr{S}$ be a set. A map

$$\tau : S \to S$$

   is called an *involution* iff

$$\tau(\tau(s)) = s, \qquad \forall s \in S.$$

   Let τ be an involution on the set S and define

$$\mathscr{F}_+(S) = \{\mathbf{f} \in \mathscr{F}(S) \mid \mathbf{f}(\tau(s)) = \mathbf{f}(s), \forall s \in S\},$$

$$\mathscr{F}_-(S) = \{\mathbf{f} \in \mathscr{F}(S) \mid \mathbf{f}(\tau(s)) = -\mathbf{f}(s), \forall s \in S\}.$$

   (a) Show that $\mathscr{F}_+(S)$ and $\mathscr{F}_-(S)$ are subspaces of $\mathscr{F}(S)$.
   (b) Show that $\mathscr{F}_+(S) \cap \mathscr{F}_-(S) = \{0\}$.
   (c) Let $\mathbf{f} \in \mathscr{F}(S)$ and define

$$\mathbf{f}_+ : S \to \mathbb{R} \quad \text{by} \quad \mathbf{f}_+(s) = \mathbf{f}(s) + \mathbf{f}(\tau(s))$$

$$\mathbf{f}_- : S \to \mathbb{R} \quad \text{by} \quad \mathbf{f}_-(s) = \mathbf{f}(s) - \mathbf{f}(\tau(s)).$$

   Show that $\mathbf{f}_+ \in \mathscr{F}_+(S)$ and $\mathbf{f}_- \in \mathscr{F}_-(S)$.

   (d) Show that $\mathscr{F}(S) = \mathscr{F}_+(S) + \mathscr{F}_-(S)$.
   (e) If S is finite show that

$$\dim \mathscr{F}_+(S) + \dim \mathscr{F}_-(S) = |S|.$$

7. Let $\mathscr{V}$ be a vector space, and $\mathscr{S}$, $\mathscr{T}$ finite-dimensional subspaces of $\mathscr{V}$. If $\mathscr{S}$ and $\mathscr{T}$ have the same dimension, and $\mathscr{S} \subset \mathscr{T}$ show that $\mathscr{S} = \mathscr{T}$.

8. Let $\mathscr{S}$, $\mathscr{T}$, $\mathscr{U}$ be subspaces of a vector space $\mathscr{V}$. Is it always true that

$$(\mathscr{S} + \mathscr{T}) \cap \mathscr{U} = (\mathscr{S} \cap \mathscr{U}) + (\mathscr{T} \cap \mathscr{U})?$$

9. Let $S = \{s, t, u\}$ and let $\tau: S \to S$ be the involution obtained by interchanging s and u and leaving t alone, that is

$$\tau(s) = u, \qquad \tau(t) = t, \qquad \tau(u) = s.$$

Find bases for $\mathscr{F}_+(S)$ and $\mathscr{F}_-(S)$.

10. Find a basis for the subspace of $\mathscr{P}_4(\mathbb{R})$ spanned by the polynomials $f \in \mathscr{P}_4(\mathbb{R})$ that satisfy the equation

$$\mathbf{f}(z) = z^4\mathbf{f}\left(\frac{1}{z}\right).$$

8 Linear transformations

In physics when vectors are used by fixing a coordinate system and employing components, it is very important to understand how the components will change if the coordinate system is changed. A change of coordinates is best expressed in terms of transformations of vectors. The simplest such transformations are the linear ones.

In the calculus when the study of curves in the plane and space are introduced the student was confronted with "vector-valued functions of a scalar." The simplest such functions were the linear ones.

In this chapter we will begin the study of linear transformations, the generalization of these linear functions to higher dimensions. A typical linear transformation from $\mathbb{R}^2$ to $\mathbb{R}^3$ is given by a system of linear equations

$$\begin{aligned} y_1 &= a_{11}x_1 + a_{12}x_2 \\ y_2 &= a_{21}x_1 + a_{22}x_2 \\ y_3 &= a_{31}x_1 + a_{32}x_2 \end{aligned}$$

where $\mathbf{X} = (x_1, x_2)$ are the components of a vector $\mathbf{X}$ of $\mathbb{R}^2$ and $\mathbf{Y} = (y_1, y_2, y_3)$ are the components of the vector of $\mathbb{R}^3$ that corresponds to $\mathbf{X}$ under the particular transformation. The numbers $a_{11}, \ldots$ are fixed and determine the transformation. The rectangular array of numbers

$$\begin{matrix} a_{11}, & a_{12} \\ a_{21}, & a_{22} \\ a_{31}, & a_{32} \end{matrix}$$

is called the matrix of the transformation (relative to the usual bases for $\mathbb{R}^2$ and $\mathbb{R}^3$). It is cumbersome and inconvenient to deal with linear transformations in terms of a system of linear equations. Instead we shall define them as functions between vector spaces with two simple properties.

Definition. Let $\mathscr{V}$ and $\mathscr{W}$ be vector spaces. A *linear transformation* T *from* $\mathscr{V}$ *to* $\mathscr{W}$, written $\mathsf{T}: \mathscr{V} \to \mathscr{W}$, is a function that assigns to each vector $\mathbf{A}$ in $\mathscr{V}$ a vector $\mathsf{T}(\mathbf{A})$ in $\mathscr{W}$ such that the following two properties hold:

(1) $\mathsf{T}(\mathbf{A} + \mathbf{B}) = \mathsf{T}(\mathbf{A}) + \mathsf{T}(\mathbf{B})$ for all $\mathbf{A}, \mathbf{B} \in \mathscr{V}$.
(2) $\mathsf{T}(a\mathbf{A}) = a\mathsf{T}(\mathbf{A})$ for all numbers a and all $\mathbf{A} \in \mathscr{V}$.

We might paraphrase this definition by saying that a linear transformation from $\mathscr{V}$ to $\mathscr{W}$ is a function that preserves vector addition and scalar multiplication.

It is to be emphasized that as soon as the dimension of $\mathscr{V}$ and $\mathscr{W}$ exceed 1 there are an enormous number of linear transformations from $\mathscr{V}$ to $\mathscr{W}$. Their variety is what makes the subject of linear algebra really interesting. The classification of linear transformations is what will concern us throughout the remainder of this course, although we will only scratch the surface of classification theory!

Let us begin by examining some of the boundless variety of linear transformations.

EXAMPLE 1. In $\mathbb{R}^2$ consider the function

$$\mathsf{T}: \mathbb{R}^2 \to \mathbb{R}^2$$

given by

$$\mathsf{T}(x, y) = (x \cos \theta - y \sin \theta, x \sin \theta + y \cos \theta)$$

where θ is a fixed number $0 \leq \theta < 2\pi$. We claim that T is a linear transformation. To prove this we must verify that T possesses the two properties characteristic of linear transformations. So suppose that $\mathbf{A} = (a_1, a_2)$, $\mathbf{B} = (b_1, b_2)$ are vectors in $\mathbb{R}^2$. Then

$$\begin{aligned}
\mathsf{T}(\mathbf{A} + \mathbf{B}) &= \mathsf{T}(a_1 + b_1, a_2 + b_2) \\
&= ((a_1 + b_1)\cos \theta - (a_2 + b_2)\sin \theta, \\
&\qquad (a_1 + b_1)\sin \theta + (a_2 + b_2)\cos \theta) \\
&= (a_1 \cos \theta - a_2 \sin \theta + b_1 \cos \theta - b_2 \sin \theta, \\
&\qquad a_1 \sin \theta + a_2 \cos \theta + b_1 \sin \theta + b_2 \cos \theta) \\
&= (a_1 \cos \theta - a_2 \sin \theta, a_1 \sin \theta + a_2 \cos \theta) \\
&\qquad + (b_1 \cos \theta - b_2 \sin \theta, b_1 \sin \theta + b_2 \cos \theta) \\
&= \mathsf{T}(\mathbf{A}) + \mathsf{T}(\mathbf{B}),
\end{aligned}$$

and hence the first characteristic property of a linear transformation is satisfied. To verify the second property we suppose that a is a number and compute

$$\begin{aligned}\mathsf{T}(a\mathbf{A}) &= \mathsf{T}(aa_1, aa_2)\\ &= (aa_1 \cos\theta - aa_2 \sin\theta, aa_1 \sin\theta + aa_2 \cos\theta)\\ &= (a(a_1 \cos\theta - a_2 \sin\theta), a(a_1 \sin\theta + a_2 \cos\theta))\\ &= a(a_1 \cos\theta - a_2 \sin\theta, a_1 \sin\theta + a_2 \cos\theta)\\ &= a\mathsf{T}(\mathbf{A}),\end{aligned}$$

so that the second characteristic property of a linear transformation holds for T.

Therefore $\mathsf{T} : \mathbb{R}^2 \to \mathbb{R}^2$ is a linear transformation.

Geometrically T has a very simple interpretation. It is a rotation of the plane $\mathbb{R}^2$ through the angle θ (in radian measure) in the counterclockwise direction as shown in Figure 8.1. In studying less concretely given linear transformations we will often seek such simple geometric interpretations.

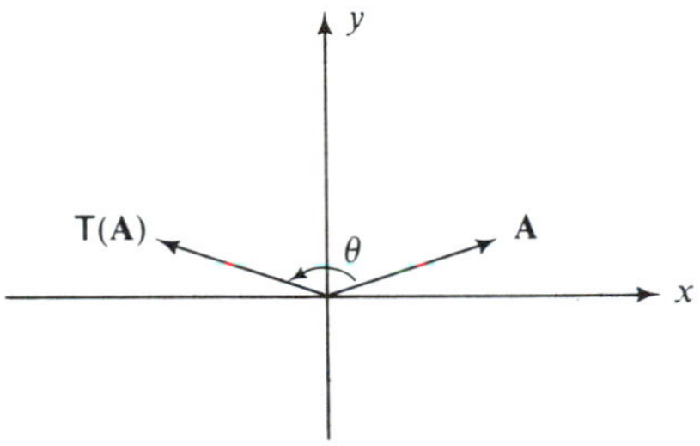

Figure 8.1

Example 2. Consider the function $\mathbb{R}^3$ to $\mathbb{R}^2$ given by

$$\mathsf{T}(x, y, z) = (x, y)$$

We claim that T is a linear transformation. To prove this we must verify that T has the two characteristic properties of a linear transformation. This is done in the following computations ($\mathbf{A} = (a_1, a_2, a_3)$, $\mathbf{B} = (b_1, b_2, b_3)$)

$$\begin{aligned}\mathsf{T}(\mathbf{A} + \mathbf{B}) &= \mathsf{T}(a_1 + b_1, a_2 + b_2, a_3 + b_3)\\ &= (a_1 + b_1, a_2 + b_2) = (a_1, a_2) + (b_1, b_2)\\ &= \mathsf{T}(\mathbf{A}) + \mathsf{T}(\mathbf{B})\end{aligned}$$

and

$$\mathsf{T}(a\mathbf{A}) = \mathsf{T}(aa_1, aa_2, aa_3) = (aa_1, aa_2) = a(a_1, a_2) = a\mathsf{T}(\mathbf{A})$$

(where a is a number).

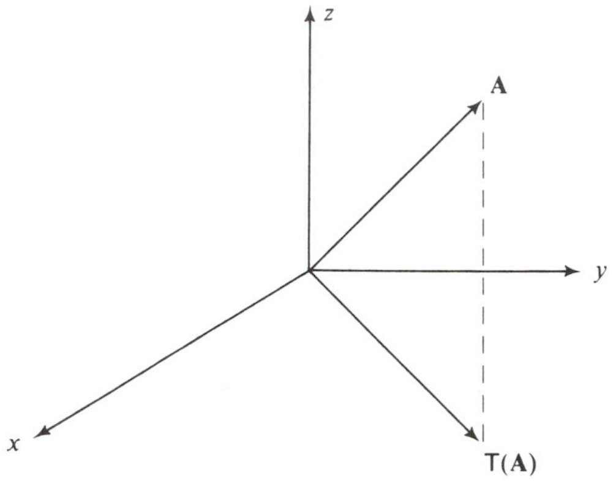

Figure 8.2

EXAMPLE 3. A slight variation of Example 2 is given by the function $\mathsf{T}: \mathbb{R}^3 \to \mathbb{R}^3$ such that

$$\mathsf{T}(x, y, z) = (x, y, 0).$$

It is routine to check that T is a linear transformation. Geometrically, T is projection onto the xy plane, as in Figure 8.2. It is possible to combine the projection of Example 3 with the rotation of Example 1 to obtain a more complex linear transformation. Many of the linear transformations from $\mathbb{R}^3$ to $\mathbb{R}^2$ are of this composite form.

EXAMPLE 4. Define a function T from $\mathbb{R}^3$ to $\mathscr{P}_2(\mathbb{R})$ by

$$\mathsf{T}(a_1, a_2, a_3) = a_1 + a_2 x + a_3 x^2$$

It is routine to verify that T is a linear transformation. Presently we will be able to use it to show that $\mathbb{R}^3$ and $\mathscr{P}_2(\mathbb{R})$ are in some sense the "same" vector space.

Before turning to some more examples of linear transformations let us see what additional properties they must have.

Proposition 8.1. *Let* $\mathsf{T}: \mathscr{V} \to \mathscr{W}$ *be a linear transformation. Then* $\mathsf{T}(\mathbf{0}) = \mathbf{0}$. *(Note that we have used the same symbol* $\mathbf{0}$ *to denote the zero vector of* $\mathscr{V}$ *and of* $\mathscr{W}$.*)*

PROOF. We have

$$\mathsf{T}(\mathbf{0}) = \mathsf{T}(0\mathbf{0}) = 0\mathsf{T}(\mathbf{0}) = \mathbf{0}$$

by (2.1). □

Proposition 8.2. *Let* $\mathsf{T}: \mathscr{V} \to \mathscr{W}$ *be a linear transformation. If* $\mathbf{A}_1, \ldots, \mathbf{A}_n$ *are vectors in* $\mathscr{V}$ *and* $a_1, \ldots, a_n$ *are numbers, then*

$$\mathsf{T}(a_1\mathbf{A}_1 + a_2\mathbf{A}_2 + \cdots + a_n\mathbf{A}_n) = a_1\mathsf{T}(\mathbf{A}_1) + a_2\mathsf{T}(\mathbf{A}_2) + \cdots + a_n\mathsf{T}(\mathbf{A}_n)$$

PROOF. We apply the fact that T preserves vector sums to write

$$\mathsf{T}(a_1\mathbf{A}_1 + \cdots + a_n\mathbf{A}_n) = \mathsf{T}(a_1\mathbf{A}_1) + \cdots + \mathsf{T}(a_n\mathbf{A}_n)$$

and now we apply the fact that T preserves scalar products to write

$$\mathsf{T}(a_1\mathbf{A}_1) + \cdots + \mathsf{T}(a_n\mathbf{A}_n) = a_1\mathsf{T}(\mathbf{A}_1) + a_2\mathsf{T}(\mathbf{A}_2) + \cdots + a_n\mathsf{T}(\mathbf{A}_n)$$

which is the desired conclusion. □

Proposition 8.3. *Let* $\mathsf{T}: \mathscr{V} \to \mathscr{W}$ *be a linear transformation. Suppose that* $\mathscr{U}$ *is a linear subspace of* $\mathscr{V}$ *and set*

$$\mathsf{T}(\mathscr{U}) = \{\mathsf{T}(\mathbf{A}) \in \mathscr{W} \mid \mathbf{A} \in \mathscr{U}\}$$

(that is $\mathsf{T}(\mathscr{U})$ *is the set of all vectors in* $\mathscr{W}$ *that are of the form* $\mathsf{T}(\mathbf{A})$ *for some vector* $\mathbf{A}$ *in* $\mathscr{V}$*). Then* $\mathsf{T}(\mathscr{U})$ *is a linear subspace of* $\mathscr{W}$.

PROOF. Suppose that $\mathbf{A}'$, $\mathbf{B}'$ are vectors in $\mathsf{T}(\mathscr{U})$. Then there are vectors $\mathbf{A}, \mathbf{B}$ in $\mathscr{U}$ such that

$$\mathbf{A}' = \mathsf{T}(\mathbf{A}) \quad \text{and} \quad \mathbf{B}' = \mathsf{T}(\mathbf{B}).$$

Then $\mathbf{A} + \mathbf{B}$ is in $\mathscr{U}$ since $\mathscr{U}$ is a linear subspace of $\mathscr{V}$. Therefore $\mathbf{C} = \mathsf{T}(\mathbf{A} + \mathbf{B})$ is in $\mathsf{T}(\mathscr{U})$. But since T is a linear transformation

$$\mathbf{C} = \mathsf{T}(\mathbf{A} + \mathbf{B}) = \mathsf{T}(\mathbf{A}) + \mathsf{T}(\mathbf{B}) = \mathbf{A}' + \mathbf{B}'$$

so we see that $\mathbf{A}' + \mathbf{B}'$ is in $\mathsf{T}(\mathscr{U})$.

Next we suppose that a is a number. Then since $\mathscr{U}$ is a linear subspace of $\mathscr{V}$, $a\mathbf{A} \in \mathscr{U}$. Therefore $\mathbf{C} = \mathsf{T}(a\mathbf{A}) \in \mathsf{T}(\mathscr{U})$. But since T is a linear transformation

$$\mathbf{C} = \mathsf{T}(a\mathbf{A}) = a\mathsf{T}(\mathbf{A}) = a\mathbf{A}'$$

so that $a\mathbf{A}'$ is in $\mathsf{T}(\mathscr{U})$ also. □

Proposition 8.4. *Suppose that* $\mathsf{T}: \mathscr{V} \to \mathscr{W}$ *is a linear transformation and* D *is a set of vectors in* $\mathscr{V}$. *Then*

$$\mathsf{T}(\mathscr{L}(D)) = \mathscr{L}(\mathsf{T}(D))$$

where $\mathsf{T}(D)$ *denotes the set of vectors in* $\mathscr{W}$ *of the form* $\mathsf{T}(\mathbf{A})$ *for some vector* $\mathbf{A}$ *in* D.

PROOF. The proof is similar to that of (8.3). We must show that each vector in $\mathsf{T}(\mathscr{L}(D))$ is in $\mathscr{L}(\mathsf{T}(D))$ and conversely. A vector in $\mathsf{T}(\mathscr{L}(D))$ is a vector $\mathsf{T}(\mathbf{A})$ for some $\mathbf{A}$ in $\mathscr{L}(D)$. Then there are vectors $\mathbf{A}_1, \ldots, \mathbf{A}_n$ in D so that

$$\mathbf{A} = a_1\mathbf{A}_1 + \cdots + a_n\mathbf{A}_n$$

for suitable numbers $a_1, \ldots, a_n$. Therefore by (8.2)

$$\mathsf{T}(\mathbf{A}) = a_1\mathsf{T}(\mathbf{A}_1) + \cdots + a_n\mathsf{T}(\mathbf{A}_n)$$

and since the right hand side belongs to $\mathscr{L}(\mathsf{T}(D))$ we see that $\mathsf{T}(\mathscr{L}(D)) \subset \mathscr{L}(\mathsf{T}(D))$. To prove the converse suppose that $\mathbf{A}'$ belongs to $\mathscr{L}(\mathsf{T}(D))$. Then there are vectors $\mathbf{A}'_1, \ldots, \mathbf{A}'_n$ in $\mathsf{T}(D)$, and numbers $a_1, \ldots, a_n$ such that

$$\mathbf{A}' = a_1\mathbf{A}'_1 + \cdots + a_n\mathbf{A}'_n.$$

There are also vectors $\mathbf{A}_1, \ldots, \mathbf{A}_n$ in D with

$$\mathbf{A}'_1 = \mathsf{T}(\mathbf{A}_1), \qquad \mathbf{A}'_2 = \mathsf{T}(\mathbf{A}_2), \qquad \ldots, \qquad \mathbf{A}'_n = \mathsf{T}(\mathbf{A}_n).$$

Let

$$\mathbf{A} = a_1\mathbf{A}_1 + a_2\mathbf{A}_2 + \cdots + a_n\mathbf{A}_n.$$

Then $\mathbf{A}$ is a vector in $\mathscr{L}(D)$ and by (8.2)

$$\begin{aligned}\mathsf{T}(\mathbf{A}) &= \mathsf{T}(a_1\mathbf{A}_1 + \cdots + a_n\mathbf{A}_n)\\ &= a_1\mathsf{T}(\mathbf{A}_1) + \cdots + a_n\mathsf{T}(\mathbf{A}_n)\\ &= a_1\mathbf{A}'_1 + \cdots + a_n\mathbf{A}'_n = \mathbf{A}'\end{aligned}$$

so that $\mathbf{A}'$ belongs to $\mathsf{T}(\mathscr{L}(D))$, that is $\mathscr{L}(\mathsf{T}(D)) \subset \mathsf{T}(\mathscr{L}(D))$ as was to be shown. □

Proposition 8.5. *Suppose that* $\mathsf{T}: \mathscr{V} \to \mathscr{W}$ and $\mathsf{S}: \mathscr{W} \to \mathscr{U}$ *are linear transformations. Then*

$$\mathsf{S}\cdot\mathsf{T}: \mathscr{V} \to \mathscr{U} \quad \textit{defined by} \quad (\mathsf{S}\cdot\mathsf{T})(\mathbf{A}) = \mathsf{S}(\mathsf{T}(\mathbf{A}))$$

is a linear transformation.

Proof. We must verify that $\mathsf{S}\cdot\mathsf{T}$ has the two characteristic properties of a linear transformation. So suppose that $\mathbf{A}, \mathbf{B}$ belong to $\mathscr{V}$. Then using the facts that S and T are linear transformations we find

$$\begin{aligned}\mathsf{S}\cdot\mathsf{T}(\mathbf{A}+\mathbf{B}) &= \mathsf{S}(\mathsf{T}(\mathbf{A}+\mathbf{B}))\\ &= \mathsf{S}(\mathsf{T}(\mathbf{A}) + \mathsf{T}(\mathbf{B}))\\ &= \mathsf{S}(\mathsf{T}(\mathbf{A})) + \mathsf{S}(\mathsf{T}(\mathbf{B}))\\ &= \mathsf{S}\cdot\mathsf{T}(\mathbf{A}) + \mathsf{S}\cdot\mathsf{T}(\mathbf{B}).\end{aligned}$$

Likewise for any number a we find

$$\begin{aligned}\mathsf{S}\cdot\mathsf{T}(a\mathbf{A}) &= \mathsf{S}(\mathsf{T}(a\mathbf{A})) = \mathsf{S}(a\mathsf{T}(\mathbf{A}))\\ &= a\mathsf{S}(\mathsf{T}(\mathbf{A})) = a\mathsf{S}\cdot\mathsf{T}(\mathbf{A})\end{aligned}$$

and hence $\mathsf{S}\cdot\mathsf{T}$ is a linear transformation. □

Definition. Suppose that $\mathsf{S}, \mathsf{T}: \mathscr{V} \to \mathscr{W}$ are both linear transformations. Let

$$\mathsf{S} + \mathsf{T}: \mathscr{V} \to \mathscr{W}$$

be the function defined by

$$(\mathsf{S}+\mathsf{T})(\mathbf{A}) = \mathsf{S}(\mathbf{A}) + \mathsf{T}(\mathbf{A}).$$

Proposition 8.6. *If* $\mathsf{S}, \mathsf{T} : \mathscr{V} \to \mathscr{W}$ *are linear transformations then so is* $\mathsf{S} + \mathsf{T} : \mathscr{V} \to \mathscr{W}$.

PROOF. We suppose that **A**, **B** belong to $\mathscr{V}$. Then we compute as follows

$$\begin{aligned}(\mathsf{S} + \mathsf{T})(\mathbf{A} + \mathbf{B}) &= \mathsf{S}(\mathbf{A} + \mathbf{B}) + \mathsf{T}(\mathbf{A} + \mathbf{B}) \\ &= \mathsf{S}(\mathbf{A}) + \mathsf{S}(\mathbf{B}) + \mathsf{T}(\mathbf{A}) + \mathsf{T}(\mathbf{B}) \\ &= \mathsf{S}(\mathbf{A}) + \mathsf{T}(\mathbf{A}) + \mathsf{S}(\mathbf{B}) + \mathsf{T}(\mathbf{B}) \\ &= (\mathsf{S} + \mathsf{T})(\mathbf{A}) + (\mathsf{S} + \mathsf{T})(\mathbf{B}).\end{aligned}$$

For any number a we have

$$\begin{aligned}(\mathsf{S} + \mathsf{T})(a\mathbf{A}) &= \mathsf{S}(a\mathbf{A}) + \mathsf{T}(a\mathbf{A}) \\ &= a\mathsf{S}(\mathbf{A}) + a\mathsf{T}(\mathbf{A}) \\ &= a(\mathsf{S}(\mathbf{A}) + \mathsf{T}(\mathbf{A})) \\ &= a((\mathsf{S} + \mathsf{T})(\mathbf{A})).\end{aligned}$$

Hence the function $\mathsf{S} + \mathsf{T} : \mathscr{V} \to \mathscr{W}$ possesses the two characteristic properties of a linear transformation. □

Definition. Suppose that $\mathsf{T} : \mathscr{V} \to \mathscr{W}$ is a linear transformation and a is a number. Define the function

$$a\mathsf{T} : \mathscr{V} \to \mathscr{W}$$

by

$$(a\mathsf{T})(\mathbf{A}) = a\mathsf{T}(\mathbf{A}).$$

Proposition 8.7. *If* $\mathsf{T} : \mathscr{V} \to \mathscr{W}$ *is a linear transformation and* a *is a number, then* $a\mathsf{T} : \mathscr{V} \to \mathscr{W}$ *is also a linear transformation.*

PROOF. The proof is similar to that of (8.6) and is left to the reader. □

Definition. If $\mathscr{V}$ and $\mathscr{W}$ are vector spaces, the *zero transformation*, denoted by $0 : \mathscr{V} \to \mathscr{W}$, is defined by $0(\mathbf{A}) = \mathbf{0}$ for all **A** in $\mathscr{V}$.

Note that taken together (8.6) and (8.7) say that we are on our way to showing that the set of all linear transformations from $\mathscr{V}$ to $\mathscr{W}$ **is again a vector space**. This is indeed true, and a very important theorem.

Theorem 8.8. *Let* $\mathscr{V}$ *and* $\mathscr{W}$ *be vector spaces and* $\mathscr{L}(\mathscr{V}, \mathscr{W})$ *the set of all linear transformations from* $\mathscr{V}$ *to* $\mathscr{W}$. *Then* $\mathscr{L}(\mathscr{V}, \mathscr{W})$ *is a vector space.*

PROOF. We must verify that Axioms 1–8 of Chapter 2 are satisfied for the procedure of adding linear transformations and multiplying them by numbers as defined above. The zero transformation defined above is the zero vector in Axiom 3. The negative of the transformation T is $(-1)\mathsf{T}$ in Axiom 4. The verifications of the details are routine. □

Note that (8.8) should finally lay to rest the idea that a vector is a quantity with direction and magnitude. The vector spaces $\mathscr{L}(\mathscr{V}, \mathscr{W})$ are enormously important and we will see more and more of them.

Definition. Suppose that $\mathsf{T}: \mathscr{V} \to \mathscr{W}$ is a linear transformation. The *image of* T, denoted by $\operatorname{Im} \mathsf{T}$, is defined by $\operatorname{Im} \mathsf{T} = \mathsf{T}(\mathscr{V})$. The *kernel of* T denoted by $\ker \mathsf{T}$ is defined by $\ker \mathsf{T} = \{\mathbf{A} \in \mathscr{V} \mid \mathsf{T}(\mathbf{A}) = \mathbf{0}\}$.

EXAMPLE 5. Let $\mathsf{T}: \mathbb{R}^4 \to \mathbb{R}^3$ be the linear transformation given by

$$\mathsf{T}(x, y, z, w) = (x + y, z + w, 0).$$

Find the image and kernel of T.

Solution. To find the image of T is easy. It is the xy-plane, that is

$$\operatorname{Im} \mathsf{T} = \{(u, v, 0) \in \mathbb{R}^3\} = \mathscr{V}_{xy}.$$

Because, given any vector $(u, v, 0)$ one checks

$$\mathsf{T}(u, 0, v, 0) = (u, v, 0),$$

showing $\operatorname{Im} \mathsf{T} \supset \mathscr{V}_{xy}$.

The last coordinate of any vector in $\operatorname{Im} \mathsf{T}$ must be zero, so $\operatorname{Im} \mathsf{T} \subset \mathscr{V}_{xy}$ and it follows that $\operatorname{Im} \mathsf{T} = \mathscr{V}_{xy}$.

The kernel of T consists of all vectors $(x, y, z, w) \in \mathbb{R}^4$ such that

$$(0, 0, 0) = (x + y, z + w, 0).$$

So one quickly sees

$$\ker \mathsf{T} = \{(a, -a, b, -b) \in \mathbb{R}^4 \mid a, b \in \mathbb{R}\}.$$

Note that both $\operatorname{Im} \mathsf{T}$ and $\ker \mathsf{T}$ are subspaces of $\mathbb{R}^3$ and $\mathbb{R}^4$ respectively. This is not a special property of this example as the next result shows.

Proposition 8.9. *Suppose that* $\mathsf{T}: \mathscr{V} \to \mathscr{W}$ *is a linear transformation. Then* $\operatorname{Im} \mathsf{T}$ *is a linear subspace of* $\mathscr{W}$ *and* $\ker \mathsf{T}$ *is a linear subspace of* $\mathscr{V}$.

PROOF. The fact that $\operatorname{Im} \mathsf{T}$ is a linear subspace of $\mathscr{W}$ is the special case of (8.3) where $\mathscr{U} = \mathscr{V}$. To show that $\ker \mathsf{T}$ is a linear subspace of $\mathscr{V}$ we suppose that $\mathbf{A}, \mathbf{B} \in \ker \mathsf{T}$. Then

$$\mathsf{T}(\mathbf{A} + \mathbf{B}) = \mathsf{T}(\mathbf{A}) + \mathsf{T}(\mathbf{B}) = \mathbf{0} + \mathbf{0} = \mathbf{0}$$

so that $\mathbf{A} + \mathbf{B} \in \ker \mathsf{T}$. Likewise

$$\mathsf{T}(a\mathbf{A}) = a\mathsf{T}(\mathbf{A}) = a \cdot \mathbf{0} = \mathbf{0}$$

for any number a, and hence $\ker \mathsf{T}$ is a linear subspace of $\mathscr{V}$. □

Proposition 8.10. *Suppose that* $\mathsf{T} : \mathscr{V} \to \mathscr{W}$ *is a linear transformation. If* $\mathscr{V}$ *is finite dimensional, then* ker T *and* Im T *are finite dimensional and*

$$\dim \mathscr{V} = \dim \operatorname{Im} \mathsf{T} + \dim \ker \mathsf{T}.$$

Remark. The proof of (8.10) is important because it indicates the process whereby one *solves* a system of linear homogeneous equations.

PROOF. Since $\mathscr{V}$ is finite dimensional ker T is finite dimensional by (6.8). Choose a basis $\mathbf{A}_1, \ldots, \mathbf{A}_s$ for ker T. By (6.7) we may find vectors $\mathbf{B}_1, \ldots, \mathbf{B}_t$ so that $\mathbf{A}_1, \ldots, \mathbf{A}_s, \mathbf{B}_1, \ldots, \mathbf{B}_t$ is a basis for $\mathscr{V}$. By (8.4) we have

$$\begin{aligned}\operatorname{Im} \mathsf{T} &= \mathsf{T}(\mathscr{V}) = \mathsf{T}(\mathscr{L}(\mathbf{A}_1, \ldots, \mathbf{A}_s, \mathbf{B}_1, \ldots, \mathbf{B}_t))\\ &= \mathscr{L}(\mathsf{T}(\mathbf{A}_1), \ldots, \mathsf{T}(\mathbf{A}_s), \mathsf{T}(\mathbf{B}_1), \ldots, \mathsf{T}(\mathbf{B}_t))\\ &= \mathscr{L}(\mathbf{0}, \ldots, \mathbf{0}, \mathsf{T}(\mathbf{B}_1), \ldots, \mathsf{T}(\mathbf{B}_t))\\ &= \mathscr{L}(\mathsf{T}(\mathbf{B}_1), \ldots, \mathsf{T}(\mathbf{B}_t))\end{aligned}$$

and hence Im T is finite dimensional. We claim that the vectors $\mathsf{T}(\mathbf{B}_1), \ldots, \mathsf{T}(\mathbf{B}_t)$ are a basis for Im T. The preceding equation shows that $\mathsf{T}(\mathbf{B}_1), \ldots, \mathsf{T}(\mathbf{B}_t)$ spans Im T and so we must show that they are linearly independent. To this end suppose that there are numbers $b_1, \ldots, b_t$, not all zero, so that

$$b_1\mathsf{T}(\mathbf{B}_1) + b_2\mathsf{T}(\mathbf{B}_2) + \cdots + b_t\mathsf{T}(\mathbf{B}_t) = 0.$$

Let

$$\mathbf{B} = b_1\mathbf{B}_1 + \cdots + b_t\mathbf{B}_t.$$

Then $\mathbf{B}$ is a vector in $\mathscr{V}$. By (8.2)

$$\mathsf{T}(\mathbf{B}) = b_1\mathsf{T}(\mathbf{B}_1) + \cdots + b_t\mathsf{T}(\mathbf{B}_t) = \mathbf{0}$$

so $\mathbf{B} \in \ker \mathsf{T}$. Since the vectors $\mathbf{A}_1, \ldots, \mathbf{A}_s$ are a basis for ker T we may find numbers $a_1, \ldots, a_s$ so that

$$\mathbf{B} = a_1\mathbf{A}_1 + \cdots + a_s\mathbf{A}_s$$

or

$$b_1\mathbf{B}_1 + \cdots + b_t\mathbf{B}_t = a_1\mathbf{A}_1 + \cdots + a_s\mathbf{A}_s$$

which may be written

$$\mathbf{0} = a_1\mathbf{A}_1 + \cdots + a_s\mathbf{A}_s - b_1\mathbf{B}_1 - \cdots - b_t\mathbf{B}_t$$

which is a linear relation between the vectors $\mathbf{A}_1, \ldots, \mathbf{A}_s, \mathbf{B}_1, \ldots, \mathbf{B}_t$. Since not all of $b_1, \ldots, b_t$ are zero this contradicts the fact that $\mathbf{A}_1, \ldots, \mathbf{A}_s, \mathbf{B}_1, \ldots, \mathbf{B}_t$ are a basis for $\mathscr{V}$. Hence $\mathsf{T}(\mathbf{B}_1), \ldots, \mathsf{T}(\mathbf{B}_t)$ is a basis for Im T. Thus

$$\begin{aligned}\dim \ker \mathsf{T} &= s\\ \dim \operatorname{Im} \mathsf{T} &= t\\ \dim \mathscr{V} &= s + t\end{aligned}$$

as was to be shown. □

EXAMPLE 6. Let $\mathsf{D} : \mathscr{P}(\mathbb{R}) \to \mathscr{P}(\mathbb{R})$ be the linear transformation defined by

$$\mathsf{D}(p(x)) = \frac{d}{dx} p(x).$$

The fact that D is a linear transformation is a consequence of elementary facts from the calculus.

Note that

$$\mathsf{D}(\mathscr{P}_n(\mathbb{R})) = \mathscr{P}_{n-1}(\mathbb{R}).$$

The kernel of D consists of all those polynomials with $(d/dx)p(x) = 0$. The only such polynomials are the constant polynomials. Thus

$$\ker \mathsf{D} = \mathscr{P}_0(\mathbb{R}).$$

Let $\mathsf{L} : \mathscr{P}(\mathbb{R}) \to \mathscr{P}(\mathbb{R})$ be the linear transformation defined by

$$\mathsf{L}(p(x)) = \int_0^x p(t)dt.$$

The fact that L is a linear transformation is also a consequence of elementary facts from the calculus. Note that

$$\mathsf{L}(\mathscr{P}_n(\mathbb{R})) = \{p(x) \in \mathscr{P}_{n+1}(\mathbb{R}) \mid p(0) = 0\},$$

that is $\mathsf{L}(\mathscr{P}_n(\mathbb{R}))$ consists of those polynomials of degree $n + 1$ whose constant term is zero. Thus

$$\operatorname{Im} \mathsf{L} = \{p(x) \mid p(0) = 0\},$$

that is the image of L consists of all the polynomials with zero constant term.

To calculate the kernel of L we recall that

$$\mathsf{D}(\mathsf{L}(p(x))) = \frac{d}{dx} \int_0^x p(t)dt = p(x).$$

Therefore, in our notation for composition of linear transformations

$$(*) \qquad \mathsf{D} \cdot \mathsf{L}(p(x)) = p(x).$$

If, therefore, $\mathsf{L}(p(x)) = \mathbf{0}$ then

$$p(x) = \mathsf{DL}(p(x)) = \mathsf{D}(\mathbf{0}) = \mathbf{0}.$$

Hence $p(x) \in \ker \mathsf{L}$ iff $p(x) = 0$, that is

$$\ker \mathsf{L} = \{\mathbf{0}\}.$$

Likewise Equation (*) shows that

$$\operatorname{Im} \mathsf{D} = \mathscr{P}(\mathbb{R}).$$

For if $p(x) \in \mathscr{P}(\mathbb{R})$ then

$$p(x) = \mathsf{D}(\mathsf{L}(x))$$

showing $p(x) \in \mathsf{D}(\mathscr{P}(\mathbb{R}))$.

EXAMPLE 7. A slight variation of the preceding example may be constructed by considering

$$\mathsf{D} : \mathscr{P}_n(\mathbb{R}) \to \mathscr{P}_n(\mathbb{R})$$

given again by

$$\mathsf{D}(p(x)) = \frac{d}{dx} p(x).$$

This is possible because $\deg \mathsf{D}(p(x)) \leq \deg p(x)$. (This is not possible for L since $\deg \mathsf{L}(p(x)) = 1 + \deg p(x)$.) We may therefore compose D with itself to form

$$\begin{aligned} \mathsf{D}^2 &= \mathsf{D} \cdot \mathsf{D} : \mathscr{P}_n(\mathbb{R}) \to \mathscr{P}_n(\mathbb{R}) \\ \mathsf{D}^3 &= \mathsf{D} \cdot \mathsf{D}^2 : \mathscr{P}_n(\mathbb{R}) \to \mathscr{P}_n(\mathbb{R}) \\ &\vdots \\ \mathsf{D}^m &= \mathsf{D} \cdot \mathsf{D}^{m-1} : \mathscr{P}_n(\mathbb{R}) \to \mathscr{P}_n(\mathbb{R}). \end{aligned}$$

Notice that

$$\begin{aligned} \ker \mathsf{D} &= \mathscr{P}_0(\mathbb{R}) \\ \operatorname{Im} \mathsf{D} &= \mathscr{P}_{n-1}(\mathbb{R}) \\ \ker \mathsf{D}^2 &= \mathscr{P}_1(\mathbb{R}) \\ \operatorname{Im} \mathsf{D}^2 &= \mathscr{P}_{n-2}(\mathbb{R}) \end{aligned}$$

and more generally

$$\begin{aligned} \ker \mathsf{D}^m &= \mathscr{P}_{m-1}(\mathbb{R}) \\ \operatorname{Im} \mathsf{D}^m &= \mathscr{P}_{n-m}(\mathbb{R}). \end{aligned}$$

In particular

$$\begin{aligned} \ker \mathsf{D}^{n+1} &= \mathscr{P}_n(\mathbb{R}) \\ \operatorname{Im} \mathsf{D}^{n+1} &= \{\mathbf{0}\}. \end{aligned}$$

That is

$$\mathsf{D}^{n+1} = \mathbf{0} : \mathscr{P}_n(\mathbb{R}) \to \mathscr{P}_n(\mathbb{R}).$$

The linear transformation D is called **nilpotent** because of this latter property. The nilpotent transformations play a central role in the more advanced theory of linear algebra.

Definition. A linear transformation $\mathsf{T}: \mathscr{V} \to \mathscr{W}$ is said to be an *isomorphism* iff there exists a linear transformation $\mathsf{S}: \mathscr{W} \to \mathscr{V}$ such that

$$\mathsf{S} \cdot \mathsf{T}(\mathbf{A}) = \mathbf{A} \quad \text{for all } \mathbf{A} \in \mathscr{V}$$
$$\mathsf{T} \cdot \mathsf{S}(\mathbf{B}) = \mathbf{B} \quad \text{for all } \mathbf{B} \in \mathscr{W}.$$

If there is a linear transformation $\mathsf{T}: \mathscr{V} \to \mathscr{W}$ that is an isomorphism, then we say that $\mathscr{V}$ and $\mathscr{W}$ are *isomorphic*.

If $\mathscr{V}$ and $\mathscr{W}$ are isomorphic vector spaces then in some sense they are the same. More precisely an isomorphism T will translate true theorems in $\mathscr{V}$ into true theorems in $\mathscr{W}$ and conversely.

EXAMPLE 8. The vector spaces $\mathbb{R}^{n+1}$ and $\mathscr{P}_n(\mathbb{R})$ are isomorphic for all non-negative integers n.

To see this we must define a linear transformation

$$\mathsf{T}: \mathbb{R}^{n+1} \to \mathscr{P}_n(\mathbb{R})$$

and a linear transformation

$$\mathsf{S}: \mathscr{P}_n(\mathbb{R}) \to \mathbb{R}^{n+1}$$

such that

$$\mathsf{ST}(\mathbf{A}) = \mathbf{A} \quad \text{for all } \mathbf{A} \in \mathbb{R}^{n+1}$$
$$\mathsf{TS}(p(x)) = p(x) \quad \text{for all } p(x) \in \mathscr{P}_n(\mathbb{R}).$$

To do this we set

$$\mathsf{T}(a_1, \ldots, a_{n+1}) = a_1 + a_2 x + a_3 x^2 + \cdots + a_{n+1} x^n$$

and

$$\mathsf{S}(b_0 + b_1 x + \cdots + b_n x^n) = (b_0, b_1, \ldots, b_n).$$

It is routine to verify that S, T are linear transformations and that

$$\mathsf{ST}(\mathbf{A}) = \mathbf{A}$$
$$\mathsf{TS}(p(x)) = p(x)$$

for all $\mathbf{A}$ in $\mathbb{R}^{n+1}$ and $p(x)$ in $\mathscr{P}_n(\mathbb{R})$.

In view of Example 8 and the theorems to follow, the reader might be puzzled why, if the two vector spaces $\mathbb{R}^{n+1}$ and $\mathscr{P}_n(\mathbb{R})$ are isomorphic, we bothered to introduce both of them rather than stick to good old $\mathbb{R}^{n+1}$. The answer is not simple and involves in part the "style" of the mathematics of the second half of this century. However to be more specific, one reason to introduce both examples is that each suggests certain natural phenomena. For example, $\mathbb{R}^{n+1}$ lends itself quite nicely to the rotation and projection type linear transformations of Examples 1 and 3. These make little geometric sense in $\mathscr{P}_n(\mathbb{R})$ directly. On the other hand, $\mathscr{P}_n(\mathbb{R})$ suggests the linear transformation D, of Example 6 which in the context of $\mathbb{R}^{n+1}$ is more than a little forced and artificial. As remarked previously the nilpotent transformations, of which D is the most natural example, are destined to play a central role in the further study of linear transformations. Finally, there is no "natural" way to construct an isomorphism from $\mathbb{R}^{n+1}$ to $\mathscr{P}_n(\mathbb{R})$. For example we leave to the reader the verification that each of the following linear transformations

$$\mathsf{T}_1(a_1, \ldots, a_{n+1}) = a_2 + a_3 x + \cdots + a_{n+1}x^{n-1} + a_1 x^n$$
$$\mathsf{T}_2(a_1, \ldots, a_{n+1}) = a_1 + (a_1 + a_2)x + \cdots + (a_1 + a_2 + \cdots + a_{n+1})x^n$$

are isomorphisms of $\mathbb{R}^{n+1}$ to $\mathscr{P}_n(\mathbb{R})$.

Proposition 8.11. *A linear transformation* $\mathsf{T} : \mathscr{V} \to \mathscr{W}$ *is an isomorphism iff* $\ker \mathsf{T} = \{\mathbf{0}\}$ *and* $\operatorname{Im} \mathsf{T} = \mathscr{W}$.

Note. We are required to construct a linear transformation

$$\mathsf{S} : \mathscr{W} \to \mathscr{V}$$

such that

$$\mathsf{ST}(\mathbf{A}) = \mathbf{A} \quad \text{for all } \mathbf{A}$$
$$\mathsf{TS}(\mathbf{B}) = \mathbf{B} \quad \text{for all } \mathbf{B}.$$

In order to do this we will first show that the linear transformation T has a special property, namely for each vector $\mathbf{B}$ in $\mathscr{W}$ there is exactly one vector $\mathbf{A}$ in $\mathscr{V}$ with $\mathsf{T}(\mathbf{A}) = \mathbf{B}$. This result is important enough to be of separate interest.

Proposition 8.12. *Suppose that* $\mathsf{T} : \mathscr{V} \to \mathscr{W}$ *is a linear transformation with* $\ker \mathsf{T} = \{\mathbf{0}\}$. *Then for each vector* $\mathbf{B}$ *in* $\operatorname{Im} \mathsf{T}$ *there is exactly one vector* $\mathbf{A}$ *in* $\mathscr{V}$ *such that* $\mathsf{T}(\mathbf{A}) = \mathbf{B}$.

PROOF. Since $\mathbf{B}$ is a vector in $\operatorname{Im} \mathsf{T}$ there is certainly *at least* one vector $\mathbf{A}$ such that $\mathsf{T}(\mathbf{A}) = \mathbf{B}$. If there were another vector $\mathbf{C}$ with $\mathsf{T}(\mathbf{C}) = \mathbf{B}$ then we would have

$$\mathsf{T}(\mathbf{A} - \mathbf{C}) = \mathsf{T}(\mathbf{A}) - \mathsf{T}(\mathbf{C}) = \mathbf{B} - \mathbf{B} = \mathbf{0}.$$

Therefore $\mathbf{A} - \mathbf{C} \in \ker \mathsf{T} = \{\mathbf{0}\}$ and hence

$$\mathbf{A} - \mathbf{C} = \mathbf{0} \qquad \text{or} \qquad \mathbf{A} = \mathbf{C}$$

so $\mathbf{C}$ could not have been different from $\mathbf{A}$ at all. □

EXAMPLE 9. Consider the linear transformation of Example 3

$$\mathsf{T} : \mathbb{R}^3 \to \mathbb{R}^3$$

defined by

$$\mathsf{T}(x, y, z) = (x, y, 0).$$

The vector (1, 1, 0) belongs to Im T and

$$(1, 1, 0) = \mathsf{T}(1, 1, 1)$$
$$(1, 1, 0) = \mathsf{T}(1, 1, -1).$$

Of course

$$\ker \mathsf{T} = \{(0, 0, z)\} \neq \{\mathbf{0}\}.$$

The preceding result is in a sense quite surprising. For if $\mathsf{T} : \mathscr{V} \to \mathscr{W}$ has the property that for each $\mathbf{B} \in \operatorname{Im} \mathsf{T}$ there is exactly one $\mathbf{A}$ in $\mathscr{V}$ with $\mathsf{T}(\mathbf{A}) = \mathbf{B}$ then $\ker \mathsf{T} = \{\mathbf{0}\}$. For $\mathbf{0} \in \operatorname{Im} \mathsf{T}$, since Im T is a subspace of $\mathscr{W}$, and $\mathsf{T}(\mathbf{0}) = \mathbf{0}$ since T is a linear transformation. Therefore if $\mathbf{A}$ is any vector in $\mathscr{V}$ with $\mathsf{T}(\mathbf{A}) = \mathbf{0}$ we must have $\mathbf{A} = \mathbf{0}$. What is surprising is that the converse holds, and this is the content of (8.12). It is a reflection of the *homogenity* of linear transformations.

PROOF OF. (8.11). Since $\ker \mathsf{T} = \{\mathbf{0}\}$ and $\operatorname{Im} \mathsf{T} = \mathscr{W}$ there is for each vector $\mathbf{A}$ in $\mathscr{W}$ exactly one vector $\mathbf{A}$ in $\mathscr{V}$ such that $\mathsf{T}(\mathbf{A}) = \mathbf{B}$. Define a function

$$\mathsf{S} : \mathscr{W} \to \mathscr{V}$$

by setting

$$\mathsf{S}(\mathbf{B}) = \mathbf{A},$$

that is

$$\mathsf{S}(\mathbf{B}) = \mathbf{A} \quad \text{iff} \quad \mathsf{T}(\mathbf{A}) = \mathbf{B}$$

Note that since there is exactly one $\mathbf{A}$ in $\mathscr{V}$ with $\mathsf{T}(\mathbf{A}) = \mathbf{B}$ this definition yields a well-defined function

$$\mathsf{S} : \mathscr{W} \to \mathscr{V}.$$

We claim that S is a linear transformation. To prove this we suppose $\mathbf{B}, \mathbf{C}$ are vectors in $\mathscr{W}$. Let $\mathbf{A}$ and $\mathbf{D}$ be the unique vectors of $\mathscr{V}$ with

$$\mathsf{T}(\mathbf{A}) = \mathbf{B}, \qquad \mathsf{T}(\mathbf{D}) = \mathbf{C}.$$

Then since T is a linear transformation

$$\mathsf{T}(\mathbf{A} + \mathbf{D}) = \mathsf{T}(\mathbf{A}) + \mathsf{T}(\mathbf{D}) = \mathbf{B} + \mathbf{C}.$$

Therefore $\mathbf{A} + \mathbf{D}$ is the unique vector of $\mathscr{V}$ with

$$\mathsf{T}(\mathbf{A} + \mathbf{D}) = \mathbf{B} + \mathbf{C}.$$

In terms of our function S this says

$$\mathsf{S}(\mathbf{B} + \mathbf{C}) = \mathbf{A} + \mathbf{D} = \mathsf{S}(\mathbf{B}) + \mathsf{S}(\mathbf{C}).$$

If now r is a number, then since T is a linear transformation

$$\mathsf{T}(r\mathbf{A}) = r\mathsf{T}(\mathbf{A}) = r\mathbf{B}.$$

Therefore

$$\mathsf{S}(r\mathbf{B}) = r\mathbf{A} = r\mathsf{S}(\mathbf{A}).$$

Thus we have shown that the function

$$\mathsf{S} : \mathscr{W} \to \mathscr{V}$$

is a linear transformation. It is immediate from the definitions that

$$\mathsf{S}(\mathsf{T}(\mathbf{A})) = \mathbf{A} \quad \text{for all } \mathbf{A} \text{ in } \mathscr{V}$$
$$\mathsf{T}(\mathsf{S}(\mathbf{B})) = \mathbf{B} \quad \text{for all } \mathbf{B} \text{ in } \mathscr{W}.$$

Therefore T is an isomorphism.

To prove the converse direction, that is, if T is an isomorphism then $\ker \mathsf{T} = \{0\}$ and $\operatorname{Im} \mathsf{T} = \mathscr{W}$, is routine and is left to the reader as an exercise. □

Let us look at some examples:

Example 10. Let $\mathsf{T} : \mathbb{R}^4 \to \mathbb{R}^4$ be the linear transformation

$$\mathsf{T}(a_1, a_2, a_3, a_4) = (a_1, a_1 + a_2, a_1 + a_2 + a_3, a_1 + a_2 + a_3 + a_4)$$

Then we claim that T is an isomorphism. To see this we *do not* have to construct a linear transformation

$$\mathsf{S} : \mathbb{R}^4 \to \mathbb{R}^4$$

such that

$$\mathsf{ST}(\mathbf{A}) = \mathbf{A} \quad \text{for all } \mathbf{A} \text{ in } \mathbb{R}^4$$
$$\mathsf{TS}(\mathbf{A}) = \mathbf{A} \quad \text{for all } \mathbf{A} \text{ in } \mathbb{R}^4$$

although such an S will certainly exist. Rather we will use (8.11). First notice that $\ker \mathsf{T} = \{\mathbf{0}\}$. For if

$$\mathsf{T}(a_1, a_2, a_3, a_4) = \mathbf{0}$$

then

$$(a_1, a_1 + a_2, a_1 + a_2 + a_3, a_1 + a_2 + a_3 + a_4) = (0, 0, 0, 0)$$

so

$$\begin{aligned} a_1 &= 0 \\ a_1 + a_2 &= 0 \\ a_1 + a_2 + a_3 &= 0 \\ a_1 + a_2 + a_3 + a_4 &= 0 \end{aligned}$$

or

$$a_1 = 0, \qquad a_2 = 0, \qquad a_3 = 0, \qquad a_4 = 0.$$

Next note that by (8.10) we have

$$4 = \dim \mathbb{R}^4 = \dim \operatorname{Im} \mathsf{T} + \dim \ker \mathsf{T} = \dim \operatorname{Im} \mathsf{T} + 0.$$

Therefore $\dim \operatorname{Im} \mathsf{T} = 4$. Thus by (6.10) $\operatorname{Im} \mathsf{T} = \mathbb{R}^4$. To summarize

$$\operatorname{Im} \mathsf{T} = \mathbb{R}^4$$
$$\ker \mathsf{T} = \{\mathbf{0}\}$$

so T is an isomorphism by (8.11).

EXAMPLE 11. Let S be a set and $T \subset S$. If

$$f : T \to \mathbb{R}$$

is a function in $\mathscr{F}(T)$ we define a new function

$$\mathsf{L}(f) : S \to \mathbb{R}$$

by

$$\mathsf{L}(f)(s) = \begin{cases} f(s) & \text{if } s \in T, \\ 0 & \text{if } s \notin T. \end{cases}$$

The assignment $f \mapsto \mathsf{L}(f)$ defines a function

$$\mathsf{L} : \mathscr{F}(T) \to \mathscr{F}(S),$$

which is a linear transformation. To see this we simply check the definitions:

$$\begin{aligned} \mathsf{L}(f + g)(s) &= (f + g)(s) = f(s) + g(s) \\ &= (\mathsf{L}(f))(s) + (\mathsf{L}(g))(s), \qquad s \in T \end{aligned}$$

and $\mathsf{L}(f + g)(s) = 0 = (\mathsf{L}(f))(s) + (\mathsf{L}(g))(s), \qquad s \notin T$, while

$$\mathsf{L}(rf)(s) = \begin{cases} rf(s) = r(\mathsf{L}(f))(s) & \text{if } s \in T \\ 0 \cdot r(\mathsf{L}(f))(s) & \text{if } s \notin T. \end{cases}$$

Let us compute the image of L. Note that for all $f \in \mathscr{F}(T)$ we have

$$(\mathsf{L}(f))(s) = 0 \quad \text{if } s \in S - T$$

where $S - T = \{s \in S \mid s \notin T\}$. Therefore $\operatorname{Im} \mathsf{L} \subset \mathscr{F}(S, S - T)$. Moreover if $g \in \mathscr{F}(S, S - T)$

$$g : S \to \mathbb{R}$$

so by restricting g to T we get

$$g|_T : T \to \mathbb{R}$$

defined by

$$g|_T(s) = g(s), \qquad s \in T.$$

Note that by definition of L

$$\mathsf{L}(g|_T)(s) = \begin{cases} g(s) & \text{if} \quad s \in T \\ 0 & \text{if} \quad s \notin T \end{cases}$$
$$= g(s)$$

because $g \in \mathscr{F}(S, S - T)$. Therefore $g \in \operatorname{Im} \mathsf{L}$, and hence $\mathscr{F}(S, S - T) \subset \operatorname{Im} \mathsf{L}$. Since we already have established the reverse inclusion we conclude $\operatorname{Im} \mathsf{L} = \mathscr{F}(S, S - T)$.

The kernel of L may be computed quite easily. If $f \in \ker \mathsf{L}$ then

$$0 = \mathsf{L}(f)(s) = \begin{cases} f(s) & \text{if } s \in T \\ 0 & \text{if } s \notin T \end{cases}$$

so that

$$f = 0 : T \to \mathbb{R}$$

and thus

$$\ker \mathsf{T} = \{\mathbf{0}\}.$$

EXAMPLE 12. Let S be a set and $T \subset S$. If $f \in \mathscr{F}(S)$ then by restricting the domain of f to T we obtain a function

$$\mathsf{R}_T(f) : T \to \mathbb{R}$$

defined by

$$\mathsf{R}_T(f)(s) = f(s).$$

The function

$$\mathsf{R}_T : \mathscr{F}(S) \to \mathscr{F}(T)$$

given by

$$f \mapsto \mathsf{R}_T(f)$$

is easily checked to be a linear transformation.

To compute the kernel of R_T observe that

$$\mathsf{R}_T f = 0 \in \mathscr{F}(T) \quad \text{iff} \quad f(s) = 0, \qquad s \in T.$$

So we see

$$\ker \ \mathsf{R}_T = \mathscr{F}(S, T).$$

Let L be the linear transformation of Example 11. By definition we have

$$\mathsf{R}_T \mathsf{L}(f)(s) = f(s), \qquad s \in T$$

so $\mathsf{R}_T\mathsf{L}(f) = f \in \mathscr{F}(T)$, therefore $\mathsf{R}_T\mathsf{L} = \mathsf{I}$, and $\operatorname{Im} \mathsf{R}_T = \mathscr{F}(T)$. (The transformation denoted by I here is the identity transformation. Precisely, $\mathsf{I} : \mathscr{V} \to \mathscr{V}$ and $\mathsf{I}(A) = A$ for all $A \in \mathscr{V}$ is called the **identity transformation** (of the vector space $\mathscr{V}$).)

Finally, by restricting R_T to the subspace $\mathscr{F}(S, S - T)$ we obtain

$$\mathsf{R}_T : \mathscr{F}(S, S - T) \to \mathscr{F}(T)$$
$$\mathsf{L} : \mathscr{F}(T) \to \mathscr{F}(S, S - T).$$

It is easy to check that these linear transformations are inverse isomorphisms, so $\mathscr{F}(T)$ is isomorphic to $\mathscr{F}(S, S - T)$.

At this point it is worthwhile to describe a very general process for constructing linear transformations.

Linear extension construction

Suppose given the following data

(1) $\mathscr{V}$, $\mathscr{W}$ vector spaces
(2) $\{\mathbf{A}_1, \ldots, \mathbf{A}_n\}$ a basis for $\mathscr{V}$
(3) $\{\mathbf{B}_1, \ldots, \mathbf{B}_n\}$ vectors in $\mathscr{W}$, they need not be a basis, **they need not even be distinct**.

We define a function $\mathsf{T} : \mathscr{V} \to \mathscr{W}$ as follows. Let $\mathbf{A}$ be a vector in $\mathscr{V}$. By (6.6) there are unique numbers $a_1, \ldots, a_n$ (the coordinates of $\mathbf{A}$ relative to $\{\mathbf{A}_1, \ldots, \mathbf{A}_n\}$) such that

$$\mathbf{A} = a_1\mathbf{A}_1 + \cdots + a_n\mathbf{A}_n.$$

Now define

$$\mathsf{T}(\mathbf{A}) = a_1\mathbf{B}_1 + \cdots + a_n\mathbf{B}_n.$$

Note that

$$\mathsf{T}(\mathbf{A}_i) = \mathbf{B}_i \qquad i = 1, 2, \ldots, n.$$

Note. The reason that T is well defined is because $\{\mathbf{A}_1, \ldots, \mathbf{A}_n\}$ is a basis for $\mathscr{V}$. For in order to define T we first must write

$$\mathbf{A} = a_1\mathbf{A}_1 + \cdots + a_n\mathbf{A}_n,$$

and we are using the fact that $\{\mathbf{A}_1, \ldots, \mathbf{A}_n\}$ spans $\mathscr{V}$. The fact that $\mathsf{T}(\mathbf{A})$ is unambiguously defined in $\mathscr{W}$ uses the fact that the coordinates of $\mathbf{A}$ are unique, which is true since $\mathbf{A}_1, \ldots, \mathbf{A}_n$ are linearly independent.

Note also, that the linear transformation T depends not only on the sets of vectors $\{\mathbf{A}_1, \ldots, \mathbf{A}_n\}$ and $\{\mathbf{B}_1, \ldots, \mathbf{B}_n\}$ but also on their *order*. To change the order of the **A**s and/or **B**s will change the transformation T.

We call the function T the *linear extension* of

$$\mathsf{T}(\mathbf{A}_i) = \mathbf{B}_i \qquad i = 1, \ldots, n.$$

The adjective **linear** is justified by:

Proposition 8.13. *Suppose that* $\mathscr{V}$ *and* $\mathscr{W}$ *are vector spaces,* $\{\mathbf{A}_1, \ldots, \mathbf{A}_n\}$ *a basis for* $\mathscr{V}$ *and* $\{\mathbf{B}_1, \ldots, \mathbf{B}_n\}$ *vectors in* $\mathscr{W}$*. Then the linear extension of*

$$\mathsf{T}(\mathbf{A}_i) = \mathbf{B}_i \qquad i = 1, \ldots, n$$

is a linear transformation

$$\mathsf{T} : \mathscr{V} \to \mathscr{W}.$$

PROOF. We must show that

$$\mathsf{T}(\mathbf{A} + \mathbf{C}) = \mathsf{T}(\mathbf{A}) + \mathsf{T}(\mathbf{C})$$
$$\mathsf{T}(r\mathbf{A}) = r\mathsf{T}(\mathbf{A})$$

for all vectors **A** and **C** in $\mathscr{V}$ and all numbers r. So suppose that

$$\mathbf{A} = a_1\mathbf{A}_1 + \cdots + a_n\mathbf{A}_n$$
$$\mathbf{C} = c_1\mathbf{A}_1 + \cdots + c_n\mathbf{A}_n.$$

Then of course

$$\mathbf{A} + \mathbf{C} = (a_1 + c_1)\mathbf{A}_1 + (a_2 + c_2)\mathbf{A}_2 + \cdots + (a_n + c_n)\mathbf{A}_n$$

by the usual sort of argument involving shuffling vectors around. Thus according to our definition of T we have

$$\mathsf{T}(\mathbf{A} + \mathbf{C}) = (a_1 + c_1)\mathbf{B}_1 + \cdots + (a_n + c_n)\mathbf{B}_n$$

hence

$$\begin{aligned}\mathsf{T}(\mathbf{A} + \mathbf{C}) &= a_1\mathbf{B}_1 + c_1\mathbf{B}_1 + \cdots + a_n\mathbf{B}_n + c_n\mathbf{B}_n \\ &= (a_1\mathbf{B}_1 + a_2\mathbf{B}_2 + \cdots + a_n\mathbf{B}_n) + (c_1\mathbf{B}_1 + \cdots + c_n\mathbf{B}_n) \\ &= \mathsf{T}(\mathbf{A}) + \mathsf{T}(\mathbf{C}).\end{aligned}$$

In a similar manner we may show

$$\mathsf{T}(r\mathbf{A}) = r\mathsf{T}(\mathbf{A})$$

for all vectors **A** in $\mathscr{V}$ and numbers r. Therefore T is a linear transformation. □

The linear extension construction has a number of surprising consequences. Here are some of them:

Theorem 8.14. *Suppose that* $\mathscr{V}$ *and* $\mathscr{W}$ *are finite-dimensional vector spaces. If* $\dim \mathscr{V} = \dim \mathscr{W}$ *then* $\mathscr{V}$ *and* $\mathscr{W}$ *are isomorphic.*

PROOF. Let $\{\mathbf{A}_1, \ldots, \mathbf{A}_n\}$ be a basis for $\mathscr{V}$ and $\{\mathbf{B}_1, \ldots, \mathbf{B}_n\}$ a basis for $\mathscr{W}$. Note that they have the same number of elements. Let $\mathsf{T} : \mathscr{V} \to \mathscr{W}$ be the linear transformation that is the linear extension of

$$\mathsf{T}(\mathbf{A}_i) = \mathbf{B}_i \qquad i = 1, 2, \ldots, n.$$

Then by (8.4) and the fact that $\{\mathbf{A}_1, \ldots, \mathbf{A}_n\}$ is a basis for

$$\begin{aligned}\operatorname{Im} \mathsf{T} = \mathsf{T}(\mathscr{L}(\mathbf{A}_1, \ldots, \mathbf{A}_n)) &= \mathscr{L}(\mathsf{T}(\mathbf{A}_1), \ldots, \mathsf{T}(\mathbf{A}_n)) \\ = \mathscr{L}(\mathbf{B}_1, \ldots, \mathbf{B}_n) \quad &= \mathscr{W},\end{aligned}$$

since $\{\mathbf{B}_1, \ldots, \mathbf{B}_n\}$ is a basis for $\mathscr{W}$. Therefore by (8.10)

$$\begin{aligned}n = \dim \mathscr{V} &= \dim \operatorname{Im} \mathsf{T} + \dim \ker \mathsf{T} = \dim \mathscr{W} + \dim \ker \mathsf{T} \\ &= n + \dim \ker \mathsf{T}\end{aligned}$$

and hence

$$\dim \ker \mathsf{T} = n - n = 0.$$

Therefore $\ker \mathsf{T} = \{\mathbf{0}\}$ and hence T is an isomorphism by (8.11). □

Notice that the converse of (8.14) is also true. That is if $\mathscr{V}$ and $\mathscr{W}$ are finite-dimensional vector spaces that are isomorphic then they must have the same dimension. To see that this is so, suppose

$$\mathsf{T} : \mathscr{V} \to \mathscr{W}$$

is a linear transformation that is an isomorphism. Let $\{\mathbf{A}_1, \ldots, \mathbf{A}_n\}$ be a basis for $\mathscr{V}$, so that $\dim \mathscr{V} = n$. Then

$$\begin{aligned}\mathscr{W} = \operatorname{Im} \mathsf{T} = \mathsf{T}(\mathscr{V}) &= \mathsf{T}(\mathscr{L}(\mathbf{A}_1, \ldots, \mathbf{A}_n)) \\ &= \mathscr{L}(\mathsf{T}(\mathbf{A}_1), \ldots, \mathsf{T}(\mathbf{A}_n))\end{aligned}$$

so that the vectors $\mathsf{T}(\mathbf{A}_1), \ldots, \mathsf{T}(\mathbf{A}_n)$ span $\mathscr{W}$ and hence satisfy half of the basis condition. It remains to show that $\{\mathsf{T}(\mathbf{A}_1), \ldots, \mathsf{T}(\mathbf{A}_n)\}$ are linearly independent, for then they are a basis and hence $\dim \mathscr{W} = n$ also. So we suppose $\{\mathsf{T}(\mathbf{A}_1), \ldots, \mathsf{T}(\mathbf{A}_n)\}$ are dependent. Then there are numbers $a_1, \ldots, a_n$, not all zero, such that

$$a_1\mathsf{T}(\mathbf{A}_1) + a_2\mathsf{T}(\mathbf{A}_2) + \cdots + a_n\mathsf{T}(\mathbf{A}_n) = \mathbf{0}.$$

We then have

$$\begin{aligned}\mathbf{0} &= a_1\mathsf{T}(\mathbf{A}_1) + a_2\mathsf{T}(\mathbf{A}_2) + \cdots + a_n\mathsf{T}(\mathbf{A}_n) \\ &= \mathsf{T}(a_1\mathbf{A}_1) + \mathsf{T}(a_2\mathbf{A}_2) + \cdots + \mathsf{T}(a_n\mathbf{A}_n) \\ &= \mathsf{T}(a_1\mathbf{A}_1 + a_2\mathbf{A}_2 + \cdots + a_n\mathbf{A}_n).\end{aligned}$$

Let $\mathsf{S} : \mathscr{W} \to \mathscr{V}$ be the inverse of T, that is the linear transformation which must exist since T is an isomorphism. Apply S to the equation

$$\mathbf{0} = \mathsf{T}(a_1\mathbf{A}_1 + a_2\mathbf{A}_2 + \cdots + a_n\mathbf{A}_n)$$

and obtain

$$\begin{aligned}\mathsf{S}(\mathbf{0}) &= \mathsf{ST}(a_1\mathbf{A}_1 + \cdots + a_n\mathbf{A}_n) \\ \mathbf{0} &= \mathrm{a}_1\mathbf{A}_1 + a_2\mathbf{A}_2 + \cdots + a_n\mathbf{A}_n,\end{aligned}$$

which says that the vectors $\{\mathbf{A}_1, \ldots, \mathbf{A}_n\}$ are linearly dependent, contrary to the fact they are a basis for $\mathscr{V}$. Thus dim $\mathscr{W} = n$ also.

To summarize: Two finite-dimensional vector spaces $\mathscr{V}$ and $\mathscr{W}$ are isomorphic iff they have the same dimension.

Corollary 8.15. *If a vector space $\mathscr{V}$ has finite dimension n then $\mathscr{V}$ is isomorphic to $\mathbb{R}^n$.* □

In view of (8.15) it is again tempting to ask if it was really worthwhile introducing abstract vector spaces if indeed up to isomorphism there are only (at least for finite-dimensional ones) the spaces $\mathbb{R}^n$. The answer lies, as in the discussion after Example 7, in the aesthetics of mathematics and the naturality of certain constructions in some examples and not in others. In particular the operator (i.e. linear transformation)

$$\mathsf{D} : \mathscr{P}(\mathbb{R}) \to \mathscr{P}(\mathbb{R})$$

is quite natural in $\mathscr{P}(\mathbb{R})$ but not in $\mathbb{R}^n$, while the rotations (see Example 1)

$$\mathsf{T} : \mathbb{R}^2 \to \mathbb{R}^2$$

make little sense in $\mathscr{P}(\mathbb{R})$, or $\mathscr{L}(\mathscr{V}, \mathscr{W})$.

Another reason is that there is no natural isomorphism between $\mathscr{V}$ and $\mathbb{R}^n$. Highly simple and natural constructions, geometric or otherwise, in $\mathscr{V}$ may go over into complex and unnatural configurations in $\mathbb{R}^n$.

Notation. Let $\mathbf{E}_i \in \mathbb{R}^n$ be the vector

$$\mathbf{E}_i = (0, \ldots, \underset{\substack{\uparrow \\ i\text{th place}}}{1}, 0, \ldots).$$

We have already seen that the vectors $\{\mathbf{E}_1, \ldots, \mathbf{E}_n\}$ are a basis for $\mathbb{R}^n$.

Proposition 8.16. *Let $\mathbf{A}_1, \ldots, \mathbf{A}_n$ be vectors in the vector space $\mathscr{V}$. Let*

$$\mathsf{T} : \mathbb{R}^n \to \mathscr{V}$$

be the linear extension of

$$\mathsf{T}(\mathbf{E}_i) = \mathbf{A}_i \qquad i = 1, 2, \ldots, n.$$

Then $\{\mathbf{A}_1, \ldots, \mathbf{A}_n\}$ is linearly independent iff ker $\mathsf{T} = \{\mathbf{0}\}$.

PROOF. Suppose that $\{\mathbf{A}_1, \ldots, \mathbf{A}_n\}$ is linearly independent. If $\mathbf{B} = (b_1, \ldots, b_n) \in \mathbb{R}^n$ then

$$\mathsf{T}(b_1, \ldots, b_n) = b_1\mathbf{A}_1 + b_2\mathbf{A}_2 + \cdots + b_n\mathbf{A}_n.$$

Suppose that

$$\mathsf{T}(b_1, \ldots, b_n) = \mathbf{0}.$$

Then

$$b_1\mathbf{A}_1 + \cdots + b_n\mathbf{A}_n = \mathbf{0}$$

which since $\{\mathbf{A}_1, \ldots, \mathbf{A}_n\}$ is linearly independent implies

$$(b_1, \ldots, b_n) = (0, \ldots, 0).$$

That is $\ker \mathsf{T} = \{\mathbf{0}\}$.

On the other hand suppose that $\ker \mathsf{T} = \{\mathbf{0}\}$. If

$$a_1\mathbf{A}_1 + \cdots + a_n\mathbf{A}_n = \mathbf{0}$$

is a linear relation between $\{\mathbf{A}_1, \ldots, \mathbf{A}_n\}$ then

$$\mathsf{T}(a_1, \ldots, a_n) = a_1\mathbf{A}_1 + \cdots + a_n\mathbf{A}_n = \mathbf{0}.$$

So $(a_1, \ldots, a_n) = (0, \ldots, 0)$ since $\ker \mathsf{T} = \{\mathbf{0}\}$. Therefore the linear relation is the trivial one, so $\{\mathbf{A}_1, \ldots, \mathbf{A}_n\}$ is linearly independent. □

Similarly we have:

Proposition 8.17. *Let $\mathbf{A}_1, \ldots, \mathbf{A}_n$ be vectors in the vector space $\mathscr{V}$. Let*

$$\mathsf{T} : \mathbb{R}^n \to \mathscr{V}$$

be the linear extension of

$$\mathsf{T}(\mathbf{E}_i) = \mathbf{A}_i \qquad i = 1, 2, \ldots, n.$$

Then $\mathbf{A}_1, \ldots, \mathbf{A}_n$ span $\mathscr{V}$ iff $\operatorname{Im} \mathsf{T} = \mathscr{V}$.

PROOF. Exercise. □

EXAMPLE 13. Let S', S'' be finite sets and $\varphi : S' \to S''$ be a function. Let

$$\mathsf{T}_\varphi : \mathscr{F}(S') \to \mathscr{F}(S'')$$

be the linear extension of

$$\mathsf{T}_\varphi(\chi_{s'}) = \chi_{\varphi(s')} \qquad s' \in S'.$$

Then T_φ is a linear transformation. If $f \in \mathscr{F}(S')$ then

$$\mathsf{T}_\varphi(f) = \mathsf{T}_\varphi\left(\sum_{s' \in S'} f(s')\chi_{s'}\right)$$
$$= \sum_{s' \in S'} f(s')\chi_{\varphi(s')}.$$

From this formula we easily see that

$$\operatorname{Im} \mathsf{T} = \mathsf{L}(\mathscr{F}(\varphi(S')))$$

where

$$\mathsf{L} : \mathscr{F}(\varphi(S')) \to \mathscr{F}(S'')$$

is the linear transformation of Example 11 for $\varphi(S') \subset S''$. In general it is not easy to compute ker T_φ, see Examples 9 and 10 in the next chapter.

Exercises

1. Show that each of the following is a linear transformation:

   (a) $\mathsf{T}: \mathbb{R}^2 \to \mathbb{R}^2$ defined by $\mathsf{T}(x, y) = (2x - y, x)$
   (b) $\mathsf{T}: \mathbb{R}^3 \to \mathbb{R}^2$ defined by $\mathsf{T}(x, y, z) = (z, x + y)$
   (c) $\mathsf{T}: \mathbb{R} \to \mathbb{R}^2$ defined by $\mathsf{T}(x) = (2x, -x)$
   (d) $\mathsf{T}: \mathbb{R}^2 \to \mathbb{R}^3$ defined by $\mathsf{T}(x, y) = (x + y, y, x)$

2. Show that each of the following is *not* a linear transformation:

   (a) $\mathsf{T}: \mathbb{R}^2 \to \mathbb{R}^2$ defined by $\mathsf{T}(x, y) = (x^2, y^2)$
   (b) $\mathsf{T}: \mathbb{R}^3 \to \mathbb{R}^2$ defined by $\mathsf{T}(x, y, z) = (x + y + z, 1)$
   (c) $\mathsf{T}: \mathbb{R} \to \mathbb{R}^2$ defined by $\mathsf{T}(x) = (1, -1)$
   (d) $\mathsf{T}: \mathbb{R}^2 \to \mathbb{R}^3$ defined by $\mathsf{T}(x, y) = (xy, y, x)$

3. Let $\mathsf{D}: \mathscr{P}_n(\mathbb{R}) \to \mathscr{P}_n(\mathbb{R})$ be the differentiation operator. That is $\mathsf{D}(p(x)) = (d/dx)p(x)$. Show that

$$\mathsf{D}^{n+1} = \underbrace{\mathsf{D} \cdot \cdots \cdot \mathsf{D}}_{(n+1 \text{ times})} = 0: \mathscr{P}_n(\mathbb{R}) \to \mathscr{P}_n(\mathbb{R}).$$

4. Let $\mathsf{T}: \mathbb{R}^3 \to \mathbb{R}$ be defined by $\mathsf{T}(x, y, z) = x - 3y + 2z$. Show that T is linear. Find a basis for the kernel of T. What are dim ker T and dim Im T?

5. Let $\mathsf{T}: \mathbb{R}^3 \to \mathbb{R}^3$ be defined by $\mathsf{T}(x, y, z) = (y, 0, z)$. Show T is linear. Find a basis for kernel T and Im T. What are their dimensions? Let $\mathscr{U} \subset \mathbb{R}^3$ be the linear subspace of vectors with $y = 0$ (that is the xz plane). Find a basis for $\mathsf{T}(\mathscr{U})$.

6. Let $\mathsf{T}: \mathscr{P}_3(\mathbb{R}) \to \mathscr{P}_2(\mathbb{R})$ be the linear transformation defined by

$$\mathsf{T}(a_0 + a_1 x + a_2 x^2 + a_3 x^3) = a_1 + a_2 x + a_3 x^2.$$

   Find a basis for ker T and Im T. What are their dimensions?

7. (a) Let $\mathsf{T}: \mathbb{R} \to \mathbb{R}$ be a linear transformation. Show that there exists a number t, depending only on T, such that $\mathsf{T}(x) = tx$ for all $x \in \mathbb{R}$.
   (b) Suppose that $\mathsf{T}: \mathbb{R} \to \mathbb{R}$ is a linear transformation such that $\mathsf{T}(3) = -4$. Calculate $\mathsf{T}(-7)$. (*Hint:* See part (a) again!)

8. (a) Let $\mathsf{T}: \mathbb{R}^2 \to \mathbb{R}$ be a linear transformation. Show that there exist two numbers a and b, depending only on T, such that $\mathsf{T}(x, y) = ax + by$ for all (x, y) in $\mathbb{R}^2$.
   (b) Let $\mathsf{T}: \mathbb{R}^2 \to \mathbb{R}$ be a linear transformation. Suppose $\mathsf{T}(1, 1) = 3$, $T(1, 0) = 4$. Calculate $\mathsf{T}(2, 1)$.

9. Let $\mathsf{S}: \mathbb{R}^n \to \mathbb{R}^n$ be the *shift operator* which is defined by $\mathsf{S}(a_1, \ldots, a_n) = (0, a_1, a_2, \ldots, a_{n-1})$. Compute the dimension of the kernel and image of S. Do the same for

$$\mathsf{S}^k = \underbrace{\mathsf{S} \cdot \cdots \cdot \mathsf{S}}_{k \text{ times}}.$$

   Show $\mathsf{S}^n = 0$.

10. A linear transformation $\mathsf{S}:\mathscr{P}(\mathbb{R}) \to \mathscr{P}(\mathbb{R})$ is defined by

$$\mathsf{S}(p(x)) = xp(x)$$

that is

$$\mathsf{S}(a_0 + a_1x + \cdots + a_nx^n) = a_0x + a_1x^2 + \cdots + a_nx^{n+1}.$$

Find ker S, Im S. Does $\mathsf{S}^n = 0$ for any n?

11. A linear transformation $\mathsf{T}:\mathscr{V} \to \mathscr{W}$ is said to be *injective* iff for each pair of vectors $\mathbf{A}, \mathbf{B}$ in $\mathscr{V}$ with $\mathbf{A} \neq \mathbf{B}$ we have $\mathsf{T}(\mathbf{A}) \neq \mathsf{T}(\mathbf{B})$. Show that T is injective iff $\mathsf{T}(\mathbf{C}) = \mathbf{0} \Rightarrow \mathbf{C} = \mathbf{0}$, that is $\ker \mathsf{T} = \{\mathbf{0}\}$.

12. Let $\mathsf{S}:\mathbb{R}^3 \to \mathbb{R}^4$, $\mathsf{T}:\mathbb{R}^4 \to \mathbb{R}^2$ be the linear transformations defined by

$$\mathsf{S}(a_1, a_2, a_3) = (a_1 + a_2, a_1 + a_3, a_2 + a_3, a_1 + a_2 + a_3)$$
$$\mathsf{T}(b_1, b_2, b_3, b_4) = (b_1 + b_2, b_3 + b_4)$$

Compute $\mathsf{T} \cdot \mathsf{S}(1, 1, 1, 1)$.

13. Let $\mathsf{S}:\mathbb{R}^3 \to \mathbb{R}^3$ and $\mathsf{P}:\mathbb{R}^3 \to \mathbb{R}^3$ be the linear transformations defined by

$$\mathsf{S}(a_1, a_2, a_3) = (0, a_1, a_2)$$
$$\mathsf{P}(a_1, a_2, a_3) = (a_1, a_2, 0).$$

Calculate dim Im S, dim ker S, dim Im P, dim ker P. Find SP(1, 0, 1), PS(1, 0, 1), (S + P)(1, 0, 1), and (P − 2S)(1, 0, 1).

14. Suppose that $\mathscr{V}$ is a finite-dimensional vector space and $\mathsf{S}, \mathsf{T}:\mathscr{V} \to \mathscr{W}$ are linear transformations. Let $\{\mathbf{A}_1, \ldots, \mathbf{A}_n\}$ be a basis for $\mathscr{V}$. Suppose further that

$$\mathsf{S}(\mathbf{A}_i) = T(\mathbf{A}_i) \qquad i = 1, 2, \ldots, n.$$

Show that S = T.

15. Suppose that $\mathscr{V}$ is a finite-dimensional vector space and $\mathsf{T}:\mathscr{V} \to \mathscr{V}$ a linear transformation. Show the following are equivalent:

(a) T is an isomorphism
(b) $\ker \mathsf{T} = \{\mathbf{0}\}$
(c) $\operatorname{Im} \mathsf{T} = \mathscr{V}$.

(*Hint*: Use (8.10) and (8.11).)

16. Let $\mathsf{T}:\mathbb{R}^3 \to \mathbb{R}^3$ be the linear extension of $\mathsf{T}(\mathbf{E}_i) = \mathbf{A}_i$, $i = 1, 2, 3$ where

$$\mathbf{A}_1 = (0, 1, 1), \qquad \mathbf{A}_2 = (-1, 0, 1), \qquad \mathbf{A}_3 = (0, 1, 2).$$

Find Im T and ker T. Also find T(1, 2, 3).

17. Let $\mathsf{T}:\mathbb{R}^n \to \mathbb{R}^n$ be the linear extension of

$$\mathsf{T}(\mathbf{E}_1) = (0, 0, \ldots, 0)$$
$$\mathsf{T}(\mathbf{E}_2) = (1, 0, \ldots, 0)$$
$$\mathsf{T}(\mathbf{E}_3) = (0, 2, 0, \ldots, 0)$$
$$\vdots$$
$$\mathsf{T}(\mathbf{E}_n) = (0, 0, \ldots, n - 1, 0).$$

Show that T is a nilpotent operator and more specifically that $\mathsf{T}^n = 0$. Find Im T and ker T.

18. Let S and T be finite sets. Show that $\mathscr{F}(S)$ and $\mathscr{F}(T)$ are isomorphic iff S and T have the same number of elements.

19. Let $S = \{a, b, c\}$ and let $\mathsf{T} : \mathscr{F}(S) \to \mathbb{R}^2$ be the linear extension of:

$$\mathsf{T}(\chi_a) = (0, 1)$$
$$\mathsf{T}(\chi_b) = (1, 1)$$
$$\mathsf{T}(\chi_c) = (1, 0).$$

Show that the kernel of T is the subspace spanned by the function

$$f : S \to \mathbb{R}$$

defined by $f(a) = 1, f(b) = -1, f(c) = 1.$

20. Let $\mathscr{V}$ and $\mathscr{W}$ be vector spaces and let $E \subset \mathscr{V}$ be a subset. Define

$$\mathsf{L}\colon \text{Fun}(E, \mathscr{W}) \to \mathscr{L}(\mathscr{L}(E), \mathscr{W})$$

by

$$\mathsf{L}(\mathbf{f})(a_1\mathbf{E}_1 + \cdots + a_n\mathbf{E}_n) = a_1\mathbf{f}(\mathrm{H}_1) + \cdots + a_n\mathbf{f}(\mathbf{E}_n).$$

Show that L is an isomorphism of vector spaces.

21. Let S be a set and define $\varphi\colon \mathscr{L}(\mathscr{F}(S), \mathbb{R}) \to \mathscr{F}(S)$ by $\varphi(\mathsf{L})(s) = \mathsf{L}(\chi_s)$, where χ_s is the characteristic function of s. Show that φ is a linear transformation. Show it is an isomorphism if S is finite. What happens when S is infinite?

22. Let S, T be sets and $\varphi\colon S \to T$ a function. Define $\varphi^*\colon \mathscr{F}(T) \to \mathscr{F}(S)$ by $\varphi^*(\mathbf{f})(s) = \mathbf{f}(\varphi(s))$.

(a) Show that φ^* is a linear transformation.
(b) Show that when φ is epic that φ^* is monic.
(c) In general show that

$$\text{Ker } \varphi^* = \mathscr{F}(T, T - \varphi(S)).$$

(d) Show that when φ is injective φ^* is epic.

23. Let $\mathscr{V}', \mathscr{V}''$ be vector subspaces of the finite-dimensional vector space $\mathscr{V}$. Show that there is an isomorphism $\mathsf{T}\colon \mathscr{V} \to \mathscr{V}$ such that T maps $\mathscr{V}'$ isomorphically onto $\mathscr{V}''$ iff $\dim \mathscr{V}' = \dim \mathscr{V}''$.

24. Consider pairs of subspaces $(\mathscr{V}'_1, \mathscr{V}'_2)$ in $\mathscr{V}$. We say that two such pairs $(\mathscr{V}'_1, \mathscr{V}'_2)$ and $(\mathscr{V}''_1, \mathscr{V}''_2)$ are *equivalent* iff there is an isomorphism $\mathsf{T}\colon \mathscr{V} \to \mathscr{V}$ such that T maps $\mathscr{V}'_i$ isomorphically onto $\mathscr{V}''_i$ for $i = 1, 2$. Try to find necessary and sufficient conditions that two pairs of subspaces are equivalent. (*Hints*: 1. Think about the formula

$$\dim(\mathscr{V}'_1 + \mathscr{V}'_2) = \dim \mathscr{V}'_1 + \dim \mathscr{V}''_1 - \dim(\mathscr{V}'_1 \cap \mathscr{V}'_2)$$

and how it is proved.

2. Think about how many different dimensions you can associate to $(\mathscr{V}'_1, \mathscr{V}'_2)$. For example, do not forget $\dim \mathscr{V}'_1 - \dim \mathscr{V}'_1 \cap \mathscr{V}''_1 = \dim \mathscr{V}'_1/\mathscr{V}'_1 \cap \mathscr{V}''_1$, etc.)

9 Linear transformations: some numerical examples

Our objective in this section is to present several numerical examples to illustrate the preceding discussion of linear transformations.

EXAMPLE 1. Which of the following are linear transformations

$$\mathsf{T} : \mathbb{R}^3 \to \mathbb{R}^3$$

and which are not?

(1) $\mathsf{T}(x, y, z) = (x + y + z, 0, 0)$
(2) $\mathsf{T}(x, y, z) = (y, z, x)$
(3) $\mathsf{T}(x, y, z) = (x + y + z, 1, -1)$
(4) $\mathsf{T}(x, y, z) = (xyz, 0, 0)$

Solution. The first and second of these are easily seen to be linear transformations. For example we will check (2).

Let $\mathbf{A} = (a_1, a_2, a_3)$, $\mathbf{B} = (b_1, b_2, b_3)$. Then we have

$$\begin{aligned} \mathsf{T}(\mathbf{A} + \mathbf{B}) &= \mathsf{T}(a_1 + b_1, a_2 + b_2, a_3 + b_3) \\ &= (a_2 + b_2, a_3 + b_3, a_1 + b_1) \\ &= (a_2, a_3, a_1) + (b_2, b_3, b_1) \\ &= \mathsf{T}(\mathbf{A}) + \mathsf{T}(\mathbf{B}). \end{aligned}$$

$$\begin{aligned} \mathsf{T}(r\mathbf{A}) &= \mathsf{T}(ra_1, ra_2, ra_3) \\ &= (ra_2, ra_3, ra_1) \\ &= r(a_2, a_3, a_1) \\ &= r\mathsf{T}(\mathbf{A}). \end{aligned}$$

The first may be checked similarly.

Suppose you tried to check the third transformation by this method. We get

$$\begin{aligned}\mathsf{T}(\mathbf{A}+\mathbf{B}) &= \mathsf{T}(a_1+b_1, a_2+b_2, a_3+b_3)\\ &= (a_1+b_1+a_2+b_2+a_3+b_3, 1, -1)\\ &= (a_1+a_2+a_3+b_1+b_2+b_3, 1, -1).\end{aligned}$$

while

$$\begin{aligned}\mathsf{T}(\mathbf{A})+\mathsf{T}(\mathbf{B}) &= (a_1+a_2+a_3, 1, -1)+(b_1+b_2+b_3, 1, -1)\\ &= (a_1+a_2+a_3+b_1+b_2+b_3, 2, -2).\end{aligned}$$

and we note that

$$\begin{aligned}&(a_1+a_2+a_3+b_1+b_2+b_3, 1, -1)\\ &\qquad\neq (a_1+a_2+a_3+b_1+b_2+b_3, 2, -2)\end{aligned}$$

because $1 \neq 2$. Therefore

$$\mathsf{T}(\mathbf{A}+\mathbf{B}) \neq \mathsf{T}(\mathbf{A})+\mathsf{T}(\mathbf{B})$$

and hence T is *not* a linear transformation.

Now actually with a little more familiarity with linear transformations it should be immediately "obvious" that T is not a linear transformation, all that is needed is a *reason why* it isn't. Here is one simple reason:

$$\mathsf{T}(0, 0, 0) = (0, 1, -1) \neq (0, 0, 0),$$

and according to (8.1)

$$\mathsf{T}(0, 0, 0) = (0, 0, 0)$$

if T is linear. Therefore T is not linear.

Let us look at the last example now. If it is a linear transformation then we must have

$$\mathsf{T}(r\mathbf{A}) = r\mathsf{T}(\mathbf{A})$$

for all numbers r and vectors $\mathbf{A} = (a_1, a_2, a_3)$. Let us therefore calculate both sides above.

$$\begin{aligned}\mathsf{T}(r\mathbf{A}) &= \mathsf{T}(ra_1, ra_2, ra_3)\\ &= (ra_1ra_2ra_3, 0, 0)\\ &= (r^3a_1a_2a_3, 0, 0)\\ &= r^3(a_1a_2a_3, 0, 0)\end{aligned}$$

while

$$r\mathsf{T}(\mathbf{A}) = r(a_1a_2a_3, 0, 0).$$

So we must ask ourselves if it is true that

$$r^3(a_1a_2a_3, 0, 0) = r(a_1a_2a_3, 0, 0)$$

for *all* numbers r and vectors $\mathbf{A} = (a_1, a_2, a_3)$. The answer is clearly no. For if we set $a_1 = a_2 = a_3 = 1$ and $r = 2$, then

$$r^3(a_1 a_2 a_3, 0, 0) = 8(1, 0, 0) = (8, 0, 0)$$
$$r(a_1 a_2 a_3, 0, 0) = 2(1, 0, 0) = (2, 0, 0)$$

and these vectors are clearly not equal, so T cannot be a linear transformation in this case.

EXAMPLE 2. Let T, S be the linear transformations

$$\mathsf{T}, \mathsf{S} : \mathbb{R}^3 \to \mathbb{R}^3$$

given by the formulas

$$\mathsf{T}(x, y, z) = (x + y + z, 0, 0)$$
$$\mathsf{S}(x, y, z) = (y, z, x).$$

Calculate

(a) $(\mathsf{T} \cdot \mathsf{S})(1, 0, 1)$
(b) $(\mathsf{S} \cdot \mathsf{T})(1, 0, 1)$
(c) $(\mathsf{S} + \mathsf{T})(1, 0, 1)$
(d) $(\mathsf{S} - \mathsf{T})(1, 0, 1)$
(e) $(\mathsf{S}(\mathsf{S} + \mathsf{T})\mathsf{T})(1, 0, 1)$.

Solution. This requires nothing but determination:

(a) $(\mathsf{T} \cdot \mathsf{S})(1, 0, 1) = \mathsf{T}(\mathsf{S}(1, 0, 1))$ (by definition of composition)
$= \mathsf{T}(0, 1, 1)$ (by definition of S)
$= (2, 0, 0)$. (by definition of T)

(b) $\mathsf{S} \cdot \mathsf{T}(1, 0, 1) = \mathsf{S}(\mathsf{T}(1, 0, 1))$
$= \mathsf{S}(2, 0, 0)$
$= (0, 0, 2)$.

(c) $(\mathsf{S} + \mathsf{T})(1, 0, 1) = \mathsf{S}(1, 0, 1) + \mathsf{T}(1, 0, 1)$
$= (0, 1, 1) + (2, 0, 0)$
$= (2, 1, 1)$.

(d) $(\mathsf{S} - \mathsf{T})(1, 0, 1) = \mathsf{S}(1, 0, 1) - \mathsf{T}(1, 0, 1)$
$= (0, 1, 1) - (2, 0, 0)$
$= (-2, 1, 1)$.

(e) $(\mathsf{S}(\mathsf{S} + \mathsf{T})\mathsf{T})(1, 0, 1) = (\mathsf{S}(\mathsf{S} + \mathsf{T}))(\mathsf{T}(1, 0, 1)$
$= (\mathsf{S}(\mathsf{S} + \mathsf{T}))(2, 0, 0)$
$= \mathsf{S}((\mathsf{S} + \mathsf{T})(2, 0, 0))$
$= \mathsf{S}(\mathsf{S}(2, 0, 0) + \mathsf{T}(2, 0, 0))$
$= \mathsf{S}((0, 0, 2) + (2, 0, 0))$
$= \mathsf{S}(2, 0, 2) = (0, 2, 2)$.

EXAMPLE 3. Let

$$\mathsf{L}: \mathbb{R}^3 \to \mathbb{R}^3$$

be the linear transformation given by the formula

$$\mathsf{L}(x, y, z) = (0, x, y).$$

Find the bases for the kernel and image of L, L^2, and L^3.

Solution. Let us work with L first. The kernel of L consists of all the vectors $\mathbf{A} = (a_1, a_2, a_3)$ such that $\mathsf{L}(\mathbf{A}) = \mathbf{0}$. That means $\mathbf{A} \in \ker \mathsf{L}$ iff

$$(0, 0, 0) = \mathsf{L}(\mathbf{A}) = (0, a_1, a_2)$$

that is

$$0 = 0, \qquad 0 = a_1, \qquad 0 = a_2,$$

and *no restriction* on a_3. Thus

$$\ker \mathsf{L} = \{(0, 0, z) \in \mathbb{R}^3\}$$

or in geometric terms the kernel of L is the z-axis. Therefore a basis for the kernel of L consists of the single vector $\{(0, 0, 1)\}$. (There are many other bases, for example, $\{(0, 0, -99/100)\}$ is also a basis for $\ker \mathsf{L}$.)

To find a basis for the image of L we note that $\mathbf{B} \in \operatorname{Im} \mathsf{L}$ iff $\mathbf{B} = \mathsf{L}(\mathbf{A})$ for some vector $\mathbf{A} \in \mathbb{R}^3$. Therefore $\mathbf{B} \in \operatorname{Im} \mathsf{L}$ iff

$$(b_1, b_2, b_3) = \mathbf{B} = \mathsf{L}(\mathbf{A}) = (0, a_1, a_2)$$

that is[1]

$$b_1 = 0, \qquad b_2 = a_1, \qquad b_3 = a_2$$

which means

$$\operatorname{Im} \mathsf{L} = \{(0, b_2, b_3)\}$$

or put geometrically, $\operatorname{Im} \mathsf{L}$ is the y, z plane. A basis for $\operatorname{Im} \mathsf{L}$ is therefore the set $\{(0, 1, 0), (0, 0, 1)\}$. Another basis is $\{(0, 1, -1), (0, 1, 0)\}$.

We must now apply the same analysis to the linear transformation L^2. First let us find a "formula" for L^2. We have

$$\begin{aligned} \mathsf{L}^2(x, y, z) &= \mathsf{L}(\mathsf{L}(x, y, z)) = \mathsf{L}(0, x, y) \\ &= (0, 0, x) \end{aligned}$$

(Remember, that L shifts the coordinates to the right one unit, inserts a 0 in the first place and strikes off the third coordinate.) So a vector $\mathbf{A}$ belongs to the kernel of L^2 iff

$$(0, 0, 0) = \mathsf{L}^2(x, y, z) = (0, 0, a_1),$$

[1] For example $(0, 2, 3) = \mathsf{L}(2, 3, 0)$ and $(0, 2, 3) = \mathsf{L}(2, 3, 1)$ etc.

that is

$$0 = 0, \qquad 0 = 0, \qquad 0 = a_1,$$

so that $\mathbf{A} \in \ker \mathsf{L}^2$ iff $a_1 = 0$, while a_2 and a_3 are arbitrary. Thus

$$\ker \mathsf{L}^2 = \{(0, a_2, a_3) \in \mathbb{R}^3\}$$

or geometrically, $\ker \mathsf{L}^2$ is the y, z plane. Hence a basis for $\ker \mathsf{L}^2$ is $\{(0, 1, 0), (0, 0, 1)\}$.

To find a basis for the image of L^2 we note that $\mathbf{B} \in \operatorname{Im} \mathsf{L}^2$ iff $\mathbf{B} = \mathsf{L}^2(\mathbf{A})$ for some vector $\mathbf{A}$ in $\mathbb{R}^3$. That is

$$(b_1, b_2, b_3) = \mathbf{B} = \mathsf{L}^2(\mathbf{A}) = (0, 0, a_1)$$

or

$$b_1 = 0, \qquad b_2 = 0, \qquad b_3 = a_1.$$

Therefore we find

$$\operatorname{Im} \mathsf{L}^2 = \{(0, 0, z) \in \mathbb{R}^3\}$$

or that $\operatorname{Im} \mathsf{L}^2$ is the z axis, which has basis $\{(0, 0, 1)\}$.

Finally we note that

$$\mathsf{L}^3(x, y, z) = (0, 0, 0)$$

so that $\operatorname{Im} \mathsf{L}^3 = \{(0, 0, 0)\}$ and $\ker \mathsf{L}^3 = \mathbb{R}^3$. A basis for the kernel of L^3 is thus any basis for $\mathbb{R}^3$, such as $\{(1, 1, 1), (1, 1, 0), (1, 0, 0)\}$, and a basis for the image of L^3 is $\varnothing$, the empty set.

EXAMPLE 4. Let S, T be the linear transformations $\mathsf{S}, \mathsf{T} : \mathbb{R}^3 \to \mathbb{R}^3$ defined by

$$\mathsf{S}(x, y, z) = (y, z, x)$$
$$\mathsf{T}(x, y, z) = (x + y + z, 0, 0).$$

Find a basis for the kernel of $\mathsf{S} + \mathsf{T}$.

Solution. Let us find a formula for $\mathsf{S} + \mathsf{T}$. Notice that

$$\begin{aligned}(\mathsf{S} + \mathsf{T})(x, y, z) &= \mathsf{S}(x, y, z) + \mathsf{T}(x, y, z)\\ &= (y, z, x) + (x + y + z, 0, 0)\\ &= (x + 2y + z, z, x).\end{aligned}$$

That is

$$(\mathsf{S} + \mathsf{T})(x, y, z) = (x + 2y + z, z, x).$$

Therefore $\mathbf{A} \in \ker (\mathsf{S} + \mathsf{T})$ iff

$$(0, 0, 0) = (\mathsf{S} + \mathsf{T})(\mathbf{A}) = (a_1 + 2a_2 + a_3, a_3, a_1)$$

or

$$0 = a_1 + 2a_2 + a_3, \qquad 0 = a_3, \qquad 0 = a_1,$$

iff

$$0 = a_2, \qquad 0 = a_3, \qquad 0 = a_1,$$

and therefore $\ker(\mathsf{S} + \mathsf{T}) = \{(0, 0, 0)\}$ so the empty set is a basis for $\ker(\mathsf{S} + \mathsf{T})$.

EXERCISE. Show that in the preceding example $\operatorname{Im}(\mathsf{S} + \mathsf{T}) = \mathbb{R}^3$.

EXAMPLE 5. Which of the following linear transformations are isomorphisms

(a) $\mathsf{T}: \mathbb{R}^3 \to \mathscr{P}_2(\mathbb{R})$ defined by

$$\mathsf{T}(a_1, a_2, a_3) = a_1 + (a_1 + a_2)x + (a_1 + a_2 + a_3)x^2,$$

(b) $\mathsf{T}: \mathbb{R}^3 \to \mathscr{P}_2(\mathbb{R})$ defined by $\mathsf{T}(a_1, a_2, a_3) = a_1 + (a_1 + a_2)x$,

(c) $\mathsf{T}: \mathbb{R}^3 \to \mathbb{R}^4$ defined by

$$\mathsf{T}(a_1, a_2, a_3) = (a_1 + a_2, a_2 + a_3, a_3 + a_1, a_1 + a_2 + a_3),$$

(d) $\mathsf{T}: \mathbb{R}^3 \to \mathbb{R}^3$ defined by

$$\mathsf{T}(a_1, a_2, a_3) = (a_1 + a_2 + a_3, a_1 + a_2, a_1 + a_2)?$$

Solution. Remember to check that T is an isomorphism you do *not* want to use the definition of an isomorphism if you can avoid it, because that requires a construction of another linear transformations S that may be quite hard. Rather we will try to use properties of linear transformations such as (8.10) and (8.11).

(a) Note that $\ker \mathsf{T} = \{\mathbf{0}\}$. Because

$$\mathbf{0} = \mathsf{T}(\mathbf{A}) = \mathsf{T}(a_1, a_2, a_3) = a_1 + (a_1 + a_2)x + (a_1 + a_2 + a_3)x^2$$

iff

$$0 = a_1, \qquad 0 = a_1 + a_2, \qquad 0 = a_1 + a_2 + a_3$$

or

$$a_1 = 0, \qquad a_2 = 0, \qquad a_3 = 0.$$

Next we apply (8.10) which gives

$$3 = \dim \mathbb{R}^3 = \dim \ker \mathsf{T} + \dim \operatorname{Im} \mathsf{T} = 0 + \dim \operatorname{Im} \mathsf{T}$$

so

$$\dim \operatorname{Im} \mathsf{T} = 3.$$

Recall that $\dim \mathscr{P}_2(\mathbb{R}) = 3$. Therefore Im T, which is a linear subspace of $\mathscr{P}_2(\mathbb{R})$ by (8.9), has the same dimension as $\mathscr{P}_2(\mathbb{R})$ and hence $\operatorname{Im} \mathsf{T} = \mathscr{P}_2(\mathbb{R})$. Therefore T is an isomorphism by (8.11).

For those who are interested, the linear transformation

$$\mathsf{S} : \mathscr{P}_2(\mathbb{R}) \to \mathbb{R}^3$$

given by the formula

$$\mathsf{S}(b_0 + b_1 x + b_2 x^2) = (b_0, b_1 - b_0, b_2 - b_1 - b_0)$$

is actually the transformation required by the definition, and whose existence is assured by (8.11).

(b) This transformation is *not* an isomorphism. We may see this by applying (8.11). Because

$$\mathsf{T}(0, 0, 1) = \mathbf{0}$$

and hence $\ker \mathsf{T} \neq \{\mathbf{0}\}$.

(c) This is *not* an isomorphism, because if it were, then by the discussion following (8.14)

$$3 = \dim \mathbb{R}^3 = \dim \mathbb{R}^4 = 4$$

which is impossible.

(d) This transformation is also *not* an isomorphism. To see this we note that $\mathbf{B} \in \operatorname{Im} \mathsf{T}$ iff there is an $\mathbf{A}$ with $\mathbf{B} = \mathsf{T}(\mathbf{A})$, so in this case

$$(b_1, b_2, b_3) = \mathbf{B} = \mathsf{T}(\mathbf{A}) = (a_1 + a_2 + a_3, a_1 + a_2, a_1 + a_2)$$

and therefore

$$\begin{aligned} b_1 &= a_1 + a_2 + a_3 \\ b_2 &= a_1 + a_2 \\ b_3 &= a_1 + a_2 \end{aligned}$$

and in particular $b_2 = b_3$. Therefore the vector (0, 1, 2) does not belong to $\operatorname{Im} \mathsf{T}$ so by (8.11) T cannot be an isomorphism since $\operatorname{Im} \mathsf{T} \neq \mathbb{R}^3$.

Example 6. Let $\mathsf{T} : \mathbb{R}^3 \to \mathscr{P}_5(\mathbb{R})$ be the linear transformation that is the linear extension of

$$\begin{aligned} \mathsf{T}(1, 1, 1) &= x^2 + x^4, \\ \mathsf{T}(1, 1, 0) &= x + x^3 + x^5, \\ \mathsf{T}(1, 0, 0) &= 1. \end{aligned}$$

Calculate $\mathsf{T}(0, 0, 1)$.

Solution. First of all we ought to note that {(1, 1, 1), (1, 1, 0), (1, 0, 0)} is a basis for $\mathbb{R}^3$ so that the transformation T is well defined by the linear extension construction. To calculate $\mathsf{T}(0, 0, 1)$ we must first write

$$(0, 0, 1) = a_1(1, 1, 1) + a_2(1, 1, 0) + a_3(1, 0, 0),$$

that is, we must find the coordinates of (0, 0, 1) relative to the basis (*ordered* basis) (1, 1, 1), (1, 1, 0), (1, 0, 0). This is just some high school algebra.

Suppose

$$(0, 0, 1) = a_1(1, 1, 1) + a_2(1, 1, 0) + a_3(1, 0, 0).$$

Then

$$\begin{aligned}(0, 0, 1) &= (a_1, a_1, a_1) + (a_2, a_2, 0) + (a_3, 0, 0)\\ &= (a_1 + a_2 + a_3, a_1 + a_2, a_1)\end{aligned}$$

so

$$0 = a_1 + a_2 + a_3, \qquad 0 = a_1 + a_2, \qquad 1 = a_1$$

or

$$a_1 = 1, \qquad a_2 = -1, \qquad a_3 = 0.$$

So

$$(0, 0, 1) = 1(1, 1, 1) - 1(1, 1, 0) + 0(1, 0, 0)$$

and therefore

$$\begin{aligned}\mathsf{T}(0, 0, 1) &= 1(x^2 + x^4) - 1(x + x^3 + x^5)\\ &= -x + x^2 - x^3 + x^4 - x^5.\end{aligned}$$

Notice that since the vectors

$$\{x^2 + x^4, x + x^3 + x^5, 1\}$$

are linearly independent in $\mathscr{P}_5(\mathbb{R})$ the kernel of the transformation T of the preceding example is $\{\mathbf{0}\}$. (Why? Because of (8.15).)

Here is another example involving linear extensions.

EXAMPLE 7. Let $\mathsf{T}: \mathbb{R}^4 \to \mathbb{R}^2$ be the linear transformation that is the linear extension of

$$\begin{aligned}\mathsf{T}(1, 0, 0, 0) &= (1, 1),\\ \mathsf{T}(1, 1, 0, 0) &= (0, 1),\\ \mathsf{T}(1, 1, 1, 0) &= (1, 0),\\ \mathsf{T}(1, 1, 1, 1) &= (-1, -1).\end{aligned}$$

Calculate

$$\mathsf{T}(4, 3, 2, 1).$$

Solution. Again, as in (6), the first step will be to find the coordinates of (4, 3, 2, 1) relative to the (*ordered*) basis

$$\{(1, 0, 0, 0), (1, 1, 0, 0), (1, 1, 1, 0), (1, 1, 1, 1)\}$$

for $\mathbb{R}^4$. O.K. So write

$$(4, 3, 2, 1) = a_1(1, 0, 0, 0) + a_2(1, 1, 0, 0) + a_3(1, 1, 1, 0) + a_4(1, 1, 1, 1)$$

and use high school algebra to find a_1, a_2, a_3, a_4. We have

$$\begin{aligned}(4, 3, 2, 1) &= (a_1, 0, 0, 0) + (a_2, a_2, 0, 0) + (a_3, a_3, a_3, 0) + (a_4, a_4, a_4, a_4)\\ &= (a_1 + a_2 + a_3 + a_4, a_2 + a_3 + a_4, a_3 + a_4, a_4)\end{aligned}$$

so

$$\begin{aligned}4 &= a_1 + a_2 + a_3 + a_4\\ 3 &= a_2 + a_3 + a_4\\ 2 &= a_3 + a_4\\ 1 &= a_4\end{aligned}$$

which implies

$$a_1 = 1, \qquad a_2 = 1, \qquad a_3 = 1, \qquad a_4 = 1.$$

Thus

$$(4, 3, 2, 1) = 1(1, 0, 0, 0) + 1(1, 1, 0, 0) + 1(1, 1, 1, 0) + 1(1, 1, 1, 1)$$

so

$$\begin{aligned}\mathsf{T}(4, 3, 2, 1) &= (1, 1) + (0, 1) + (1, 0) + (-1, -1)\\ &= (1 + 0 + 1 - 1, 1 + 1 + 0 - 1)\\ &= (1, 1).\end{aligned}$$

EXAMPLE 8. Let $\mathscr{V}$ be the subspace

$$\mathscr{V} = \{(x, y, z) | x + y + z = 0\}$$

of $\mathbb{R}^3$. Is $\mathscr{V}$ isomorphic to $\mathbb{R}^2$?

Solution. Yes. Note $\dim \mathscr{V} = 2$ since $\{(1, -1, 0), (0, 1, -1)\}$ is a basis for $\mathscr{V}$. Now apply (8.14).

EXAMPLE 9. Let S be the set $\{u, v, w\}$ and define

$$\mathsf{E} : \mathscr{F}(S) \to \mathbb{R}$$

by

$$\mathsf{E}(\mathbf{f}) = \mathbf{f}(u) + \mathbf{f}(v) + \mathbf{f}(w).$$

Then E is a linear transformation. To compute the kernel of E note that

$$\mathsf{E}(\mathbf{f}) = \mathbf{0} \quad \text{iff} \quad \mathbf{f}(u) + \mathbf{f}(v) + \mathbf{f}(w) = 0.$$

Let us define $\mathbf{g}_1, \mathbf{g}_2 \in \mathscr{F}(S)$, that is, $g_1, g_2 : S \to \mathbb{R}$ by

$$\mathbf{g}_1(u) = 1, \qquad \mathbf{g}_1(v) = -1, \qquad \mathbf{g}_1(w) = 0,$$

and

$$\mathbf{g}_2(u) = 0, \qquad \mathbf{g}_2(v) = -1, \qquad \mathbf{g}_2(w) = 1.$$

Then $\mathbf{g}_1, \mathbf{g}_2 \in \ker \mathsf{E}$, for

$$\mathsf{E}(\mathbf{g}_1) = \mathbf{g}_1(u) + \mathbf{g}_1(v) + \mathbf{g}_1(w) = 1 + (-1) + 0 = 0$$
$$\mathsf{E}(\mathbf{g}_2) = \mathbf{g}_2(u) + \mathbf{g}_2(v) + \mathbf{g}_2(w) = 0 + (-1) + 1 = 0.$$

The vectors $\mathbf{g}_1$ and $\mathbf{g}_2$ are linearly independent because

$$a_1\mathbf{g}_1 + a_2\mathbf{g}_2 = \mathbf{0}$$

implies

$$\begin{aligned} 0 = (a_1\mathbf{g}_1 + a_2\mathbf{g}_2)(u) &= a_1\mathbf{g}_1(u) + a_2\mathbf{g}_2(u) \\ &= a_1 + a_2 \cdot 0 = a_1, \\ 0 = (a_1\mathbf{g}_1 + a_2\mathbf{g}_2)(w) &= a_1\mathbf{g}_1(w) + a_2\mathbf{g}_2(w) \\ &= a_1 \cdot 0 + a_2 = a_2, \end{aligned}$$

so $a_1 = 0$ and $a_2 = 0$.

Finally if $\mathbf{f} \in \ker \mathsf{E}$ then

$$\mathbf{f}(u) + \mathbf{f}(v) + \mathbf{f}(w) = 0$$

so

$$-\mathbf{f}(w) = \mathbf{f}(u) + \mathbf{f}(v).$$

Therefore

$$\begin{aligned} (\mathbf{f} - \mathbf{f}(u)\mathbf{g}_1)(u) &= 0 = (\mathbf{f}(w)\mathbf{g}_2)(u) \\ (\mathbf{f} - \mathbf{f}(u)\mathbf{g}_1)(v) &= \mathbf{f}(v) + \mathbf{f}(u) = -\mathbf{f}(w) = (\mathbf{f}(w)\mathbf{g}_2)(v) \\ (\mathbf{f} - \mathbf{f}(u)\mathbf{g}_1)(w) &= \mathbf{f}(w) = (\mathbf{f}(w)\mathbf{g}_2)(w) \end{aligned}$$

and so

$$\mathbf{f} - \mathbf{f}(u)\mathbf{g}_1 = \mathbf{f}(w)\mathbf{g}_2,$$

whence

$$\mathbf{f} = \mathbf{f}(u)\mathbf{g}_1 + \mathbf{f}(w)\mathbf{g}_2$$

so that $\mathbf{g}_1, \mathbf{g}_2$ span $\ker \mathsf{E}$; being linearly independent they are a basis for $\ker \mathsf{E}$.

Example 10. Let S' and S'' be the sets $S' = \{u, v, w\}$, $S'' = \{a, b\}$, and let $\varphi : S' \to S''$ be defined by

$$\varphi(u) = a, \qquad \varphi(v) = a, \qquad \varphi(w) = b.$$

Then $\varphi(S') = S''$ so by Example 13 of the preceding chapter the transformation

$$\mathsf{T}_\varphi : \mathscr{F}(S') \to \mathscr{F}(S'')$$

is onto. To compute $\ker \mathsf{T}_\varphi$ note that by (8.10)

$$\dim \mathscr{F}(S') = \dim \ker \mathsf{T}_\varphi + \dim \operatorname{Im} \mathsf{T}_\varphi = \dim \ker \mathsf{T}_\varphi + \dim \mathscr{F}(S'')$$

so dim ker $\mathsf{T}_\varphi = 1$ since dim $\mathscr{F}(S') = 3$ and dim $\mathscr{F}(S'') = 2$. Therefore all we need to do to obtain a basis for ker T_φ is to find a single nonzero vector in ker T_φ. Clearly $\boldsymbol{\chi}_u - \boldsymbol{\chi}_v$ is such a vector, so ker T_φ is spanned by the function $\mathbf{k} = \boldsymbol{\chi}_u - \boldsymbol{\chi}_v$.

EXERCISES

Let $k(x)$ be a *fixed* polynomial. Define functions

$$\mathsf{M}_{k(x)} : \mathscr{P}(\mathbb{R}) \to \mathscr{P}(\mathbb{R})$$
$$\mathsf{L}_{k(x)} : \mathscr{P}(\mathbb{R}) \to \mathscr{P}(\mathbb{R})$$

by

$$\mathsf{M}_{k(x)}(p(x)) = k(x)p(x)$$

$$\mathsf{L}_{k(x)}(p(x)) = \int_0^x p(t)k(t)dt.$$

1. Show that $\mathsf{M}_{k(x)}$ and $\mathsf{L}_{k(x)}$ are linear transformations for all polynomials $k(x)$. Calculate

$$\mathsf{M}_{x^2+2x^3}(x + x^4) \quad \text{and} \quad \mathsf{L}_{x+x^2}(1 + x^3).$$

2. Calculate $\mathsf{DM}_x - \mathsf{M}_x\mathsf{D} : \mathscr{P}(\mathbb{R}) \to \mathscr{P}(\mathbb{R})$ where D is defined in Example 6 of Chapter 8. That is find a "formula" for

$$(\mathsf{DM}_x - \mathsf{M}_x\mathsf{D})(p(x)).$$

3. Show that $\mathsf{LM}_{k(x)} = \mathsf{L}_{k(x)} : \mathscr{P}(\mathbb{R}) \to \mathscr{P}(\mathbb{R})$.

4. Show that $\mathsf{T} : \mathbb{R}^2 \to \mathbb{R}^2$ defined by

$$\mathsf{T}(x, y) = (ax + by, cx + dy),$$

a, b, c, d fixed real numbers, is a linear transformation and that T is an isomorphism iff $ad \neq bc$.

5. Show that $\mathsf{T} : \mathbb{R}^3 \to \mathbb{R}^3$ defined by

$$\mathsf{T}(x, y, z) = (a_1x + a_2y + a_3z, b_1x + b_2y + b_3z, c_1x + c_2y + c_3z)$$

is a linear transformation. Any linear transformation from $\mathbb{R}^3$ to $\mathbb{R}^3$ takes this form, where a_i, b_i, c_i are scalars.

6. Let $\mathsf{T} : \mathbb{R}^2 \to \mathbb{R}^2$ be defined by $\mathsf{T}(x, y) = (x - y, 2x + y)$ and let $\mathsf{S} : \mathbb{R}^2 \to \mathbb{R}^2$ be defined by $\mathsf{S}(x, y) = (y - 2x, x + y)$.

(a) Find $\mathsf{T} \cdot \mathsf{S}(1, 0)$, $(\mathsf{T} + \mathsf{S})(1, 0)$, $\mathsf{S} \cdot \mathsf{T}(1, 0)$, $(2\mathsf{T} - \mathsf{S})(1, 0)$ and $\mathsf{T}^2(1, 0)$.
(b) Find $\mathsf{T} \cdot \mathsf{S}(x, y)$ and $\mathsf{S} \cdot \mathsf{T}(x, y)$.
(c) What are the vectors (x, y) satisfying $\mathsf{T}(x, y) = (1, 0)$?

7. Let $\mathsf{T} : \mathbb{R}^2 \to \mathbb{R}^2$ be defined by $\mathsf{T}(x, y) = (x, x)$. What is the kernel of T, and the image of T?

8. Let $\mathsf{T}: \mathbb{R}^3 \to \mathbb{R}^3$ be defined by $\mathsf{T}(x, y, z) = (x + y, y + z, z + x)$. Is T an isomorphism? Find $\mathsf{T}(1, 0, 0)$, $\mathsf{T}(0, 1, 0)$, and $\mathsf{T}(0, 0, 1)$.

9. Let $\mathsf{T}: \mathbb{R}^3 \to \mathbb{R}^3$ be a linear transformation satisfying the condition

$$\mathsf{T}(x, y, z) = (0, 0, 0) \quad \text{whenever } 2x - y + z = 0.$$

Let $\mathsf{T}(0, 0, 1) = (1, 2, 3)$. Find $\mathsf{T}(1, 0, 0)$ and $\mathsf{T}(0, 1, 0)$. What is dim Im T?

10. Let $\mathsf{T}: \mathbb{R}^3 \to \mathbb{R}^3$ be the linear transformation given by the formula

$$\mathsf{T}(x, y, z) = (y + z, x + z, x + y).$$

Show that T is an isomorphism and find an inverse for T.

11. Let $\mathscr{V}$ be a finite-dimensional vector space over $\mathbb{R}$, and $\mathbf{E}_1, \ldots, \mathbf{E}_n$ be a basis for $\mathscr{V}$. Define for each i, $1 \leq i \leq n$ a function

$$\mathsf{E}_i^*: \mathscr{V} \to \mathbb{R}$$

by the formula

$$\mathsf{E}_i^*(\mathbf{A}) = a_i, \qquad 1 \leq i \leq n,$$

where

$$\mathbf{A} = a_1\mathbf{E}_1 + \cdots + a_n\mathbf{E}_n.$$

(a) Show that $\mathbf{E}_1^*, \ldots, \mathbf{E}_n^*$ are linear transformations.
(b) Find a basis for Ker E_i.
(c) Let $\mathsf{S}: \mathscr{V} \to \mathscr{L}(\mathscr{V}, \mathbb{R})$ be the linear extension of the map

$$\mathsf{E}_i \to \mathsf{E}_i^*, \qquad i = 1, \ldots, n.$$

Show that S is an isomorphism.

(d) Compute $\mathrm{Dim}_{\mathbb{R}}\, \mathscr{L}(\mathscr{V}, \mathbb{R})$.

12. Let $\mathscr{V}$ be a vector space and define $\mathscr{V}^* = \mathscr{L}(\mathscr{V}; \mathbb{R})$

$$\mathsf{B}: \mathscr{V} \to \mathscr{V}^{**} \qquad (= \mathscr{L}(\mathscr{V}^*, \mathbb{R}) = \mathscr{L}(\mathscr{L}(\mathscr{V}; \mathbb{R}), \mathbb{R}))$$

by

$$\mathsf{B}(\mathbf{A})(\mathbf{f}) = \mathbf{f}(\mathbf{A}).$$

(a) Show that B is a linear transformation.
(b) Show that Ker $\mathsf{B} = \{0\}$.
(c) Show that B is an isomorphism if $\mathscr{V}$ is finite dimensional.
(d) Show by an example that Im $\mathsf{B} \neq \mathscr{V}^{**}$ when $\mathscr{V}$ is not finite dimensional.

13. Let $\mathscr{V}$ be a finite-dimensional vector space with basis $\mathbf{E} = \{\mathbf{E}_1, \ldots, \mathbf{E}_n\}$. Show that the linear extension construction gives an isomorphism

$$\mathsf{L}: \mathscr{F}(E) \to \mathscr{V}$$

by

$$\mathsf{L}(\mathbf{f}) = \mathbf{f}(\mathbf{E}_1) \cdot E_1 + \cdots + \mathbf{f}(\mathbf{E}_n) \cdot E_n.$$

14. Let $\mathscr{V}$ be a finite-dimensional vector space with basis $\mathbf{E} = \{\mathbf{E}_1, \ldots, \mathbf{E}_n\}$. Show that the map (remember $\mathscr{V}^* = \mathscr{L}(\mathscr{V}; \mathbb{R})$)

$$\mathsf{C}: \mathscr{V}^* \to \mathscr{F}(E)$$

defined by

$$\mathsf{C}(\mathsf{T})(\mathbf{E}_i) = \mathsf{T}(\mathbf{E}_i), \qquad i = 1, \ldots, n$$

is a vector space isomorphism.

15. Define a function

$$l: \mathscr{P}_n(\mathbb{R}) \to \mathscr{P}_n(\mathbb{R})$$

by

$$l(p(x)) = p(x + 1).$$

Determine if l is a linear transformation or not.

16. Let $\mathscr{V}$ and $\mathscr{W}$ be finite-dimensional vector spaces. Suppose $\{\mathbf{E}_1, \ldots, \mathbf{E}_n\}$ is a basis for $\mathscr{V}$ and $\{\mathbf{F}_1, \ldots, \mathbf{F}_m\}$ a basis for $\mathscr{W}$. For each ordered pair (i, j) show that

$$\mathsf{M}_{i,j}: \mathscr{V} \to \mathscr{W},$$

$$\mathsf{M}_{i,j}(\mathbf{A}) = a_j \mathbf{F}_i, \qquad A = a_1\mathbf{E}_1 + \cdots + a_n\mathbf{E}_n$$

is a linear transformation.

(a) Show $\{\mathsf{M}_{i,j} | 1 \le i \le n, 1 \le j \le m\}$ is a basis for $\mathscr{L}(\mathscr{V}, \mathscr{W})$.
(b) Show $\operatorname{Dim} \mathscr{L}(\mathscr{V}, \mathscr{W}) = nm$.

17. Let $\mathscr{V}$ be a vector space and suppose that $\mathsf{S}, \mathsf{T} \in \mathscr{V}^* = \mathscr{L}(\mathscr{V}; \mathbb{R})$. Define

$$\mathsf{L}: \mathscr{V} \to \mathbb{R}^2$$

by

$$\mathsf{L}(\mathbf{A}) = (\mathsf{S}(\mathbf{A}), \mathsf{T}(\mathbf{A})).$$

(a) Show that L is a linear transformation.
(b) Show that $\operatorname{Ker} \mathsf{L} = \operatorname{Ker} \mathsf{S} \cap \operatorname{Ker} \mathsf{T}$.
(c) Show that $\operatorname{Im} \mathsf{L} = \mathbb{R}^2$ iff $\mathsf{S} \ne 0 \ne \mathsf{T}$.

18. Let $\mathscr{V}$ be a vector space and define

$$\mathsf{C}_i: \mathscr{L}(\mathscr{V}; \mathbb{R}) \to \mathscr{L}(\mathscr{V}; \mathbb{R}^2), \qquad i = 1, 2$$

by

$$\left.\begin{aligned} \mathsf{C}_1(\mathsf{T})(\mathbf{A}) &= (\mathsf{T}(\mathbf{A}), 0), \\ \mathsf{C}_2(\mathsf{T})(\mathbf{A}) &= (0, \mathsf{T}(\mathbf{A})), \end{aligned}\right\} \qquad \mathbf{A} \in \mathscr{V}.$$

(a) Show that C_i is a linear transformation for $i = 1, 2$.
(b) Show that $\operatorname{Ker} \mathsf{C}_i = \{0\}$, $i = 1, 2$.
(c) Show that $\operatorname{Im} \mathsf{C}_1 \cap \operatorname{Im} \mathsf{C}_2 = \{0\}$.
(d) Let $\mathscr{L}_i = \operatorname{Im} \mathsf{C}_i$, $i = 1, 2$ and show that

$$\mathscr{L}(\mathscr{V}; \mathbb{R}^2) = \mathscr{L}_1 + \mathscr{L}_2.$$

19. Let $\mathscr{V}$ be a vector space over $\mathbb{R}$. A *complex structure* on $\mathscr{V}$ is a linear transformation

$$\mathsf{J}: \mathscr{V} \to \mathscr{V}$$

such that for all **A** in $\mathscr{V}$

$$\mathsf{J}^2(\mathbf{A}) = -\mathbf{A}.$$

(a) Show that

$$\mathsf{J}(x, y) = (-y, x)$$

is a complex structure on $\mathbb{R}^2$.

(b) Show that

$$\mathsf{J}(x, y) = (y, -x)$$

is a different complex structure on $\mathbb{R}^2$.

For a real vector space $\mathscr{V}$ with complex structure J define a multiplication of vectors in $\mathscr{V}$ by *complex* scalars by the formula

$$(a + bi) \cdot \mathbf{A} = a\mathbf{A} + b\mathsf{J}(\mathbf{A})$$

(c) Show that this definition of complex scalar multiplication makes $\mathscr{V}$ into a complex vector space.

(d) Show that on $\mathbb{R}^3$ there is no complex structure.

20. A linear transformation $\mathsf{T}: \mathscr{V} \to \mathscr{V}$ is called *invertible* iff it is an isomorphism.

(a) If $\mathscr{V}$ is finite dimensional show that T is invertible iff $\operatorname{Ker} \mathsf{T} = \{0\}$.

(b) If S, T are invertible show $\mathsf{S} \cdot \mathsf{T}$ is also. What is $(\mathsf{S} \cdot \mathsf{T})^{-1}$?

10 Matrices and linear transformations

We come now to the connecting link between linear transformations and matrices. Our approach will be to consider first the case of a linear transformation

$$\mathsf{T} : \mathbb{R}^3 \to \mathbb{R}^3$$

in some detail and then abstract the salient features to the general case. Let us therefore suppose given a fixed linear transformation $\mathsf{T} : \mathbb{R}^3 \to \mathbb{R}^3$. As usual we will denote by $\mathbf{E}_1, \mathbf{E}_2, \mathbf{E}_3$ the standard basis vectors (1, 0, 0), (0, 1, 0), (0, 0, 1) in $\mathbb{R}^3$. Let

$$\begin{aligned}
\mathsf{T}(\mathbf{E}_1) &= (a_{11}, a_{21}, a_{31}) \\
\mathsf{T}(\mathbf{E}_2) &= (a_{12}, a_{22}, a_{32}) \\
\mathsf{T}(\mathbf{E}_3) &= (a_{13}, a_{23}, a_{33}).
\end{aligned}$$

Notice that the subscript ij on an a means that it is the ith coordinate of $\mathsf{T}(E_j)$, and that i, j assume the values 1, 2, 3. Of course using the vectors $\mathbf{E}_1, \mathbf{E}_2, \mathbf{E}_3$ we can write these equations as

$$\begin{aligned}
\mathsf{T}(\mathbf{E}_1) &= a_{11}\mathbf{E}_1 + a_{21}\mathbf{E}_2 + a_{31}\mathbf{E}_3 \\
\mathsf{T}(\mathbf{E}_2) &= a_{12}\mathbf{E}_1 + a_{22}\mathbf{E}_2 + a_{32}\mathbf{E}_3 \\
\mathsf{T}(\mathbf{E}_3) &= a_{13}\mathbf{E}_1 + a_{23}\mathbf{E}_2 + a_{33}\mathbf{E}_3.
\end{aligned}$$

If we now recall the "$\sum$ notation" for sums from the calculus we see that we may sum up these three equations by the one equation

$$\mathsf{T}(\mathbf{E}_j) = \sum_{i=1}^{3} a_{ij}\mathbf{E}_i \qquad j = 1, 2, 3.$$

Suppose now that $\mathbf{V} = (v_1, v_2, v_3)$ is any vector in $\mathbb{R}^3$. Then of course

$$\mathbf{V} = v_1\mathbf{E}_1 + v_2\mathbf{E}_2 + v_3\mathbf{E}_3.$$

Therefore

$$\begin{aligned}\mathsf{T}(\mathbf{V}) &= \mathsf{T}(v_1\mathbf{E}_1 + v_2\mathbf{E}_2 + v_3\mathbf{E}_3)\\ &= v_1\mathsf{T}(\mathbf{E}_1) + v_2\mathsf{T}(\mathbf{E}_2) + v_3\mathsf{T}(\mathbf{E}_3)\end{aligned}$$

because T is a linear transformation. Now referring to the preceding list of equations for $\mathsf{T}(\mathbf{E}_j)$, $j = 1, 2, 3$, we find

$$\begin{aligned}\mathsf{T}(\mathbf{V}) = v_1(a_{11}\mathbf{E}_1 + a_{21}\mathbf{E}_2 + a_{31}\mathbf{E}_3) + v_2(a_{12}\mathbf{E}_1 + a_{22}\mathbf{E}_2 + a_{32}\mathbf{E}_3)\\ + v_3(a_{13}\mathbf{E}_1 + a_{23}\mathbf{E}_2 + a_{33}\mathbf{E}_3)\end{aligned}$$

or after some algebraic manipulations, that

$$\begin{aligned}\mathsf{T}(\mathbf{V}) = (a_{11}v_1 + a_{12}v_2 + a_{13}v_3)\mathbf{E}_1 + (a_{21}v_1 + a_{22}v_2 + a_{33}v_3)\mathbf{E}_2\\ + (a_{31}v_1 + a_{32}v_2 + a_{33}v_3)\mathbf{E}_3,\end{aligned}$$

which in the more compact $\sum$ notation may be written

$$\mathsf{T}(\mathbf{V}) = \sum_{i=1}^{3}\sum_{j=1}^{3} a_{ij}v_j\mathbf{E}_i.$$

Thus we see that from the array of 9 numbers

$$(a_{ij})_{\substack{i=1,2,3\\ j=1,2,3}}$$

we may calculate the value of the linear transformation T on any vector $\mathbf{V}$ whatsoever. Notice that it is not just the collection of 9 numbers that matters but the pattern (which is 3×3) that they fit into. Notice also that we have tacitly agreed to *fix* the ordered basis $\{\mathbf{E}_1, \mathbf{E}_2, \mathbf{E}_3\}$ for $\mathbb{R}^3$ and calculate all coordinates with respect to this ordered basis. Let us formalize these ideas with a definition.

Definition. Let $\mathsf{T} : \mathbb{R}^3 \to \mathbb{R}^3$ be a linear transformation. Suppose that

$$\mathsf{T}(\mathbf{E}_j) = \sum_{i=1}^{3} a_{ij}\mathbf{E}_i \qquad j = 1, 2, 3,$$

for the unique numbers $(a_{ij})_{i,j=1,2,3}$. The *matrix of* T *relative to the ordered basis* $\mathbf{E}_1, \mathbf{E}_2, \mathbf{E}_3$ is the 3×3 array of numbers

$$\mathbf{A} = \begin{pmatrix} a_{11} & a_{12} & a_{13}\\ a_{21} & a_{22} & a_{23}\\ a_{31} & a_{32} & a_{33}\end{pmatrix}.$$

Notice that the number a_{ij} occurs as the entry in the ith row and jth column of the matrix of T. In order to write down the matrix of T we must calculate the coordinates of $\mathsf{T}(\mathbf{E}_j)$ relative to the ordered basis $\{\mathbf{E}_1, \mathbf{E}_2, \mathbf{E}_3\}$ and write the resulting numbers as the jth *column* of the matrix of T.

EXAMPLE 1. Let $\mathsf{T} : \mathbb{R}^3 \to \mathbb{R}^3$ be the linear transformation given by

$$\mathsf{T}(x, y, z) = (x + 2z, y - x, z + y).$$

Calculate the matrix of T relative to the standard basis $\mathbf{E}_1, \mathbf{E}_2, \mathbf{E}_3$ of $\mathbb{R}^3$.

Solution. We must first calculate the coordinates of $\mathsf{T}(\mathbf{E}_1)$, $\mathsf{T}(\mathbf{E}_2)$, $\mathsf{T}(\mathbf{E}_3)$ and write them down according to the rule explained above. We have

$$\begin{aligned}
\mathsf{T}(\mathbf{E}_1) &= \mathsf{T}(1, 0, 0) = (1, -1, 0) = 1\mathbf{E}_1 + (-1)\mathbf{E}_2 + 0\mathbf{E}_3\\
\mathsf{T}(\mathbf{E}_2) &= \mathsf{T}(0, 1, 0) = (0, 1, 1) = 0\mathbf{E}_1 + (1)\mathbf{E}_2 + (1)\mathbf{E}_3\\
\mathsf{T}(\mathbf{E}_3) &= \mathsf{T}(0, 0, 1) = (2, 0, 1) = 2\mathbf{E}_1 + 0\mathbf{E}_2 + 1\mathbf{E}_3.
\end{aligned}$$

Therefore the matrix of T relative to the basis $\mathbf{E}_1, \mathbf{E}_2, \mathbf{E}_3$ is

$$\begin{pmatrix} 1 & 0 & 2\\ -1 & 1 & 0\\ 0 & 1 & 1 \end{pmatrix}.$$

EXAMPLE 2. Let $\mathsf{S} : \mathbb{R}^3 \to \mathbb{R}^3$ be the linear transformation given by[1]

$$\mathsf{S}(x, y, z) = (0, x, y).$$

Calculate the matrix of S relative to the standard basis $\mathbf{E}_1, \mathbf{E}_2, \mathbf{E}_3$ of $\mathbb{R}^3$.

Solution. We have

$$\begin{aligned}
\mathsf{S}(\mathbf{E}_1) &= \mathsf{S}(1, 0, 0) = (0, 1, 0) = 0\mathbf{E}_1 + \mathbf{E}_2 + 0\mathbf{E}_3\\
\mathsf{S}(\mathbf{E}_2) &= \mathsf{S}(0, 1, 0) = (0, 0, 1) = 0\mathbf{E}_1 + 0\mathbf{E}_2 + \mathbf{E}_3\\
\mathsf{S}(\mathbf{E}_3) &= \mathsf{S}(0, 0, 1) = (0, 0, 0) = 0\mathbf{E}_1 + 0\mathbf{E}_2 + 0\mathbf{E}_3
\end{aligned}$$

so the matrix we seek is

$$\begin{pmatrix} 0 & 0 & 0\\ 1 & 0 & 0\\ 0 & 1 & 0 \end{pmatrix}.$$

In order for the matrix of a linear transformation $\mathsf{T} : \mathbb{R}^3 \to \mathbb{R}^3$ to be of any real use to us it should be possible to answer questions about T in terms of its matrix, and for each operation on T to have a corresponding operation of the matrix of T. To illustrate this latter point let us suppose given two linear transformations

$$\mathsf{T}, \mathsf{S} : \mathbb{R}^3 \to \mathbb{R}^3$$

which have matrices

$$\mathbf{A} = \begin{pmatrix} a_{11} & a_{12} & a_{13}\\ a_{21} & a_{22} & a_{23}\\ a_{31} & a_{32} & a_{33} \end{pmatrix}, \qquad \mathbf{B} = \begin{pmatrix} b_{11} & b_{12} & b_{13}\\ b_{21} & b_{22} & b_{23}\\ b_{31} & b_{32} & b_{33} \end{pmatrix}$$

[1] The transformation S is called the *shift* operator.

relative to the standard basis $\mathbf{E}_1, \mathbf{E}_2, \mathbf{E}_3$ of $\mathbb{R}^3$. There is then also the linear transformation

$$\mathsf{S} \cdot \mathsf{T} : \mathbb{R}^3 \to \mathbb{R}^3$$

that is the composition of S and T. The question we pose is what is the matrix of $\mathsf{S} \cdot \mathsf{T}$ relative to the standard basis of $\mathbb{R}^3$? To answer this question we simply follow the procedure of the two preceding examples, taking care not to become too entangled in the notations. Let us therefore calculate $\mathsf{S} \cdot \mathsf{T}(\mathbf{E}_1)$, $\mathsf{S} \cdot \mathsf{T}(\mathbf{E}_2)$ and $\mathsf{S} \cdot \mathsf{T}(\mathbf{E}_3)$ in terms of what has been given to us, namely the matrices $\mathbf{A}$ and $\mathbf{B}$. We have

$$\begin{aligned}
\mathsf{S} \cdot \mathsf{T}(\mathbf{E}_1) &= \mathsf{S}(\mathsf{T}(\mathbf{E}_1)) \\
&= \mathsf{S}(a_{11}\mathbf{E}_1 + a_{21}\mathbf{E}_2 + a_{31}\mathbf{E}_3) \\
&= a_{11}\mathsf{S}(\mathbf{E}_1) + a_{21}\mathsf{S}(\mathbf{E}_2) + a_{31}\mathsf{S}(\mathbf{E}_3) \\
&= a_{11}(b_{11}\mathbf{E}_1 + b_{21}\mathbf{E}_2 + b_{31}\mathbf{E}_3) \\
&\quad + a_{21}(b_{12}\mathbf{E}_1 + b_{22}\mathbf{E}_2 + b_{32}\mathbf{E}_3) \\
&\quad + a_{31}(b_{13}\mathbf{E}_1 + b_{23}\mathbf{E}_2 + b_{33}\mathbf{E}_3)
\end{aligned}$$

which after a little algebraic manipulation gives

$$\begin{aligned}
\mathsf{S} \cdot \mathsf{T}(\mathbf{E}_1) = (b_{11}a_{11} + b_{12}a_{21} + b_{13}a_{31})\mathbf{E}_1 \\
+ (b_{21}a_{11} + b_{22}a_{21} + b_{23}a_{31})\mathbf{E}_2 \\
+ (b_{31}a_{11} + b_{32}a_{21} + b_{33}a_{31})\mathbf{E}_3.
\end{aligned}$$

Thus the first column of the matrix we are seeking is

$$\begin{aligned}
&b_{11}a_{11} + b_{12}a_{21} + b_{13}a_{31} \\
&b_{21}a_{11} + b_{22}a_{21} + b_{23}a_{31} \\
&b_{31}a_{11} + b_{32}a_{21} + b_{33}a_{31}.
\end{aligned}$$

In a similar manner we compute $\mathsf{S} \cdot \mathsf{T}(\mathbf{E}_2)$, $\mathsf{S} \cdot \mathsf{T}(\mathbf{E}_3)$ and we find that

$$\begin{aligned}
\mathsf{S} \cdot \mathsf{T}(\mathbf{E}_2) = (b_{11}a_{12} + b_{12}a_{22} + b_{13}a_{32})\mathbf{E}_1 \\
+ (b_{21}a_{12} + b_{22}a_{22} + b_{23}a_{32})\mathbf{E}_2 \\
+ (b_{31}a_{12} + b_{32}a_{22} + b_{33}a_{33})\mathbf{E}_3,
\end{aligned}$$

and

$$\begin{aligned}
\mathsf{S} \cdot \mathsf{T}(\mathbf{E}_3) = (b_{11}a_{13} + b_{12}a_{23} + b_{13}a_{33})\mathbf{E}_1 \\
+ (b_{21}a_{13} + b_{22}a_{23} + b_{23}a_{33})\mathbf{E}_2 \\
+ (b_{31}a_{13} + b_{32}a_{23} + b_{33}a_{33})\mathbf{E}_3.
\end{aligned}$$

Therefore the matrix of $\mathsf{S} \cdot \mathsf{T}$ relative to the standard basis of $\mathbb{R}^3$ is given by (the commas have been inserted so that the columns may be easily distinguished)

$$\mathbf{C} = \begin{pmatrix}
b_{11}a_{11} + b_{12}a_{21} + b_{13}a_{31}, & b_{11}a_{12} + b_{12}a_{22} + b_{13}a_{32}, & b_{11}a_{13} + b_{12}a_{23} + b_{13}a_{33} \\
b_{21}a_{11} + b_{22}a_{21} + b_{23}a_{31}, & b_{21}a_{12} + b_{22}a_{22} + b_{23}a_{32}, & b_{21}a_{13} + b_{22}a_{23} + b_{23}a_{33} \\
b_{31}a_{11} + b_{32}a_{21} + b_{33}a_{31}, & b_{31}a_{12} + b_{32}a_{22} + b_{33}a_{32}, & b_{31}a_{13} + b_{32}a_{23} + b_{33}a_{33}
\end{pmatrix}.$$

While the above formulas are quite formidable to contemplate there is really a very simple rule underlying the computation of the matrix **C**. Write

$$\mathbf{C} = \begin{pmatrix} c_{11} & c_{12} & c_{13} \\ c_{21} & c_{22} & c_{23} \\ c_{31} & c_{32} & c_{33} \end{pmatrix}.$$

The entry c_{ij} is calculated by taking the ith *row* of **B**

$$b_{i1}, b_{i2}, b_{i3}$$

and the jth *column* of **A**

$$\begin{matrix} a_{1j} \\ a_{2j} \\ a_{3j} \end{matrix}$$

multiplying the elements in the corresponding positions and adding them up, that is

$$c_{ij} = b_{i1}a_{1j} + b_{i2}a_{2j} + b_{i3}a_{3j} = \sum_{k=1}^{3} b_{ik}a_{kj}.$$

This row-by-column method of multiplication for matrices that perhaps some of you have seen before is not artificial, but a natural consequence of the attempt to solve the problem we posed; namely, to calculate the matrix of $\mathsf{S} \cdot \mathsf{T}$ from **A** and **B**.

EXAMPLE 3. Let S and T be the linear transformations introduced in Examples 1 and 2. Calculate the matrix of the linear transformation $\mathsf{S} \cdot \mathsf{T} : \mathbb{R}^3 \to \mathbb{R}^3$ relative to the standard basis $\mathbf{E}_1, \mathbf{E}_2, \mathbf{E}_3$ of $\mathbb{R}^3$.

Solution. Recall that we found the matrices

$$\begin{pmatrix} 0 & 0 & 0 \\ 1 & 0 & 0 \\ 0 & 1 & 0 \end{pmatrix}, \quad \begin{pmatrix} 1 & 0 & 2 \\ -1 & 1 & 0 \\ 0 & 1 & 1 \end{pmatrix}$$

to be the matrices of S and T respectively. Therefore using the row by column multiplication procedure described above we find the required matrix is:

$$\begin{pmatrix} 0 & 0 & 0 \\ 1 & 0 & 2 \\ -1 & 1 & 0 \end{pmatrix}.$$

EXAMPLE 4. With S and T as in Examples 1 and 2 calculate the matrix of $\mathsf{T} \cdot \mathsf{S} : \mathbb{R}^3 \to \mathbb{R}^3$ relative to the standard basis of $\mathbb{R}^3$.

Solution. We have to employ row-by-column multiplication in the opposite order from Example 3. We write

$$\begin{pmatrix} 1 & 0 & 2 \\ -1 & 1 & 0 \\ 0 & 1 & 1 \end{pmatrix}, \quad \begin{pmatrix} 0 & 0 & 0 \\ 1 & 0 & 0 \\ 0 & 1 & 0 \end{pmatrix}$$

and so the required matrix is:

$$\begin{pmatrix} 0 & 2 & 0 \\ 1 & 0 & 0 \\ 1 & 1 & 0 \end{pmatrix}$$

which is quite a bit different from Example 3.

The matrix of a linear transformation $\mathsf{T} : \mathbb{R}^3 \to \mathbb{R}^3$ gives a very convenient way to specify the transformation. This is particularly so if the matrix has a large number of zero entries. However this may not be the case for a particular T if we only allow ourselves to use the standard basis $\mathbf{E}_1, \mathbf{E}_2, \mathbf{E}_3$ for $\mathbb{R}^3$. This suggests that we introduce a matrix for $\mathsf{T} : \mathbb{R}^3 \to \mathbb{R}^3$ relative to *any* ordered basis $\mathbf{A}_1, \mathbf{A}_2, \mathbf{A}_3$ for $\mathbb{R}^3$, or more generally still, relative to a *pair* of ordered bases $\mathbf{A}_1, \mathbf{A}_2, \mathbf{A}_3$ and $\mathbf{B}_1, \mathbf{B}_2, \mathbf{B}_3$ for $\mathbb{R}^3$. Of course there is nothing sacred about $\mathbb{R}^3$ and so we might as well consider the general case of a linear transformation $\mathsf{T} : \mathscr{V} \to \mathscr{W}$ where $\mathscr{V}$ and $\mathscr{W}$ are finite-dimensional vector spaces. We will take up this subject in Chapter 12 after some elementary matrix notions in the next chapter.

Exercises

1. For each of the following linear transformations of $\mathbb{R}^3$ to $\mathbb{R}^3$ calculate the matrix relative to the standard basis:
   (a) $\mathsf{T} : \mathbb{R}^3 \to \mathbb{R}^3$ by $\mathsf{T}(x, y, z) = (x + y + z, 0, 0)$
   (b) $\mathsf{Q} : \mathbb{R}^3 \to \mathbb{R}^3$ by $\mathsf{Q}(x, y, z) = (x, x + y, x + y + z)$
   (c) $\mathsf{F} : \mathbb{R}^3 \to \mathbb{R}^3$ by $\mathsf{F}(x, y, z) = (x + 2y + 3z, 2x + y, z - x)$
   (d) $\mathsf{G} : \mathbb{R}^3 \to \mathbb{R}^3$ by $\mathsf{G}(x, y, z) = (y - z, x + y, z - 2x)$.

2. Calculate the matrix of the linear transformations

$$\begin{aligned} \mathsf{T} \cdot \mathsf{Q} &: \mathbb{R}^3 \to \mathbb{R}^3 \\ \mathsf{F} \cdot \mathsf{G} &: \mathbb{R}^3 \to \mathbb{R}^3 \\ \mathsf{Q} \cdot \mathsf{F} \cdot \mathsf{G} &: \mathbb{R}^3 \to \mathbb{R}^3 \end{aligned}$$

   where T, Q, F and G are as in Exercise (1).

3. Suppose that the matrix of the linear transformation $\mathsf{S} : \mathbb{R}^3 \to \mathbb{R}^3$ is

$$\begin{pmatrix} 1 & 2 & 3 \\ 0 & 0 & 1 \\ 1 & -1 & 2 \end{pmatrix}$$

   relative to the standard basis of $\mathbb{R}^3$. Calculate $\mathsf{S}(1, 2, 3)$.

4. Let $\mathsf{P} : \mathbb{R}^3 \to \mathbb{R}^3$ be the linear transformation given by

$$\mathsf{P}(x, y, z) = (x, y, 0).$$

Calculate the matrix for P and for P^2 relative to the standard basis of $\mathbb{R}^3$.

5. Let $\mathsf{S}, \mathsf{T} : \mathbb{R}^3 \to \mathbb{R}^3$ be linear transformations with matrices **A** and **B** relative to the standard basis of $\mathbb{R}^3$. What is the matrix of $\mathsf{S} + \mathsf{T}$ relative to this basis?

6. Let $\mathsf{T} : \mathbb{R}^3 \to \mathbb{R}^3$ be the linear extension of

$$\mathsf{T}(\mathbf{E}_1) = (-1, 0, 2)$$
$$\mathsf{T}(\mathbf{E}_2) = (1, 1, -1)$$
$$\mathsf{T}(\mathbf{E}_3) = (1, -3, 4).$$

Calculate the matrix of T relative to the standard basis.

7. Let $\mathsf{T} : \mathbb{R}^2 \to \mathbb{R}^2$ be the linear extension of

$$\mathsf{T}(1, 1) = (1, -1)$$
$$\mathsf{T}(1, -2) = (2, 2).$$

Calculate the matrix of the linear transformation T relative to the standard basis.

8. Let $\mathsf{T} : \mathbb{R}^3 \to \mathbb{R}^3$ be the linear transformation defined by

$$\mathsf{T}(x, y, z) = (0, y, z).$$

Calculate the matrix of T relative to the standard basis of $\mathbb{R}^3$.

9. Let $\mathsf{T} : \mathbb{R}^4 \to \mathbb{R}^4$ be the linear transformation defined by

$$\mathsf{T}(x, y, z, w) = (0, x, y, z).$$

Calculate the matrix of T relative to the standard basis. Calculate the matrix of $\mathsf{T}^2, \mathsf{T}^3, \mathsf{T}^4$ relative to the standard basis.

10. Let $\mathsf{T} : \mathbb{R}^3 \to \mathbb{R}^3$ be the linear transformation whose matrix relative to the standard basis is

$$\begin{pmatrix} 1 & 3 & 2 \\ 0 & 1 & 1 \\ 2 & -1 & 0 \end{pmatrix}.$$

(a) Calculate $\mathsf{T}(\mathbf{E}_1)$, $\mathsf{T}(\mathbf{E}_2)$, $\mathsf{T}(\mathbf{E}_3)$.
(b) Let r be a real number, calculate the matrix of $r\mathsf{T}$, relative to the standard basis.

11. Let $\mathsf{T} : \mathbb{R}^2 \to \mathbb{R}^2$ be the linear transformation whose matrix relative to the standard basis is

$$\begin{pmatrix} 1 & 1 \\ 1 & 2 \end{pmatrix}$$

(a) Calculate $\mathsf{T}(\mathbf{E}_1)$ and $\mathsf{T}(\mathbf{E}_2)$.
(b) Find $\mathbf{A}_1, \mathbf{A}_2$ satisfying $\mathsf{T}(\mathbf{A}_i) = \mathbf{E}_i, i = 1, 2.$

12. Let $\mathsf{T}: \mathbb{R}^3 \to \mathbb{R}^3$ be the linear transformation whose matrix relative to the standard basis is

$$\begin{pmatrix} 2 & 0 & -1 \\ 1 & 1 & 0 \\ -1 & 1 & 1 \end{pmatrix}.$$

(a) Calculate $\mathsf{T}(\mathbf{E}_1)$, $\mathsf{T}(\mathbf{E}_2)$, $\mathsf{T}(\mathbf{E}_3)$, and $\mathsf{T}(1, 2, 3)$.
(b) Is T surjective? (Check if $(1, 0, 0)$ belongs to $\operatorname{Im} \mathsf{T}$.)

13. Let $\mathbf{A}_1, \mathbf{A}_2, \mathbf{A}_3 \in \mathbb{R}^3$ and write

$$\mathbf{A}_j = (a_{1j}, a_{2j}, a_{3j}), \qquad j = 1, 2, 3.$$

Let $\mathsf{T}: \mathbb{R}^3 \to \mathbb{R}^3$ be the linear extension of the function $\mathsf{T}(\mathbf{E}_j) = \mathbf{A}_j$, $j = 1, 2, 3$. Show that the matrix of T relative to the standard basis of $\mathbb{R}^3$ is (a_{ij}).

14. Let $\mathsf{T}: \mathbb{R}^3 \to \mathbb{R}^3$ be a linear transformation, and $\mathbf{A} = (a_{ij})$ its matrix with respect to the standard basis of $\mathbb{R}^3$. Regard the columns of $\mathbf{A}$ as vectors in $\mathbb{R}^3$. Call these three vectors $\mathbf{A}_1, \mathbf{A}_2, \mathbf{A}_3$. Show $\operatorname{Im} \mathsf{T} = \mathscr{L}\{\mathbf{A}_1, \mathbf{A}_2, \mathbf{A}_3\}$.

15. If $\mathsf{T}: \mathbb{R}^3 \to \mathbb{R}^3$ is the linear transformation with matrix

$$\begin{pmatrix} 0 & 1 & 1 \\ 1 & 0 & 1 \\ 1 & 1 & 0 \end{pmatrix}$$

with respect to the standard basis of $\mathbb{R}^3$ show that T is an isomorphism.

11 Matrices

In the last chapter we saw that a linear transformation $\mathsf{T}: \mathbb{R}^3 \to \mathbb{R}^3$ could be represented (that is, was completely determined by) 9 numbers arranged in a 3×3 array. In this chapter we will study such arrays, which are called matrices. We will return to the connection between matrices and linear transformations in the next chapter.

Definition. A rectangular array of numbers composed of m rows and n columns

$$\mathbf{A} = \begin{pmatrix} a_{11} & a_{12} & \cdots & a_{1n} \\ a_{21} & a_{22} & \cdots & a_{2n} \\ \vdots & & & \\ a_{m1} & a_{m2} & \cdots & a_{mn} \end{pmatrix}$$

is called an $m \times n$ *matrix* (read *m by n* matrix[1]). If there is a possibility of confusing entries from two adjacent columns as a product we will insert commas between the entries of a given row to carefully distinguish which entry belongs to which column.

The elements $a_{i1}, a_{i2}, \ldots, a_{in}$ form the ith *row of* **A** and the elements

$$\begin{matrix} a_{1j}, \\ a_{2j}, \\ \vdots \\ a_{mj}, \end{matrix}$$

[1] We also say that the matrix **A** is of, or has, *size* $m \times n$.

form the jth *column* of A. We will often write

$$\mathbf{A} = (a_{ij})_{\substack{1 \le i \le m \\ 1 \le j \le n}}$$

for **A**, or simply

$$\mathbf{A} = (a_{ij})$$

when m and n are understood from context. Note that the order of the subscripts is important; the first subscript denotes the row and the second subscript the column to which an entry belongs.

Just as with vectors in $\mathbb{R}^n$, two matrices are equal iff they have the same entries. That is:

Definition. If $\mathbf{A} = (a_{ij})$ and $\mathbf{B} = (b_{ij})$ are $m \times n$ matrices, then $\mathbf{A} = \mathbf{B}$ iff $a_{ij} = b_{ij}$ for $i = 1, 2, \ldots, m$ and $j = 1, \ldots, n$.

Our study of linear transformations suggests the following definitions.

Definition. If $\mathbf{A} = (a_{ij})$ and $\mathbf{B} = (b_{ij})$ are two $m \times n$ matrices their *sum*, $\mathbf{A} + \mathbf{B}$, is the matrix $\mathbf{C} = (c_{ij})$ where $c_{ij} = a_{ij} + b_{ij}$, $i = 1, 2, \ldots, m$, $j = 1, 2, \ldots, n$.

Definition. If $\mathbf{A} = (a_{ij})$ is an $m \times n$ matrix and r is a number then rA, *the scalar multiple of* **A** *by* r, is the matrix $\mathbf{C} = (c_{ij})$ where $c_{ij} = ra_{ij}$, $i = 1, \ldots, m$ and $j = 1, \ldots, n$.

The following result is a routine verification of definitions:

Proposition 11.1. *The matrices of size $m \times n$ form a vector space under the operations of matrix addition and scalar multiplication. We denote this vector space by $\mathcal{M}_{mn}$.* □

The dimension of the vector space $\mathcal{M}_{mn}$ is not hard to compute. We take our lead from the method we used to show that $\dim \mathbb{R}^n = n$. Introduce the $m \times n$ matrix $\mathbf{E}_{rs} = (e_{ij})$ by the requirement

$$e_{ij} = \begin{cases} 0 & \text{if } i \ne r, j \ne s, \\ 1 & \text{if } i = r, j = s. \end{cases}$$

For example the 6×4 matrix $\mathbf{E}_{32}$ is

$$\mathbf{E}_{32} = \begin{pmatrix} 0 & 0 & 0 & 0 \\ 0 & 0 & 0 & 0 \\ 0 & 1 & 0 & 0 \\ 0 & 0 & 0 & 0 \\ 0 & 0 & 0 & 0 \\ 0 & 0 & 0 & 0 \end{pmatrix}.$$

It is then a routine verification to prove:

Proposition 11.2. *The vectors* $\{\mathbf{E}_{rs} | r = 1, 2, \ldots, m, s = 1, 2, \ldots, n\}$ *form a basis for* $\mathscr{M}_{mn}$. *Therefore* $\dim \mathscr{M}_{mn} = mn$. □

Recall now that if $\mathscr{V}$ and $\mathscr{W}$ are vector spaces we introduced the vector space $\mathscr{L}(\mathscr{V}, \mathscr{W})$ of linear transformations from $\mathscr{V}$ to $\mathscr{W}$. If $\mathscr{V}$ and $\mathscr{W}$ are finite dimensional then we will see that the innocent looking (11.2) implies the useful fact $\dim \mathscr{L}(\mathscr{V}, \mathscr{W}) = \dim \mathscr{V} \dim \mathscr{W}$.

EXAMPLE 1.

$$\begin{pmatrix} 3 & 1 & 4 \\ 2 & 0 & -1 \\ -2 & -1 & 0 \end{pmatrix} + \begin{pmatrix} 0 & -1 & 3 \\ 2 & -9 & 4 \\ 7 & 6 & 1 \end{pmatrix}$$

$$= \begin{pmatrix} 3+0 & 1-1 & 4+3 \\ 2+2 & 0-9 & -1+4 \\ -2+7 & -1+6 & 0+1 \end{pmatrix} = \begin{pmatrix} 3 & 0 & 7 \\ 4 & -9 & 3 \\ 5 & 5 & 1 \end{pmatrix}.$$

EXAMPLE 2.

$$\begin{pmatrix} 1 & 4 & 6 & 0 & 1 \\ 2 & 0 & -1 & 7 & 9 \end{pmatrix} + \begin{pmatrix} 0 & 1 & -1 & 4 & 7 \\ -1 & -2 & -3 & 7 & 9 \end{pmatrix}$$

$$= \begin{pmatrix} 1+0 & 4+1 & 6-1 & 0+4 & 1+7 \\ 2-1 & 0-2 & -1-3 & 7+7 & 9+9 \end{pmatrix}$$

$$= \begin{pmatrix} 1 & 5 & 5 & 4 & 8 \\ 1 & -2 & -4 & 14 & 18 \end{pmatrix}.$$

EXAMPLE 3.

$$4\begin{pmatrix} 3 & 1 \\ 7 & 4 \\ 6 & -4 \end{pmatrix} = \begin{pmatrix} 12 & 4 \\ 28 & 16 \\ 24 & -16 \end{pmatrix}.$$

The discussion of Chapter 10 suggests the following:

Definition. If $\mathbf{A} = (a_{ij})$ is an $m \times n$ matrix and $\mathbf{B} = (b_{ij})$ is an $n \times p$ matrix their *matrix product* $\mathbf{A} \cdot \mathbf{B}$ is the $m \times p$ matrix $\mathbf{AB} = (c_{ij})$ where

$$c_{ij} = \sum_{k=1}^{n} a_{ik} b_{kj}$$

where $i = 1, \ldots, m$, $j = 1, \ldots, p$.

Thus the entry of the ith row and jth column of the product $\mathbf{A} \cdot \mathbf{B}$ is obtained by taking the ith row

$$a_{i1}, a_{i2}, \ldots, a_{in}$$

of the matrix $\mathbf{A}$ and the jth column of the matrix $\mathbf{B}$

$$\begin{matrix} b_{1j} \\ b_{2j} \\ \vdots \\ b_{nj}, \end{matrix}$$

multiplying the corresponding entries together and adding the resulting products, i.e.

$$a_{i1}b_{1j} + a_{i2}b_{2j} + \cdots + a_{ik}b_{kj} + \cdots + a_{in}b_{nj},$$

$i = 1, 2, \ldots, m$, $j = 1, 2, \ldots, p$.

Note that for the product of $\mathbf{A}$ *and* $\mathbf{B}$ *to be defined the number of columns of* $\mathbf{A}$ *must be equal to the number of rows of* $\mathbf{B}$. Thus the *order* in which the product of $\mathbf{A}$ and $\mathbf{B}$ is taken is very important, for $\mathbf{A} \cdot \mathbf{B}$ can be defined without $\mathbf{B} \cdot \mathbf{A}$ being defined.

EXAMPLE 4. Compute the matrix product

$$(1 \quad 2 \quad 3) \cdot \begin{pmatrix} 4 \\ 5 \\ 6 \end{pmatrix}.$$

Solution. Note the answer is a 1×1 matrix.

$$(1 \quad 2 \quad 3)\begin{pmatrix} 4 \\ 5 \\ 6 \end{pmatrix} = (4 + 10 + 18) = (32).$$

Note that the product in the reverse order

$$\begin{pmatrix} 4 \\ 5 \\ 6 \end{pmatrix} \cdot (1 \quad 2 \quad 3) = \begin{pmatrix} 4 & 8 & 12 \\ 5 & 10 & 15 \\ 6 & 12 & 18 \end{pmatrix}$$

does not even have the same size. So the products in different orders are certainly not equal.

EXAMPLE 5. Compute the matrix product

$$\begin{pmatrix} 0 & 1 & 1 \\ 0 & 0 & 1 \\ 0 & 0 & 0 \end{pmatrix} \cdot \begin{pmatrix} 0 & 1 & 1 \\ 0 & 0 & 1 \\ 0 & 0 & 0 \end{pmatrix}.$$

Answer.

$$\begin{pmatrix} 0 & 0 & 1 \\ 0 & 0 & 0 \\ 0 & 0 & 0 \end{pmatrix}.$$

EXAMPLE 6. Let

$$\mathbf{A} = \begin{pmatrix} 1 & 2 & 3 \\ 4 & 5 & 6 \end{pmatrix}$$

and

$$\mathbf{B} = \begin{pmatrix} 1 & 2 & 3 & 4 \\ 1 & 2 & 3 & 4 \\ 1 & 2 & 3 & 4 \end{pmatrix}.$$

Calculate the product $\mathbf{A} \cdot \mathbf{B}$.

Solution. We have

$$\begin{aligned} &\begin{pmatrix} 1 & 2 & 3 \\ 4 & 5 & 6 \end{pmatrix}\begin{pmatrix} 1 & 2 & 3 & 4 \\ 1 & 2 & 3 & 4 \\ 1 & 2 & 3 & 4 \end{pmatrix} \\ &= \begin{pmatrix} 1+2+3 & 2+4+6, & 3+6+9, & 4+8+12 \\ 4+5+6 & 8+10+12, & 12+15+18, & 16+20+24 \end{pmatrix} \\ &= \begin{pmatrix} 6 & 12 & 18 & 24 \\ 15 & 30 & 45 & 60 \end{pmatrix}. \end{aligned}$$

Note that the product $\mathbf{B} \cdot \mathbf{A}$ is *not* defined.

Definition. A matrix $\mathbf{A}$ is said to be a *square matrix of size n* iff it has n rows and n columns (that is the number of rows equals the number of columns equals n).

If $\mathbf{A}$ and $\mathbf{B}$ are *square* matrices of size n then the products $\mathbf{AB}$ and $\mathbf{BA}$ are both defined, and are square matrices of size n. However they may not be equal.

EXAMPLE 7. Let

$$\mathbf{A} = \begin{pmatrix} 1 & 0 \\ 0 & 3 \end{pmatrix} \quad \text{and} \quad \mathbf{B} = \begin{pmatrix} 3 & 0 \\ 2 & 1 \end{pmatrix}.$$

Compute the matrix products $\mathbf{AB}$ and $\mathbf{BA}$.

Solution. We have

$$\mathbf{AB} = \begin{pmatrix} 1 & 0 \\ 0 & 3 \end{pmatrix}\begin{pmatrix} 3 & 0 \\ 2 & 1 \end{pmatrix} = \begin{pmatrix} 3 & 0 \\ 6 & 3 \end{pmatrix}, \qquad \mathbf{BA} = \begin{pmatrix} 3 & 0 \\ 2 & 1 \end{pmatrix}\begin{pmatrix} 1 & 0 \\ 0 & 3 \end{pmatrix} = \begin{pmatrix} 3 & 0 \\ 2 & 3 \end{pmatrix}$$

and so we see that $\mathbf{AB} \neq \mathbf{BA}$.

As the preceding example shows even if **AB** and **BA** are defined we should not expect that $\mathbf{AB} = \mathbf{BA}$.

Notation. If **A** is a square matrix then **AA** is defined and is denoted by $\mathbf{A}^2$. Similarly

$$\underset{n \text{ times}}{\mathbf{A} \cdots \mathbf{A}}$$

is defined and denoted $\mathbf{A}^n$.

EXAMPLE 8. Let

$$\mathbf{A} = \begin{pmatrix} 0 & 0 \\ 1 & 0 \end{pmatrix}.$$

Calculate $\mathbf{A}^2$.

Solution. We have

$$\mathbf{A}^2 = \begin{pmatrix} 0 & 0 \\ 1 & 0 \end{pmatrix}\begin{pmatrix} 0 & 0 \\ 1 & 0 \end{pmatrix} = \begin{pmatrix} 0 & 0 \\ 0 & 0 \end{pmatrix}.$$

Thus not only does matrix multiplication behave strangely in that it is not commutative, it is also possible for the square of a matrix with nonzero entries to have only zero entries. We may summarize the basic rules of matrix operations in the following formulas: (assume that the indicated operations are defined, that is, that the sizes are correct for the operations to make sense).

$$\begin{aligned} \mathbf{A} + \mathbf{B} &= \mathbf{B} + \mathbf{A} \\ \mathbf{A} + (\mathbf{B} + \mathbf{C}) &= \mathbf{A} + (\mathbf{B} + \mathbf{C}) \\ r(\mathbf{A} + \mathbf{B}) &= r\mathbf{A} + r\mathbf{B} \\ (r + s)\mathbf{A} &= r\mathbf{A} + s\mathbf{A} \\ 0\mathbf{A} &= \mathbf{0} \qquad (\text{where } \mathbf{0} = (0_{ij}) \text{ and } 0_{ij} = 0) \\ \mathbf{A} + (-1)\mathbf{A} &= \mathbf{0} \\ \mathbf{A} + \mathbf{0} &= \mathbf{A} \\ (\mathbf{A} + \mathbf{B}) \cdot \mathbf{C} &= \mathbf{A} \cdot \mathbf{C} + \mathbf{B} \cdot \mathbf{C} \\ \mathbf{C} \cdot (\mathbf{A} + \mathbf{B}) &= \mathbf{C} \cdot \mathbf{A} + \mathbf{C} \cdot \mathbf{B} \\ \mathbf{0} \cdot \mathbf{A} &= \mathbf{0} = \mathbf{A} \cdot \mathbf{0} \\ \mathbf{A} \cdot (\mathbf{B} \cdot \mathbf{C}) &= (\mathbf{A} \cdot \mathbf{B}) \cdot \mathbf{C}. \end{aligned}$$

Of course the first few of these rules were needed to prove (11.1).

In discussing matrices it is convenient to distinguish certain special types of matrices.

Special types of matrices

The identity matrix. The *identity matrix of size n* is the square $n \times n$ matrix denoted by **I**, where

$$\mathbf{I} = \begin{pmatrix} 1 & 0 & & \cdots & 0 \\ 0 & 1 & & \cdots & 0 \\ 0 & 0 & 1 & \cdots & 0 \\ \vdots & & & & \\ 0 & 0 & & \cdots & 1 \end{pmatrix} = (i_{ij}) \quad \text{where } i_{ij} = \begin{cases} 1 & \text{if } i = j, \\ 0 & \text{if } i \neq j. \end{cases}$$

For example, the identity matrices of size 1, 2, 3, and 4 are

$$(1), \quad \begin{pmatrix} 1 & 0 \\ 0 & 1 \end{pmatrix}, \quad \begin{pmatrix} 1 & 0 & 0 \\ 0 & 1 & 0 \\ 0 & 0 & 1 \end{pmatrix}, \quad \begin{pmatrix} 1 & 0 & 0 & 0 \\ 0 & 1 & 0 & 0 \\ 0 & 0 & 1 & 0 \\ 0 & 0 & 0 & 1 \end{pmatrix}.$$

The following important facts are easily verified:

$$\mathbf{IB} = \mathbf{B} \quad \text{for any } n \times p \text{ matrix } \mathbf{B},$$
$$\mathbf{AI} = \mathbf{A} \quad \text{for any } m \times n \text{ matrix } \mathbf{A}.$$

Scalar matrices. A square matrix $\mathbf{A} = (a_{ij})$ is called a *scalar matrix* iff $\mathbf{A} = r\mathbf{I}$ for some number r. For example

$$\begin{pmatrix} 3 & 0 & 0 \\ 0 & 3 & 0 \\ 0 & 0 & 3 \end{pmatrix} = 3\begin{pmatrix} 1 & 0 & 0 \\ 0 & 1 & 0 \\ 0 & 0 & 1 \end{pmatrix}$$

is a scalar matrix but

$$\begin{pmatrix} 3 & 0 & 0 \\ 0 & -3 & 0 \\ 0 & 0 & 3 \end{pmatrix}$$

is not a scalar matrix. The following formulas are easily checked

$$(a\mathbf{I})\mathbf{B} = a\mathbf{B} \quad \text{for any } n \times p \text{ matrix } \mathbf{B}.$$
$$\mathbf{A}(a\mathbf{I}) = a\mathbf{A} \quad \text{for any } m \times n \text{ matrix } \mathbf{B}.$$

For example

$$\begin{pmatrix} 3 & 0 & 0 \\ 0 & 3 & 0 \\ 0 & 0 & 3 \end{pmatrix}\begin{pmatrix} 1 \\ 2 \\ 3 \end{pmatrix} = \begin{pmatrix} 3 \\ 6 \\ 9 \end{pmatrix} = 3\begin{pmatrix} 1 \\ 2 \\ 3 \end{pmatrix}.$$

Diagonal matrices. For any square matrix $\mathbf{A} = (a_{ij})$ of size n the entries

$$a_{11}, a_{22}, \ldots, a_{nn}$$

are called the *diagonal entries* of $\mathbf{A}$. For example, the diagonal entries of

$$\begin{pmatrix} 3 & 2 & 1 \\ 6 & 5 & 4 \\ 9 & 8 & 7 \end{pmatrix}$$

are 3, 5, 7. A square matrix is said to be a *diagonal matrix* iff its only nonzero entries are on the diagonal. That is $\mathbf{A} = (a_{ij})$ is a diagonal matrix iff $a_i = 0$ for $i \neq j$.

For example $\mathbf{I}$ and $a\mathbf{I}$ are diagonal matrices as is

$$\begin{pmatrix} 3 & 0 & 0 \\ 0 & -3 & 0 \\ 0 & 0 & 3 \end{pmatrix}.$$

Note that the diagonal entries themselves need not be nonzero. For example

$$\begin{pmatrix} 0 & 0 & 0 \\ 0 & 1 & 0 \\ 0 & 0 & 0 \end{pmatrix} \quad \text{and} \quad \begin{pmatrix} 0 & 0 & 0 \\ 0 & 0 & 0 \\ 0 & 0 & 0 \end{pmatrix}$$

are also diagonal matrices.

In general a diagonal matrix looks like

$$\mathbf{A} = \begin{pmatrix} a_{11} & & & \text{\Large 0} \\ & a_{22} & & \\ & & \ddots & \\ \text{\Large 0} & & & a_{nn} \end{pmatrix}$$

where the giant 0s mean that all other entries are zero. If $\mathbf{A}$ and $\mathbf{B}$ are diagonal matrices of size n then so are $\mathbf{AB}$ and $\mathbf{BA}$. Indeed if

$$\mathbf{A} = \begin{pmatrix} a_{11} & & \text{\Large 0} \\ & \ddots & \\ \text{\Large 0} & & a_{nn} \end{pmatrix} \quad \text{and} \quad \mathbf{B} = \begin{pmatrix} b_{11} & & \text{\Large 0} \\ & \ddots & \\ \text{\Large 0} & & b_{nn} \end{pmatrix}$$

then

$$\mathbf{AB} = \begin{pmatrix} a_{11}b_{11} & & & \text{\Large 0} \\ & a_{22}b_{22} & & \\ & & \ddots & \\ \text{\Large 0} & & & a_{nn}b_{nn} \end{pmatrix} = \mathbf{BA}.$$

Triangular matrices. A square matrix $\mathbf{A}$ is said to be *lower triangular* iff $\mathbf{A} = (a_{ij})$ where $a_{ij} = 0$ if $j > i$. For example

$$\begin{pmatrix} 1 & 0 & 0 \\ 0 & 2 & 0 \\ 3 & -1 & -3 \end{pmatrix}$$

is a lower triangular matrix. A triangular matrix $\mathbf{A} = (a_{ij})$ for which $a_{ii} = 0$, $i = 1, \ldots, n$ (that is, all of whose diagonal entries are 0) is said to be *strictly triangular*. An example of a strictly triangular matrix is

$$\begin{pmatrix} 0 & 0 & 0 \\ 1 & 0 & 0 \\ 2 & 3 & 0 \end{pmatrix}$$

The Zero matrix. The *zero matrix* of size $m \times n$ is the $m \times n$ matrix $\mathbf{0}$ all of those entries are 0.

Idempotent matrices. A square matrix $\mathbf{A}$ is said to be *idempotent* iff $\mathbf{A}^2 = \mathbf{A}$. There are lots of idempotent matrices. Here are a few examples

$$\begin{pmatrix} 1 & -1 \\ 0 & 0 \end{pmatrix}, \quad \begin{pmatrix} 1 & 0 & 0 \\ 0 & 1 & 0 \\ 0 & 0 & 0 \end{pmatrix}$$

as may be easily checked by explicit computation.

Nilpotent matrices. A square matrix $\mathbf{A}$ is said to be *nilpotent* iff there is an integer q such $\mathbf{A}^q = \mathbf{0}$. (The smallest such integer q is called the *index of nilpotence* of $\mathbf{A}$.). For example if $\mathbf{A}$ is the matrix of the shift operator on $\mathbb{R}^3$, that is

$$\mathbf{A} = \begin{pmatrix} 0 & 0 & 0 \\ 1 & 0 & 0 \\ 0 & 1 & 0 \end{pmatrix}$$

then

$$\mathbf{A}^2 = \begin{pmatrix} 0 & 0 & 0 \\ 1 & 0 & 0 \\ 0 & 1 & 0 \end{pmatrix}\begin{pmatrix} 0 & 0 & 0 \\ 1 & 0 & 0 \\ 0 & 1 & 0 \end{pmatrix} = \begin{pmatrix} 0 & 0 & 0 \\ 0 & 0 & 0 \\ 1 & 0 & 0 \end{pmatrix}$$

and

$$\mathbf{A}^3 = \begin{pmatrix} 0 & 0 & 0 \\ 0 & 0 & 0 \\ 1 & 0 & 0 \end{pmatrix}\begin{pmatrix} 0 & 0 & 0 \\ 1 & 0 & 0 \\ 0 & 1 & 0 \end{pmatrix} = \begin{pmatrix} 0 & 0 & 0 \\ 0 & 0 & 0 \\ 0 & 0 & 0 \end{pmatrix}$$

so that A is nilpotent of index 3.

As a matter of fact there is a very simple geometric explanation for why $\mathbf{A}^3 = \mathbf{0}$. Indeed any strictly lower triangular matrix

$$\mathbf{B} = \begin{pmatrix} 0 & 0 & 0 \\ a & 0 & 0 \\ b & c & 0 \end{pmatrix}$$

will be nilpotent of index at most 3. Of course we could prove this by an orgy of calculation but that is not what we want to do. Rather we are going to exploit the relation between matrices and linear transformations that we started to discuss in the preceding section. Let us therefore construct a linear transformation

$$\mathsf{T} : \mathbb{R}^3 \to \mathbb{R}^3$$

whose matrix is $\mathbf{B}$. To do this we must recall that if there is such a T then it must satisfy

$$\begin{aligned} \mathsf{T}(1, 0, 0) &= (0, a, b) \\ \mathsf{T}(0, 1, 0) &= (0, 0, c) \\ \mathsf{T}(0, 0, 1) &= (0, 0, 0) \end{aligned}$$

if its matrix is going to be $\mathbf{B}$. (Remember how to compute the columns of $\mathbf{B}$!) Therefore

$$\begin{aligned} \mathsf{T}(x, y, z) &= \mathsf{T}(x(1, 0, 0) + y(0, 1, 0) + z(0, 0, 1)) \\ &= x\mathsf{T}(1, 0, 0) + y\mathsf{T}(0, 1, 0) + z\mathsf{T}(0, 0, 1) \\ &= x(0, a, b) + y(0, 0, c) + z(0, 0, 0) \\ &= (0, ax, bx + cy), \end{aligned}$$

that is,

$$\mathsf{T}(x, y\ z) = (0, ax, bx + cy)$$

is the only linear transformation

$$\mathsf{T} : \mathbb{R}^3 \to \mathbb{R}^3$$

whose matrix is $\mathbf{B}$. Let us now compute

$$\mathsf{T}^2 : \mathbb{R}^3 \to \mathbb{R}^3, \qquad \mathsf{T}^3 : \mathbb{R}^3 \to \mathbb{R}^3.$$

We have

$$\begin{aligned} \mathsf{T}^2(x, y, z) &= \mathsf{T}(\mathsf{T}(x, y, z)) = \mathsf{T}(0, ax, bx + cy) \\ &= (0, 0, cax) \end{aligned}$$

and

$$\mathsf{T}^3(x, y, z) = \mathsf{T}(\mathsf{T}^2(x, y, z)) = \mathsf{T}(0, 0, cax) = (0, 0, 0).$$

Therefore

$$\mathsf{T}^3 = 0 : \mathbb{R}^3 \to \mathbb{R}^3.$$

Hence the matrix of T^3 relative to the standard basis of $\mathbb{R}^3$ is

$$\mathbf{0} = \begin{pmatrix} 0 & 0 & 0 \\ 0 & 0 & 0 \\ 0 & 0 & 0 \end{pmatrix}.$$

But now remember that our definition of matrix multiplication was rigged up so that $\mathbf{B}^3$ would be the matrix of T^3 relative to the standard basis of $\mathbb{R}^3$ and therefore we have shown that

$$\mathbf{B}^3 = \mathbf{0},$$

and hence that $\mathbf{B}$ is nilpotent of index at most 3.

Thus we see the geometric reason behind the fact that $\mathbf{B}^3 = \mathbf{0}$ is that $\mathbf{B}$ is the matrix of a linear transformation that is basically a shift on $\mathbb{R}^3$. (Actually T is what is called a *weighted* shift, the weights being a, b, c.) This discussion should show that it is sometimes possible, and indeed advantageous to discover properties of a matrix by examining a linear transformation that it represents. We will see more examples of this later.

Nonsingular matrices. A square matrix $\mathbf{A}$ is said to be *invertible* or *nonsingular* iff there exists a matrix $\mathbf{B}$ such that

$$\mathbf{AB} = \mathbf{I} \quad \text{and} \quad \mathbf{BA} = \mathbf{I}.$$

If $\mathbf{A}$ is nonsingular then the matrix $\mathbf{B}$ with $\mathbf{AB} = \mathbf{I} = \mathbf{BA}$ is called the *inverse matrix* of $\mathbf{A}$ and is denoted by $\mathbf{A}^{-1}$.

It is a theorem that we will prove in the next chapter that if there exists a matrix $\mathbf{B}$ such that

$$\mathbf{AB} = \mathbf{I}$$

then also

$$\mathbf{BA} = \mathbf{I}.$$

Thus to check that $\mathbf{B} = \mathbf{A}^{-1}$ we need only calculate one of the two products $\mathbf{AB}$ and $\mathbf{BA}$ and see if it is $\mathbf{I}$. For example if

$$\mathbf{A} = \begin{pmatrix} 0 & 1 & 0 \\ 1 & 0 & 0 \\ 0 & 0 & 1 \end{pmatrix}$$

then

$$\mathbf{A}^{-1} = \begin{pmatrix} 0 & 1 & 0 \\ 1 & 0 & 0 \\ 0 & 0 & 1 \end{pmatrix}.$$

For we have

$$\mathbf{AA} = \begin{pmatrix} 0 & 1 & 0 \\ 1 & 0 & 0 \\ 0 & 0 & 1 \end{pmatrix}\begin{pmatrix} 0 & 1 & 0 \\ 1 & 0 & 0 \\ 0 & 0 & 1 \end{pmatrix} = \begin{pmatrix} 1 & 0 & 0 \\ 0 & 1 & 0 \\ 0 & 0 & 1 \end{pmatrix} = \mathbf{I}$$

and therefore $\mathbf{A} = \mathbf{A}^{-1}$. There is actually a simple geometric explanation for why $\mathbf{A} = \mathbf{A}^{-1}$. For a moment's reflection on our discussion of matrices of linear transformations shows that $\mathbf{A}$ is the matrix of the linear transformation $\mathsf{T} : \mathbb{R}^3 \to \mathbb{R}^3$ defined by

$$\mathsf{T}(x, y, z) = (y, x, z)$$

that switches the first two coordinates, relative to the standard bases for $\mathbb{R}^3$. Clearly switching the first two coordinates twice will change nothing, that is

$$\mathsf{T}^2(x, y, z) = \mathsf{T}(\mathsf{T}(x, y, z)) = \mathsf{T}(y, x, z) = (x, y, z).$$

Therefore the matrix of

$$\mathsf{T}^2 : \mathbb{R}^3 \to \mathbb{R}^3$$

relative to the standard basis of $\mathbb{R}^3$ is

$$\mathbf{I} = \begin{pmatrix} 1 & 0 & 0 \\ 0 & 1 & 0 \\ 0 & 0 & 1 \end{pmatrix}.$$

But now remember that the matrix of T^2 is also $\mathbf{A}^2$, because that is how we defined matrix multiplication. Thus $\mathbf{A}^2 = \mathbf{I}$ for the simple geometric reason that it is the matrix of the transformation T that switches the first two coordinates.

The preceding discussion illustrates again that it is sometimes possible to extract information about a matrix by examining the corresponding linear transformation.

A matrix $\mathbf{A}$ with the property that $\mathbf{A} = \mathbf{A}^{-1}$ is called *involutory* or an *involution*. The example above shows that there are nontrivial involutions.

Another example of a nonsingular matrix is

$$\mathbf{B} = \begin{pmatrix} 0 & 1 & 0 \\ 0 & 0 & 1 \\ 1 & 0 & 0 \end{pmatrix}.$$

The inverse of $\mathbf{B}$ is the matrix

$$\mathbf{C} = \begin{pmatrix} 0 & 0 & 1 \\ 1 & 0 & 0 \\ 0 & 1 & 0 \end{pmatrix}$$

as we may certainly compute

$$\mathbf{BC} = \begin{pmatrix} 0 & 1 & 0 \\ 0 & 0 & 1 \\ 1 & 0 & 0 \end{pmatrix}\begin{pmatrix} 0 & 0 & 1 \\ 1 & 0 & 0 \\ 0 & 1 & 0 \end{pmatrix} = \begin{pmatrix} 1 & 0 & 0 \\ 0 & 1 & 0 \\ 0 & 0 & 1 \end{pmatrix} = \mathbf{I}.$$

An example of a matrix that is *not* invertible is

$$\begin{pmatrix} 0 & 0 & 0 \\ 1 & 0 & 0 \\ 0 & 1 & 0 \end{pmatrix},$$

and more generally we have:

A nilpotent matrix is not invertible. For suppose that $\mathbf{A}$ is a nilpotent matrix that is invertible. Let $\mathbf{B}$ be an inverse for $\mathbf{A}$. Since $\mathbf{A}$ is nilpotent there is an integer q such that

$$\mathbf{A}^q = \mathbf{0}.$$

Then

$$\mathbf{0} = \mathbf{A}^q\mathbf{B} = \mathbf{A}^{q-1}\mathbf{A}\mathbf{B} = \mathbf{A}^{q-1}\mathbf{I} = \mathbf{A}^{q-1}$$

so

$$\mathbf{A}^{q-1} = \mathbf{0}.$$

We may then repeat the above trick to show

$$\mathbf{A}^{q-2} = \mathbf{0}.$$

If we repeat this trick $q - 1$ times we will get

$$\mathbf{A} = \mathbf{0}.$$

But then

$$\mathbf{I} = \mathbf{A}\mathbf{B} = \mathbf{0}\mathbf{B} = \mathbf{0}$$

which is impossible.

We may also show:

The only invertible idempotent matrix is $\mathbf{I}$. For if $\mathbf{A}$ is an idempotent matrix then $\mathbf{A}^2 = \mathbf{A}$. If in addition $\mathbf{A}$ is invertible with inverse $\mathbf{B}$ then

$$\mathbf{A}^2 = \mathbf{A}$$

implies

$$\mathbf{A} = \mathbf{I}\mathbf{A} = \mathbf{B}\mathbf{A}\mathbf{A} = \mathbf{B}\mathbf{A}^2 = \mathbf{B}\mathbf{A} = \mathbf{I}$$

so $\mathbf{A} = \mathbf{I}$ as claimed.

Symmetric and skew-symmetric matrices. A square matrix $\mathbf{A} = (a_{ij})$ is said to be *symmetric* iff $a_{ij} = a_{ji}$ for $i, j = 1, \ldots, n$; it is said to be *skew-symmetric* iff $a_{ij} = -a_{ji}$ for $i, j = 1, \ldots, n$. For example

$$\begin{pmatrix} 1 & 0 & 1 \\ 0 & 0 & 0 \\ 1 & 0 & 3 \end{pmatrix} \quad \text{and} \quad \begin{pmatrix} 0 & 1 & 2 & 3 \\ 1 & 4 & 5 & 6 \\ 2 & 5 & 7 & 8 \\ 3 & 6 & 8 & 9 \end{pmatrix}$$

are symmetric matrices, and

$$\begin{pmatrix} 0 & -1 & 2 \\ 1 & 0 & -3 \\ -2 & 3 & 0 \end{pmatrix} \quad \text{and} \quad \begin{pmatrix} 0 & 1 \\ -1 & 0 \end{pmatrix}$$

are skew-symmetric matrices. Notice that the matrix

$$\begin{pmatrix} 1 & 0 & 1 \\ 0 & 0 & 0 \\ -1 & 0 & 3 \end{pmatrix}$$

is not skew-symmetric because

$$a_{11} = 1 \neq -1 = -a_{11}.$$

That is to say, if a matrix $\mathbf{A} = (a_{ij})$ is skew-symmetric then the equations $a_{11} = -a_{11}$, $a_{22} = -a_{22}, \ldots, a_{nn} = -a_{nn}$ certainly imply that $a_{11} = 0$, $a_{22} = 0, \ldots, a_{nn} = 0$, that is, a skew-symmetric matrix has all its diagonal entries equal to 0.

The skew-symmetric matrix

$$\mathbf{A} = \begin{pmatrix} 0 & 1 \\ -1 & 0 \end{pmatrix}$$

is interesting because it is also nonsingular since

$$\begin{pmatrix} 0 & 1 \\ -1 & 0 \end{pmatrix}\begin{pmatrix} 0 & -1 \\ 1 & 0 \end{pmatrix} = \begin{pmatrix} 1 & 0 \\ 0 & 1 \end{pmatrix}.$$

Thus various combinations of the preceding concepts may occur simultaneously. By now the student must be wondering how to tell when a matrix has an inverse, and how to calculate it when it does. This is a topic we will take up in Chapter 13, for the moment we will content ourselves with the 2×2 case.

Proposition 11.3. A 2×2 matrix

$$\mathbf{A} = \begin{pmatrix} a & b \\ c & d \end{pmatrix}$$

is nonsingular iff $ad - bc \neq 0$. If $ad - bc \neq 0$ then

$$\mathbf{A}^{-1} = \frac{1}{ad - cb}\begin{pmatrix} d & -b \\ -c & a \end{pmatrix}.$$

PROOF. Suppose that $ad - bc \neq 0$. Let

$$\mathbf{B} = \frac{1}{ad - bc}\begin{pmatrix} d & -b \\ -c & a \end{pmatrix}$$

then

$$\mathbf{BA} = \frac{1}{ad - bc}\begin{pmatrix} d & -b \\ -c & a \end{pmatrix}\begin{pmatrix} a & b \\ c & d \end{pmatrix}$$
$$= \frac{1}{ad - bc}\begin{pmatrix} da - bc & bd - bd \\ -ca + ac & -cb + ad \end{pmatrix}$$
$$= \frac{1}{ad - bc}\begin{pmatrix} ad - bc & 0 \\ 0 & ad - bc \end{pmatrix}$$
$$= \begin{pmatrix} 1 & 0 \\ 0 & 1 \end{pmatrix} = \mathbf{I},$$

and therefore $\mathbf{A}$ is nonsingular with

$$\mathbf{A}^{-1} = \frac{1}{ad - bc}\begin{pmatrix} d & -b \\ -c & a \end{pmatrix}$$

as claimed.

Suppose conversely that $\mathbf{A}$ is nonsingular, but that $ad - bc = 0$. We will deduce a contradiction. Let

$$\mathbf{C} = \begin{pmatrix} d & -b \\ -c & a \end{pmatrix}.$$

Then computing as above

$$\mathbf{CA} = \begin{pmatrix} ad - bc & 0 \\ 0 & ad - bc \end{pmatrix} = (ad - bc)\mathbf{I} = \mathbf{0}$$

This gives the equation

$$\mathbf{C} = \mathbf{CI} = \mathbf{C}(\mathbf{AA}^{-1}) = (\mathbf{CA})\mathbf{A}^{-1} = \mathbf{0A}^{-1} = \mathbf{0}.$$

Therefore

$$\mathbf{C} = \begin{pmatrix} d & -b \\ -c & a \end{pmatrix} = \begin{pmatrix} 0 & 0 \\ 0 & 0 \end{pmatrix}$$

so that

$$a = 0, \qquad b = 0, \qquad c = 0, \qquad d = 0.$$

But then $\mathbf{A} = 0$ also, so

$$\mathbf{I} = \mathbf{AA}^{-1} = \mathbf{0A}^{-1} = \mathbf{0}$$

so

$$\begin{pmatrix} 1 & 0 \\ 0 & 1 \end{pmatrix} = \begin{pmatrix} 0 & 0 \\ 0 & 0 \end{pmatrix}$$

and hence $1 = 0$ which is impossible. □

The student may wonder how the above result was originally discovered. Did somebody just make a lucky guess? Perhaps, but a more logical development may be found in Chapter 13.

EXERCISES

1. Perform the following matrix computations:

(a)
$$2\begin{pmatrix} 3 & 1 & 4 & -1 \\ 2 & 0 & -1 & 2 \end{pmatrix} - 3\begin{pmatrix} 1 & 0 & -1 & 7 \\ -1 & -2 & 0 & -4 \end{pmatrix}$$

(b)
$$2\begin{pmatrix} 1 \\ 2 \\ 3 \end{pmatrix} - 4\begin{pmatrix} 3 \\ 2 \\ 1 \end{pmatrix} + 8\begin{pmatrix} 1 \\ 0 \\ 1 \end{pmatrix}.$$

2. Perform the following matrix multiplications:

$$(1 \quad 2 \quad 3)\begin{pmatrix} 1 \\ 2 \\ 3 \end{pmatrix},$$

$$\begin{pmatrix} 0 & 1 & 0 \\ 1 & 1 & 0 \\ 0 & 0 & 2 \end{pmatrix}\begin{pmatrix} 1 & 0 \\ 0 & 1 \\ 1 & 0 \end{pmatrix},$$

$$\begin{pmatrix} 1 & 0 & 1 & 0 \\ 0 & 1 & 0 & 1 \end{pmatrix}\begin{pmatrix} 1 & 0 \\ 0 & 1 \\ 1 & 0 \\ 0 & 1 \end{pmatrix}\begin{pmatrix} 2 & 2 \\ 2 & 2 \end{pmatrix},$$

$$\begin{pmatrix} 1 & 2 & 3 \\ 4 & 5 & 6 \\ 7 & 8 & 9 \end{pmatrix}\begin{pmatrix} 0 & 0 & 0 \\ 1 & 0 & 0 \\ 0 & 0 & 0 \end{pmatrix},$$

$$\begin{pmatrix} 0 & 0 & 0 \\ 1 & 0 & 0 \\ 0 & 0 & 0 \end{pmatrix}\begin{pmatrix} 1 & 2 & 3 \\ 4 & 5 & 6 \\ 7 & 8 & 9 \end{pmatrix}.$$

3. Which of the following matrices are nonsingular, involutory, idempotent, nilpotent, symmetric, or skew-symmetric?

$$\mathbf{A} = \begin{pmatrix} 1 & -1 \\ 0 & 0 \end{pmatrix} \qquad \mathbf{F} = \begin{pmatrix} 0 & 1 \\ -1 & 0 \end{pmatrix}$$

$$\mathbf{B} = \begin{pmatrix} 1 & 2 \\ 3 & 4 \end{pmatrix} \qquad \mathbf{G} = \begin{pmatrix} 1 & 0 \\ 0 & 0 \end{pmatrix}$$

$$\mathbf{C} = \begin{pmatrix} 1 & -1 \\ -1 & 1 \end{pmatrix} \qquad \mathbf{H} = \begin{pmatrix} 1 & 0 \\ -1 & 0 \end{pmatrix}$$

$$\mathbf{D} = \begin{pmatrix} 1 & 1 \\ 1 & 1 \end{pmatrix} \qquad \mathbf{J} = \begin{pmatrix} 4 & 1 \\ 0 & 2 \end{pmatrix}$$

$$\mathbf{E} = \begin{pmatrix} 1 & 1 \\ -1 & 1 \end{pmatrix} \qquad \mathbf{K} = \begin{pmatrix} 0 & 0 \\ 1 & 0 \end{pmatrix}$$

Find the inverse for those that are invertible.

4. Show that a diagonal matrix

$$\mathbf{A} = \begin{pmatrix} a_{11} & 0 & 0 \\ 0 & a_{22} & 0 \\ 0 & 0 & a_{33} \end{pmatrix}$$

is nonsingular iff $a_{11}a_{22}a_{33} \neq 0$. That is $a_{11} \neq 0, a_{22} \neq 0, a_{33} \neq 0$. If A has an inverse what is it?

5. Show that a diagonal matrix

$$\mathbf{A} = \begin{pmatrix} a_{11} & & & 0 \\ & a_{22} & & \\ & & \ddots & \\ 0 & & & a_{nn} \end{pmatrix}$$

is nonsingular iff *all* its diagonal entries are nonzero.

6. If $\mathbf{A} = (a_{ij})$ is a matrix we define the *transpose of* $\mathbf{A}$ to be the matrix $\mathbf{A}^t = (b_{ij})$ where $b_{ij} = a_{ji}$. Find the transpose of each of the following matrices:

$$\begin{pmatrix} 1 & 2 & 3 \\ 4 & 5 & 6 \end{pmatrix} \qquad (1 \quad 2 \quad 3)$$

$$\begin{pmatrix} 1 & 0 & 1 \\ 0 & 1 & 0 \\ 1 & 0 & 1 \end{pmatrix} \quad \begin{pmatrix} 1 & 2 \\ 3 & 4 \\ 5 & 6 \end{pmatrix}$$

Show for any matrix $\mathbf{A}$ that $(\mathbf{A}^t)^t = \mathbf{A}$.

7. Let $\mathbf{A}$ be a square matrix, show that $\mathbf{A}$ is symmetric iff $\mathbf{A} = \mathbf{A}^t$, $\mathbf{A}$ is skew-symmetric iff $\mathbf{A} = -\mathbf{A}^t$.

8. Show that the product of two lower triangular matrices is again lower triangular. If you cannot work the general case do the 2×2 and 3×3 cases.

9. For any square matrix $\mathbf{A}$ show that $\mathbf{A} + \mathbf{A}^t$ is symmetric and $\mathbf{A} - \mathbf{A}^t$ is skew-symmetric.

10. Let $\mathbf{A}$ and $\mathbf{B}$ be 3×3 matrices. Show that $\mathbf{A}^t\mathbf{B}^t = (\mathbf{BA})^t$.

11. Let $\mathbf{A}$ be an idempotent matrix. Show that $\mathbf{I} - \mathbf{A}$ is also idempotent.

12. Show that a matrix $\mathbf{A}$ is involutory iff $(\mathbf{I} - \mathbf{A})(\mathbf{I} + \mathbf{A}) = \mathbf{0}$.

13. Let $\mathbf{A} = (a_{ij})$ be a 3×3 matrix. Compute $\mathbf{AE}_{rs}$ and $\mathbf{E}_{rs}\mathbf{A}$.

14. A square matrix $\mathbf{A}$ is said to commute with a matrix $\mathbf{B}$ iff $\mathbf{AB} = \mathbf{BA}$. When does a 3×3 matrix $\mathbf{A}$ commute with the matrix $\mathbf{E}_{rs}$?

15. Show that if a 3×3 matrix $\mathbf{A}$ commutes with every 3×3 matrix $\mathbf{B}$ then $\mathbf{A}$ is a scalar matrix. (*Hint*: If $\mathbf{A}$ commutes with every matrix $\mathbf{B}$ it commutes with the 9 matrices $\mathbf{E}_{rs}$ $r, s = 1, 2, 3$. Now use (14).)

16. Find all 2×2 matrices that commute with

$$\begin{pmatrix} 1 & 1 \\ 0 & 1 \end{pmatrix}.$$

17. Construct a 3×3 matrix $\mathbf{A}$ such that $\mathbf{A}^3 = \mathbf{I}$. (Try to think of a simple linear transformation $\mathsf{T} : \mathbb{R}^3 \to \mathbb{R}^3$ with $\mathsf{T}^3 = \mathbf{I}$ and use its matrix relative to the standard basis.)

18. Let $\mathbf{A}$ be a 3×3 matrix, $\mathbf{D}$ be the diagonal matrix

$$\mathbf{D} = \begin{pmatrix} d_1 & 0 & 0 \\ 0 & d_2 & 0 \\ 0 & 0 & d_3 \end{pmatrix}$$

(a) Compute $\mathbf{D} \cdot \mathbf{A}$.
(b) Compute $\mathbf{A} \cdot \mathbf{D}$.

19. Let $\mathbf{A}$ be a 3×3 matrix, Compute $\mathbf{E}_{12} \cdot \mathbf{A}$, $\mathbf{A} \cdot \mathbf{E}_{12}$, $\mathbf{E}_{21} \cdot \mathbf{A}$, $\mathbf{A} \cdot \mathbf{E}_{21}$. What conclusion can you obtain in general for $\mathbf{E}_{rs} \cdot \mathbf{A}$ and $\mathbf{A} \cdot \mathbf{E}_{rs}$?

20. If $\mathbf{A}$ is an idempotent square matrix show $\mathbf{I} - 2\mathbf{A}$ is invertible. (*Hint*: Idempotents correspond to projections. Interpret $\mathbf{I} - 2\mathbf{A}$ as a reflection. Try the 2×2 case first, then try to generalize.)

21. Let $\mathbf{A}$, $\mathbf{B}$ and $\mathbf{A} + \mathbf{B}$ be invertible $n \times n$ matrices. Show that $\mathbf{A}^{-1} + \mathbf{B}^{-1}$ is invertible. (*Hint*: If you cannot think of anything else try to show

$$(\mathbf{A}^{-1} + \mathbf{B}^{-1})^{-1} = \mathbf{A}(\mathbf{A} + \mathbf{B})^{-1}\mathbf{B}.)$$

22. Find all the 2×2 matrices A satisfying the equation $\mathbf{A}^2 + 1 = 0$.

23. Find the inverse of the matrix

$$\begin{pmatrix} 0 & 1 & 1 \\ 1 & 0 & 1 \\ 1 & 1 & 0 \end{pmatrix}.$$

24. Show that

$$\begin{pmatrix} 1 & 2 \\ -2 & 1 \end{pmatrix}$$

is invertible. Find its inverse. Solve the equation

$$\begin{pmatrix} 1 & 2 \\ -2 & 1 \end{pmatrix}\begin{pmatrix} x \\ y \end{pmatrix} = \begin{pmatrix} -1 \\ 1 \end{pmatrix}$$

for x and y.

25. Let $\mathbf{A} \in \mathbb{R}^n$. Show that $\mathbf{A} = 0$ iff the matrix product $\mathbf{A}\mathbf{A}^t$ is the zero matrix.

12 Representing linear transformations by matrices

Let us return now to the ideas we developed in Chapter 10 for representing a linear transformation $\mathsf{T}: \mathbb{R}^3 \to \mathbb{R}^3$ by a 3×3 matrix. There is of course nothing sacred about $\mathbb{R}^3$ and its standard basis as the following discussion will show.

Let us suppose given a linear transformation

$$\mathsf{T}: \mathscr{V} \to \mathscr{W}$$

between the finite-dimensional vector spaces $\mathscr{V}$ and $\mathscr{W}$. Let $\{\mathbf{A}_1, \ldots, \mathbf{A}_n\}$ be an *ordered* basis (that is, a basis whose vectors are placed in a specified order) for $\mathscr{V}$ and $\{\mathbf{B}_1, \ldots, \mathbf{B}_m\}$ an ordered basis for $\mathscr{W}$. It is then possible to find unique numbers a_{ij}, $i = 1, \ldots, n$; $j = 1, \ldots, m$ such that

$$\begin{aligned}
\mathsf{T}(\mathbf{A}_1) &= a_{11}\mathbf{B}_1 + a_{21}\mathbf{B}_2 + \cdots + a_{m1}\mathbf{B}_m, \\
\mathsf{T}(\mathbf{A}_2) &= a_{12}\mathbf{B}_1 + a_{22}\mathbf{B}_2 + \cdots + a_{m2}\mathbf{B}_m, \\
&\vdots \\
\mathsf{T}(\mathbf{A}_n) &= a_{1n}\mathbf{B}_1 + a_{2n}\mathbf{B}_2 + \cdots + a_{mn}\mathbf{B}_m,
\end{aligned}$$

which we have seen may be written more compactly as

$$\mathsf{T}(\mathbf{A}_j) = \sum_{i=1}^{m} a_{ij}\mathbf{B}_i \qquad j = 1, 2, \ldots, n.$$

The $m \times n$ matrix $\mathbf{A} = (a_{ij})$ is called the *matrix of* T *relative* to the ordered bases $\{\mathbf{A}_1, \ldots, \mathbf{A}_n\}$, $\{\mathbf{B}_1, \ldots, \mathbf{B}_m\}$. Note that just as in the case of $\mathscr{V} = \mathbb{R}^3 = \mathscr{W}$ that the *columns* of $\mathbf{A}$ are the coordinates of the vectors $\mathsf{T}(\mathbf{A}_j)$ relative to the basis $\{\mathbf{B}_1, \ldots, \mathbf{B}_m\}$. Note also that in saying "$\mathbf{A}$ is the matrix of T" you **must** also specify relative to which pairs of ordered bases.

EXAMPLE 1. Let

$$\mathsf{T}: \mathbb{R}^2 \to \mathbb{R}^2$$

be the linear transformation given by

$$\mathsf{T}(x, y) = (y, x).$$

Calculate the matrix of T relative to the standard basis of $\mathbb{R}^2$.

Solution. By the "standard basis of $\mathbb{R}^2$" we mean the basis $\mathbf{E}_1 = (1, 0)$, $\mathbf{E}_2 = (0, 1)$, and that we are to use this ordered basis in both the domain and range of T. We have

$$\mathsf{T}(1, 0) = (0, 1) \qquad \mathsf{T}(0, 1) = (1, 0)$$

so the matrix we seek is

$$\mathbf{A} = \begin{pmatrix} 0 & 1 \\ 1 & 0 \end{pmatrix}.$$

EXAMPLE 2. With T as in Example 1 find the matrix of T **relative** to the pair of ordered bases $\{\mathbf{E}_1, \mathbf{E}_2\}$ and $\{\mathbf{F}_1, \mathbf{F}_2\}$ where $\mathbf{F}_1 = (0, 1)$, $\mathbf{F}_2 = (1, 0)$. (Here $\{\mathbf{E}_1, \mathbf{E}_2\}$ is the basis in the domain of T and $\{\mathbf{F}_1, \mathbf{F}_2\}$ the basis for the range of T.)

Solution. We still have

$$\mathsf{T}(1, 0) = (0, 1) \qquad \mathsf{T}(0, 1) = (1, 0)$$

but now we must write these equations as

$$\mathsf{T}(E_1) = 1\mathbf{F}_1 + 0\mathbf{F}_2 \qquad \mathsf{T}(E_2) = 0\mathbf{F}_1 + 1\mathbf{F}_2$$

so that

$$\mathbf{B} = \begin{pmatrix} 1 & 0 \\ 0 & 1 \end{pmatrix}$$

is the matrix that we now seek.

The lesson to be learned from the preceding example is that appearance can be deceiving! It also suggests that we might profitably inquire into when two matrices represent the same linear transformation relative to different ordered bases. Before doing so let us see a few more examples.

EXAMPLE 3. Calculate the matrix of the differentiation operator

$$\mathsf{D}: \mathscr{P}_n(\mathbb{R}) \to \mathscr{P}_n(\mathbb{R})$$

relative to the usual basis $\{1, x, x^2, \ldots, x^n\}$ for $\mathscr{P}_n(\mathbb{R})$.

Solution. We have for $m = 1, 2, \ldots$

$$\mathsf{D}x^m = mx^{m-1} = 0 + 0x + \cdots + mx^{m-1} + 0x^m + \cdots + 0x^n$$

and

$$\mathsf{D}(1) = 0.$$

Thus the matrix that we seek is

$$\mathbf{N} = \begin{pmatrix} 0 & 1 & 0 & \cdots & 0 \\ 0 & 0 & 2 & \cdots & 0 \\ \vdots & \vdots & \vdots & & \vdots \\ 0 & 0 & 0 & \cdots & n \\ 0 & 0 & 0 & & 0 \end{pmatrix}.$$

For example

$$\mathbf{N} = \begin{pmatrix} 0 & 1 \\ 0 & 0 \end{pmatrix} \qquad \text{when } n = 1$$

$$\mathbf{N} = \begin{pmatrix} 0 & 1 & 0 \\ 0 & 0 & 2 \\ 0 & 0 & 0 \end{pmatrix} \qquad \text{when } n = 2$$

$$\mathbf{N} = \begin{pmatrix} 0 & 1 & 0 & 0 \\ 0 & 0 & 2 & 0 \\ 0 & 0 & 0 & 3 \\ 0 & 0 & 0 & 0 \end{pmatrix} \qquad \text{when } n = 3$$

etc. Notice that all the nonzero entries are along the *superdiagonal*. (The superdiagonal of a square matrix $\mathbf{A} = (a_{ij})$ are the entries $a_{1,2}, a_{2,3}, \ldots, a_{n-1,n}$ of $\mathbf{A}$.)

There is of course no reason to restrict ourselves to square matrices. For example we have the linear transformation

$$\mathbf{D} : \mathscr{P}_n(\mathbb{R}) \to \mathscr{P}_{n-1}(\mathbb{R})$$

and we can ask for its matrix relative to the standard bases of $\mathscr{P}_n(\mathbb{R})$ and $\mathscr{P}_{n-1}(\mathbb{R})$. (What size is this matrix? Answer: $n \times (n + 1)$.) Computing as before we see the required matrix is

$$\mathbf{N}' = \begin{pmatrix} 0 & 1 & 0 & 0 \\ 0 & 0 & 2 & 0 \\ \vdots & \vdots & \vdots & \vdots \\ 0 & 0 & 0 & n \end{pmatrix}.$$

For example

$$\mathbf{N}' = (0 \quad 1) \qquad \text{when } n = 1$$

$$\mathbf{N}' = \begin{pmatrix} 0 & 1 & 0 \\ 0 & 0 & 2 \end{pmatrix} \qquad \text{when } n = 2$$

$$\mathbf{N}' = \begin{pmatrix} 0 & 1 & 0 & 0 \\ 0 & 0 & 2 & 0 \\ 0 & 0 & 0 & 3 \end{pmatrix} \qquad \text{when } n = 3$$

etc.

For a related example, try to calculate the matrix of

$$\mathsf{D} : \mathscr{P}_n(\mathbb{R}) \to \mathscr{P}_{n+1}(\mathbb{R})$$

($(n + 1)$ not $n - 1$). Before you write anything down decide what size it should be.

EXAMPLE 4. Calculate the matrix of the linear transformation:

$$\mathsf{T} : \mathbb{R}^4 \to \mathscr{P}_1(\mathbb{R})$$

given by

$$\mathsf{T}(a_1, a_2, a_3, a_4) = a_1 + a_3 + (a_2 + a_4)x$$

relative to the ordered bases

$$\begin{aligned} \mathbf{A}_1 &= (1, 1, 1, 1) \\ \mathbf{A}_2 &= (1, 1, 1, 0) \\ \mathbf{A}_3 &= (1, 1, 0, 0) \\ \mathbf{A}_4 &= (1, 0, 0, 0) \end{aligned}$$

for $\mathbb{R}^4$ and

$$\begin{aligned} \mathbf{B}_1 &= 1 + x \\ \mathbf{B}_2 &= 1 - x \end{aligned}$$

for $\mathscr{P}_1(\mathbb{R})$.

Solution. We have (note $1 = \frac{1}{2}(\mathbf{B}_1 + \mathbf{B}_2)$, $x = \frac{1}{2}(\mathbf{B}_1 - \mathbf{B}_2)$)

$$\begin{aligned} \mathsf{T}(\mathbf{A}_1) &= \mathsf{T}(1, 1, 1, 1) = 2 + 2x = 2\mathbf{B}_1 + 0\mathbf{B}_2 \\ \mathsf{T}(\mathbf{A}_2) &= \mathsf{T}(1, 1, 1, 0) = 2 + x \;\;= \tfrac{3}{2}\mathbf{B}_1 + \tfrac{1}{2}\mathbf{B}_2 \\ \mathsf{T}(\mathbf{A}_3) &= \mathsf{T}(1, 1, 0, 0) = 1 + x \;\;= \mathbf{B}_1 + 0\mathbf{B}_2 \\ \mathsf{T}(\mathbf{A}_4) &= \mathsf{T}(1, 0, 0, 0) = 1 \qquad\;\; = \tfrac{1}{2}\mathbf{B}_1 + \tfrac{1}{2}\mathbf{B}_2 \end{aligned}$$

so the matrix that we seek is

$$\begin{pmatrix} 2 & \frac{3}{2} & 1 & \frac{1}{2} \\ 0 & \frac{1}{2} & 0 & \frac{1}{2} \end{pmatrix}.$$

EXAMPLE 5. Let

$$\mathsf{T} : \mathbb{R}^3 \to \mathbb{R}^2$$

be given by

$$\mathsf{T}(x, y, z) = (x + y, y + z).$$

Calculate the matrix of T relative to (a) the standard bases of $\mathbb{R}^3$ and $\mathbb{R}^2$, (b) the bases

$$\begin{aligned} \mathbf{A}_1 &= (1, 0, 0) \qquad & \mathbf{B}_1 &= (1, 0) \\ \mathbf{A}_2 &= (0, 0, 1) \qquad & \mathbf{B}_2 &= (0, 1). \\ \mathbf{A}_3 &= (1, -1, 1) \end{aligned}$$

Solution. We have

$$\begin{aligned} \mathsf{T}(1, 0, 0) &= (1, 0) \\ \mathsf{T}(0, 1, 0) &= (1, 1) \\ \mathsf{T}(0, 0, 1) &= (0, 1) \end{aligned}$$

so that

$$\mathbf{A} = \begin{pmatrix} 1 & 1 & 0 \\ 0 & 1 & 1 \end{pmatrix}$$

is the matrix for Part (a). On the other hand

$$\mathsf{T}(1, -1, 1) = (0, 0)$$

so that

$$\mathbf{B} = \begin{pmatrix} 1 & 0 & 0 \\ 0 & 1 & 0 \end{pmatrix}$$

is the matrix for Part (b).

EXAMPLE 6. Let $\mathsf{T} : \mathscr{P}_3(\mathbb{R}) \to \mathscr{P}_5(\mathbb{R})$ be the linear transformation given by the formula

$$\mathsf{T}(p(x)) = (1 + x - x^2)p(x).$$

Find the matrix of T relative to the standard bases of $\mathscr{P}_3(\mathbb{R})$ and $\mathscr{P}_5(\mathbb{R})$.

Solution. We compute

$$\begin{aligned} \mathsf{T}(1) &= 1 + x - x^2 \\ \mathsf{T}(x) &= x + x^2 - x^3 \\ \mathsf{T}(x^2) &= x^2 + x^3 - x^4 \\ \mathsf{T}(x^3) &= x^3 + x^4 - x^5 \end{aligned}$$

and recalling that the columns of T are the vectors $\mathsf{T}(1)$, $\mathsf{T}(x)$, $\mathsf{T}(x^2)$, $\mathsf{T}(x^3)$ expressed in terms of the basis $\{1, x, x^2, x^3, x^4, x^5\}$ we obtain

$$\begin{pmatrix} 1 & 0 & 0 & 0 \\ 1 & 1 & 0 & 0 \\ -1 & 1 & 1 & 0 \\ 0 & -1 & 1 & 1 \\ 0 & 0 & -1 & 1 \\ 0 & 0 & 0 & -1 \end{pmatrix}$$

as the required matrix.

As a last example consider:

Example 7. Let $S = \{u, v, w\}$ and define

$$\mathsf{E} : \mathscr{F}(S) \to \mathbb{R}$$

by

$$\mathsf{E}(\mathbf{f}) = \mathbf{f}(u) + \mathbf{f}(v) + \mathbf{f}(w), \qquad \mathbf{f} \in \mathscr{F}(S).$$

Determine the matrix of $\mathbf{E}$ relative to the ordered bases $\{\chi_u, \chi_v, \chi_w\}$, $\{1\}$ for $\mathscr{F}(S)$ and $\mathbb{R}$ respectively.

Solution. We compute

$$\begin{aligned} \mathsf{E}(\chi_u) &= \chi_u(u) + \chi_u(v) + \chi_u(w) = 1 + 0 + 0 = 1 \\ \mathsf{E}(\chi_v) &= \chi_v(u) + \chi_v(v) + \chi_v(w) + 0 + 1 + 0 = 1 \\ \mathsf{E}(\chi_w) &= \chi_w(u) + \chi_w(v) + \chi_w(w) = 0 + 0 + 1 = 1 \end{aligned}$$

so the required matrix is

$$(1 \quad 1 \quad 1)$$

for E.

The preceding examples illustrate (I hope!) that the matrix of a linear transformation $\mathsf{T} : \mathscr{V} \to \mathscr{W}$ can be exceedingly complex if some care is not exercised in the choice of the basis relative to which the matrix is computed. Indeed finding a basis relative to which the matrix is as simple as possible should clearly be one's goal if matrices are to simplify our numerical computations with linear transformations.

With these numerical examples behind us we turn now to a more careful investigation of the relation between linear transformations and matrices.

Theorem 12.1. *Let $\mathscr{V}$ and $\mathscr{W}$ be finite-dimensional vector spaces. Suppose that $\{\mathbf{A}_1, \ldots, \mathbf{A}_n\}$, $\{\mathbf{B}_1, \ldots, \mathbf{B}_m\}$ are ordered bases for $\mathscr{V}$ and $\mathscr{W}$ respectively. Then assigning to each linear transformation $\mathsf{T} : \mathscr{V} \to \mathscr{W}$ its matrix relative to these ordered bases defines an isomorphism*

$$\mathsf{M} : \mathscr{L}(\mathscr{V}, \mathscr{W}) \to \mathscr{M}_{mn}$$

of the vector space of linear transformations from $\mathscr{V}$ to $\mathscr{W}$ with the vector space of $m \times n$ matrices.

PROOF. It is clear that M is a linear transformation. Indeed our definition of matrix addition and scalar multiplication was rigged up so that this would be so. To show that M is an isomorphism we will actually construct a linear map

$$\mathsf{L} : \mathscr{M}_{mn} \to \mathscr{L}(\mathscr{V}, \mathscr{W})$$

such that

$$\mathsf{L} \cdot \mathsf{M}(\mathsf{T}) = \mathsf{T} \quad \text{for all } \mathsf{T} \text{ in } \mathscr{L}(\mathscr{V}, \mathscr{W})$$
$$\mathsf{M} \cdot \mathsf{L}(\mathbf{A}) = \mathbf{A} \quad \text{for all } \mathbf{A} \text{ in } \mathscr{M}_{mn}.$$

So suppose that $\mathbf{A} = (a_{ij})$ is an $m \times n$ matrix. Define $\mathsf{T} : \mathscr{V} \to \mathscr{W}$ to be the linear extension of

$$\begin{aligned} \mathsf{T}(\mathbf{A}_1) &= a_{11}\mathbf{B}_1 + a_{21}\mathbf{B}_2 + \cdots + a_{m1}\mathbf{B}_m, \\ \mathsf{T}(\mathbf{A}_2) &= a_{12}\mathbf{B}_1 + a_{22}\mathbf{B}_2 + \cdots + a_{m2}\mathbf{B}_m, \\ &\vdots \\ \mathsf{T}(\mathbf{A}_n) &= a_{1n}\mathbf{B}_1 + a_{2n}\mathbf{B}_n + \cdots + a_{mn}\mathbf{B}_m. \end{aligned}$$

Thus if

$$\mathbf{C} = c_1\mathbf{A}_1 + c_2\mathbf{A}_2 + \cdots + c_n\mathbf{A}_n$$

is an arbitrary vector of $\mathscr{V}$ we have the horrendous formula:

$$\begin{aligned} \mathsf{T}(\mathbf{C}) &= (a_{11}c_1 + a_{12}c_2 + \cdots + a_{1n}c_n)\mathbf{B}_1 + \cdots \\ &\quad + (a_{m1}c_1 + a_{m2}c_2 + \cdots + a_{mn}c_n)\mathbf{B}_m \\ &= \sum_{i=1}^{m} \sum_{j=1}^{n} a_{ij}c_j\mathbf{B}_i. \end{aligned}$$

It is immediately clear that the assignment of the linear transformation T to the matrix $\mathbf{A}$ defines a linear transformation

$$\mathsf{L} : \mathscr{M}_{mn} \to \mathscr{L}(\mathscr{V}, \mathscr{W}).$$

The matrix of $\mathsf{L}(\mathbf{A})$ is seen by definition to be $\mathbf{A}$. That is

$$\mathsf{ML}(\mathbf{A}) = \mathbf{A} \quad \text{for all } \mathbf{A} \text{ in } \mathscr{M}_{mn}.$$

On the other hand suppose that $\mathsf{T} : \mathscr{V} \to \mathscr{W}$ is in $\mathscr{L}(\mathscr{V}, \mathscr{W})$. Then T and LM(T) are given on the basis $\mathbf{A}_1, \ldots, \mathbf{A}_n$ by

$$\mathsf{T}(\mathbf{A}_j) = \sum_{i=1}^{m} a_{ij}\mathbf{B}_i = \mathsf{LM}(\mathsf{T}) \qquad j = 1, \ldots, n.$$

Since $\mathbf{A}_1, \ldots, \mathbf{A}_n$ is a basis for $\mathscr{V}$ one has for any vector a unique expression

$$\mathbf{C} = \sum_{j=1}^{n} c_j \mathbf{A}_j,$$

so we have

$$\mathsf{T}(\mathbf{C}) = \sum_{i=1}^{m} \sum_{j=1}^{n} a_{ij} c_j \mathbf{B}_i = (\mathsf{LM})(\mathsf{T})(\mathbf{C}),$$

and hence $\mathsf{T} = \mathsf{LM}(\mathsf{T})$, that is

$$\mathsf{LM}(\mathsf{T}) = \mathsf{T} \quad \text{for all } \mathsf{T} \text{ in } \mathscr{L}(\mathscr{V}, \mathscr{W}).$$

Therefore M is an isomorphism. □

Before examining several interesting consequences of (12.1) let it be noted that the proof of (12.1) is of interest in its own right. Namely, given the following data:

(1) a finite-dimensional vector space $\mathscr{V}$ with basis $\mathbf{A}_1, \ldots, \mathbf{A}_n$
(2) a finite-dimensional vector space $\mathscr{W}$ with basis $\mathbf{B}_1, \ldots, \mathbf{B}_m$
(3) an $m \times n$ matrix $\mathbf{A} = (a_{ij})$

we may manufacture a linear transformation

$$\mathsf{L}(\mathbf{A}) = \mathsf{T} : \mathscr{V} \to \mathscr{W}$$

by the formula

$$(*) \qquad \mathsf{T}(\mathbf{C}) = \sum_{i=1}^{m} \sum_{j=1}^{n} a_{ij} c_j \mathbf{B}_i$$

where $\mathbf{C} = \sum c_j \mathbf{A}_j$, or what is the same thing, *by requiring that the matrix of* T *relative to the ordered bases* $\{\mathbf{A}_1, \ldots, \mathbf{A}_n\}$, $\{\mathbf{B}_1, \ldots, \mathbf{B}_m\}$ *be* $\mathbf{A}$.

EXAMPLE 8. Find the value of the linear transformation $\mathsf{T} : \mathbb{R}^3 \to \mathscr{P}_2(\mathbb{R})$ whose matrix relative to the bases $\{\mathbf{E}_1, \mathbf{E}_2, \mathbf{E}_3\}$, $\{1, x, x^2\}$ is

$$\mathbf{A} = \begin{pmatrix} 1 & 0 & -1 \\ 2 & 4 & -3 \\ 3 & 0 & 2 \end{pmatrix}$$

on the vector $\mathbf{C} = (1, 1, -1)$ of $\mathbb{R}^3$.

Solution. We have

$$\begin{aligned} \mathsf{T}(C) &= \mathsf{T}(1\mathbf{E}_1 + 1\mathbf{E}_2 - 1\mathbf{E}_3) \\ &= \mathsf{T}(\mathbf{E}_1) + \mathsf{T}(\mathbf{E}_2) - \mathsf{T}(\mathbf{E}_3) \\ &= (1 + 2x + 3x^2) + (4x) - (-1 - 3x + 2x^2) \\ &= 1 + 1 + 2x + 4x + 3x + 3x^2 - 2x^2 \\ &= 2 + 9x + x^2. \end{aligned}$$

A moment's (an hour?) reflection on the above example and Equation (*) will show that the coordinates of T(**C**) relative to the ordered basis $\{\mathbf{B}_1, \ldots, \mathbf{B}_n\}$ appear as the entries in the ***column*** vector

$$A\begin{pmatrix} c_1 \\ \vdots \\ c_n \end{pmatrix} = \begin{pmatrix} a_{11} & \cdots & a_{1n} \\ \vdots & & \\ a_{m1} & \cdots & a_{mn} \end{pmatrix}\begin{pmatrix} c_1 \\ \vdots \\ c_n \end{pmatrix}$$

Thus we may solve problems such as Example 8 *by matrix multiplications.*

Example 9. Let $\mathsf{T} : \mathbb{R}^3 \to \mathbb{R}^4$ be the linear transformation whose matrix relative to the standard bases of $\mathbb{R}^3$ and $\mathbb{R}^4$ is

$$\mathbf{A} = \begin{pmatrix} 0 & 1 & 0 \\ 3 & 4 & 1 \\ -1 & -5 & -1 \\ 2 & 2 & 0 \end{pmatrix}$$

Calculate the value of T on $(1, 2, -4)$.

Solution. We have

$$\begin{pmatrix} 0 & 1 & 0 \\ 3 & 4 & 1 \\ -1 & -5 & -1 \\ 2 & 2 & 0 \end{pmatrix}\begin{pmatrix} 1 \\ 2 \\ -4 \end{pmatrix} = \begin{pmatrix} 2 \\ 7 \\ -7 \\ 6 \end{pmatrix}$$

Therefore $\mathsf{T}(1, 2, 4) = (2, 7, -7, 6)$.

While it may appear strange that we calculate the value of T(**C**) by using the coordinates of **C** to make a column matrix and that our answer appears as a column of numbers instead of a row, let us just remark that we could very well have agreed to write vectors in $\mathbb{R}^n$ as columns of numbers instead of rows. The reason for not doing so is that English reads naturally from left to right. The interchange of rows and columns in the mathematical formalism is an unfortunate, though not accidental occurrence, and is tied up with the difference between covariance and contravariance in physics. In Chinese, column vectors would be perfectly natural.

Example 10. Let $\mathsf{T} : \mathbb{R}^3 \to \mathbb{R}$ be the linear transformation whose matrix is

$$(1, -2, 1)$$

relative to the standard bases. Find $\mathsf{T}(6, -4, 9)$.

Solution. We have

$$\mathsf{T}(6, -4, 9) = (1, -2, 1)\begin{pmatrix} 6 \\ -4 \\ 9 \end{pmatrix} = (6 + 8 + 9) = (23).$$

EXAMPLE 11. Let $\mathsf{T} : \mathbb{R}^3 \to \mathbb{R}^4$ be the linear transformation with matrix

$$\begin{pmatrix} 1 & 0 & 0 \\ 0 & 1 & 1 \\ 1 & 1 & 0 \\ 0 & 0 & 1 \end{pmatrix}$$

relative to the standard bases of $\mathbb{R}^3$ and $\mathbb{R}^4$. Find bases for the kernel and image of T.

Solution. If $(x, y, z) \in \mathbb{R}^3$ then by matrix multiplication

$$\begin{pmatrix} 1 & 0 & 0 \\ 0 & 1 & 1 \\ 1 & 1 & 0 \\ 0 & 0 & 1 \end{pmatrix} \begin{pmatrix} x \\ y \\ z \end{pmatrix} = \begin{pmatrix} x \\ y + z \\ x + y \\ z \end{pmatrix}$$

so

$$\mathsf{T}(x, y, z) = (x, y + z, x + y, z),$$

and

$$(x, y, z) \in \ker \mathsf{T} \quad \text{iff} \quad \mathbf{0} = (x, y + z, x + y, z)$$

so

$$\left.\begin{aligned} 0 &= x \\ 0 &= x + y \end{aligned}\right\} \qquad x = 0 \quad \text{and} \quad y = 0$$

$$\left.\begin{aligned} 0 &= y + z \\ 0 &= z \end{aligned}\right\} \qquad z = 0 \quad \text{and} \quad y = 0.$$

Therefore $\ker \mathsf{T} = \{\mathbf{0}\}$. To find a basis for the image of T, note that since $\ker \mathsf{T} = \{\mathbf{0}\}$ the image has dimension 3. Therefore the vectors

$$\mathsf{T}(1, 0, 0), \qquad \mathsf{T}(0, 1, 0), \qquad \mathsf{T}(0, 0, 1)$$

are a basis for the image. By the definition of the matrix of a linear transformation the components of these vectors are the columns of the matrix, so

$$(1, 0, 1, 0), \qquad (0, 1, 1, 0), \qquad (0, 1, 0, 1)$$

is a basis for $\operatorname{Im} \mathsf{T}$.

Let us return now to the consequences of (12.1) we hinted at previously. First we have:

Corollary 12.2. *Let $\mathscr{V}$ and $\mathscr{W}$ be finite-dimensional vector spaces. Then $\mathscr{L}(\mathscr{V}, \mathscr{W})$ is also finite dimensional and moreover*

$$\dim \mathscr{L}(\mathscr{V}, \mathscr{W}) = \dim \mathscr{V} \dim \mathscr{W}.$$

PROOF. This is immediate from the fact that $\mathscr{L}(\mathscr{V}, \mathscr{W})$ and $\mathscr{M}_{mn}$ are isomorphic where $m = \dim \mathscr{V}$ and $n = \dim \mathscr{W}$. □

To state our next corollary we will need the following very important result.

Proposition 12.3. *Let $\mathscr{U}$, $\mathscr{V}$, and $\mathscr{W}$ be finite-dimensional vector spaces with bases $\{\mathbf{C}_1, \ldots, \mathbf{C}_p\}$, $\{\mathbf{A}_1, \ldots, \mathbf{A}_n\}$, and $\{\mathbf{B}_1, \ldots, \mathbf{B}_m\}$. Suppose that*

$$\mathsf{S} : \mathscr{U} \to \mathscr{V}, \qquad \mathsf{T} : \mathscr{V} \to \mathscr{W}$$

are linear transformations. Let $\mathbf{B} = (b_{ij})$ be the matrix of S relative to the bases $\{\mathbf{C}_1, \ldots, \mathbf{C}_p\}$, $\{\mathbf{A}_1, \ldots, \mathbf{A}_n\}$ and $\mathbf{A} = (a_{jk})$ the matrix of T relative to the bases $\{\mathbf{A}_1, \ldots, \mathbf{A}_n\}$, $\{\mathbf{B}_1, \ldots, \mathbf{B}_m\}$. Then the matrix of $\mathsf{T} \cdot \mathsf{S}$ relative to the bases $\{\mathbf{C}_1, \ldots, \mathbf{C}_p\}$, $\{\mathbf{B}_1, \ldots, \mathbf{B}_m\}$ is the matrix product $\mathbf{AB}$.

PROOF. This is an immediate consequence of the definition of the product of two matrices. □

EXAMPLE 12. Let

$$\mathsf{S} : \mathbb{R}^3 \to \mathbb{R}^4, \qquad \mathsf{T} : \mathbb{R}^4 \to \mathbb{R}^2$$

be the linear transformations whose matrices relative to the standard bases are

$$\mathbf{B} = \begin{pmatrix} 1 & 0 & -1 \\ -2 & -1 & -2 \\ 0 & 2 & 5 \\ 4 & 3 & 0 \end{pmatrix} \qquad \mathbf{A} = \begin{pmatrix} -1 & 0 & 3 & -1 \\ 2 & 4 & 1 & 2 \end{pmatrix}$$

Find the matrix of the transformation $\mathsf{T} \cdot \mathsf{S} : \mathbb{R}^3 \to \mathbb{R}^2$ relative to the standard bases.

Solution. According to (12.3) we need only calculate the matrix product $\mathbf{AB}$ which is

$$\begin{pmatrix} -1 & 0 & 3 & -1 \\ 2 & 4 & 1 & 2 \end{pmatrix} \begin{pmatrix} 1 & 0 & -1 \\ -2 & -1 & -2 \\ 0 & 2 & 5 \\ 4 & 3 & 0 \end{pmatrix} = \begin{pmatrix} -5 & 3 & 16 \\ 2 & 4 & -5 \end{pmatrix}$$

and so the matrix that we seek is

$$\begin{pmatrix} -5 & 3 & 16 \\ 2 & 4 & -5 \end{pmatrix}.$$

Corollary 12.4. *A linear transformation $\mathsf{T} : \mathscr{V} \to \mathscr{V}$ is an isomorphism iff its matrix $\mathbf{A}$ is invertible. (The matrix $\mathbf{A}$ is to be computed relative to any pair $\{\mathbf{A}_1, \ldots, \mathbf{A}_n\}$, $\{\mathbf{B}_1, \ldots, \mathbf{B}_n\}$ of ordered bases for $\mathscr{V}$.)*

PROOF. Suppose that T is an isomorphism. Let

$$\mathsf{S} : \mathscr{V} \to \mathscr{V}$$

be the linear transformation inverse to T. Let **B** be the matrix of S relative to the basis pair $\{\mathbf{B}_1, \ldots, \mathbf{B}_n\}$, $\{\mathbf{A}_1, \ldots, \mathbf{A}_n\}$. (N.B. We have interchanged the role of the bases $\{\mathbf{B}_1, \ldots, \mathbf{B}_n\}$, and $\{\mathbf{A}_1, \ldots, \mathbf{A}_n\}$. Thus if $\mathbf{B} = (b_{ij})$ then $\mathsf{S}(\mathbf{B}_j) = \sum b_{ij}\mathbf{A}_i$.) According to (12.3) the matrix product **AB** is the matrix of the linear transformation $\mathsf{T} \cdot \mathsf{S} : \mathscr{V} \to \mathscr{V}$ relative to the basis pair $\{\mathbf{B}_1, \ldots, \mathbf{B}_n\}$, $\{\mathbf{B}_1, \ldots, \mathbf{B}_n\}$. But $\mathsf{T} \cdot \mathsf{S}(\mathbf{C}) = \mathbf{C}$ for all $\mathbf{C} \in \mathscr{V}$ since T and S are inverse isomorphisms. In particular

$$\mathsf{T} \cdot \mathsf{S}(\mathbf{B}_j) = 0\mathbf{B}_1 + 0\mathbf{B}_2 + \cdots + 0\mathbf{B}_{j-1} + 1\mathbf{B}_j + 0\mathbf{B}_{j+1} + \cdots + 0\mathbf{B}_n$$

and hence the matrix of $\mathsf{T} \cdot \mathsf{S}$ relative to the bases $\{\mathbf{B}_1, \ldots, \mathbf{B}_n\}$, $\{\mathbf{B}_1, \ldots, \mathbf{B}_n\}$ is

$$\mathbf{I} = \begin{pmatrix} 1 & & 0 \\ & 1 & \\ & & \ddots & \\ 0 & & & 1 \end{pmatrix}.$$

Therefore $\mathbf{AB} = \mathbf{I}$. Likewise, according to (12.3) the matrix product **BA** is the matrix of the linear transformation $\mathsf{ST} : \mathscr{V} \to \mathscr{V}$ relative to the basis pair $\{\mathbf{A}_1, \ldots, \mathbf{A}_n\}$, $\{\mathbf{A}_1, \ldots, \mathbf{A}_n\}$. But $\mathsf{ST}(\mathbf{C}) = \mathbf{C}$ for all **C** in $\mathscr{V}$ because S and T are inverse isomorphisms, and hence as before we find $\mathbf{BA} = \mathbf{I}$. This shows that if

$$\mathsf{T} : \mathscr{V} \to \mathscr{V}$$

is an isomorphism then a matrix **A** for T is always invertible.

To prove the converse, we suppose that the matrix **A** of T is invertible. Let **B** be a matrix such that

$$\mathbf{AB} = \mathbf{I} = \mathbf{BA}.$$

Let $\mathsf{S} : \mathscr{V} \to \mathscr{V}$ be the linear transformation whose matrix relative to the ordered bases $\{\mathbf{B}_1, \ldots, \mathbf{B}_n\}$, $\{\mathbf{A}_1, \ldots, \mathbf{A}_n\}$ is **B**. (*Note*: We have again interchanged the roles of $\{\mathbf{B}_1, \ldots, \mathbf{B}_n\}$ and $\{\mathbf{A}_1, \ldots, \mathbf{A}_n\}$.) Then the matrix of $\mathsf{ST} : \mathscr{V} \to \mathscr{V}$ relative to the ordered bases $\{\mathbf{A}_1, \ldots, \mathbf{A}_n\}$, $\{\mathbf{A}_1, \ldots, \mathbf{A}_n\}$ is

$$\mathbf{I} = \begin{pmatrix} 1 & & 0 \\ & 1 & \\ & & \ddots & \\ 0 & & & 1 \end{pmatrix}$$

Therefore $\mathsf{S} \cdot \mathsf{T}$ and I have the same matrix relative to the bases $\{\mathbf{A}_1, \ldots, \mathbf{A}_n\}$, $\{\mathbf{A}_1, \ldots, \mathbf{A}_n\}$ so that by (12.1) $\mathsf{S} \cdot \mathsf{T} = \mathsf{I}$, that is

$$\mathsf{S} \cdot \mathsf{T}(\mathbf{C}) = \mathbf{C} \quad \text{for all } \mathbf{C} \text{ in } \mathscr{V}.$$

Likewise we see that

$$\mathsf{T} \cdot \mathsf{S}(\mathbf{C}) = \mathbf{C} \quad \text{for all } \mathbf{C} \text{ in } \mathscr{V}$$

so that S and T are inverse isomorphisms. □

Note that in the proof of (12.4) we used the fact that the matrix of the transformation $\mathsf{I} : \mathscr{V} \to \mathscr{V}$ defined by $\mathsf{I}(\mathbf{C}) = \mathbf{C}$ for all $\mathbf{C}$ relative to the bases $\{\mathbf{A}_1, \ldots, \mathbf{A}_n\}$, $\{\mathbf{A}_1, \ldots, \mathbf{A}_n\}$ is the identity matrix. This is not the case if we have two *different* bases (or even different orderings on the same basis) in $\mathscr{V}$.

EXAMPLE 13. Find the matrix of the identity linear transformation $\mathsf{I} : \mathbb{R}^3 \to \mathbb{R}^3$ relative to the ordered bases

$$\begin{array}{lll} \mathbf{E}_1 = (1, 0, 0), & \mathbf{E}_2 = (0, 1, 0), & \mathbf{E}_3 = (0, 0, 1) \\ \mathbf{F}_1 = (1, 1, 1), & \mathbf{F}_2 = (1, 1, 0), & \mathbf{F}_3 = (1, 0, 0) \end{array}$$

of $\mathbb{R}^3$.

Solution. We have

$$\begin{aligned} \mathsf{I}(E_1) &= \mathsf{I}(1, 0, 0) = (1, 0, 0) = 0\mathbf{F}_1 + 0\mathbf{F}_2 + 1\mathbf{F}_3 \\ \mathsf{I}(E_2) &= \mathsf{I}(0, 1, 0) = (0, 1, 0) = 0\mathbf{F}_1 + 1\mathbf{F}_2 + (-1)\mathbf{F}_3 \\ \mathsf{I}(E_3) &= \mathsf{I}(0, 0, 1) = (0, 0, 1) = 1\mathbf{F}_1 + (-1)\mathbf{F}_2 + 0\mathbf{F}_3. \end{aligned}$$

So the matrix we seek is

$$\mathbf{A} = \begin{pmatrix} 0 & 0 & 1 \\ 0 & 1 & -1 \\ 1 & -1 & 0 \end{pmatrix}$$

The moral of the example is that appearances are **really** deceiving.

In view of Example 13 it is reasonable to expect that when we calculate with matrices of transformations $\mathsf{T} : \mathscr{V} \to \mathscr{V}$ we insist upon using the same ordered basis $\{\mathbf{A}_1, \ldots, \mathbf{A}_n\}$ twice to do the calculation, rather than work with distinct ordered bases $\{\mathbf{A}_1, \ldots, \mathbf{A}_n\}$, $\{\mathbf{B}_1, \ldots, \mathbf{B}_n\}$. Other reasons for insisting on using only one ordered basis $\{\mathbf{A}_1, \ldots, \mathbf{A}_n\}$ when we study a transformation $\mathsf{T} : \mathscr{V} \to \mathscr{V}$ from the same space to itself will appear in Chapter 14.

We turn now to another consequence of (12.1) which substantiates a remark we made concerning inverses of matrices in the last chapter.

Corollary 12.5. *Suppose that* $\mathbf{A}$ *and* $\mathbf{B}$ *are square matrices of size* n *and* $\mathbf{AB} = \mathbf{I}$. *Then* $\mathbf{BA} = \mathbf{I}$. (*Thus to check that a square matrix* $\mathbf{B}$ *is the inverse of a square matrix* $\mathbf{A}$ *we need only check one of the two conditions* $\mathbf{AB} = \mathbf{I}$ *and* $\mathbf{BA} = \mathbf{I}$.)

PROOF. Let $\mathsf{T}, \mathsf{S} : \mathbb{R}^n \to \mathbb{R}^n$ be the linear transformations whose matrices with respect to the standard basis of $\mathbb{R}^n$ are **A** and **B** respectively. That is

$$\mathsf{L}(\mathbf{A}) = \mathsf{T}, \qquad \mathsf{L}(\mathbf{B}) = \mathsf{S}$$

or what is the same thing

$$\mathsf{M}(\mathsf{T}) = \mathbf{A}, \qquad \mathsf{M}(\mathsf{S}) = \mathbf{B}.$$

Then by (12.3)

$$\mathsf{M}(\mathsf{T} \cdot \mathsf{S}) = \mathsf{M}(\mathsf{T})\mathsf{M}(\mathsf{S}) = \mathbf{AB} = \mathbf{I} = \mathsf{M}(\mathsf{I}).$$

Therefore since M is an isomorphism (see 8.12)

$$\mathsf{T} \cdot \mathsf{S} = \mathsf{I}.$$

From this we may conclude that $\mathsf{S} : \mathbb{R}^n \to \mathbb{R}^n$ is an isomorphism as follows. Suppose $\mathbf{C} \in \ker \mathsf{S}$. Then

$$\mathsf{S}(\mathbf{C}) = \mathbf{0}$$

so

$$\mathbf{0} = \mathsf{T}(\mathbf{0}) = \mathsf{T}(\mathsf{S}(\mathbf{C})) = (\mathsf{T} \cdot \mathsf{S})(\mathbf{C}) = \mathsf{I}(\mathbf{C}) = \mathbf{C}.$$

Therefore S has kernel $\mathbf{0}$ so by (8.10)

$$n = \dim \mathbb{R}^n = \dim \operatorname{Im} \mathsf{S} + \dim \ker \mathsf{S} = \dim \operatorname{Im} \mathsf{S}.$$

Therefore $\operatorname{Im} \mathsf{S} = \mathbb{R}^n$ by (6.10) and hence S is an isomorphism by (8.11). Therefore there is a transformation

$$\bar{\mathsf{T}} : \mathbb{R}^n \to \mathbb{R}^n$$

such that

$$\mathsf{S}\bar{\mathsf{T}} = \mathsf{I}.$$

Then

$$\mathsf{T}(\mathsf{S}\bar{\mathsf{T}}) = \mathsf{T}(\mathsf{I}) = \mathsf{T}$$
$$(\mathsf{TS})\bar{\mathsf{T}} = \mathsf{I}\bar{\mathsf{T}} = \bar{\mathsf{T}}$$

so that $\mathsf{T} = \bar{\mathsf{T}}$, that is

$$\mathsf{ST} = \mathsf{I}.$$

Hence

$$\mathbf{I} = \mathsf{M}(\mathsf{I}) = \mathsf{M}(\mathsf{ST}) = \mathsf{M}(\mathsf{S})\mathsf{M}(\mathsf{T}) = \mathbf{B} \cdot \mathbf{A}$$

as required. □

While the proof of (12.5) may seem complicated we invite the reader to try and prove the result using only matrices. Even in the 3×3 case such a proof will be most painful! The corollaries (12.2) and (12.5) illustrate an important point, namely some results concerning linear transformations are best proved using matrices, and conversely some results concerning matrices are best proved using linear transformations.

In the study of linear transformations we will very often make use of matrix representations to prove theorems and make computations. The matrix representation for a linear transformation $\mathsf{T} : \mathscr{V} \to \mathscr{W}$ depends on a choice of bases for $\mathscr{V}$ and $\mathscr{W}$. Our initial choice of these bases may be bad and unsuited to the problem at hand in that we obtain a matrix which does not convey the information we seek. It is therefore quite natural to change the bases to obtain a "better" representation for T, because to change the bases does nothing to the linear transformation, but it does change the matrix representation. Several natural and important questions therefore arise. First of all, what is the relation (numerically, that is) between two matrix representations of T computed with respect to different pairs of bases in $\mathscr{V}$ and $\mathscr{W}$? Secondly, if we are given two $m \times n$ matrices **A** and **B**, when (that is, what numerical relations must hold between them) can we conclude that they represent one and the same linear transformation T, but relative to different bases? As it turns out, both questions may be answered simultaneously, however before doing so we consider one more example of the phenomena under discussion.

EXAMPLE 14. Let $\mathsf{T} : \mathbb{R}^3 \to \mathbb{R}^3$ be the linear transformation given by

$$\mathsf{T}(x, y, z) = (y + z, x + z, y + x).$$

Calculate the matrix of T relative to

(a) the standard basis of $\mathbb{R}^3$,
(b) the basis $\{(1, 1, 1), (1, -1, 0), (1, 1, -2)\}$ used twice.

Solution. To do Part (a) we compute as follows:

$$\begin{aligned} \mathsf{T}(1, 0, 0) &= (0, 1, 1) \\ \mathsf{T}(0, 1, 0) &= (1, 0, 1) \\ \mathsf{T}(0, 0, 1) &= (1, 1, 0) \end{aligned}$$

Thus the desired matrix is

$$\mathbf{A} = \begin{pmatrix} 0 & 1 & 1 \\ 1 & 0 & 1 \\ 1 & 1 & 0 \end{pmatrix}$$

To do the computations of Part (b) let us set $\mathbf{F}_1 = (1, 1, 1)$, $\mathbf{F}_2 = (1, -1, 0)$, $\mathbf{F}_3 = (1, 1, -2)$. Then we find

$$\begin{aligned} \mathsf{T}(\mathbf{F}_1) &= \mathsf{T}(1, 1, 1) = (2, 2, 2) = 2\mathbf{F}_1 + 0\mathbf{F}_2 + 0\mathbf{F}_3 \\ \mathsf{T}(\mathbf{F}_2) &= \mathsf{T}(1, -1, 0) = (-1, 1, 0) = 0\mathbf{F}_1 - \mathbf{F}_2 + 0\mathbf{F}_3 \\ \mathsf{T}(\mathbf{F}_3) &= \mathsf{T}(1, 1 - 2) = (-1, -1, 2) = 0\mathbf{F}_1 + 0\mathbf{F}_2 - \mathbf{F}_3. \end{aligned}$$

So the matrix for Part (b) is

$$\mathbf{B} = \begin{pmatrix} 2 & 0 & 0 \\ 0 & -1 & 0 \\ 0 & 0 & -1 \end{pmatrix}$$

This should serve to illustrate that the matrix of a transformation depends heavily on the bases we use to compute it.

Theorem 12.6. *Let* $\mathbf{A}$ *and* $\mathbf{B}$ *be* $m \times n$ *matrices,* $\mathscr{V}$ *an* n*-dimensional vector space and* $\mathscr{W}$ *an* m*-dimensional vector space. Then* $\mathbf{A}$ *and* $\mathbf{B}$ *represent the same linear transformation* $\mathsf{T} : \mathscr{V} \to \mathscr{W}$ *relative to (perhaps) different pairs of ordered bases iff there exist nonsingular matrices* P *and* Q *such that*

$$\mathbf{A} = \mathbf{PBQ}^{-1}.$$

(Note that $\mathbf{P}$ is $m \times m$ and $\mathbf{Q}$ is $n \times n$.)

PROOF. There are two things we must prove. First, if $\mathbf{A}$ and $\mathbf{B}$ represent the same linear transformation relative to different bases of $\mathscr{V}$ and $\mathscr{W}$ we must construct invertible matrices $\mathbf{P}$ and $\mathbf{Q}$ such that

$$\mathbf{A} = \mathbf{PBQ}^{-1}.$$

Secondly if we are given invertible matrices $\mathbf{P}$ and $\mathbf{Q}$ such that

$$\mathbf{A} = \mathbf{PBQ}^{-1}$$

we must construct a linear transformation

$$\mathsf{T} : \mathscr{V} \to \mathscr{W}$$

and pairs of ordered bases for $\mathscr{V}$ and $\mathscr{W}$ so that $\mathbf{A}$ represents T relative to one pair and $\mathbf{B}$ relative to the other. Consider the first of these. We suppose given bases $\{\mathbf{A}_1, \ldots, \mathbf{A}_n\}$, $\{\mathbf{B}_1, \ldots, \mathbf{B}_m\}$ such that the matrix of T relative to these bases is $\mathbf{A}$, and bases $\{\mathbf{C}_1, \ldots, \mathbf{C}_n\}$, $\{\mathbf{D}_1, \ldots, \mathbf{D}_m\}$ such that the matrix of T relative to these bases is $\mathbf{B}$. Let $\mathbf{P}$ be the matrix of (remember Example 13)

$$\mathsf{I} : \mathscr{W} \to \mathscr{W}$$

relative to the bases $\{\mathbf{D}_1, \ldots, \mathbf{D}_m\}$, $\{\mathbf{B}_1, \ldots, \mathbf{B}_m\}$. Then by (12.4) $\mathbf{P}$ is invertible. Let $\mathbf{Q}$ be the matrix of

$$\mathsf{I} : \mathscr{V} \to \mathscr{V}$$

relative to the bases $\{\mathbf{C}_1, \ldots, \mathbf{C}_n\}$, $\{\mathbf{A}_1, \ldots, \mathbf{A}_n\}$. Then by (12.4) $\mathbf{Q}$ is invertible and by (12.3) $\mathbf{Q}^{-1}$ represents the matrix of

$$\mathsf{I} : \mathscr{V} \to \mathscr{V}$$

relative to the bases $\{\mathbf{A}_1, \ldots, \mathbf{A}_n\}$, $\{\mathbf{C}_1, \ldots, \mathbf{C}_n\}$.

Therefore by (12.4) $\mathbf{PB}$ is the matrix of $\mathsf{T} : \mathscr{V} \to \mathscr{W}$ relative to the bases $\{\mathbf{C}_1, \ldots, \mathbf{C}_n\}$, $\{\mathbf{B}_1, \ldots, \mathbf{B}_m\}$. If we apply (12.4) again we see that $\mathbf{PBQ}^{-1}$ is the matrix of T relative to the bases $\{\mathbf{A}_1, \ldots, \mathbf{A}_n\}$ and $\{\mathbf{B}_1, \ldots, \mathbf{B}_m\}$. But $\mathbf{A}$

is also the matrix of T relative to the bases $\{\mathbf{A}_1, \ldots, \mathbf{A}_n\}$ and $(\mathbf{B}_1, \ldots, \mathbf{B}_m\}$ so that by (12.1)

$$\mathbf{A} = \mathbf{PBQ}^{-1}$$

as required.

To prove the converse suppose given invertible matrices $\mathbf{P}$ and $\mathbf{Q}$ such that $\mathbf{A} = \mathbf{PBQ}^{-1}$. Choose bases $\{\mathbf{A}_1, \ldots, \mathbf{A}_n\}$, $\{\mathbf{B}_1, \ldots, \mathbf{B}_m\}$ for $\mathscr{V}$ and $\mathscr{W}$ respectively. Let $\mathsf{T} : \mathscr{V} \to \mathscr{W}$ be the linear transformation whose matrix is $\mathbf{A}$ relative to these bases. Let

$$\begin{aligned} \mathbf{C}_1 &= \mathbf{P}(\mathbf{A}_1), \ldots, \mathbf{C}_n = \mathbf{P}(\mathbf{A}_n) \\ \mathbf{D}_1 &= \mathbf{Q}(\mathbf{B}_1), \ldots, \mathbf{D}_m = \mathbf{Q}(\mathbf{B}_m). \end{aligned}$$

Since $\mathbf{P}$ and $\mathbf{Q}$ are isomorphisms, the collections $\{\mathbf{C}_1, \ldots, \mathbf{C}_n\}$, $\{\mathbf{D}_1, \ldots, \mathbf{D}_m\}$ are bases for $\mathscr{V}$ and $\mathscr{W}$ respectively. (See for example (8.15) and (8.16).) A brute force computation now shows that $\mathbf{B}$ is the matrix of T relative to the bases $\{\mathbf{C}_1, \ldots, \mathbf{C}_n\}$, $\{\mathbf{D}_1, \ldots, \mathbf{D}_m\}$. □

We will return to (12.6) in chapter 15 where we will discuss in more detail when $\mathbf{A}$ and $\mathbf{B}$ represent the same linear transformation relative to distinct bases.

EXAMPLE 15. Let us see how (12.6) applies to Example 14. Recall that we are given the linear transformation $\mathsf{T} : \mathbb{R}^3 \to \mathbb{R}^3$ defined by

$$\mathsf{T}(x, y, z) = (y + z, x + z, y + x)$$

and

$$\mathbf{A} = \begin{pmatrix} 0 & 1 & 1 \\ 1 & 0 & 1 \\ 1 & 1 & 0 \end{pmatrix}$$

is the matrix of T relative to the standard basis of $\mathbb{R}^3$, while

$$\mathbf{B} = \begin{pmatrix} 2 & 0 & 0 \\ 0 & -1 & 0 \\ 0 & 0 & -1 \end{pmatrix}$$

is the matrix of T relative to the ordered basis $\{(1, 1, 1), (1, -1, 0), (1, 1, -2)\}$ of $\mathbb{R}^3$.

According to (12.6) there are invertible matrices $\mathbf{P}$ and $\mathbf{Q}^{-1}$ such that $\mathbf{A} = \mathbf{PBQ}^{-1}$. Our task is to compute $\mathbf{P}$ and $\mathbf{Q}^{-1}$. The proof of (12.6) tells us how. Namely:

(1) $\mathbf{P}$ is the matrix of $\mathsf{I} : \mathbb{R}^3 \to \mathbb{R}^3$ relative to the basis pair $\{(1, 1, 1), (1, -1, 0), (1, 1, -2)\}$ and $\{(1, 0, 0), (0, 1, 0), (0, 0, 1)\}$.
(2) $\mathbf{Q}^{-1}$ is the matrix of $\mathsf{I} : \mathbb{R}^3 \to \mathbb{R}^3$ relative to the basis pair $\{(1, 0, 0), (0, 1, 0), (0, 0, 1)\}$ and $\{(1, 1, 1), (1, -1, 0), (1, 1, -2)\}$.

The computation of $\mathbf{P}$ is easy and gives us

$$\mathbf{P} = \begin{pmatrix} 1 & 1 & 1 \\ 1 & -1 & 1 \\ 1 & 0 & -2 \end{pmatrix}.$$

The computation of $\mathbf{Q}^{-1}$ is not hard and depends on the following equations

$$\begin{aligned} (1, 0, 0) &= \tfrac{1}{3}(1, 1, 1) + \tfrac{1}{2}(1, -1, 0) + \tfrac{1}{6}(1, 1, -2) \\ (0, 1, 0) &= \tfrac{1}{3}(1, 1, 1) - \tfrac{1}{2}(1, -1, 0) + \tfrac{1}{6}(1, 1, -2) \\ (0, 0, 1) &= \tfrac{1}{3}(1, 1, 1) + 0(1, -1, 0) - \tfrac{1}{3}(1, 1, -2) \end{aligned}$$

so that

$$\mathbf{Q}^{-1} = \begin{pmatrix} \frac{1}{3} & \frac{1}{3} & \frac{1}{3} \\ \frac{1}{2} & -\frac{1}{2} & 0 \\ \frac{1}{6} & \frac{1}{6} & -\frac{1}{3} \end{pmatrix}.$$

A tedious computation shows

$$\mathbf{A} = \mathbf{PBQ}^{-1},$$

that is,

$$\begin{pmatrix} 0 & 1 & 1 \\ 1 & 0 & 1 \\ 1 & 1 & 0 \end{pmatrix} = \begin{pmatrix} 1 & 1 & 1 \\ 1 & -1 & 1 \\ 1 & 0 & -2 \end{pmatrix} \begin{pmatrix} 2 & 0 & 0 \\ 0 & -1 & 0 \\ 0 & 0 & -1 \end{pmatrix} \begin{pmatrix} \frac{1}{3} & \frac{1}{2} & \frac{1}{6} \\ \frac{1}{3} & -\frac{1}{2} & \frac{1}{6} \\ \frac{1}{3} & 0 & -\frac{1}{3} \end{pmatrix}.$$

Exercises

1. Find the matrix of the following linear transformations relative to the standard bases for $\mathbb{R}^n$

   (a) $\mathsf{T} : \mathbb{R}^3 \to \mathbb{R}^5$ given by

   $$\mathsf{T}(a_1, a_2, a_3) = (a_1, a_1 - a_3, a_2 + a_3, a_3, a_1 + a_2)$$

   (b) $\mathsf{T} : \mathbb{R}^4 \to \mathbb{R}^4$ given by

   $$\mathsf{T}(a_1, a_2, a_3, a_4) = (a_1, a_1 + a_2, a_1 + a_2 + a_3, a_1 + a_2 + a_3 + a_4)$$

   (c) $\mathsf{T} : \mathbb{R}^3 \to \mathbb{R}$ given by

   $$\mathsf{T}(a_1, a_2, a_3) = a_1 + a_2 + a_3$$

   (d) $\mathsf{T} : \mathbb{R}^2 \to \mathbb{R}^4$ given by

   $$\mathsf{T}(a_1, a_2) = (a_1 + a_2, 2a_1, 3a_2 - a_1, 2a_1 - a_2).$$

2. Find the matrix of the following linear transformations relative to the usual bases of $\mathscr{P}_n(\mathbb{R})$.

   (a) $\mathsf{D} : \mathscr{P}_3(\mathbb{R}) \to \mathscr{P}_3(\mathbb{R})$ (*Hint*: think about size!)
   (b) $\mathsf{D} : \mathscr{P}_3(\mathbb{R}) \to \mathscr{P}_2(\mathbb{R})$
   (c) $\mathsf{T} : \mathscr{P}_2(\mathbb{R}) \to \mathscr{P}_2(\mathbb{R})$ given by $\mathsf{T}(p(x)) = p(x + 1)$
   (d) $\mathsf{T} : \mathscr{P}_2(\mathbb{R}) \to \mathscr{P}_3(\mathbb{R})$ given by $\mathsf{T}(p(x)) = xp(x)$.

3. Let $\mathsf{T} : \mathbb{R}^4 \to \mathbb{R}^7$ be the linear transformation whose matrix relative to the standard bases is

$$\begin{pmatrix} 1 & 2 & -1 & 0 \\ 0 & -1 & 4 & 1 \\ 2 & 0 & 2 & 0 \\ 0 & -1 & 3 & 1 \\ 3 & 2 & -5 & 0 \\ 0 & 0 & 7 & 1 \\ 4 & 0 & 1 & 0 \end{pmatrix}$$

Find $\mathsf{T}(1, 2, 3, 4)$.

4. Let $\mathsf{S} : \mathbb{R}^7 \to \mathbb{R}^3$ be the linear transformation whose matrix relative to the standard bases is

$$\begin{pmatrix} 1 & 0 & 0 & 0 & 0 & 0 & 0 \\ 0 & 1 & 0 & 0 & 0 & 0 & 0 \\ 1 & 0 & 0 & 0 & 0 & 0 & 0 \end{pmatrix}$$

and let T be the linear transformation of Problem 3. Find the matrix of $\mathsf{ST} : \mathbb{R}^4 \to \mathbb{R}^3$ relative to the standard bases. What is $\mathsf{S} \cdot \mathsf{T}(1, -1, 1, -1)$?

5. Let $\mathsf{S}, \mathsf{T} : \mathbb{R}^3 \to \mathbb{R}^2$ be the linear transformations with matrices

$$\begin{pmatrix} 6 & -1 & 2 \\ 2 & 4 & 1 \end{pmatrix} \quad \text{and} \quad \begin{pmatrix} -5 & 0 & -7 \\ 2 & -1 & 9 \end{pmatrix}$$

respectively. What is the matrix of the linear transformation

$$3\mathsf{S} - 7\mathsf{T} : \mathbb{R}^3 \to \mathbb{R}^2?$$

Find $(3\mathsf{S} - 7\mathsf{T})(1, 2, 3)$.

6. Let $\mathscr{V}$ and $\mathscr{W}$ be finite-dimensional vector spaces and $\mathsf{T} : \mathscr{V} \to \mathscr{W}$ a linear transformation. Show that there are bases $\{\mathbf{A}_1, \ldots, \mathbf{A}_n\}$, $\{\mathbf{B}_1, \ldots, \mathbf{B}_m\}$ for $\mathscr{V}$ and $\mathscr{W}$ such that the matrix of T is

$$\begin{pmatrix} 1 & & & & & \\ & \ddots & & & 0 & \\ & & 1 & & & \\ & & & 0 & & \\ & 0 & & & \ddots & \\ & & & & & 0 \end{pmatrix}.$$

(*Hint*: Choose a basis $\mathbf{A}'_1, \ldots, \mathbf{A}'_r$ for $\ker \mathsf{T}$. Extend this to a basis $\mathbf{A}''_1, \ldots, \mathbf{A}''_s$, $\mathbf{A}'_1, \ldots, \mathbf{A}'_r$ for $\mathscr{V}$. *Show* that the vectors

$$\mathbf{B}_1 = \mathsf{T}(\mathbf{A}''_1), \quad \ldots, \quad \mathbf{B}_s = \mathsf{T}(\mathbf{A}''_s)$$

are linearly independent in $\mathscr{W}$. Extend them to a basis $\mathbf{B}_1, \ldots, \mathbf{B}_s, \mathbf{B}_{s+1}, \ldots, \mathbf{B}_m$ for $\mathscr{W}$. Let $\mathbf{A}_1 = \mathbf{A}''_1, \ldots, \mathbf{A}_s = \mathbf{A}''_s$, $\mathbf{A}_{s+1} = \mathbf{A}'_1, \ldots, \mathbf{A}_n = \mathbf{A}'_r$. Calculate the matrix of T relative to these bases.)

7. Let

$$\mathbf{A} = \begin{pmatrix} 0 & 0 & 0 & 0 \\ a & 0 & 0 & 0 \\ b & d & 0 & 0 \\ c & e & f & 0 \end{pmatrix}$$

Show that $\mathbf{A}^4 = 0$. (*Hint*: Recall how we did an analogous problem in the 3×3 case using the linear transformation $\mathbf{T}: \mathbb{R}^3 \to \mathbb{R}^3$ whose matrix relative to the standard basis was the analog of $\mathbf{A}$.) Can you generalize your result to $n \times n$?

8. Let $\mathsf{T}: \mathbb{R}^2 \to \mathbb{R}^2$ be the linear transformation

$$\mathsf{T}(x, y) = (x + 3y, 3x + y).$$

Find the matrix of T relative to

(a) the standard basis of $\mathbb{R}^2$
(b) the basis $\{\mathbf{F}_1 = (1, 1), \mathbf{F}_2 = (1, -1)\}$ used twice.
(c) $\{\mathbf{F}_1, \mathbf{F}_2\}$ in the domain and the standard basis in the range.

9. Let $\mathsf{T}: \mathbb{R}^3 \to \mathbb{R}^3$ be the linear transformation

$$\mathsf{T}(x, y, z) = (3y + 4z, 3x, 4x).$$

Calculate the matrix of T relative to

(a) the standard basis of $\mathbb{R}^3$
(b) the basis $\{(0, 4, -3), (5, 3, 4), (5, -3, -4)\}$ used twice.

10. Find the matrix of the shift $\mathbf{S}: \mathbb{R}^n \to \mathbb{R}^n$ relative to the standard basis used twice.

11. Let $\mathsf{T}: \mathbb{R}^3 \to \mathbb{R}^3$ be the linear transformation whose matrix relative to the standard basis is

$$\begin{pmatrix} 1 & 1 & 1 \\ -1 & 1 & 0 \\ 1 & 1 & 0 \end{pmatrix}.$$

Find the matrix of T relative to the basis $\{\mathbf{F}_1, \mathbf{F}_2, \mathbf{F}_3\}$ used twice, where $\mathbf{F}_1 = (1, 1, 1)$, $\mathbf{F}_2 = (1, 1, 0)$, $\mathbf{F}_3 = (1, 0, 0)$.

12. Let $\mathsf{T}: \mathbb{R}^3 \to \mathbb{R}^3$ be the linear transformation whose matrix relative to the basis of Problem 11, $\{\mathbf{F}_1, \mathbf{F}_2, \mathbf{F}_3\}$ used twice is

$$\begin{pmatrix} 1 & 0 & 0 \\ 0 & -1 & 1 \\ 0 & 0 & 1 \end{pmatrix}.$$

Find the matrix of T relative to the standard basis.

13. Let $\mathsf{T}: \mathbb{R}^3 \to \mathbb{R}^3$ be the linear transformation $\mathsf{T}(x, y, z) = (x + y, y + z, z + x)$.

(a) Find the matrix of T relative to the standard basis.
(b) Find $\mathbf{A}_1, \mathbf{A}_2, \mathbf{A}_3$ satisfying $\mathsf{T}(\mathbf{A}_i) = \mathbf{E}_i$, $i = 1, 2, 3$.
(c) Show that $\mathbf{A}_1, \mathbf{A}_2, \mathbf{A}_3$ form a set of linearly independent vectors.
(d) Show that $\mathsf{T}(\mathbf{E}_1), \mathsf{T}(\mathbf{E}_2), \mathsf{T}(\mathbf{E}_3)$ form a set of linearly independent vectors.
(e) Consider $\mathsf{S}: \mathbb{R}^3 \to \mathbb{R}^3$ the linear extension of $\mathsf{S}(\mathbf{E}_1) = \mathbf{A}_1$, $\mathsf{S}(\mathbf{E}_2) = \mathbf{A}_2$, $\mathsf{S}(\mathbf{E}_3) = \mathbf{A}_3$, where $\mathbf{A}_i$ are that of (b).
(f) Find the matrix of S relative to the standard basis.
(g) Show that the matrices obtained in (a) and (f) are inverse matrices of each other.

14. Let $\mathsf{T}: \mathscr{P}_3(\mathbb{R}) \to \mathbb{R}^2$ be the linear transformation defined by

$$\mathsf{T}(p(x)) = \left(p(0), \int_0^1 p(x)\, dx\right).$$

Find the matrix of T relative to the ordered basis pair $\{1, x, x^2, x^3\}$ and $\{(1, 0), (0, 1)\}$.

15. Let σ: $\{a, b, c\} \to \{a, b, c\}$ be given by

$$\sigma(a) = b, \qquad \sigma(b) = c, \qquad \sigma(c) = a.$$

Find the matrix of

$$\sigma^* = \mathscr{F}(\mathbf{S}) \to \mathbf{F}(\mathbf{S}): \mathbf{S} = \{a, b, c\}$$

with respect to the basis of characteristic functions.

16. Determine if

$$\begin{pmatrix} 1 & 1 \\ 1 & 1 \end{pmatrix}, \quad \begin{pmatrix} 1 & 1 \\ -1 & 1 \end{pmatrix}$$

can represent the same linear transformation T: $\mathbb{R}^2 \to \mathbb{R}^2$ with respect to different pairs of ordered bases.

17. Find a pair of 2×2 matrices $\mathbf{A}, \mathbf{B}$ such that

$$\mathbf{AB} - \mathbf{BA} = \begin{pmatrix} 1 & 0 \\ 0 & 1 \end{pmatrix}.$$

18. Show that a 3×3 matrix

$$\begin{pmatrix} a & 0 & 0 \\ 0 & b & c \\ 0 & d & e \end{pmatrix}$$

is invertible iff

$$a(be - cd) \neq 0.$$

19. Show that the 3×3 matrix

$$\begin{pmatrix} a & 0 & c \\ 0 & b & 0 \\ c & 0 & a \end{pmatrix}$$

is invertible iff $abc \neq 0$.

20. Determine if the matrices

$$\begin{pmatrix} 1 & 0 & 1 \\ 0 & 2 & 0 \\ 1 & 0 & 1 \end{pmatrix}$$

and

$$\begin{pmatrix} 1 & 0 & 0 \\ 0 & 1 & 1 \\ 0 & 1 & 1 \end{pmatrix}$$

represent the same linear transformation with respect to different ordered pairs of bases of $\mathbb{R}^3$.

21. Let $\mathsf{M}: \mathbb{C}^2 \to \mathbb{C}$ be the linear transformation defined by

$$\mathsf{M}(u, v) = u \cdot v,$$

the dot denoting multiplication of complex numbers. Regard $\mathbb{C}$ and $\mathbb{C}^2$ as *real* vector spaces using $\{1, i\}$ as a basis for $\mathbb{C}$ and $\{(1, 0), (i, 0), (0, 1), (0, i)\}$ as a basis for $\mathbb{C}^2$. Find the matrix of M with respect to these ordered bases.

22. Show that any non-zero linear combination of the matrices

$$\begin{pmatrix} 0 & 1 \\ -1 & 0 \end{pmatrix}, \quad \begin{pmatrix} 0 & -1 \\ 1 & 0 \end{pmatrix}$$

is invertible.

23. A vector $\mathbf{A} \in \mathscr{U}$ is called a *fixed point* of a linear transformation $\mathsf{T}: \mathscr{U} \to \mathscr{U}$ iff $\mathsf{T}(\mathbf{A}) = \mathbf{A}$. It is called an *antifixed point* if $\mathsf{T}(\mathbf{A}) = -\mathbf{A}$. Let $\mathsf{T}: \mathbb{R}^2 \to \mathbb{R}^2$ be represented by each of the following matrices in turn and find the fixed points and antifixed points if any

$$\begin{pmatrix} 0 & 1 \\ 1 & 0 \end{pmatrix}, \quad \begin{pmatrix} 0 & 1 \\ -1 & 0 \end{pmatrix}, \quad \begin{pmatrix} 1 & 1 \\ 1 & 1 \end{pmatrix}.$$

24. Let $\mathscr{V}$ and $\mathscr{W}$ be finite-dimensional vector spaces with ordered bases $\{\mathbf{E}_1, \ldots, \mathbf{E}_n\}$ and $\{\mathbf{F}_1, \ldots, \mathbf{F}_m\}$. Let $\mathsf{T}: \mathscr{V} \to \mathscr{W}$ be a linear transformation with matrix $\mathbf{M}$ with respect to this basis pair. Let $\mathscr{V}^* = \mathscr{L}(\mathscr{V}; \mathbb{R})$ and $\mathscr{W}^* = \mathscr{L}(\mathscr{W}; \mathbb{R})$.

(a) Define $\mathsf{T}^*: \mathscr{W}^* \to \mathscr{V}^*$ by

$$\mathsf{T}^*(\mathbf{f})(\mathbf{A}) = \mathbf{f}(\mathsf{T}(\mathbf{A})), \qquad \mathbf{A} \in \mathscr{V}, \mathbf{f} \in \mathscr{W}^*.$$

Show T is a linear transformation. (It is called the *dual* transformation.)

(b) Define $\mathbf{E}_1^*, \ldots, \mathbf{E}_n^* \in \mathscr{V}^*$ to be the linear extension of the functions

$$\mathbf{E}_i^*(\mathbf{E}_j) = \begin{cases} 1, & i = k, \\ 0, & \text{otherwise.} \end{cases}$$

Show that $\{\mathbf{E}_1^*, \ldots, \mathbf{E}_n^*\}$ is a basis for $\mathscr{V}^*$, called the *dual* basis to $\{\mathbf{E}_1, \ldots, \mathbf{E}_n\}$.

(c) Show that the matrix of T^* with respect to the dual basis $\{\mathbf{F}_1^*, \ldots, \mathbf{F}_m^*\}$, $\{\mathbf{E}_1^*, \ldots, \mathbf{E}_n^*\}$ is the transpose of $\mathbf{M}$.

25. Let $\mathsf{T}\colon \mathbb{R}^3 \to \mathbb{R}^3$ be the linear transformation whose matrix with respect to the standard basis of $\mathbb{R}^3$ is

$$\begin{pmatrix} 0 & 1 & 1 \\ 1 & 0 & 1 \\ 1 & 1 & 0 \end{pmatrix}$$

Find bases for the kernel and image of T. Use this to show T is invertible. Find the matrix of T^{-1} relative to the standard basis.

More on representing linear transformations by matrices 12bis

Our purpose in this chapter is to develop further the theory and technique of representing linear transformations by matrices. We will touch on several scattered topics and techniques. It is to be emphasized that we are only scratching the surface of an iceberg!

EXAMPLE 1. *Projections.*

One of the simplest type of linear transformation is a projection. For example in $\mathbb{R}^3$ consider the plane $\mathscr{V} = \{(x, y, z) | x + y + z = 0\}$ (Note $\mathscr{V} \subset \mathbb{R}^3$ is a 2-dimensional subspace.) and the linear transformation $\mathsf{P} : \mathbb{R}^3 \to \mathbb{R}^3$ that sends a vector $\mathbf{A}$ into its projection on the plane $\mathscr{V}$. See Figure $12^{\text{bis}}.1$. A formula for $\mathsf{P} : \mathbb{R}^3 \to \mathbb{R}^3$ is not hard to find. Notice that for any vector $\mathbf{A}$ that lies in $\mathscr{V}$ we must have $\mathsf{P}(\mathbf{A}) = \mathbf{A}$. For example

$$\mathsf{P}(1, -1, 0) = (1, -1, 0)$$
$$\mathsf{P}(0, 1, -1) = (0, 1, -1),$$

and notice that $\{(1, -1, 0), (0, 1, -1)\}$ is a basis for $\mathscr{V}$. Notice next that for any vector $\mathbf{B}$ lying on the line through the origin perpendicular to $\mathscr{V}$ we must have $\mathsf{P}(\mathbf{B}) = \mathbf{0}$. For example

$$\mathsf{P}(1, 1, 1) = \mathbf{0}.$$

If we now notice that $\{(1, -1, 0), (0, 1, -1), (1, 1, 1)\}$ is a basis for $\mathbb{R}^3$ we can easily crank out several formulas for P. For example, the matrix of P relative to the basis $\{(1, -1, 0), (0, 1, -1), (1, 1, 1)\}$ is just

$$\mathbf{M} = \begin{pmatrix} 1 & 0 & 0 \\ 0 & 1 & 0 \\ 0 & 0 & 0 \end{pmatrix}.$$

Of course you may object that this is cheating somewhat, and what you really want is the matrix of P relative to the standard basis for $\mathbb{R}^3$. This isn't hard either. Notice that

$$\begin{aligned}(1, 0, 0) &= \tfrac{2}{3}(1, -1, 0) + \tfrac{1}{3}(0, 1, -1) + \tfrac{1}{3}(1, 1, 1)\\ (0, 1, 0) &= -\tfrac{1}{3}(1, -1, 0) + \tfrac{1}{3}(0, 1, -1) + \tfrac{1}{3}(1, 1, 1)\\ (0, 0, 1) &= -\tfrac{1}{3}(1, -1, 0) - \tfrac{2}{3}(0, 1, -1) + \tfrac{1}{3}(1, 1, 1).\end{aligned}$$

Therefore we find that

$$\begin{aligned}\mathsf{P}(1, 0, 0) &= \tfrac{2}{3}\mathsf{P}(1, -1, 0) + \tfrac{1}{3}\mathsf{P}(0, 1, -1) + \tfrac{1}{3}\mathsf{P}(1, 1, 1)\\ &= \tfrac{2}{3}(1, -1, 0) + \tfrac{1}{3}(0, 1, -1) = (\tfrac{2}{3}, -\tfrac{1}{3}, -\tfrac{1}{3})\\ \mathsf{P}(0, 1, 0) &= -\tfrac{1}{3}\mathsf{P}(1, -1, 0) + \tfrac{1}{3}\mathsf{P}(0, 1, -1) + \tfrac{1}{3}\mathsf{P}(1, 1, 1)\\ &= -\tfrac{1}{3}(1, -1, 0) + \tfrac{1}{3}(0, 1, -1) = (-\tfrac{1}{3}, \tfrac{2}{3}, -\tfrac{1}{3})\\ \mathsf{P}(0, 0, 1) &= -\tfrac{1}{3}\mathsf{P}(1, -1, 0) - \tfrac{2}{3}\mathsf{P}(0, 1, -1) + \tfrac{1}{3}\mathsf{P}(1, 1, 1)\\ &= -\tfrac{1}{3}(1, -1, 0) - \tfrac{2}{3}(0, 1, -1) = (-\tfrac{1}{3}, -\tfrac{1}{3}, \tfrac{2}{3})\end{aligned}$$

so that the matrix of P relative to the standard basis is

$$\mathbf{N} = \begin{pmatrix} \frac{2}{3} & -\frac{1}{3} & -\frac{1}{3}\\ -\frac{1}{3} & \frac{2}{3} & -\frac{1}{3}\\ -\frac{1}{3} & -\frac{1}{3} & \frac{2}{3}\end{pmatrix} = \frac{1}{3}\begin{pmatrix} 2 & -1 & -1\\ -1 & 2 & -1\\ -1 & -1 & 2\end{pmatrix}$$

or if one enjoys such things

$$\mathsf{P}(x, y, z) = (\tfrac{2}{3}x - \tfrac{1}{3}y - \tfrac{1}{3}z, -\tfrac{1}{3}x + \tfrac{2}{3}y - \tfrac{1}{3}z, -\tfrac{1}{3}x - \tfrac{1}{3}y + \tfrac{2}{3}z).$$

Now the first question to ask is: Which representation of P tells us the most about it? I am sure you will agree that the matrix representation $\mathbf{M}$ **together with the basis** $\{(1, -1, 0), (0, 1, -1), (1, 1, 1)\}$ relative to which we computed M tells us everything we could possibly want to know about P. For it clearly tells us that P is projection onto the plane spanned by the vectors $(1, -1, 0)$ and $(0, 1, -1)$ which is, of course, the plane $\mathscr{V} = \{(x, y, z): x + y + z = 0\}$. On the other hand the matrix representation $\mathbf{N}$

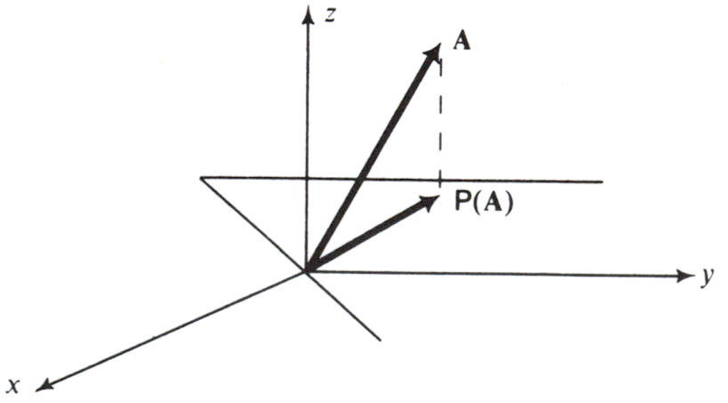

Figure 12bis.1

of P relative to the standard basis conveys relatively little about the simple geometric nature of P. This brings us to another question: How does one recognize when a linear transformation $\mathsf{T}: \mathbb{R}^3 \to \mathbb{R}^3$ is a projection? Well to deal with this question we had better begin by defining what we mean by a projection. (Remember a definition is either useful or useless, not true or false!) A little experimentation shows that the following is a useful choice.

Definition. A linear transformation

$$\mathsf{P}: \mathscr{W} \to \mathscr{W}$$

is called a *projection* iff $\mathsf{P}^2 = \mathsf{P}$, that is iff $\mathsf{P}(\mathsf{P}(\mathbf{A})) = \mathsf{P}(\mathbf{A})$ for every vector $\mathbf{A}$ in $\mathscr{W}$.

Now it pays to be warned that there are linear transformations

$$\mathsf{S}: \mathbb{R}^3 \to \mathbb{R}^3$$

that are projections in the sense of this definition that are not projections onto a plane (or line) in the sense we have spoken of so far. Let us illustrate by an example.

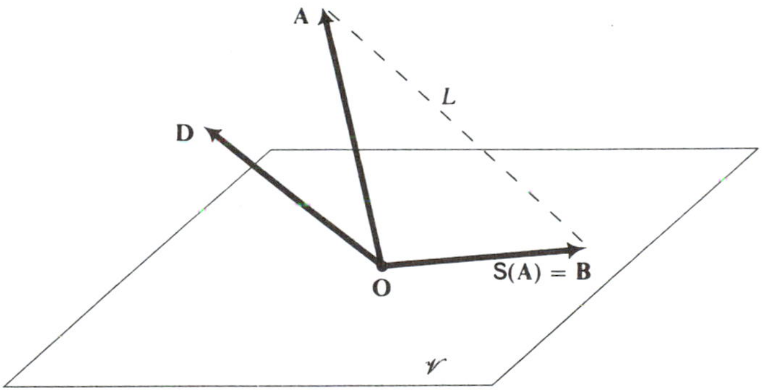

Figure 12bis.2

Let $\mathscr{V} = \{(x, y, z) | x + y + z = 0\}$ be the plane in $\mathbb{R}^3$ we have considered already. Let $\mathbf{D} = (1, 1, 0)$. Note that $\mathbf{D}$ is not a vector in $\mathscr{V}$. We are going to describe a sort of *skew* projection onto $\mathscr{V}$. Let $\mathbf{A}$ be a vector in $\mathbb{R}^3$. Draw a line L parallel to $\mathbf{D}$ passing through the tip of the vector $\mathbf{A}$. This line will meet the plane $\mathscr{V}$ in a unique point (why?) which determines a unique vector lying in the plane $\mathscr{V}$ which we call $\mathbf{B}$. Set $\mathsf{S}(\mathbf{A}) = \mathbf{B}$ as in Figure 12bis.2. It is rather easily checked from the definition that $\mathsf{S}: \mathbb{R}^3 \to \mathbb{R}^3$ is indeed a linear transformation. Note that

$$\begin{aligned} \mathsf{S}(\mathbf{A}) &= \mathbf{A} \quad \text{if } \mathbf{A} \in \mathscr{V} \\ \mathsf{S}(\mathbf{A}) &= \mathbf{0} \quad \text{if } \mathbf{A} \in \mathscr{L}(\mathbf{D}). \end{aligned}$$

Since $\mathbf{D}$ is not in the plane $\mathscr{V}$, if we choose a basis $\{\mathbf{F}, \mathbf{E}\}$ for $\mathscr{V}$ then $\{\mathbf{F}, \mathbf{E}, \mathbf{D}\}$ is a basis for $\mathbb{R}^3$. If $\mathbf{A}$ is any vector in $\mathbb{R}^3$ then

$$\mathbf{A} = a_1\mathbf{F} + a_2\mathbf{E} + a_3\mathbf{D}$$

so

$$\begin{aligned}\mathsf{S}(\mathbf{A}) &= a_1\mathsf{S}(\mathbf{F}) + a_2\mathsf{S}(\mathbf{E}) + a_3\mathsf{S}(\mathbf{D})\\ &= a_1\mathbf{F} + a_2\mathbf{E}\end{aligned}$$

since $\mathbf{E}$ and $\mathbf{F}$ belong to $\mathscr{V}$ and $\mathbf{D}$ belongs to $\mathscr{L}(\mathbf{D})$. Therefore

$$\begin{aligned}\mathsf{S}^2(\mathbf{A}) &= \mathsf{S}(a_1\mathbf{F} + a_2\mathbf{E}) = a_1\mathsf{S}(\mathbf{F}) + a_2\mathsf{S}(\mathbf{E})\\ &= a_1\mathbf{F} + a_2\mathbf{E}.\end{aligned}$$

Thus $\mathsf{S}^2 = \mathsf{S}$ so S is a projection.

It is a fact we are going to prove that every projection

$$\mathsf{S} : \mathbb{R}^3 \to \mathbb{R}^3$$

such that $\dim(\operatorname{Im} \mathsf{S}) = 2$ is a skew projection of the preceding type. The more usual projections we have dealt with up until now are what are called *orthogonal* projections ($\mathbf{D}$ is orthogonal to $\mathscr{V}$) or *self-adjoint* projections (see, for example Chapter 15).

Let us return now to the question that started all this, namely how does one recognize a projection? The definition we have proposed is going to be useful if and only if we can somehow show that every projection $\mathsf{S} : \mathbb{R}^3 \to \mathbb{R}^3$ is a skew projection in the preceding sense. In fact more is true.

Theorem 12bis.1. *Let $\mathscr{W}$ be a finite-dimensional vector space of dimension n and $\mathsf{S} : \mathscr{W} \to \mathscr{W}$ a projection. Then there is a basis $\{\mathbf{A}_1, \ldots, \mathbf{A}_n\}$ for $\mathscr{W}$ such that*

$$\mathsf{S}(\mathbf{A}_i) = \begin{cases}\mathbf{A}_i & \text{if } 1 \le i \le r\\ 0 & \text{if } r + 1 \le i \le n\end{cases}$$

where $r = \dim(\operatorname{Im} \mathsf{S})$, and hence the matrix of S relative to the basis $\{\mathbf{A}_1, \ldots, \mathbf{A}_n\}$ is

$$\begin{pmatrix} \overbrace{\begin{matrix}1 & & \\ & \ddots & \\ & & 1\end{matrix}}^{r} & 0 \\ 0 & \overbrace{\begin{matrix}0 & & \\ & \ddots & \\ & & 0\end{matrix}}^{n-r}\end{pmatrix}$$

PROOF. Let $\{\mathbf{B}_1, \ldots, \mathbf{B}_r\}$ be a basis for $\operatorname{Im} \mathsf{S}$ and $\{\mathbf{C}_1, \ldots, \mathbf{C}_s\}$ be a basis for $\ker S$. Notice that by (8.10) we have

$$n = r + s.$$

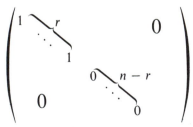

Therefore the collection $\{\mathbf{B}_1, \ldots, \mathbf{B}_r, \mathbf{C}_1, \ldots, \mathbf{C}_s\}$ contains n vectors, the correct number of vectors to be a basis for $\mathscr{W}$. Let us show that indeed it is a basis for $\mathscr{W}$. To this end notice the following two facts:

$$\begin{aligned} \mathsf{S}(\mathbf{C}_i) &= \mathbf{0} \quad \text{for } i = 1, \ldots, s \\ \mathsf{S}(\mathbf{B}_i) &= \mathbf{B}_i \quad \text{for } i = 1, \ldots, r. \end{aligned}$$

The first of these is clear from the definition of $\{\mathbf{C}_1, \ldots, \mathbf{C}_s\}$ while the second requires some proof. (Be sensible, we still have not used the fact that S is a projection!) We note that since $\mathbf{B}_i$ belongs to Im S we may find a vector $\mathbf{D}_i$ in $\mathscr{W}$ such that $\mathsf{S}(\mathbf{D}_i) = \mathbf{B}_i$. Therefore since S is a projection

$$\mathsf{S}(\mathbf{B}_i) = \mathsf{S}(\mathsf{S}(\mathbf{D}_i)) = \mathsf{S}^2(\mathbf{D}_i) = \mathsf{S}(\mathbf{D}_i) = \mathbf{B}_i,$$

which verifies the second set of equations preceding.

Now since there are n vectors in the collection $\{\mathbf{B}_1, \ldots, \mathbf{B}_r, \mathbf{C}_1, \ldots, \mathbf{C}_s\}$ we may apply (6.9) to conclude that we need only show $\{\mathbf{B}_1, \ldots, \mathbf{C}_s\}$ is linearly independent for them to be a basis, that is, we must verify only one-half of the basis condition.

Suppose the vectors $\{\mathbf{B}_1, \ldots, \mathbf{B}_r, \mathbf{C}_1, \ldots, \mathbf{C}_s\}$ are linearly dependent. Then we may find numbers $b_1, \ldots, b_r, c_1, \ldots, c_s$, not all zero, so that

$$(*) \qquad b_1\mathbf{B}_1 + \cdots + b_r\mathbf{B}_r + c_1\mathbf{C}_1 + \cdots + c_s\mathbf{C}_s = \mathbf{0}.$$

Apply S to both sides of this equation to get

$$\begin{aligned} \mathbf{0} &= \mathsf{S}(b_1\mathbf{B}_1 + \cdots + b_r\mathbf{B}_r + c_1\mathbf{C}_1 + \cdots + c_s\mathbf{C}_s) \\ &= b_1\mathsf{S}(\mathbf{B}_1) + \cdots + b_r\mathsf{S}(\mathbf{B}_r) + c_1\mathsf{S}(\mathbf{C}_1) + \cdots + c_s\mathsf{S}(\mathbf{C}_s) \\ &= b_1\mathbf{B}_1 + \cdots + b_r\mathbf{B}_r. \end{aligned}$$

But since $\mathbf{B}_1, \ldots, \mathbf{B}_r$ are a basis for Im S they are linearly independent and hence $b_1 = b_2 = \cdots = b_r = 0$. Thus our original relation $(*)$ reads

$$c_1\mathbf{C}_1 + \cdots + c_s\mathbf{C}_s = \mathbf{0}.$$

But since $\mathbf{C}_1, \ldots, \mathbf{C}_s$ are a basis for ker S they are linearly independent also, so $c_1 = c_2 = \cdots = c_s = 0$. Therefore the collection $\{\mathbf{B}_1, \ldots, \mathbf{B}_r, \mathbf{C}_1, \ldots, \mathbf{C}_s\}$ is not linearly dependent, so must be linearly independent. Thus as we remarked earlier $\{\mathbf{B}_1, \ldots, \mathbf{B}_r, \mathbf{C}_1, \ldots, \mathbf{C}_s\}$ being linearly independent is a basis for $\mathscr{W}$. Set

$$\begin{aligned} \mathbf{A}_1 &= \mathbf{B}_1 \\ \mathbf{A}_2 &= \mathbf{B}_2 \\ &\vdots \\ \mathbf{A}_r &= \mathbf{B}_r \qquad \text{(remember } r + s = n.\text{)} \\ \mathbf{A}_{r+1} &= \mathbf{C}_1 \\ &\vdots \\ \mathbf{A}_n &= \mathbf{C}_s. \end{aligned}$$

Then we will have

$$\begin{aligned} \mathsf{P}(\mathbf{A}_i) &= \mathsf{P}(\mathbf{B}_i) = \mathbf{B}_i = \mathbf{A}_i \quad &&\text{for } 1 \le i \le r \\ &= \mathsf{P}(\mathbf{C}_{i-r}) = \mathbf{0} &&\text{for } r+1 \le i \le n \end{aligned}$$

as required. □

It is now a rather easy consequence that every linear transformation $\mathsf{S}: \mathbb{R}^3 \to \mathbb{R}^3$ such that

$$\mathsf{S}^2 = \mathsf{S} \quad \text{and} \quad \dim(\operatorname{Im} \mathsf{S}) = 2$$

is a skew projection as described preceding (12bis.1).

The reader should turn ahead to (14.1) and examine the fundamental difference between it and (12bis.1).

For a given projection $\mathsf{S}: \mathscr{W} \to \mathscr{W}$ the actual computation of a (not *the*!) basis as in (12bis.1) may be a numerical horror, but still the proof gives you a method for doing so in terms of the type of problem we have successfully handled before.

EXAMPLE 2. *Nilpotent transformations.*

A linear transformation $\mathsf{T}: \mathscr{V} \to \mathscr{V}$ is said to be *nilpotent of index k* iff $\mathsf{T}^k = 0$ and $\mathsf{T}^{k-1} \neq 0$, that is, $\mathsf{T}^k(\mathbf{A}) = \mathbf{0}$ for all vectors $\mathbf{A}$ in $\mathscr{V}$ but there is at least one vector $\mathbf{B}$ in $\mathscr{V}$ for which $\mathsf{T}^{k-1}(\mathbf{B}) \neq \mathbf{0}$. There is a sort of canonical example of a nilpotent transformation namely the differentiation operator

$$\mathsf{D}: \mathscr{P}_n(\mathbb{R}) \to \mathscr{P}_n(\mathbb{R})$$

is nilpotent of index $n + 1$. Another family of simple examples is provided by the *shift operator*

$$\mathsf{S}: \mathbb{R}^n \to \mathbb{R}^n$$

defined by

$$\mathsf{S}(a_1, a_2, \ldots, a_n) = (0, a_1, a_2, \ldots, a_{n-1}).$$

Thus if $\{\mathbf{E}_1, \ldots, \mathbf{E}_n\}$ denotes the usual basis for $\mathbb{R}^n$ then

$$\begin{aligned} \mathsf{S}(\mathbf{E}_1) &= \mathbf{E}_2 \\ \mathsf{S}(\mathbf{E}_2) &= \mathbf{E}_3 \\ &\vdots \\ \mathsf{S}(\mathbf{E}_n) &= \mathbf{0} \end{aligned}$$

so that the matrix of S relative to the usual basis is

$$\begin{pmatrix} 0 & 0 & & & 0 \\ 1 & 0 & & & 0 \\ 0 & 1 & & & 0 \\ \vdots & & \ddots & & \vdots \\ 0 & 0 \cdots 0 & & 1 & 0 \end{pmatrix}.$$

In this case S is nilpotent of index n. Notice in both cases the index of nilpotence is the dimension of the vector space on which the transformation is defined. This is a very special situation. In fact we can determine all such nilpotent transformations. We will need the following very important fact.

Proposition 12bis.2. *Let $\mathsf{T} : \mathscr{V} \to \mathscr{V}$ be a linear transformation that is nilpotent of index k. Let $\mathbf{B} \in \mathscr{V}$ be a vector such that $\mathsf{T}^{k-1}(\mathbf{B}) \neq \mathbf{0}$. Then the vectors $\{\mathbf{B}, \mathsf{T}(\mathbf{B}), \ldots, \mathsf{T}^{k-1}(\mathbf{B})\}$ are linearly independent.*

Proof. Let us suppose that

$$(*) \qquad b_0\mathbf{B} + b_1\mathsf{T}\mathbf{B} + \cdots + b_{k-1}\mathsf{T}^{k-1}\mathbf{B} = \mathbf{0}$$

is a linear relation among $\{\mathbf{B}, \mathsf{T}(\mathbf{B}), \ldots, \mathsf{T}^{k-1}(\mathbf{B})\}$. What we must show is that $b_0 = b_1 = \cdots = b_{k-1} = 0$. Now notice that since $\mathsf{T}^k(\mathbf{B}) = \mathbf{0}$ we must have

$$\mathbf{0} = \mathsf{T}^k(\mathbf{B}) = \mathsf{T}^{k+1}(\mathbf{B}) = \cdots$$

by (8.1). Apply T^{k-1} to both sides of $(*)$ to get

$$\begin{aligned} \mathbf{0} &= b_0\mathsf{T}^{k-1}(\mathbf{B}) + b_1\mathsf{T}^k(\mathbf{B}) + \cdots + b_{k-1}\mathsf{T}^{2k-2}(\mathbf{B}) \\ &= b_0\mathsf{T}^{k-1}(\mathbf{B}) + b_1\mathbf{0} + \cdots + b_{k-1}\mathbf{0} \\ &= b_0\mathsf{T}^{k-1}(\mathbf{B}). \end{aligned}$$

Therefore since $\mathsf{T}^{k-1}(\mathbf{B}) \neq \mathbf{0}$ we must have $b_0 = 0$, and our linear relation $(*)$ has become

$$(**) \qquad b_1\mathsf{T}(\mathbf{B}) + \cdots + b_{k-1}\mathsf{T}^{k-1}(\mathbf{B}) = \mathbf{0}.$$

Now apply T^{k-2} to both sides getting

$$\begin{aligned} \mathbf{0} &= b_1\mathsf{T}^{k-1}(\mathbf{B}) + b_2\mathsf{T}^k(\mathbf{B}) + \cdots + b_{k-1}\mathsf{T}^{2k-3}(\mathbf{B}) \\ &= b_1\mathsf{T}^{k-1}(\mathbf{B}) + b_2\mathbf{0} + \cdots + b_{k-1}\mathbf{0} \\ &= b_1\mathsf{T}^{k-1}(\mathbf{B}) \end{aligned}$$

and so $b_1 = 0$. Continuing in this way we may show that $0 = b_2 = \cdots = b_{k-1}$. Therefore $\{\mathbf{B}, \mathsf{T}(\mathbf{B}), \ldots, \mathsf{T}^{k-1}(\mathbf{B})\}$ is linearly independent. □

Corollary 12bis.3. *Suppose that $\mathsf{T} : \mathscr{V} \to \mathscr{V}$ is a nilpotent linear transformation and the index of T equals the dimension of $\mathscr{V}$. Call this common number k. Then there is a vector $\mathbf{B} \in \mathscr{V}$ such that $\{\mathbf{B}, \mathsf{T}(\mathbf{B}), \ldots, \mathsf{T}^{k-1}(\mathbf{B})\}$ is a basis for $\mathscr{V}$ and $\mathsf{T}^k(\mathbf{B}) = \mathbf{0}$.*

Proof. According to (12bis.2) the vectors $\{\mathbf{B}, \mathsf{T}(\mathbf{B}), \ldots, \mathsf{T}^{k-1}(\mathbf{B})\}$ are linearly independent. There are k of them and since $k = \dim \mathscr{V}$ it follows from (6.9) that they are basis for $\mathscr{V}$. Since k is also the index of T, $\mathsf{T}^k(\mathbf{B}) = \mathbf{0}$. □

Therefore if $\mathsf{T} : \mathscr{V} \to \mathscr{V}$ is nilpotent ***and*** index $\mathsf{T} = k = \dim \mathscr{V}$ then we may choose a vector $\mathbf{B}$ in $\mathscr{V}$ so that $\{\mathbf{B}, \mathsf{T}(\mathbf{B}), \ldots, \mathsf{T}^{k-1}(\mathbf{B})\}$ is a basis for $\mathscr{V}$ and $\mathsf{T}^k\mathbf{B} = \mathbf{0}$. The matrix of T relative to this basis is therefore

$$\begin{pmatrix} 0 & 0 & & & 0 & 0 \\ 1 & 0 & & & 0 & 0 \\ 0 & 1 & \ddots & & 0 & 0 \\ \vdots & \vdots & & \ddots & \vdots & \vdots \\ 0 & 0 & \cdots & 0 & 1 & 0 \end{pmatrix}$$

and if we use this basis to identify $\mathscr{V}$ with $\mathbb{R}^n$ as in (8.13–8.15) we find that T is identified with the shift operator. Thus the structure of nilpotent transformations $\mathsf{T}: \mathscr{V} \to \mathscr{V}$ *of index equal to dimension* $\mathscr{V}$ is completely determined. The structure of more general nilpotent transformations is more complicated and a subject for a more advanced course in linear algebra.

Example 3. *Cyclic transformations*:

A linear transformation $\mathsf{T}: \mathscr{V} \to \mathscr{V}$ is called *cyclic* iff there exists a vector $\mathbf{A}$ in $\mathscr{V}$ such that the collection $\{\mathbf{A}, \mathsf{T}(\mathbf{A}), \mathsf{T}^2(\mathbf{A}), \ldots\}$ spans $\mathscr{V}$. The vector $\mathbf{A}$ is called a *cyclic vector* for T. Examples of cyclic transformations are plentiful. It follows from (12bis.3) that any nilpotent transformation $\mathsf{T}: \mathscr{V} \to \mathscr{V}$ whose index is equal to the dimension of $\mathscr{V}$ is cyclic. In dimension 2 essentially all nontrivial linear transformations are cyclic. We begin with a preliminary result of independent interest.

Proposition 12bis.4. *Suppose that $\mathsf{T}: \mathscr{V} \to \mathscr{V}$ is a cyclic linear transformation. Let the dimension of $\mathscr{V}$ be n and let $\mathbf{A} \in \mathscr{V}$ be a cyclic vector for T. Then the vectors $\mathbf{A}, \mathsf{T}(\mathbf{A}), \ldots, \mathsf{T}^{n-1}(\mathbf{A})$ are a basis for $\mathscr{V}$.*

Proof. There are n vectors in the set $\{\mathbf{A}, \mathsf{T}(\mathbf{A}), \ldots, \mathsf{T}^{n-1}(\mathbf{A})\}$ so by (6.9) to show they are a basis in the n-dimensional space we need only show that they are linearly independent. Suppose to the contrary that they are linearly dependent. By (6.2) there will be a positive integer $m < n$ such that

$$\mathsf{T}^m(\mathbf{A}) \in \mathscr{L}(\mathbf{A}, \mathsf{T}(\mathbf{A}), \ldots, \mathsf{T}^{m-1}(\mathbf{A})).$$

Therefore by applying T

$$\mathsf{T}^{m+1}(\mathbf{A}) \in \mathscr{L}(\mathsf{T}(\mathbf{A}), \mathsf{T}^2(\mathbf{A}), \ldots, \mathsf{T}^m(\mathbf{A})) \subset \mathscr{L}(\mathbf{A}, \mathsf{T}(\mathbf{A}), \ldots, \mathsf{T}^{m-1}(\mathbf{A}))$$

since all of the vectors $\mathsf{T}(\mathbf{A}), \mathsf{T}^2(\mathbf{A}), \ldots, \mathsf{T}^m(\mathbf{A})$ belong to $\mathscr{L}(\mathbf{A}, \mathsf{T}(\mathbf{A}), \ldots, \mathsf{T}^{m-1}(\mathbf{A}))$. Applying T again we find

$$\mathsf{T}^{m+2}(\mathbf{A}) \in \mathscr{L}(\mathsf{T}(\mathbf{A}), \mathsf{T}^2(\mathbf{A}), \ldots, \mathsf{T}^m(\mathbf{A})) \subset \mathscr{L}(\mathbf{A}, \mathsf{T}(\mathbf{A}), \ldots, \mathsf{T}^{m-1}(\mathbf{A}))$$

and continuing in this way we show that

$$\mathsf{T}^m(\mathbf{A}), \mathsf{T}^{m+1}(\mathbf{A}), \mathsf{T}^{m+2}(\mathbf{A}), \ldots, \in \mathscr{L}(\mathbf{A}, \mathsf{T}(\mathbf{A}), \ldots, \mathsf{T}^{m-1}(\mathbf{A}))$$

and therefore that

$$\mathscr{V} = \mathscr{L}(A, \mathsf{T}(\mathbf{A}), \ldots, \mathsf{T}^m(\mathbf{A}), \mathsf{T}^{m+1}(\mathbf{A}), \ldots) \subset \mathscr{L}(\mathbf{A}, \mathsf{T}(\mathbf{A}), \ldots, \mathsf{T}^{m-1}(\mathbf{A}))$$

so that $\dim \mathscr{V} \leq m < n$ which is impossible. Therefore the vectors $\mathbf{A}, \mathsf{T}(\mathbf{A}), \ldots, \mathsf{T}^{n-1}(\mathbf{A})$ must be linearly independent. □

Proposition 12bis.5. *A linear transformation $\mathsf{T} : \mathbb{R}^2 \to \mathbb{R}^2$ is cyclic iff $\mathsf{T} \neq e\mathsf{I}$, where e is a number and $\mathsf{I} : \mathbb{R}^2 \to \mathbb{R}^2$ the identity transformation.*

Proof. Suppose that $\mathsf{T} : \mathbb{R}^2 \to \mathbb{R}^2$. A vector $\mathbf{A} \in \mathbb{R}^2$ is a cyclic vector for T iff $\mathsf{T}(\mathbf{A}) \notin \mathscr{L}(\mathbf{A})$, that is, iff $\mathsf{T}(\mathbf{A})$ does not belong to the line through the origin spanned by $\mathbf{A}$. For then $\{\mathbf{A}, \mathsf{T}(\mathbf{A})\}$ will be a basis for $\mathbb{R}^2$ so that $\mathbb{R}^2 = \mathscr{L}(\mathbf{A}, \mathsf{T}(\mathbf{A}), \ldots)$ as required. So suppose that T does not have a cyclic vector. Then $\mathsf{T}(\mathbf{A}) \in \mathscr{L}(\mathbf{A})$ for every $\mathbf{A} \in \mathbb{R}^2$. In particular

$$\begin{aligned} \mathsf{T}(1, 0) \in \mathscr{L}(1, 0) \quad &\Rightarrow \quad \mathsf{T}(1, 0) = e_1(1, 0) \\ \mathsf{T}(0, 1) \in \mathscr{L}(0, 1) \quad &\Rightarrow \quad \mathsf{T}(0, 1) = e_2(0, 1). \end{aligned}$$

Now we claim $e_1 = e_2$ for we also have

$$\mathsf{T}(1, 1) \in \mathscr{L}(1, 1) \quad \Rightarrow \quad \mathsf{T}(1, 1) = e(1, 1)$$

so

$$\begin{aligned} e(1, 1) = \mathsf{T}(1, 1) &= \mathsf{T}((1, 0) + (0, 1)) \\ &= \mathsf{T}(1, 0) + \mathsf{T}(0, 1) \\ &= e_1(1, 0) + e_2(0, 1) \end{aligned}$$

so

$$(e, e) = (e_1, e_2)$$

and hence $e_1 = e = e_2$. Therefore

$$\begin{aligned} \mathsf{T}(1, 0) &= (e, 0) \\ \mathsf{T}(0, 1) &= (0, e) \end{aligned}$$

so the matrix of T relative to the standard basis of $\mathbb{R}^2$ is

$$\begin{pmatrix} e & 0 \\ 0 & e \end{pmatrix} = e\mathbf{I}$$

and hence $\mathsf{T} = e\mathsf{I}$. □

Suppose that $\mathsf{T} : \mathscr{V} \to \mathscr{V}$ is a cyclic transformation with cyclic vector $\mathbf{A}$. Then the matrix of T relative to the basis $\{\mathbf{A}, \mathsf{T}(\mathbf{A}), \ldots, \mathsf{T}^{n-1}(\mathbf{A})\}$ is

$$\begin{pmatrix} 0 & & & 0 & a_0 \\ 1 & & & & a_1 \\ \vdots & \ddots & 1 & & \vdots \\ 0 & \cdots & 0 & 1 & a_{n-1} \end{pmatrix}$$

where $\mathsf{T}^n(\mathbf{A}) = a_0\mathbf{A} + a_1\mathsf{T}(\mathbf{A}) + \cdots + a_{n-1}\mathsf{T}^{n-1}(\mathbf{A})$. This matrix is particularly simple and is one reason for looking for a cyclic vector. In general however a linear transformation need not have any cyclic vectors at all! Again this is a subject for study in a more advanced course on linear algebra.

Exercises

1. Find the matrix of the projection $\mathsf{P} : \mathbb{R}^3 \to \mathbb{R}^3$, relative to the standard basis, where P is the projection onto the plane $\mathscr{V} = \{(x, y, z) | x + y = 0\}$.

2. Repeat Exercise 1 for $\mathscr{V} = \{(x, y, z) | z = 0\}$.

3. Find the basis $\{\mathbf{A}_1, \mathbf{A}_2, \mathbf{A}_3\}$ for $\mathbb{R}^3$ such that the projection P of Exercise 1 relative to the basis is of the form

$$\begin{pmatrix} 1 & 0 & 0 \\ 0 & 1 & 0 \\ 0 & 0 & 0 \end{pmatrix}$$

4. Repeat Exercise 3 for the projection of Exercise 2.

5. Show that the differentiation operator $\mathsf{D} : \mathscr{P}_3(\mathbb{R}) \to \mathscr{P}_3(\mathbb{R})$ is nilpotent of index 4. Find a basis of $\mathscr{P}_3(\mathbb{R})$ of the form

$$\mathscr{E} = \{\mathbf{B}, \mathsf{D}(\mathbf{B}), \mathsf{D}^2(\mathbf{B}), \mathsf{D}^3(\mathbf{B})\}$$

where $\mathbf{B} \in \mathscr{P}_3(\mathbb{R})$. Calculate the matrix of D relative to the basis $\mathscr{E}$.

6. Let $\mathsf{T} : \mathbb{R}^3 \to \mathbb{R}^3$ be the linear transformation

$$\mathsf{T}(x, y, z) = (0, x, 2y).$$

Show that T is nilpotent of index 3. Find a vector $\mathbf{B}$ so that

$$\{\mathbf{B}, \mathsf{T}(\mathbf{B}), \mathsf{T}^2(\mathbf{B})\}$$

is a basis for $\mathbb{R}^3$. Calculate the matrix of T relative to this basis.

Calculate the matrix of T relative to the standard basis.

7. Let $\mathsf{T} : \mathbb{R}^3 \to \mathbb{R}^3$ be the linear transformation

$$\mathsf{T}(x, y, z) = (x + y, y + z, x).$$

Show that T is a cyclic linear transformation. (*Hint*: find $\mathbf{B}$ such that $\{\mathbf{B}, \mathsf{T}(\mathbf{B}), \mathsf{T}^2(\mathbf{B})\}$ is a set of independent vectors.)

8. Let $\mathsf{T} : \mathbb{R}^3 \to \mathbb{R}^3$ be the linear transformation defined by

$$\mathsf{T}(x, y, z) = (y, x, z).$$

Show that T is not cyclic.

9. Let $\mathsf{T}: \mathbb{R}^3 \to \mathbb{R}^3$ be the linear transformation defined by

$$\mathsf{T}(x, y, z) = (y, z, x).$$

Show that T is cyclic. Find the basis $\{\mathbf{B}, \mathsf{T}(\mathbf{B}), \mathsf{T}^2(\mathbf{B})\}$ and calculate the matrix of T relative to $\{\mathbf{B}, \mathsf{T}(\mathbf{B}), \mathsf{T}^2(\mathbf{B})\}$.

10. Is the product of two projections P and Q in $\mathscr{V}$ again a projection? What if P and Q commute with each other?

11. Is the product of two nilpotent transformations again nilpotent?

12. A linear transformation $\mathsf{T}: \mathscr{V} \to \mathscr{V}$ is called an involution iff $\mathsf{T}^2 = \mathsf{I}$. Which of the following matrices represent involutions in $\mathbb{R}^3$ with respect to the standard bases:

$$\begin{pmatrix} 1 & 0 & 0 \\ 0 & -1 & 0 \\ 0 & 0 & 1 \end{pmatrix}, \quad \begin{pmatrix} 0 & 0 & 1 \\ 0 & 1 & 0 \\ 1 & 0 & 0 \end{pmatrix}$$

$$\begin{pmatrix} 1 & 0 & 1 \\ 0 & 1 & 0 \\ 1 & 0 & 1 \end{pmatrix}, \quad \begin{pmatrix} 0 & 1 & 0 \\ 1 & 0 & 0 \\ 0 & 0 & 1 \end{pmatrix}$$

13. Is the product of two involutions again in an involution?

14. Two transformations $\mathsf{S}, \mathsf{T}: \mathscr{V} \to \mathscr{V}$ are said to be mutually annihilating if $\mathsf{ST} = 0$ and $\mathsf{TS} = 0$. Show that the sum of two mutually annihilating projections is again a projection.

15. A linear transformation $\mathsf{T}: \mathscr{V} \to \mathscr{V}$ is the n-dimensional vector space, $\mathscr{V}$ is called a *reflection* if there is an $(n - 1)$-dimensional subspace $\mathscr{H}$ in $\mathscr{V}$, the reflecting hyperplane, and a non-zero vector $\mathbf{v}$ in $\mathscr{V}$ such that $\mathbf{v}$ does not belong to $\mathscr{H}$ and:

$$\mathsf{T}(\mathbf{h}) = \mathbf{h} \quad \text{for all } h \in \mathscr{H},$$
$$\mathsf{T}(\mathbf{v}) = -\mathbf{v}.$$

Show that with respect to a suitable basis the matrix of a reflection takes the form

$$\begin{pmatrix} -1 & 0 & 0 & \cdots & 0 \\ 0 & 1 & 0 & \cdots & 0 \\ 0 & & & \ddots & 0 \\ \vdots & 0 & & & 0 \\ 0 & & \cdots & 0 & 1 \end{pmatrix}$$

16. Let $\mathsf{S}: \mathscr{P}_n \to \mathscr{P}_n$ be the linear transformation defined by the equation

$$\mathsf{S}(p(x)) = p(x + 1)$$

Find the matrix of S with respect to the standard basis $\{1, x, \ldots, x^n\}$ of $\mathscr{P}_n$.

17. Let $\mathbf{A}$ be the 2×2 matrix

$$\begin{pmatrix} 1 & 1 \\ 1 & 1 \end{pmatrix}$$

and let $\mathsf{L}_{\mathbf{A}}: \mathscr{M}_2 \to \mathscr{M}$ be the linear transformation given by

$$\mathsf{L}_{\mathbf{A}}(\mathbf{M}) = \mathbf{A} \cdot \mathbf{M}$$

for $\mathbf{M}$ in $\mathscr{M}$. Find the matrix of $\mathsf{L}_{\mathbf{A}}$ with respect to the ordered basis

$$\begin{pmatrix} 1 & 0 \\ 0 & 0 \end{pmatrix}, \quad \begin{pmatrix} 0 & 1 \\ 0 & 0 \end{pmatrix}, \quad \begin{pmatrix} 0 & 0 \\ 1 & 0 \end{pmatrix}, \quad \begin{pmatrix} 0 & 0 \\ 0 & 1 \end{pmatrix}$$

for $\mathscr{M}_i$.

18. Let $\mathsf{T}: \mathscr{M}_i \to \mathscr{M}_i$ be the linear transformation given by

$$\mathsf{T}(\mathbf{M}) = \mathbf{M}^t.$$

Find the matrix of T with respect to the same ordered basis of $\mathscr{M}_i$ as in Exercise 17.

19. If P is a projection show that $\mathsf{T} = 2\mathsf{P} - \mathsf{I}$ is an involution.

20. If T is an involution show that $\mathsf{P} = \frac{1}{2}(\mathsf{T} + \mathsf{I})$ is a projection.

21. Use Exercises 19 and 20 to show that with respect to a suitable basis the matrix of an involution $\mathsf{T}: \mathscr{V} \to \mathscr{V}$ is of the form

$$\begin{pmatrix} 1 & & & & & \\ & \ddots & & & \text{\Large 0} & \\ & & 1 & & & \\ & & & -1 & & \\ & \text{\Large 0} & & & \ddots & \\ & & & & & -1 \end{pmatrix}.$$

22. Find all possible 2×2 matrices that represent a nilpotent transformation $\mathsf{T}: \mathbb{R}^2 \to \mathbb{R}^2$ of index 2 with respect to some basis of $\mathbb{R}^2$.

23. (a) If $\mathbf{A}$ is a matrix show $\mathbf{A} + \mathbf{A}^t$ is symmetric and $\mathbf{A} - \mathbf{A}^t$ is skew-symmetric.
(b) Show every matrix is the sum of a symmetric and a skew-symmetric matrix.

Systems of linear equations 13

In the historical development of linear algebra the geometry of linear transformations and the algebra of systems of linear equations played significant and important rolls. A system of linear equations has the form

$$\text{(L)} \qquad \begin{aligned} a_{11}x_1 + a_{12}x_2 + \cdots + a_{1n}x_n &= b_1 \\ a_{21}x_1 + a_{22}x_2 + \cdots + a_{2n}x_n &= b_2 \\ &\vdots \\ a_{m1}x_1 + a_{m2}x_2 + \cdots + a_{mn}x_n &= b_n. \end{aligned}$$

Here $x_1, x_2, \ldots, x_n$, denote the unknowns, which are to be determined. The mn numbers a_{ij}, $i = 1, \ldots, m$, $j = 1, \ldots, n$ are called the *coefficients* of the linear system (L) and are, of course, fixed. A *solution* of the system (L) is an ordered n-tuple of numbers $(s_1, \ldots, s_n)$ such that the m equations

$$\begin{aligned} a_{11}s_1 + a_{12}s_2 + \cdots + a_{1n}s_n &= b_1 \\ a_{21}s_1 + a_{22}s_2 + \cdots + a_{2n}s_n &= b_2 \\ &\vdots \\ a_{m1}s_1 + a_{m2}s_2 + \cdots + a_{mn}s_n &= b_n \end{aligned}$$

are all true. To solve the system (L) means to find all solutions of (L).

EXAMPLE 1. Solve the linear system

$$\begin{aligned} x_1 + x_2 + x_3 &= 3 \\ 2x_2 + x_3 &= 4 \\ x_1 - x_2 \qquad &= -1. \end{aligned}$$

Solution. From previous experience we recall that this system may be solved as follows:

$$\left.\begin{array}{llr} (1) & x_1 + x_2 + x_3 = & 3 \\ (2) & \quad 2x_2 + x_3 = & 4 \\ (3) & x_1 - x_2 \qquad = & -1 \end{array}\right\}$$

subtract (1) from (3) to get

$$\left.\begin{aligned} x_1 + x_2 + x_3 &= 3 \\ 2x_2 + x_3 &= 4 \\ -2x_2 - x_3 &= -4 \end{aligned}\right\}.$$

Erase the third equation, it follows from the second, to get

$$\begin{aligned} x_1 + x_2 + x_3 &= 3 \\ 2x_2 + x_3 &= 4 \end{aligned} \quad \text{or} \quad \begin{aligned} x_1 &= 3 - x_3 - x_2 \\ x_3 &= 4 - 2x_2, \end{aligned}$$

or

$$\begin{aligned} x_1 &= x_2 - 1 \\ x_3 &= 4 - 2x_2. \end{aligned}$$

Therefore the solutions are all triples of numbers

$$(s - 1, s, 4 - 2s)$$

where s is arbitrary. For example

$$(0, 1, 2), (-1, 0, 4)$$

are solutions but

$$(1, 1, 2)$$

is not.

On the other hand past experience should have taught us that not every system of linear equations has a solution. For example the linear system

$$\begin{aligned} x_1 + x_2 &= 1 \\ x_1 + x_2 &= -1 \end{aligned}$$

clearly can have no solutions. Thus our first order of business in the study of linear equations should be to determine when a linear system has solutions, and only afterwards take up the discussion of actual techniques of solution.

Our study of matrices in the preceding chapters will come in handy here. In fact it is through the study of linear systems that matrices most frequently appear in modern scientific investigations. First let us introduce the matrices

$$\mathbf{A} = \begin{pmatrix} a_{11} & a_{12} & \cdots & a_{1n} \\ a_{21} & a_{22} & \cdots & a_{2n} \\ \vdots & & & \\ a_{m1} & a_{m2} & \cdots & a_{mn} \end{pmatrix}, \quad \mathbf{X} = \begin{pmatrix} x_1 \\ x_2 \\ \vdots \\ x_n \end{pmatrix}, \quad \mathbf{B} = \begin{pmatrix} b_1 \\ b_2 \\ \vdots \\ b_m \end{pmatrix}.$$

The matrix $\mathbf{A}$ is called the *coefficient matrix* of the linear system and $\mathbf{B}$ the matrix of *constants* of the linear system. We may then write the *system* of linear equations (L) in a more compact form as the single *matrix* equation

$$\text{(M)} \qquad \mathbf{AX} = \mathbf{B}.$$

Solving this matrix equation may be given a very simple interpretation in terms of linear transformations. We let

$$\mathsf{T} : \mathbb{R}^n \to \mathbb{R}^m$$

be the linear transformation whose matrix relative to the standard bases is $\mathbf{A}$. Thus

$$\mathsf{T}(x_1, \quad \ldots, x_n) = \left(\sum_{j=1}^{n} a_{1j}x_j, \sum_{j=1}^{n} a_{2j}x_j, \ldots, \sum_{j=1}^{n} a_{mj}x_j \right).$$

Let $\mathbf{B} = (b_1, \ldots, b_m)$, which is a vector in $\mathbb{R}^m$. A *solution* of (M) is then a vector $(s_1, s_2, \ldots, s_n)$ in $\mathbb{R}^n$ such that

$$\mathsf{T}(s_1, s_2, \ldots, s_n) = (b_1, b_2, \ldots, b_m),$$

and to find all solutions of (M) or equivalently (L), means to find all vectors $(s_1, \ldots, s_n)$ in $\mathbb{R}^n$ such that

$$\mathsf{T}(s_1, s_2, \ldots, s_n) = (b_1, \ldots, b_m).$$

The following result is therefore clear.

Proposition 13.1. *With the preceding notations, the linear system* (L) *has a solution iff the vector* $(b_1, b_2, \ldots, b_m)$ *lies in the image of the transformation* $\mathsf{T} : \mathbb{R}^n \to \mathbb{R}^m$, *that is, iff* $\mathbf{B} \in \operatorname{Im} \mathsf{T}$.

Now (13.1) is all well and good from a theoretical point of view, but how does one tell if $\mathbf{B} \in \operatorname{Im} \mathsf{T}$ from the coefficient matrix $\mathbf{A}$? The answer is really quite simple. Let us introduce the vectors

$$\begin{aligned} \mathbf{A}_{(1)} &= (a_{11}, a_{21}, \ldots, a_{m1}) \\ &\vdots \\ \mathbf{A}_{(n)} &= (a_{1n}, a_{2n}, \ldots, a_{mn}) \end{aligned}$$

in $\mathbb{R}^m$. Notice that

$$\begin{aligned} \mathbf{A}_{(1)} &= \mathsf{T}(\mathbf{E}_1) \\ &\vdots \\ \mathbf{A}_{(n)} &= \mathsf{T}(\mathbf{E}_n). \end{aligned}$$

Therefore by (8.4)

$$\operatorname{Im} \mathsf{T} = \mathscr{L}(\mathsf{T}(\mathbf{E}_1), \ldots, \mathsf{T}(\mathbf{E}_n)) = \mathscr{L}(\mathbf{A}_{(1)}, \ldots, \mathbf{A}_{(n)}),$$

that is to say, the image of T is the linear span of the vectors $\mathbf{A}_{(1)}, \ldots, \mathbf{A}_{(n)}$. This may all be summed up in a definition and a theorem.

Definition. If $\mathbf{A} = (a_{ij})$ is an $m \times n$ matrix, the *column vectors* of $\mathbf{A}$ are the n vectors

$$\begin{aligned} \mathbf{A}_{(1)} &= (a_{11}, a_{21}, \ldots, a_{m1}) \\ \mathbf{A}_{(2)} &= (a_{12}, a_{22}, \ldots, a_{m2}) \\ &\vdots \\ \mathbf{A}_{(n)} &= (a_{1n}, a_{2n}, \ldots, a_{mn}) \end{aligned}$$

in $\mathbb{R}^m$. The *column space* of $\mathbf{A}$ is the linear span of its column vectors in $\mathbb{R}^m$.

EXAMPLE 2. What are the column vectors and what is the column space of the matrix

$$\mathbf{A} = \begin{pmatrix} 1 & 0 \\ -1 & -2 \\ 0 & 2 \end{pmatrix}?$$

Solution. The column vectors are the two vectors

$$(1, -1, 0), \qquad (0, -2, 2)$$

in $\mathbb{R}^3$. The column space of $\mathbf{A}$ is the linear span in $\mathbb{R}^3$ of these two vectors. A moment's computation shows this to be the plane

$$x + y + z = 0$$

in $\mathbb{R}^3$.

From our preceding discussion we now obtain:

Theorem 13.2. Let

$$\begin{aligned} a_{11}x_1 + a_{12}x_2 + \cdots + a_{1n}x_n &= b_1 \\ &\vdots \\ a_{m1}x_1 + a_{m2}x_2 + \cdots + a_{mn}x_n &= b_m \end{aligned}$$

be a system of linear equations. Let

$$\mathbf{A} = \begin{pmatrix} a_{11}, a_{12}, \ldots, a_{1n} \\ \vdots \\ a_{m1}, a_{m2}, \ldots, a_{mn} \end{pmatrix}, \qquad \mathbf{X} = \begin{pmatrix} x_1 \\ \vdots \\ x_n \end{pmatrix}, \qquad \mathbf{B} = \begin{pmatrix} b_1 \\ \vdots \\ b_n \end{pmatrix}.$$

Then the system has a solution iff the vector $\mathbf{B} = (b_1, \ldots, b_m)$ in $\mathbb{R}^m$ is in the column space of $\mathbf{A}$, that is, the vector $(b_1, \ldots, b_m)$ can be written as a linear combination of the column vectors of $\mathbf{A}$.

PROOF. Let $\mathsf{T} : \mathbb{R}^n \to \mathbb{R}^m$ be the linear transformation whose matrix is $\mathbf{A}$. Then the image of T is the column space of $\mathbf{A}$. Now apply (13.1). □

Example 3. Does the linear system

$$\begin{aligned} x_1 + x_2 + x_3 &= 1 \\ x_1 + x_3 &= 1 \\ 2x_1 + x_2 + 2x_3 &= 0 \end{aligned}$$

have any solutions?

Solution. The coefficient matrix of this linear system is

$$\mathbf{A} = \begin{pmatrix} 1 & 1 & 1 \\ 1 & 0 & 1 \\ 2 & 1 & 2 \end{pmatrix}$$

and its matrix of constants is

$$\mathbf{B} = \begin{pmatrix} 1 \\ 1 \\ 0 \end{pmatrix}.$$

Thus the column vectors of **A** are

$$(1, 1, 2), \quad (1, 0, 1), \quad (1, 1, 2).$$

Thus the column space of **A** is spanned by

$$(1, 1, 2) \quad \text{and} \quad (1, 0, 1)$$

while the vector **B** is

$$(1, 1, 0).$$

Now suppose

$$(1, 1, 0) = a(1, 1, 2) + b(1, 0, 1).$$

Then

$$\begin{aligned} 1 &= a + b \\ 1 &= a \\ 0 &= 2a + b \end{aligned}$$

which implies

$$\begin{aligned} b &= 0 \\ a &= 1 \\ b &= -2 \end{aligned}$$

which is impossible (compare the first and third equations). Therefore **B** does not belong to the column space of **A** and the system has no solutions at all.

The simplest type of a linear system is one whose matrix of constants is the zero matrix. That is, where

$$\mathbf{B} = \begin{pmatrix} 0 \\ \vdots \\ 0 \end{pmatrix} = \mathbf{0}.$$

This case merits a special name.

Definition. A system of linear equations (in matrix notation)

$$\mathbf{AX} = \mathbf{B}$$

is called *homogeneous* iff

$$\mathbf{B} = \begin{pmatrix} 0 \\ \vdots \\ 0 \end{pmatrix} = \mathbf{0}.$$

Theorem 13.3. *Let (in matrix notation)*

$$\mathbf{AX} = \mathbf{0}$$

be a homogeneous system of linear equations. Then the set $\mathscr{S} = \{(s_1, \ldots, s_n)\}$ *of all solutions to this linear system is a linear subspace of* $\mathbb{R}^n$. *In fact, if*

$$\mathsf{T} : \mathbb{R}^n \to \mathbb{R}^m$$

is the linear transformation whose matrix relative to the standard bases is $\mathbf{A}$, *then* $\mathscr{S} = \ker \mathsf{T}$.

PROOF. This theorem takes longer to state than to prove. Simply note that $(s_1, \ldots, s_n)$ is a solution of $\mathbf{AX} = \mathbf{0}$ iff $\mathbf{T}(s_1, \ldots, s_n) = (0, \ldots, 0)$. □

Remark. You might go back and look at Example 6 of Chapter 4 again.

Thus in order to solve a homogeneous system

$$\mathbf{AX} = \mathbf{0}$$

we have merely to find a basis for the kernel of the linear transformation $\mathsf{T} : \mathbb{R}^n \to \mathbb{R}^m$ whose matrix relative to the standard basis is $\mathbf{A}$. For by (13.3) any solution will be a linear combination of these basis vectors.

Before we turn to the case of a general linear system we pause to collect some preliminary facts.

Definition. Let $\mathscr{V}$ be a vector space, let $\mathscr{U}$ be a linear subspace of $\mathscr{V}$, and let $\mathbf{A} \in \mathscr{V}$. Denote by $\mathscr{A} = \mathbf{A} + \mathscr{U}$ the set of all vectors of the form $\mathbf{A} + \mathbf{X}$ where $\mathbf{X} \in \mathscr{U}$. The set $\mathbf{A} + \mathscr{U}$ is called a *parallel* of $\mathscr{U}$ in $\mathscr{V}$. It is said to result from *parallel translation* of $\mathscr{U}$ by the vector $\mathbf{A}$. A parallel of some linear subspace of $\mathscr{V}$ is called an *affine subspace* of $\mathscr{V}$.

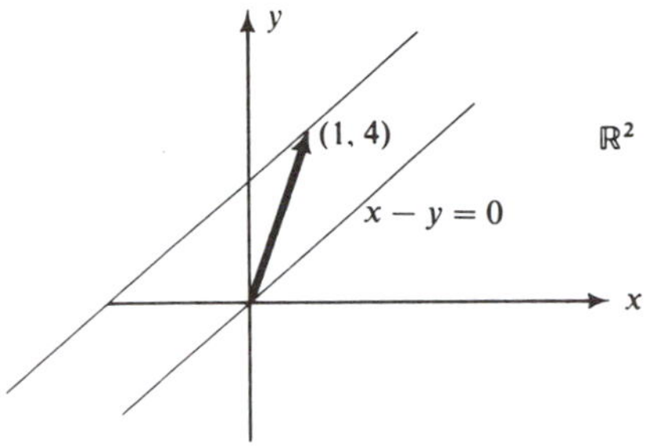

Figure 13.1

EXAMPLE 4. We let $\mathscr{V} = \mathbb{R}^2$ and $\mathscr{U}$ be the linear subspace $x - y = 0$. If $\mathbf{A} = (1, 4)$ then $\mathbf{A} + \mathscr{U}$ is the line through the point $(1, 4)$ that is parallel to $x - y = 0$. (See Figure 13.1).

Proposition 13.4. *If $\mathscr{U}$ is a linear subspace of a vector space $\mathscr{V}$ then*

(1) $\mathbf{A} \in \mathbf{A} + \mathscr{U}$ *for any* $\mathbf{A} \in \mathscr{V}$.
(2) *If* $\mathbf{B} \in \mathbf{A} + \mathscr{U}$ *then* $\mathbf{B} + \mathscr{U} = \mathbf{A} + \mathscr{U}$.
(3) *Two parallels of $\mathscr{U}$ either coincide or have no vector in common.*
(4) *If* $\mathbf{B}, \mathbf{C} \in \mathbf{A} + \mathscr{U}$ *then* $\mathbf{B} - \mathbf{C} \in \mathscr{U}$.

PROOF. Since $\mathbf{A} = \mathbf{A} + \mathbf{0}$, and $\mathbf{0} \in \mathscr{U}$ since $\mathscr{U}$ is a subspace, it follows that $\mathbf{A} \in \mathbf{A} + \mathscr{U}$ which proves (1). To prove (2) suppose that $\mathbf{B} \in \mathbf{A} + \mathscr{U}$. Then

$$\mathbf{B} = \mathbf{A} + \mathbf{X}$$

for some vector $\mathbf{X}$ in $\mathscr{U}$. Thus if $\mathbf{C} \in \mathbf{B} + \mathscr{U}$ we may find $\mathbf{Y}$ in $\mathscr{U}$ such that

$$\mathbf{C} = \mathbf{B} + \mathbf{Y} = (\mathbf{A} + \mathbf{X}) + \mathbf{Y} = \mathbf{A} + (\mathbf{X} + \mathbf{Y}) = \mathbf{A} + \mathbf{W}.$$

Now note that $\mathbf{W} = \mathbf{X} + \mathbf{Y} \in \mathscr{U}$ because $\mathscr{U}$ is a linear subspace of $\mathscr{V}$. Therefore $\mathbf{C} \in \mathbf{A} + \mathscr{U}$, and hence $\mathbf{B} + \mathscr{U} \subset \mathbf{A} + \mathscr{U}$. If we now write the equation $\mathbf{B} = \mathbf{A} + \mathbf{X}$ in the form

$$\mathbf{A} = \mathbf{B} + (-\mathbf{X})$$

we may apply exactly the same argument to show that $\mathbf{A} + \mathscr{U} \subset \mathbf{B} + \mathscr{U}$ and hence we may conclude that $\mathbf{A} + \mathscr{U} = \mathbf{B} + \mathscr{U}$ as desired. To prove (3) we suppose tht $\mathbf{A} + \mathscr{U}$ and $\mathbf{B} + \mathscr{U}$ are two parallels of $\mathscr{U}$ in $\mathscr{V}$. Suppose that they have a vector $\mathbf{C}$ in common, that is, $\mathbf{C} \in \mathbf{A} + \mathscr{U}$ and $\mathbf{C} \in \mathbf{B} + \mathscr{U}$. Then by (2) we find

$$\mathbf{C} + \mathscr{U} = \mathbf{A} + \mathscr{U}$$
$$\mathbf{C} + \mathscr{U} = \mathbf{B} + \mathscr{U}$$

so that $\mathbf{A} + \mathscr{U} = \mathbf{B} + \mathscr{U}$ as claimed. Finally to prove Part (4) we suppose that $\mathbf{B}, \mathbf{C} \in \mathbf{A} + \mathscr{U}$. Then we may find vectors $\mathbf{X}$ and $\mathbf{Y}$ in $\mathscr{U}$ such that

$$\mathbf{B} = \mathbf{A} + \mathbf{X}$$
$$\mathbf{C} = \mathbf{A} + \mathbf{Y}$$

from which we obtain

$$\mathbf{B} - \mathbf{C} = \mathbf{X} - \mathbf{Y}.$$

But $\mathbf{X} - \mathbf{Y}$ belongs to $\mathscr{U}$ since $\mathscr{U}$ is a linear subspace of $\mathscr{V}$. □

It follows from (13.4) that an affine subspace $\mathscr{A}$ of $\mathscr{V}$ is completely determined by a basis for $\mathscr{U}$ and a *single* vector $\mathbf{A}$ in the affine subspace $\mathscr{A}$.

Affine subspaces may be tied in neatly with linear equations with the aid of (13.1) and the following:

Proposition 13.5. *Let* $\mathsf{T}: \mathscr{V} \to \mathscr{W}$ *be a linear transformation and* $\mathbf{C} \in \operatorname{Im} \mathsf{T}$. *Let* $\mathscr{A} = \{\mathbf{V} \in \mathscr{V} \mid \mathsf{T}(\mathbf{V}) = \mathbf{C}\}$, *that is,* $\mathscr{A}$ *is the set of all vectors* $\mathbf{V}$ *in* $\mathscr{V}$ *such that* $\mathsf{T}(\mathbf{V}) = \mathbf{C}$. *Then* $\mathscr{A}$ *is an affine subspace of* $\mathscr{V}$. *In fact if* A *is any vector in* $\mathscr{V}$ *such that* $\mathsf{T}(\mathbf{A}) = \mathbf{C}$ *then* $\mathscr{A} = \mathbf{A} + \ker \mathsf{T}$.

PROOF. Let A be any vector in $\mathscr{V}$ such that $\mathsf{T}(\mathbf{A}) = \mathbf{C}$. If $\mathbf{B} \in \mathbf{A} + \ker \mathsf{T}$ then $\mathbf{B} = \mathbf{A} + \mathbf{X}$ for some $\mathbf{X} \in \ker \mathsf{T}$. Hence

$$\mathsf{T}(\mathbf{B}) = \mathsf{T}(\mathbf{A} + \mathbf{X}) = \mathsf{T}(\mathbf{A}) + \mathsf{T}(\mathbf{X}) = \mathsf{T}(\mathbf{A}) + \mathbf{0} = \mathsf{T}(\mathbf{A}) = \mathbf{C}$$

showing that $\mathbf{B} \in \mathscr{A}$. Therefore $\mathbf{A} + \ker \mathsf{T} \subset \mathscr{A}$. Conversely, if $\mathbf{B} \in \mathscr{A}$ then

$$\mathsf{T}(\mathbf{B} - \mathbf{A}) = \mathsf{T}(\mathbf{B}) - \mathsf{T}(\mathbf{A}) = \mathbf{C} - \mathbf{C} = \mathbf{0}$$

so that $\mathbf{B} - \mathbf{A} = \mathbf{X}$ belongs to $\ker \mathsf{T}$. Then

$$\mathbf{B} = \mathbf{A} + \mathbf{X}$$

with $\mathbf{X} \in \ker \mathsf{T}$, showing $\mathbf{B} \in \mathbf{A} + \ker \mathsf{T}$. Therefore $\mathscr{A} \subset \mathbf{A} + \ker \mathsf{T}$. Combining this with the preceding inclusion yields $\mathscr{A} = \mathbf{A} + \ker \mathsf{T}$ as desired. □

Theorem 13.6. *Suppose (in matrix notation) that*

$$\mathbf{AX} = \mathbf{B}$$

is a system of linear equations. Then the sets of all solutions $\mathscr{S} = \{(s_1, \ldots, s_n)\}$ *to this linear system is either empty or is an affine subspace of* $\mathbb{R}^n$. *In fact, if* $\mathsf{T}: \mathbb{R}^n \to \mathbb{R}^m$ *is the linear transformation whose matrix relative to the standard bases is* $\mathbf{A}$, *then* $\mathscr{S}$ *is a parallel translate of* $\ker \mathsf{T}$ *by any vector* $\mathbf{S}$ *in* $\mathscr{S}$, *or* $\mathscr{S} = \varnothing$.

Thus we see that in order to specify all solutions of the linear system

$$\mathbf{AX} = \mathbf{B}$$

we must specify a certain affine subspace of $\mathbb{R}^n$. This we may do (according to (13.6)) by finding a basis for the solution space of the system

$$\mathbf{AX} = \mathbf{0}$$

(called the *associated homogeneous* system) and finding a single particular solution to the equation (if there is one)

$$\mathbf{AX} = \mathbf{B}.$$

EXAMPLE 5. Solve the system of linear equations

$$\begin{aligned} 2x_1 + x_2 - 2x_3 + 3x_4 &= 4 \\ 3x_1 + 2x_2 - x_3 + 2x_4 &= 6 \\ 3x_1 + 3x_2 + 3x_3 - 3x_4 &= 6. \end{aligned}$$

Solution. The associated homogeneous system is

$$(\mathrm{L}_1) \qquad 2x_1 + x_2 - 2x_3 + 3x_4 = 0$$

$$(\mathrm{L}_2) \qquad 3x_1 + 2x_2 - x_3 + 2x_4 = 0$$

$$(\mathrm{L}_3) \qquad 3x_1 + 3x_2 + 3x_3 - 3x_4 = 0.$$

And this may be solved as follows: Multiply the first equation by 3 and the second by 2 to get

$$(\mathrm{L}_4) \qquad 6x_1 + 3x_2 - 6x_3 + 9x_4 = 0$$

$$(\mathrm{L}_5) \qquad 6x_1 + 4x_2 - 2x_3 + 4x_4 = 0.$$

Subtract (L_4) from (L_5) to get

$$(\mathrm{L}_6) \qquad x_2 + 4x_3 - 5x_4 = 0.$$

Subtract (L_6) from (L_1) to get

$$(\mathrm{L}_7) \qquad 2x_1 - 6x_3 + 8x_4 = 0.$$

Equations (L_7) and (L_6) yield

$$\begin{aligned} x_1 &= 3x_3 - 4x_4 \\ x_2 &= -4x_3 + 5x_4. \end{aligned}$$

A basis for the solution space to the associated homogeneous system is provided by the two vectors

$$(3, -4, 1, 0) \qquad (-4, 5, 0, 1).$$

Similar manipulations show that a solution to the original system of equations is

$$(1, 1, 1, 1).$$

Thus the solution space is the affine subspace

$$(1, 1, 1, 1) + \mathscr{L}((3, -4, 1, 0), (-4, 5, 0, 1))$$

of $\mathbb{R}^4$, or what is the same, all solutions to the original system are of the form

$$(1, 1, 1, 1) + a(3, -4, 1, 0) + b(-4, 5, 0, 1)$$

where a and b are arbitrary numbers.

It is time now to take up in more detail explicit methods for solving systems of linear equations.

Reduction to echelon form

The first method we will describe for solving systems of linear equations is by far the simplest of the many methods available. It will always work if one perseveres enough. However it is usually far from the shortest method.

Definition. A matrix $\mathbf{A} = (a_{ij})$ is said to be an *echelon matrix* iff the first nonzero entry in any row is a 1 and it appears to the right of the first nonzero entry of the preceding row. If in addition the first nonzero entry in a given row is the only nonzero entry in its column we say that the matrix $\mathbf{A}$ is in *reduced echelon form.*

EXAMPLE 6. The matrices

$$\begin{pmatrix} 1 & 0 & 3 & 1 \\ 0 & 0 & 1 & 0 \\ 0 & 0 & 0 & 0 \end{pmatrix}, \qquad \begin{pmatrix} 1 & 1 & 1 & 1 & 1 & 1 \\ 0 & 0 & 1 & 0 & 0 & 0 \\ 0 & 0 & 0 & 1 & 2 & 3 \\ 0 & 0 & 0 & 0 & 1 & 1 \\ 0 & 0 & 0 & 0 & 0 & 1 \end{pmatrix}$$

are echelon matrices, but not reduced echelon matrices. The following two matrices are reduced echelon matrices

$$\begin{pmatrix} 0 & 1 & 0 & 3 & 4 \\ 0 & 0 & 1 & 5 & 6 \end{pmatrix}, \qquad \begin{pmatrix} 1 & 0 & 1 & 0 & 1 & 0 \\ 0 & 1 & 1 & 0 & 2 & 1 \\ 0 & 0 & 0 & 1 & 3 & 0 \\ 0 & 0 & 0 & 0 & 0 & 0 \end{pmatrix}.$$

Notice that an echelon matrix or reduced echelon matrix is a very special type of upper triangular matrix. If the coefficient matrix of a linear system

$$\mathbf{AX} = \mathbf{B}$$

is in *reduced echelon form* then it is possible by inspection to read off all the solutions.

EXAMPLE 7. Solve the linear system

$$\begin{aligned} x_1 + 0x_2 + x_3 + 0x_4 + x_5 + 0x_6 &= 1 \\ 0x_1 + x_2 + x_3 + 0x_4 + 2x_5 + x_6 &= 2 \\ 0x_1 + 0x_2 + 0x_3 + x_4 + 3x_5 + 0x_6 &= 3. \end{aligned}$$

Solution. The coefficient matrix is in reduced echelon form. The associated homogeneous system is

$$\begin{aligned} x_1 + 0x_2 + x_3 + 0x_4 + x_5 + 0x_6 &= 0 \\ 0x_1 + x_2 + x_3 + 0x_4 + 2x_5 + x_6 &= 0 \\ 0x_1 + 0x_2 + 0x_3 + x_4 + 3x_5 + 0x_6 &= 0. \end{aligned}$$

Thus working our way up from the bottom we find

$$\begin{aligned} x_4 &= -3x_5 \\ x_2 &= -x_3 - 2x_5 - x_6 \\ x_1 &= -x_3 - x_5. \end{aligned}$$

A basis for the solution space of the associated homogeneous system is therefore seen to be

$$\begin{aligned} &(-1, -1, 1, 0, 0, 0) \\ &(-1, -2, 0, -3, 1, 0) \\ &(0, -1, 0, 0, 0, 1) \end{aligned}$$

and a solution to the original system is gotten from

$$\begin{aligned} x_4 &= 3 - 3x_5 \\ x_2 &= 2 - x_3 - 2x_5 - x_6 \\ x_1 &= 1 - x_3 - x_5 \end{aligned}$$

by setting $x_3 = x_5 = x_6 = 0$ and is seen to be

$$(1, 2, 0, 3, 0, 0).$$

Thus the general solution to the original system is seen to be

$$\begin{aligned} (s_1, s_2, s_3, s_4, s_5, s_6) = {} & (1, 2, 0, 3, 0, 0) + a(-1, -1, 1, 0, 0, 0) \\ & + b(-1, -2, 0, -3, 1, 0) + c(0, -1, 0, 0, 0, 1). \end{aligned}$$

(By the *general solution* we mean that any (and all) solution to the original system is obtained for a suitable choice of the three numbers (parameters) a, b, c.)

In *solving* a system of equations

$$\mathbf{AX} = \mathbf{B}$$

how do we proceed? Well the idea is to exchange the given system for a new system

$$\overline{\mathbf{A}}\mathbf{X} = \overline{\mathbf{B}}$$

which on the one hand has the same solution set as the original system but where the coefficient matrix has a more useful distribution of zeros. In fact as the preceding example suggests we would like for the new coefficient matrix to be in reduced echelon form. What kinds of operations can we perform on the original system to accomplish this end? Clearly the following three operations and any combination of them will be allowed, because they will not change the solution space.

(1) Interchange any two equations.
(2) Multiply any equation by a *nonzero* number.
(3) Add one equation to another.

Thus the following theorem says that by careful and perhaps lengthy application of these operations we may solve any linear system.

Theorem 13.7. *Let (in matrix notation)*

$$\mathbf{AX} = \mathbf{B}$$

be a system of linear equations. Then by a combination of the operations of interchanging two equations, multiplying any equation by a nonzero number, and adding one equation to another we may obtain a system of equations

$$\overline{\mathbf{A}}\mathbf{X} = \overline{\mathbf{B}}$$

whose coefficient matrix is in reduced echelon form.

The proof of (13.7) is an orgy of manipulation with indices and is omitted. Rather we will illustrate the technique of the proof by working some numerical examples.

Definition. If $\mathbf{AX} = \mathbf{B}$ is a system of linear equations then the matrix

$$\begin{pmatrix} a_{11} & a_{12} & \cdots & a_{1n} & b_1 \\ a_{21} & a_{22} & \cdots & a_{2n} & b_2 \\ \vdots & & & & \\ a_{m1} & a_{m2} & \cdots & a_{mn} & b_m \end{pmatrix}$$

is called the *augmented matrix* of the linear system.

Operations (1), (2), (3) above to be performed on the linear equations can instead be performed on the augmented matrix in order to save space. We illustrate this in the following examples also.

EXAMPLE 8. Solve the linear system

$$\begin{aligned} x + 2y - 3z &= 4 \\ x + 3y + z &= 11 \\ 2x + 5y - 4z &= 13 \\ 2x + 6y + 2z &= 22. \end{aligned}$$

Solution. The augmented matrix for this system is

$$\begin{pmatrix} 1 & 2 & -3 & 4 \\ 1 & 3 & 1 & 11 \\ 2 & 5 & -4 & 13 \\ 2 & 6 & 2 & 22 \end{pmatrix}.$$

We proceed to apply the three basic operations to reduce this matrix to row echelon form

$$\begin{matrix} \\ L_2 - L_1 \\ L_3 - 2L_1 \\ L_4 - 2L_1 \end{matrix} \begin{pmatrix} 1 & 2 & -3 & 4 \\ 0 & 1 & 4 & 7 \\ 0 & 1 & 2 & 5 \\ 0 & 2 & 8 & 14 \end{pmatrix} \rightsquigarrow \begin{matrix} L_1 - 2L_2 \\ \\ L_3 - L_2 \\ L_4 - 2L_2 \end{matrix} \begin{pmatrix} 1 & 0 & -11 & -10 \\ 0 & 1 & 4 & 7 \\ 0 & 0 & -2 & -2 \\ 0 & 0 & 0 & 0 \end{pmatrix}$$

$$\rightsquigarrow -\tfrac{1}{2}L_3 \begin{pmatrix} 1 & 0 & -11 & -10 \\ 0 & 1 & 4 & 7 \\ 0 & 0 & 1 & 1 \\ 0 & 0 & 0 & 0 \end{pmatrix} \rightsquigarrow \begin{matrix} L_1 + 11L_3 \\ L_2 - 4L_3 \\ \\ \\ \end{matrix} \begin{pmatrix} 1 & 0 & 0 & 1 \\ 0 & 1 & 0 & 3 \\ 0 & 0 & 1 & 1 \\ 0 & 0 & 0 & 0 \end{pmatrix}$$

which is the augmented matrix of the system

$$\begin{aligned} x_1 &= 1 \\ x_2 &= 3 \\ x_3 &= 1 \end{aligned}$$

having the same solutions as the original system. Thus the original system has the unique solution (1, 3, 1).

EXAMPLE 9. Solve the linear system

$$\begin{aligned} 2x + y - 2z + 3w &= 1 \\ 3x + 2y - z + 2w &= 4 \\ 3x + 3y + 3z - 3w &= 5. \end{aligned}$$

Solution. The augmented matrix for this system is

$$\begin{pmatrix} 2 & 1 & -2 & 3 & 1 \\ 3 & 2 & -1 & 2 & 4 \\ 3 & 3 & 3 & -3 & 5 \end{pmatrix}$$

which reduces to echelon form as follows:

$$\begin{matrix} 3L_1 \\ 2L_2 \\ 2L_3 \end{matrix} \begin{pmatrix} 6 & 3 & -6 & 9 & 3 \\ 6 & 4 & -2 & 4 & 8 \\ 6 & 6 & 6 & -6 & 10 \end{pmatrix} \rightsquigarrow \begin{pmatrix} 6 & 3 & -6 & 9 & 3 \\ 0 & 1 & 4 & -5 & 5 \\ 0 & 3 & 12 & -15 & 7 \end{pmatrix} \begin{matrix} \\ L_2 - L_1 \\ L_3 - L_1 \end{matrix}$$

$$\rightsquigarrow 3L_2 \begin{pmatrix} 6 & 3 & -6 & 9 & 3 \\ 0 & 3 & 12 & -15 & 15 \\ 0 & 3 & 12 & -15 & 7 \end{pmatrix}$$

$$\rightsquigarrow \begin{matrix} L_1 - L_2 \\ \\ L_3 - L_2 \end{matrix} \begin{pmatrix} 6 & 0 & -18 & 24 & -12 \\ 0 & 3 & 12 & -15 & 15 \\ 0 & 0 & 0 & 0 & -8 \end{pmatrix}$$

$$\rightsquigarrow \begin{matrix} \frac{1}{6}L_1 \\ \frac{1}{3}L_2 \\ -\frac{1}{8}L_3 \end{matrix} \begin{pmatrix} 1 & 0 & -3 & 4 & -2 \\ 0 & 1 & 4 & -5 & 5 \\ 0 & 0 & 0 & 0 & 1 \end{pmatrix}$$

which is the augmented matrix of the linear system

$$\begin{aligned} x \phantom{{}+y} - 3z + 4w &= -2 \\ y + 4z - 5w &= 5 \\ 0 &= 1. \end{aligned}$$

The impossibility of the last equation shows that this linear system has *no* solutions, hence the original system has no solutions. Note that the column space of the coefficient matrix for the new system is spanned by

$$(1, 0, 0), \quad (0, 1, 0), \quad (-3, 4, 0), \quad (4, -5, 0)$$

which is the xy-plane in $\mathbb{R}^3$. While the vector $(-2, 5, 1)$ does not lie in this plane.

As a last example of this method we work:

EXAMPLE 10. Solve the linear system

$$\begin{aligned} x_2 + 3x_3 + x_4 - x_5 &= 2 \\ x_1 - x_2 + 3x_3 - 4x_4 + 2x_5 &= 6 \\ x_1 + x_2 - x_3 + 2x_4 + x_5 &= 1 \\ x_1 - x_3 \phantom{{}+2x_4} x_5 &= 1. \end{aligned}$$

Solution. The augmented matrix is

$$\begin{pmatrix} 0 & 1 & 3 & 1 & -1 & 2 \\ 1 & -1 & 3 & -4 & 2 & 6 \\ 1 & 1 & -1 & 2 & 1 & 1 \\ 1 & 0 & -1 & 0 & 1 & 1 \end{pmatrix}.$$

Reducing to echelon form interchange L_1 and L_2

$$\begin{pmatrix} 1 & -1 & 3 & -4 & 2 & 6 \\ 0 & 1 & 3 & 1 & -1 & 2 \\ 1 & 1 & -1 & 2 & 1 & 1 \\ 1 & 0 & -1 & 0 & 1 & 1 \end{pmatrix}$$

$$\begin{array}{r} \\ \\ L_3 - L_1 \\ \rightsquigarrow L_4 - L_1 \end{array} \begin{pmatrix} 1 & -1 & 3 & -4 & 2 & 6 \\ 0 & 1 & 3 & 1 & -1 & 2 \\ 0 & 2 & -4 & 6 & -1 & -5 \\ 0 & 1 & -4 & 4 & -1 & -5 \end{pmatrix}$$

$$\begin{array}{r} L_1 + L_2 \\ \rightsquigarrow \\ L_3 - 2L_2 \\ L_4 - L_2 \end{array} \begin{pmatrix} 1 & 0 & 6 & -3 & 1 & 8 \\ 0 & 1 & 3 & 1 & -1 & 2 \\ 0 & 0 & -10 & 4 & 1 & -9 \\ 0 & 0 & -7 & 3 & 0 & -7 \end{pmatrix}$$

$$\begin{array}{r} \\ \rightsquigarrow \\ L_3 \times (-7) \\ L_4 \times 10 \end{array} \begin{pmatrix} 1 & 0 & 6 & -3 & 1 & 8 \\ 0 & 1 & 3 & 1 & -1 & 2 \\ 0 & 0 & 70 & -28 & -7 & 63 \\ 0 & 0 & -70 & 30 & 0 & -70 \end{pmatrix}$$

$$\begin{array}{r} \\ \rightsquigarrow \\ \\ L_4 + L_3 \end{array} \begin{pmatrix} 1 & 0 & 6 & -3 & 1 & 8 \\ 0 & 1 & 3 & 1 & -1 & 2 \\ 0 & 0 & 70 & -28 & -7 & 63 \\ 0 & 0 & 0 & 2 & -7 & -7 \end{pmatrix}$$

$$\begin{array}{r} \\ \rightsquigarrow \\ L_3/70 \\ L_4/2 \end{array} \begin{pmatrix} 1 & 0 & 6 & -3 & 1 & 8 \\ 0 & 1 & 3 & 1 & -1 & 2 \\ 0 & 0 & 1 & -\frac{28}{70} & -\frac{1}{10} & \frac{63}{70} \\ 0 & 0 & 0 & 1 & -\frac{7}{2} & -\frac{7}{2} \end{pmatrix}$$

which is the augmented matrix of the system

$$\begin{aligned} x_1 \quad + 6x_3 - 3x_4 + x_5 &= 8 \\ x_2 + 3x_3 + x_4 - x_5 &= 2 \\ x_3 - \tfrac{2}{5}x_4 - \tfrac{1}{10}x_5 &= \tfrac{9}{10} \\ x_4 - \tfrac{7}{2}x_5 &= -\tfrac{7}{2}. \end{aligned}$$

The associated homogeneous system is

$$\begin{aligned} x_1 \quad + 6x_3 - 3x_4 + x_5 &= 0 \\ x_2 + 3x_3 + x_4 - x_5 &= 0 \\ x_3 - \tfrac{2}{5}x_4 - \tfrac{1}{10}x_5 &= 0 \\ x_4 - \tfrac{7}{2}x_5 &= 0 \end{aligned}$$

or

$$x_4 = \frac{7}{2}x_5$$

$$x_3 = \frac{2x_4}{5} + \frac{x_5}{10}$$

$$x_2 = -3x_3 - x_4 + x_5$$

$$x_1 = -6x_3 + 3x_4 - x_5$$

and so a basis for the solution space consists of the single vector (set $x_5 = 1$)

$$\left(\frac{1}{2}, -7, \frac{3}{2}, \frac{7}{2}, 1\right)$$

or if you don't like fractions

$$(1, -14, 3, 7, 2).$$

To find a particular solution to the nonhomogeneous system we use the equations

$$x_4 = -\frac{7}{2} + \frac{7}{2}x_5$$

$$x_3 = \frac{9}{10} + \frac{2}{5}x_4 + \frac{x_5}{10}$$

$$x_2 = 2 - 3x_3 - x_4 + x_5$$

$$x_1 = 8 - 6x_3 + 3x_4 - x_5.$$

Setting $x_5 = 0$ gives the particular solution

$$x_5 = 0$$

$$x_4 = -\frac{7}{2}$$

$$x_3 = \frac{9}{10} + \frac{2}{5}\left(-\frac{7}{2}\right) = \frac{9}{10} - \frac{14}{10} = \frac{-5}{10} = -\frac{1}{2}$$

$$x_2 = 2 - 3\left(-\frac{1}{2}\right) + \frac{7}{2} = 2 + \frac{3}{2} + \frac{7}{2} = 7$$

$$x_1 = 8 - 6\left(-\frac{1}{2}\right) + 3\left(-\frac{7}{2}\right) = 8 + 3 - \frac{21}{2} = \frac{1}{2}$$

that is, $(\frac{1}{2}, 7, -\frac{1}{2}, -\frac{7}{2}, 0)$ solves the nonhomogeneous system. So the general solution to the original system is

$$(s_1, s_2, s_3, s_4, s_5) = (\tfrac{1}{2}, 7, -\tfrac{1}{2}, -\tfrac{7}{2}, 0) + a(1, -14, 3, 7, 2).$$

Setting $x_5 = 1$ gives another particular solution, namely $(1, 0, 1, 0, 1)$, so the general solution could equally well be written

$$(s_1, s_2, s_3, s_4, s_5) = (1, 0, 1, 0, 1) + a(1, -14, 3, 7, 2)$$

where a is an arbitrary number.

The simplex method

The method of solving a linear system

$$\mathbf{AX} = \mathbf{B}$$

by reducing the coefficient matrix $\mathbf{A}$ to reduced echelon form is a system of trading off the given linear system for a new and simpler linear system, whose solution sets are the same,

$$\bar{\mathbf{A}}\mathbf{X} = \bar{\mathbf{B}}$$

and from which the solutions may be read off by a purely mechanical process. There is another general approach to the solution of linear systems which involves a more direct kind of horse trading of "knowns and unknowns." To be more specific let us examine our linear system in longhand again. It looks like

$$\begin{aligned}
a_{11}x_1 + a_{12}x_2 + \cdots + a_{1n}x_n &= b_1 \\
a_{21}x_1 + a_{22}x_2 + \cdots + a_{2n}x_n &= b_2 \\
\vdots \qquad \vdots \qquad\qquad \vdots \quad &\quad \vdots \\
a_{m1}x_1 + a_{m2}x_2 + \cdots + a_{mn}x_n &= b_m.
\end{aligned}$$

Now we may regard this system as expressing the known quantities $b_1, \ldots, b_m$ as linear combinations of the unknown quantities $x_1, \ldots, x_n$. To solve the system means to reverse this process, that is to express the unknowns $x_1, \ldots, x_n$ as linear combinations of the knowns $b_1, \ldots, b_m$. For example suppose that $a_{ij} \neq 0$. Then we may solve the equation

$$a_{i1}x_1 + a_{i2}x_2 + \cdots + a_{ij}x_j + \cdots + a_{in}x_n = b_i$$

for x_j obtaining

$$x_j = -\frac{1}{a_{ij}}[a_{i1}x_1 + a_{i2}x_2 + \cdots + a_{ij-1}x_{j-1} - b_i + a_{ij+1}x_{j+1} + \cdots + a_{in}x_n].$$

(Note that x_j does *not* appear on the right in this formula.) We may now take this formula for x_j and put it into the other equations of the system obtaining the new system

$$\begin{aligned}
&\left(a_{11} - \frac{a_{1j}a_{i1}}{a_{ij}}\right)x_1 + \cdots + \frac{a_{1j}}{a_{ij}}\, b_i + \cdots + \left(a_{1n} - \frac{a_{1j}a_{in}}{a_{ij}}\right)x_n = b_1 \\
&\quad\vdots \\
&-\frac{a_{i1}}{a_{ij}}\, x_1 - \cdots + \frac{b_i}{a_{ij}} - \cdots - \frac{a_{in}}{a_{ij}}\, x_n = x_j \\
&\quad\vdots \\
&\left(a_{m1} - \frac{a_{mj}a_{i1}}{a_{ij}}\right)x_1 + \cdots + \frac{a_{mj}}{a_{ij}}\, b_i + \cdots + \left(a_{mn} - \frac{a_{mj}a_{in}}{a_{ij}}\right)x_n = b_m.
\end{aligned}$$

This process of obtaining a new system of equations is called a *pivot operation on a_{ij}*. Note that in this new system the unknown x_j does not appear in any of the equations on the left. It has been traded for the known quantity b_i. By repeated application of this process we may be able to obtain a linear system in which each right-hand side is an x_i and each left-hand side is a linear combination of the b_j and those x_k which do not appear on the left. In this way we will have solved the system. (Note that each step is reversible.) In order to apply this method (known as the *simplex method*) to solve systems of equations we must have available a simple way to write down the coefficients of the system obtained by pivoting at a_{ij}. To do this we require a simple definition.[1]

Definition. Suppose that

$$\mathbf{C} = \begin{pmatrix} c_{11} & c_{12} \\ c_{21} & c_{22} \end{pmatrix}$$

is a 2×2 matrix. The *determinant* of $\mathbf{C}$, denoted by det $\mathbf{C}$, is the number

$$\det \mathbf{C} = c_{11}c_{22} - c_{21}c_{12}.$$

EXAMPLE 11. Find the determinant of the matrix

$$\begin{pmatrix} 1 & 4 \\ 3 & 4 \end{pmatrix}.$$

Solution. We have

$$\det \begin{pmatrix} 1 & 4 \\ 3 & -4 \end{pmatrix} = 4 - 4(3) = -12.$$

[1] Recall (11.3) our discussion of inverting 2×2 matrices.

Example 12. Find the determinant of the matrix

$$\begin{pmatrix} 1 & -3 \\ 1 & -3 \end{pmatrix}.$$

Solution. We have

$$\det\begin{pmatrix} 1 & -3 \\ 1 & -3 \end{pmatrix} = 1(-3) - (-3)1 = -3 + 3 = 0.$$

Notice that a matrix with all positive entries can have a negative determinant and a matrix with all nonzero entries a zero determinant.

Return now to our linear system

$$\mathbf{AX} = \mathbf{B}.$$

If we suppose that $a_{ij} \neq 0$ then we may apply a pivot operation at a_{ij} to obtain a *new* linear system

$$\overline{\mathbf{A}}\overline{\mathbf{X}} = \overline{\mathbf{B}}$$

where ($\overline{\mathbf{A}} = (\bar{a}_{rs})$)

$$\bar{a}_{rs} = \begin{cases} \dfrac{1}{a_{ij}} \det\begin{pmatrix} a_{rs} & a_{rj} \\ a_{is} & a_{ij} \end{pmatrix}, & \text{if } r \neq i \text{ and } s \neq j \\[2ex] \dfrac{a_{rj}}{a_{ij}} & \text{if } r \neq i \text{ and } s = j \\[2ex] -\dfrac{a_{is}}{a_{ij}} & \text{if } r = i \text{ and } s \neq j \\[2ex] \dfrac{1}{a_{ij}} & \text{if } r = i \text{ and } s = j \end{cases}$$

and

$$\overline{\mathbf{B}} = \begin{pmatrix} b_1 \\ \vdots \\ b_{i-1} \\ x_j \\ b_{i+1} \\ \vdots \\ b_m \end{pmatrix} \qquad \overline{\mathbf{X}} = \begin{pmatrix} x_1 \\ \vdots \\ x_{j-1} \\ b_j \\ x_{j+1} \\ \vdots \\ x_n \end{pmatrix}.$$

With these formulas in mind let us run through a few examples by this method.

Note that we change not only the coefficient matrix **A** and constant matrix **B** *but also the matrix of unknowns*. Our notations should keep track of this fact.

Example 13. Solve the linear system

$$\begin{aligned} 2x - y + z &= 2 \\ x + 2y - z &= 3 \\ 3x + y + 2z &= -1. \end{aligned}$$

Solution: We represent our linear system by the diagram

x	y	z	
2	−1	1	2
1	2	−1	3
3	1	2	−1

It is always easiest to pivot on a 1 so let us pivot on the a_{13} position where there is a 1. Our new system is then represented by the diagram

x	y	2	
−2	1	1	z
3	1	−1	3
−1	3	2	−1

(Since we are pivoting on a 1 in the (1, 3) position the first row changes sign and the third column changes not at all.) For example ($i = 1, j = 3$)

$$\bar{a}_{21} = \det\begin{pmatrix} a_{21} & a_{23} \\ a_{11} & a_{13} \end{pmatrix} = \det\begin{pmatrix} 1 & -1 \\ 2 & 1 \end{pmatrix} = 1 + 2 = 3$$

$$\bar{a}_{22} = \det\begin{pmatrix} a_{22} & a_{23} \\ a_{12} & a_{13} \end{pmatrix} = \det\begin{pmatrix} 2 & -1 \\ -1 & 1 \end{pmatrix} = 2 - 1 = 1$$

$$\bar{a}_{31} = \det\begin{pmatrix} a_{31} & a_{33} \\ a_{11} & a_{13} \end{pmatrix} = \det\begin{pmatrix} 3 & 2 \\ 2 & 1 \end{pmatrix} = 3 - 4 = -1$$

$$\bar{a}_{32} = \det\begin{pmatrix} a_{32} & a_{33} \\ a_{12} & a_{13} \end{pmatrix} = \det\begin{pmatrix} 1 & 2 \\ -1 & 1 \end{pmatrix} = 1 + 2 = 3.$$

This new system has a handy 1 in the (2, 2) spot, so let's perform a pivot there. The resulting system is represented by the diagram

x	3	2	
−5	1	2	z
−3	1	1	y
−10	3	5	−1

Then $i = 2$, $j = 2$ and for example

$$\bar{\bar{a}}_{11} = \det\begin{pmatrix} a_{11} & a_{12} \\ a_{21} & a_{22} \end{pmatrix} = \det\begin{pmatrix} -2 & 1 \\ 3 & 1 \end{pmatrix} = -2 - 3 = -5$$

$$\bar{\bar{a}}_{13} = \det\begin{pmatrix} a_{13} & a_{12} \\ a_{23} & a_{22} \end{pmatrix} = \det\begin{pmatrix} 1 & 1 \\ -1 & 1 \end{pmatrix} = 1 + 1 = 2$$

$$\bar{\bar{a}}_{31} = \det\begin{pmatrix} a_{31} & a_{32} \\ a_{21} & a_{22} \end{pmatrix} = \det\begin{pmatrix} -1 & 3 \\ 3 & 1 \end{pmatrix} = -1 - 9 = -10$$

$$\bar{\bar{a}}_{33} = \det\begin{pmatrix} a_{33} & a_{32} \\ a_{23} & a_{22} \end{pmatrix} = \det\begin{pmatrix} 2 & 3 \\ -1 & 1 \end{pmatrix} = 2 + 3 = 5.$$

Finally we pivot on the (1, 3) position to obtain

-1	3	2	
$\frac{1}{2}$	$-\frac{1}{2}$	$-\frac{1}{2}$	z
$\frac{3}{10}$	$\frac{1}{10}$	$-\frac{1}{2}$	y
$\frac{-1}{10}$	$\frac{3}{10}$	$\frac{5}{10}$	x

$$\bar{\bar{\bar{a}}}_{12} = \frac{-1}{10}\det\begin{pmatrix} 1 & -5 \\ 3 & -10 \end{pmatrix} = \frac{-1}{10}(-10 + 15) = \frac{-5}{10} = -\frac{1}{2}$$

$$\bar{\bar{\bar{a}}}_{13} = \frac{-1}{10}\det\begin{pmatrix} 2 & -5 \\ 5 & -10 \end{pmatrix} = \frac{-1}{10}(-20 + 25) = -\frac{1}{2}$$

$$\bar{\bar{\bar{a}}}_{22} = \frac{-1}{10}\det\begin{pmatrix} 1 & -3 \\ 3 & -10 \end{pmatrix} = \frac{-1}{10}(-10 + 9) = \frac{1}{10}$$

$$\bar{\bar{\bar{a}}}_{23} = \frac{-1}{10}\det\begin{pmatrix} 1 & -3 \\ 5 & -10 \end{pmatrix} = \frac{-1}{10}(-10 + 15) = \frac{-5}{10} = -\frac{1}{2}$$

and thus we find

$$z = \frac{1}{2}(-1) - \frac{1}{2}(3) - \frac{1}{2}(2) = -\frac{1}{2} - \frac{3}{2} - 1 = -3$$

$$y = \frac{3}{10}(-1) + \frac{1}{10}(3) - \frac{1}{2}(2) = -1$$

$$x = \frac{-1}{10}(-1) + \frac{3}{10}(3) + \frac{5}{10}(2) = \frac{20}{10} = 2.$$

So there is only one solution and it is

$$x = 2, \quad y = -1, \quad z = -3.$$

As a second example consider the following:

EXAMPLE 14. Solve the linear system

$$\begin{aligned} x - y + z &= 3 \\ 2x + y - z &= 6 \\ -x + 2y + 2z &= 1 \\ 3x - 2y - 2z &= -1. \end{aligned}$$

Solution. We represent our system by the diagram

x	y	z	
1	−1	1	3
2	1	−1	6
−1	2	2	1
3	−2	−2	−1

We will pivot on the a_{11} position. Our new system is then represented by the diagram

3	y	z	
1	1	−1	x
2	3	−3	6
−1	1	3	1
3	1	−5	−1

(since we are pivoting on a 1 in position a_{11} the first row changes sign and the first column is unchanged).

For example ($i = 1, j = 1$)

$$\bar{a}_{22} = \det\begin{pmatrix} a_{22} & a_{21} \\ a_{12} & a_{11} \end{pmatrix} = \det\begin{pmatrix} 1 & 2 \\ -1 & 1 \end{pmatrix} = 1 - (-1)(2) = 3$$

$$\bar{a}_{23} = \det\begin{pmatrix} a_{23} & a_{21} \\ a_{13} & a_{11} \end{pmatrix} = \det\begin{pmatrix} -1 & 2 \\ 1 & 1 \end{pmatrix} = -1 - 2 = -3$$

$$\bar{a}_{32} = \det\begin{pmatrix} a_{32} & a_{31} \\ a_{12} & a_{11} \end{pmatrix} = \det\begin{pmatrix} 2 & -1 \\ -1 & 1 \end{pmatrix} = 2 - (-1)(-1) = 1$$

$$\bar{a}_{33} = \det\begin{pmatrix} a_{33} & a_{31} \\ a_{13} & a_{11} \end{pmatrix} = \det\begin{pmatrix} 2 & -1 \\ 1 & 1 \end{pmatrix} = 2 + 1 = 3$$

$$\bar{a}_{42} = \det\begin{pmatrix} a_{42} & a_{41} \\ a_{12} & a_{11} \end{pmatrix} = \det\begin{pmatrix} -2 & 3 \\ -1 & 1 \end{pmatrix} = -2 - (-1)(3) = 1$$

$$\bar{a}_{43} = \det\begin{pmatrix} a_{43} & a_{41} \\ a_{13} & a_{11} \end{pmatrix} = \det\begin{pmatrix} -2 & 3 \\ 1 & 1 \end{pmatrix} = -2 - 3 = -5.$$

Let us now pivot on the a_{32} position where there is a 1 in the new matrix. This will give us a system represented by the diagram

3	1	z	
2	1	-4	x
5	3	-12	6
1	1	-3	y
4	1	-8	-6

We now have $i = 3, j = 2$ and

$$\bar{\bar{a}}_{11} = \det\begin{pmatrix} \bar{a}_{11} & \bar{a}_{12} \\ \bar{a}_{31} & \bar{a}_{32} \end{pmatrix} = \det\begin{pmatrix} 1 & 1 \\ -1 & 1 \end{pmatrix} = 1 - (-1) = 2$$

$$\bar{\bar{a}}_{21} = \det\begin{pmatrix} 2 & 3 \\ -1 & 1 \end{pmatrix} = 2 - (-3) = 5$$

$$\bar{\bar{a}}_{41} = \det\begin{pmatrix} 3 & 1 \\ -1 & 1 \end{pmatrix} = 3 + 1 = 4$$

$$\bar{\bar{a}}_{13} = \det\begin{pmatrix} -1 & 1 \\ 3 & 1 \end{pmatrix} = -1 - 3 = -4$$

$$\bar{\bar{a}}_{23} = \det\begin{pmatrix} -3 & 3 \\ 3 & 1 \end{pmatrix} = -3 - 9 = -12$$

$$\bar{\bar{a}}_{43} = \det\begin{pmatrix} -5 & 1 \\ 3 & 1 \end{pmatrix} = -5 - 3 = -8.$$

If we pivot once more we can get rid of the z and thereby solve the system. The most convenient place to pivot is on the -8 in the (4, 4) position. This will give us

3	1	-6	
		$\frac{1}{2}$	x
8	-12	$\frac{3}{2}$	6
		$\frac{3}{8}$	y
$\frac{1}{2}$	$\frac{1}{8}$	$-\frac{1}{8}$	z

There is no need to calculate further as we have found the contradiction

$$24 - 12 - 9 = 6$$

which shows the original system has no solutions.

As our final example we consider:

EXAMPLE 15. Solve the linear system

$$\begin{aligned} x + 2y - 3z &= 6 \\ 2x - y + 4z &= 2 \\ 4x + 3y - 2z &= 14. \end{aligned}$$

Solution. Our system is represented by

x	y	z	
1	2	−3	6
2	−1	4	2
4	3	−2	14

Pivoting on the 1 in the (1, 1) position gives the system

6	y	z	
1	−2	3	x
2	−5	10	2
4	−5	10	14

Let us pivot now on the 10 in the (3, 3) position to obtain

6	y	14	
$-\frac{6}{5}$	$-\frac{1}{2}$	$\frac{3}{10}$	x
−2	0	1	2
$-\frac{2}{5}$	$\frac{1}{2}$	$\frac{1}{10}$	z

from which we find

$$\begin{aligned} x &= -\frac{6}{5} - \frac{y}{2} + \frac{3(14)}{10} = \frac{42 - 12}{10} - \frac{y}{2} \\ &= 3 - \frac{y}{2} \end{aligned}$$

$$-12 + 14 = 2$$

$$z = -\frac{12}{5} + \frac{1}{2}y + \frac{14}{10} = \frac{14 - 24}{10} + \frac{1}{2}y = \frac{1}{2}y - 1$$

or

$$x = 3 - \frac{y}{2}$$

$$z = \frac{1}{2}y - 1$$

so the solution space is

$$(3, 0, -1) + \mathscr{L}(-1, 2, 1).$$

Exercises

1. Solve the following systems of linear equations.

(a) $$\begin{cases} x - 3y + z = -2 \\ 2x + y - z = 6 \\ x + 2y + 2z = 2 \end{cases}$$

(b) $$\begin{cases} x - y + z = 2 \\ x + y = 1 \\ x + y + z = 8 \end{cases}$$

(c) $$\begin{cases} x + 2y - 3z = 4 \\ -x + y + z = 0 \\ 4x - 2y + z = 9 \end{cases}$$

(d) $$\begin{cases} 3x - 6y + z = 7 \\ x + 2y + z = 5 \\ -2x + 5y - 2z = -1 \end{cases}$$

2. Find a system of linear equations in three unknowns whose solution space is the line through (1, 1, 1) and the origin.

3. Find the solution space of following systems of linear homogeneous equations.

(a) $$\begin{cases} x - y + z - w = 0 \\ 2x + y - z + 2w = 0 \\ 2y + 3z + w = 0 \end{cases}$$

(b) $$\begin{cases} 3x_1 + 2x_2 + x_3 - x_4 - x_5 = 0 \\ x_1 - x_2 - x_3 - x_4 + 2x_5 = 0 \\ -x_1 + 2x_2 + 3x_3 + x_4 - x_5 = 0 \end{cases}$$

(c) $$\begin{cases} 2x + 3y - z = 0 \\ x - y + z = 0 \\ x + 9y - 5z = 0 \end{cases}$$

(d) $$\begin{cases} 11x - 8y + 5z = 0 \\ 3x - y - 2z = 0 \\ -x - 12y + 13z = 0 \end{cases}$$

4. Find the solution space of the following systems.

(a) $$\begin{cases} 2x + 6y - z + w = -3 \\ x - y + z - w = 2 \\ -x - 3y + 3z + 2w = 9 \end{cases}$$

(b) $$\begin{cases} \phantom{x_1 - {}} x_2 + 2x_3 - x_4 + x_5 = 5 \\ x_1 - x_2 + x_3 + 2x_4 - x_5 = 6 \\ x_1 + x_2 - x_3 \phantom{{}+ 2x_4} + x_5 = -2 \\ x_1 \phantom{{}- x_2} + x_3 - 3x_4 - 2x_5 = -3 \end{cases}$$

(c) $$\begin{cases} x - 2y + z = 5 \\ 2x + y - 2z = 7 \\ x - 7y + 5z = 8 \end{cases}$$

(d) $$\begin{cases} 3x + y - z = 10 \\ x - 2y - z = -2 \\ -x + y + z = 0 \\ 2x - y - 3z = 7 \end{cases}$$

5. Solve the linear systems:

$$\begin{pmatrix} 0 & 1 & 1 \\ 1 & 0 & 1 \\ 1 & 1 & 0 \end{pmatrix} \begin{pmatrix} x \\ y \\ z \end{pmatrix} = \begin{pmatrix} 1 \\ 1 \\ 1 \end{pmatrix},$$

$$\begin{pmatrix} -1 & 1 & 1 \\ 1 & -1 & 1 \\ 1 & 1 & -1 \end{pmatrix} \begin{pmatrix} x \\ y \\ z \end{pmatrix} = \begin{pmatrix} 1 \\ -1 \\ 1 \end{pmatrix}.$$

6. Write a flow chart for a computer program that reduces an $m \times n$ matrix to a reduced echelon form. If you have access to a machine write, debug, and run the program.

7. For which value of a does the following system of equations have a solution

$$\begin{aligned} y + z &= 2, \\ x + y + z &= a, \\ x + y \phantom{{}+ z} &= 2? \end{aligned}$$

8. For which values of a does the following system of equations have a solution

$$\begin{aligned} y + z &= 2, \\ x + ay + z &= 2, \\ x + y \phantom{{}+ z} &= 2? \end{aligned}$$

The elements of eigenvalue and eigenvector theory 14

Suppose that

$$\mathsf{T} : \mathscr{V} \to \mathscr{V}$$

is a linear transformation of the vector space $\mathscr{V}$ to itself. Such linear transformations have a special name (because their domain and range space are the same), they are called *endomorphisms of* $\mathscr{V}$.

Proposition 14.1. *Let* $\mathsf{T} : \mathscr{V} \to \mathscr{V}$ *be an endomorphism of the finite-dimensional vector space* $\mathscr{V}$. *Then there exist bases* $\{\mathbf{A}_1, \ldots, \mathbf{A}_n\}$ *and* $\{\mathbf{B}_1, \ldots, \mathbf{B}_n\}$ *for* $\mathscr{V}$ *such that the matrix of* T *is*

$$\left(\begin{array}{ccc|ccc} 1 & & & & & \\ & \ddots & & & 0 & \\ & & 1 & & & \\ \hline & & & 0 & & \\ & 0 & & & \ddots & \\ & & & & & 0 \end{array}\right) \begin{array}{l} \left.\vphantom{\begin{array}{c}1\\ \ddots\\ 1\end{array}}\right\} k \\ \left.\vphantom{\begin{array}{c}0\\ \ddots\\ 0\end{array}}\right\} n-k \end{array}$$

for some integer k. (*The integer k is usually called the rank of the linear transformation* T. *In fact* $k = \dim \operatorname{Im} \mathsf{T}$, *as our proof will show.*) (*Compare to* (12bis.1).)

Proof. Let $\mathbf{C}_1, \ldots, \mathbf{C}_m$ be a basis for $\ker \mathsf{T}$. By (6.7) we may find vectors $\mathbf{C}_{m+1}, \ldots, \mathbf{C}_n$ so that $\mathbf{C}_1, \ldots, \mathbf{C}_m, \mathbf{C}_{m+1}, \ldots, \mathbf{C}_n$ is a basis for $\mathscr{V}$. Thus $n = \dim \mathscr{V}$. Now we will require:

Lemma. *The vectors* $\mathsf{T}(\mathbf{C}_{m+1}), \mathsf{T}(\mathbf{C}_{m+2}), \ldots, \mathsf{T}(\mathbf{C}_n)$ *are linearly independent.*

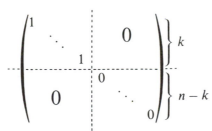

PROOF. Suppose to the contrary that the vectors $\mathsf{T}(\mathbf{C}_{m+1}), \ldots, \mathsf{T}(\mathbf{C}_n)$ are linearly dependent. Then we may find numbers $c_{m+1}, \ldots, c_n$, not all zero, so that

$$c_{m+1}\mathsf{T}(\mathbf{C}_{m+1}) + \cdots + c_n\mathsf{T}(\mathbf{C}_n) = \mathbf{0}.$$

Therefore

$$\mathsf{T}(c_{m+1}\mathbf{C}_{m+1} + \cdots + c_n\mathbf{C}_n) = c_{m+1}\mathsf{T}(\mathbf{C}_{m+1}) + \cdots + c_n\mathsf{T}(\mathbf{C}_n) = \mathbf{0}.$$

Hence the vector $c_{m+1}\mathbf{C}_{m+1} + \cdots + c_n\mathbf{C}_n$ belongs to the kernel of T. Remember that we chose the vectors $\mathbf{C}_1, \ldots, \mathbf{C}_m$ to be a basis for $\ker \mathsf{T}$ and therefore

$$c_{m+1}\mathbf{C}_{m+1} + \cdots + c_n\mathbf{C}_n = c_1\mathbf{C}_1 + \cdots + c_m\mathbf{C}_m$$

for suitable numbers $c_1, \ldots, c_n$. But then

$$(-c_1)\mathbf{C}_1 + (-c_2)\mathbf{C}_2 + \cdots + (-c_m)\mathbf{C}_m + c_{m+1}\mathbf{C}_{m+1} + \cdots + c_n\mathbf{C}_n = \mathbf{0}$$

which (since not all of $c_1, \ldots, c_n$ are zero) means that $\mathbf{C}_1, \ldots, \mathbf{C}_n$ are linearly dependent contrary to the fact that they are a basis for $\mathscr{V}$. Therefore the vectors $\mathsf{T}(\mathbf{C}_{m+1}), \ldots, \mathsf{T}(\mathbf{C}_n)$ are linearly independent. □

PROOF OF 14.1, CONTINUED. Note that

$$\begin{aligned}\operatorname{Im} \mathsf{T} &= \mathscr{L}(\mathsf{T}(\mathbf{C}_1), \ldots, \mathsf{T}(\mathbf{C}_n)) = \mathscr{L}(\mathbf{0}, \ldots, \mathbf{0}, \mathsf{T}(\mathbf{C}_{m+1}), \ldots, \mathsf{T}(\mathbf{C}_n)) \\ &= \mathscr{L}(\mathsf{T}(\mathbf{C}_{m+1}), \ldots, \mathsf{T}(\mathbf{C}_n))\end{aligned}$$

and hence $\mathsf{T}(\mathbf{C}_{m+1}), \ldots, \mathsf{T}(\mathbf{C}_n)$ is a basis for $\operatorname{Im} \mathsf{T}$.

Apply (6.7) again to choose vectors $\mathbf{D}_1, \ldots, \mathbf{D}_s$ so that $\{\mathbf{D}_1, \ldots, \mathbf{D}_s, \mathsf{T}(\mathbf{C}_{m+1}), \ldots, \mathsf{T}(\mathbf{C}_n)\}$ is a basis for $\mathscr{V}$. Now note $s = m$ since $s + n - m = \dim \mathscr{V} = n$. Set

$$\begin{array}{ll} \mathbf{A}_1 = \mathbf{C}_{m+1} & \mathbf{B}_1 = \mathsf{T}(\mathbf{C}_{m+1}) \\ \vdots & \vdots \\ \mathbf{A}_{n-m} = \mathbf{C}_n & \mathbf{B}_{n-m} = \mathsf{T}(\mathbf{C}_n) \\ \mathbf{A}_{n-m+1} = \mathbf{C}_1 & \mathbf{B}_{n-m+1} = \mathbf{D}_1 \\ \vdots & \vdots \\ \mathbf{A}_n = \mathbf{C}_m & \mathbf{B}_n = \mathbf{D}_s. \end{array}$$

Then we have

$$\begin{aligned} &\mathsf{T}(\mathbf{A}_1) = \mathbf{B}_1 \\ &\mathsf{T}(\mathbf{A}_2) = \mathbf{B}_2 \\ &\vdots \\ &\mathsf{T}(\mathbf{A}_{n-m}) = \mathbf{B}_{n-m} \\ &\mathsf{T}(\mathbf{A}_{n-m+1}) = \mathbf{0} \\ &\vdots \\ &\mathsf{T}(\mathbf{A}_n) = \mathbf{0}. \end{aligned}$$

If we put $k = n - m$ then the matrix of T relative to the basis pair $\{\mathbf{A}_1, \ldots, \mathbf{A}_n\}$, $\{\mathbf{B}_1, \ldots, \mathbf{B}_n\}$ is

$$\begin{matrix} k\left\{ \right. \\ \\ m = n - k\left\{ \right. \end{matrix} \begin{pmatrix} 1 & & & & & \\ & \ddots & & & 0 & \\ & & 1 & & & \\ & & & 0 & & \\ & 0 & & & \ddots & \\ & & & & & 0 \end{pmatrix}$$

as required. □

It follows from (14.1) that if when we study an endomorphism

$$\mathsf{T} : \mathscr{V} \to \mathscr{V}$$

on a finite-dimensional vector space through its matrix representatives we will perhaps learn nothing more useful about T than its rank if we are free to choose different bases in the domain and range of T. Since the domain and range of T are the same it is reasonable, in seeking to force the matrix to reveal more of the structure of T, to demand that we use the *same* basis in both the domain and range of T. If in this way we were to obtain a *diagonal* matrix then the structure of T will be completely revealed. Now suppose that $\{\mathbf{E}_1, \ldots, \mathbf{E}_n\}$ is a basis for $\mathscr{V}$ such that the matrix of T relative to this basis (*used in both domain and range*) is diagonal, say

$$\begin{pmatrix} e_1 & & & 0 \\ & e_2 & & \\ & & \ddots & \\ 0 & & & e_n \end{pmatrix}.$$

What does this mean? It means that

$$\mathsf{T}(\mathbf{E}_i) = e_i \mathbf{E}_i \qquad i = 1, \ldots, n.$$

That is, T is represented by a diagonal matrix iff there is a basis $\{\mathbf{E}_1, \ldots, \mathbf{E}_n\}$ for $\mathscr{V}$ and numbers $e_1, \ldots, e_n$ such that

$$\mathsf{T}(\mathbf{E}_i) = e_i \mathbf{E}_i \qquad i = 1, \ldots, n.$$

This discussion suggests that we introduce the following definition:

Definition. Let $\mathsf{T} : \mathscr{V} \to \mathscr{V}$ be an endomorphism of $\mathscr{V}$. A number e is called an *eigenvalue of* T iff there exists a **nonzero** vector $\mathbf{E}$ such that

$$\mathsf{T}(\mathbf{E}) = e\mathbf{E}.$$

Such a vector $\mathbf{E}$ is then called an *eigenvector of* T *associated* to the eigenvalue e.

Throughout our study of endomorphisms we will insist that our matrix representatives be constructed by using the same basis for both the domain and range. If $\mathsf{T} : \mathscr{V} \to \mathscr{V}$ is an endomorphism and $\{\mathbf{A}_1, \ldots, \mathbf{A}_n\}$ a basis for $\mathscr{V}$ we will use the phrase "the matrix of T relative to $\{\mathbf{A}_1, \ldots, \mathbf{A}_n\}$" (and similar such expressions) to mean the matrix $\mathbf{A} = (a_{ij})$ representing T obtained by using the basis $\{\mathbf{A}_1, \ldots, \mathbf{A}_n\}$ in both the domain and range of T, that is,

$$\mathsf{T}(\mathbf{A}_j) = \sum_{i=1}^{n} a_{ij}\mathbf{A}_i$$

for $j = 1, \ldots, n$. In this way if $\mathbf{A}_i$ happens to be an eigenvector of T corresponding to the eigenvalue e_i then

$$\mathsf{T}(A_i) = 0\mathbf{A}_1 + \cdots + 0\mathbf{A}_{i-1} + e_i\mathbf{A}_i + 0\mathbf{A}_{i+1} + \cdots + 0\mathbf{A}_n$$

so that in the ith column of $\mathbf{A}$ we will find

$$\begin{matrix} 0 \\ \vdots \\ 0 \\ e_i \\ 0 \\ \vdots \\ 0 \end{matrix} \qquad \begin{matrix} \\ \\ \\ i\text{th row.} \\ \\ \\ \\ \end{matrix}$$

In fact, we have the following important result,

Proposition 14.2. *Let* $\mathsf{T} : \mathscr{V} \to \mathscr{V}$ *be an endomorphism of the finite-dimensional vector space* $\mathscr{V}$*. Then* T *is represented by a diagonal matrix using the same basis in domain and range iff* $\mathscr{V}$ *has a basis composed of eigenvectors of* T.

Definition. An endomorphism $\mathsf{T} : \mathscr{V} \to \mathscr{V}$ of the finite-dimensional vector space $\mathscr{V}$ is said to be *diagonalizable* iff there exists a basis of $\mathscr{V}$ such that T is represented by a diagonal matrix relative to this basis.

Note that in view of (14.2) an endomorphism $\mathsf{T} : \mathscr{V} \to \mathscr{V}$ is diagonalizable iff $\mathscr{V}$ has a basis composed of eigenvectors of T. Thus if $\mathscr{V}$ is n-dimensional T is diagonalizable iff there exist n linearly independent eigenvectors of T. To *diagonalize* T means to find a basis of $\mathscr{V}$ composed of eigenvectors of T and the matrix of T with respect to this basis.

Example 1. Let $\mathsf{P} : \mathscr{V} \to \mathscr{V}$ be a projection, that is, $\mathsf{P}^2 = \mathsf{P}$ (see Chapter 12bis). Then the only eigenvalues of P are 0 and 1.

Proof. Suppose that e is an eigenvalue of P and $\mathbf{E} \in \mathscr{V}$ a corresponding eigenvector so that

$$\mathsf{P}(\mathbf{E}) = e\mathbf{E}.$$

Then computing $\mathsf{P}^2(\mathbf{E})$ in two different ways gives

$$\mathsf{P}^2(\mathbf{E}) = \mathsf{P}(\mathbf{E}) = e\mathbf{E}$$
$$\mathsf{P}^2(\mathbf{E}) = \mathsf{P}(\mathsf{P}(\mathbf{E})) = \mathsf{P}(e\mathbf{E}) = e\mathsf{P}(\mathbf{E}) = e(e\mathbf{E}) = e^2\mathbf{E}.$$

So equating yields

$$e\mathbf{E} = e^2\mathbf{E}$$

or

$$(e^2 - e)\mathbf{E} = \mathbf{0}.$$

Since the vector $\mathbf{E} \neq \mathbf{0}$ this implies

$$0 = e^2 - e = e(e - 1).$$

So $e = 0$ or $e = 1$.

In fact in (12bis.1) we even showed that a projection has the diagonal form

$$\begin{pmatrix} \overbrace{\begin{matrix} 1 & & & \\ & 1 & & \\ & & \ddots & \\ & & & 1 \end{matrix}}^{s} & \mathbf{0} \\ \mathbf{0} & \overbrace{\begin{matrix} 0 & & \\ & \ddots & \\ & & 0 \end{matrix}}^{r} \end{pmatrix}$$

where $r = \dim(\ker \mathsf{T})$ and $s = \dim(\operatorname{Im} \mathsf{P})$.

EXAMPLE 2. Let $\mathsf{T} : \mathbb{R}^2 \to \mathbb{R}^2$ be the linear transformation given by

$$\mathsf{T}(x, y) = (x + 2y, 3x + 2y).$$

Find all the eigenvalues and eigenvectors of T.

Solution. Note that relative to the standard basis of $\mathbb{R}^2$ T is represented by the matrix

$$\begin{pmatrix} 1 & 2 \\ 3 & 2 \end{pmatrix}$$

which is hardly diagonal.

To discover the eigenvalues and eigenvectors of T we must look for numbers e and nonzero vectors $\mathbf{E}$ such that $\mathsf{T}(\mathbf{E}) = e\mathbf{E}$. That is, if $\mathbf{E} = (x, y)$,

$$\mathsf{T}(x, y) = (ex, ey).$$

On the other hand

$$\mathsf{T}(x, y) = (x + 2y, 3x + 2y)$$

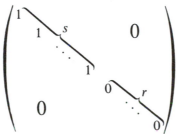

by definition. So we must look at when the system of equations

$$\begin{aligned} x + 2y &= ex \\ 3x + 2y &= ey \end{aligned}$$

has a nonzero solution. That is, for what values of e does the homogeneous linear system

$$\begin{aligned} (1 - e)x + 2y &= 0 \\ 3x + (2 - e)y &= 0 \end{aligned}$$

have a nontrivial solution? According to (13.3) this system has a nontrivial solution iff the matrix

$$\begin{pmatrix} 1 - e & 2 \\ 3 & 2 - e \end{pmatrix}$$

is *not* invertible. (That is, the linear transformation represented by this matrix is by (12.4) *not* an isomorphism.) But (11.3) says that a 2×2 matrix $\mathbf{A}$ is not an isomorphism iff $\det \mathbf{A} = 0$. Putting all this together we find that e is an eigenvalue of T iff

$$\det \begin{pmatrix} 1 - e & 2 \\ 3 & 2 - e \end{pmatrix} = 0.$$

But

$$\begin{aligned} \det \begin{pmatrix} 1 - e & 2 \\ 3 & 2 - e \end{pmatrix} &= (1 - e)(2 - e) - 6 \\ &= 2 - 3e + e^2 - 6 \\ &= e^2 - 3e - 4 \\ &= (e - 4)(e + 1). \end{aligned}$$

Thus the eigenvalues of T are those numbers e such that

$$0 = (e - 4)(e + 1),$$

that is

$$e = 4, -1$$

are the eigenvalues of T. Now we must find the eigenvectors of T associated to 4 and -1 respectively. That means we must solve the linear equations

$$\begin{aligned} e = 4: \qquad -3x + 2y &= 0 \\ 3x - 2y &= 0, \\ e = -1: \qquad 2x + 2y &= 0 \\ 3x + 3y &= 0. \end{aligned}$$

The first set has as solution space the linear span of the vector $\mathbf{E} = (2, 3)$ and the second set has as solution space the linear span of $(1, -1) = \mathbf{F}$.

Thus we see the eigenvalues of T are 4 and -1 and all eigenvectors associated to 4 are multiples of (2, 3) and all eigenvectors associated to -1 are multiples of $(1, -1)$.

While the preceding example is long and tedious it is at least partly so because we really do not have enough tools yet to work such examples comfortably. In fact the example suggests a few useful results.

Proposition 14.3. *Let $\mathsf{T} : \mathscr{V} \to \mathscr{V}$ be an endomorphism of the finite-dimensional vector space $\mathscr{V}$. A number e is an eigenvalue of* T *iff the endomorphism*

$$\mathsf{T} - e\mathsf{I} : \mathscr{V} \to \mathscr{V}$$

is not an isomorphism.

Proof. Suppose that e is an eigenvalue of T. Let **E** be an eigenvector of T associated to e. Then

$$\mathsf{T}(\mathbf{E}) = e\mathbf{E}.$$

Therefore

$$(\mathsf{T} - e\mathsf{I})(\mathbf{E}) = \mathsf{T}(\mathbf{E}) - e\mathbf{E} = e\mathbf{E} - e\mathbf{E} = \mathbf{0}$$

so $\mathbf{0} \neq \mathbf{E} \in \ker(\mathsf{T} - e\mathsf{I})$. Hence by (8.11) $\mathsf{T} - e\mathsf{I}$ is not an isomorphism. Conversely, suppose that $\mathsf{T} - e\mathsf{I}$ is not an isomorphism. By (8.11) it follows that either

$$\ker(\mathsf{T} - e\mathsf{I}) \neq \{\mathbf{0}\}$$

or

$$\operatorname{Im}(\mathsf{T} - e\mathsf{I}) \neq \mathscr{V}.$$

From (8.10) we have the equation

$$n = \dim \ker(\mathsf{T} - e\mathsf{I}) + \dim \operatorname{Im}(\mathsf{T} - e\mathsf{I}).$$

Hence in either case $\dim \ker(\mathsf{T} - e\mathsf{I}) > 0$, that is, we must have

$$\ker(\mathsf{T} - e\mathsf{I}) \neq \{\mathbf{0}\}$$

and

$$\operatorname{Im}(\mathsf{T} - e\mathsf{I}) \neq \mathscr{V}.$$

At any rate, $\ker(\mathsf{T} - e\mathsf{I}) \neq \{\mathbf{0}\}$ and so we may select $\mathbf{E} \in \ker(\mathsf{T} - e\mathsf{I})$, $\mathbf{E} \neq 0$. Then

$$\mathbf{0} = (\mathsf{T} - e\mathsf{I})(\mathbf{E}) = \mathsf{T}(\mathbf{E}) - e(\mathbf{E})$$

so

$$\mathsf{T}(\mathbf{E}) = e\mathbf{E}$$

and hence **E** is an eigenvector of T associated to the eigenvalue e. □

Definition. Let $\mathsf{T}: \mathscr{V} \to \mathscr{V}$ be an endomorphism of the vector space $\mathscr{V}$. Suppose that e is an eigenvalue of T and set

$$\mathscr{V}_e = \{\mathbf{E} \mid \mathsf{T}(\mathbf{E}) = e\mathbf{E}\}.$$

Thus $\mathscr{V}_e$ is the set of all eigenvectors of T associated with the eigenvalue e, **together with the zero vector**. $\mathscr{V}_e$ is called the *eigenspace* of T associated to the eigenvalue e.

Note that by definition an eigenspace $\mathscr{V}_e$ always contains a nonzero vector, namely an eigenvector (at least one), and therefore dim $\mathscr{V}_e$ is always positive. Recall that $\ker \mathsf{T} = \{\mathbf{E} \in \mathscr{V} \mid \mathsf{T}(\mathbf{E}) = \mathbf{0}\}$, and so the number 0 is an eigenvalue of $\mathsf{T}: \mathscr{V} \to \mathscr{V}$ iff $\ker \mathsf{T} \neq \{\mathbf{0}\}$ in which case $\mathscr{V}_0 = \ker \mathsf{T}$. The following example shows that there are many nonzero endomorphisms for which 0 is the only eigenvalue.

EXAMPLE 3. Let $\mathsf{T}: \mathscr{V} \to \mathscr{V}$ be nilpotent. Then the only eigenvalue of T is 0.

PROOF. Since T is nilpotent there exists an integer k such that $\mathsf{T}^{k-1} \neq 0$. Let k be the smallest such integer. Then $\mathsf{T}^{k-1} \neq 0$, so there is a vector $\mathbf{E} \in \mathscr{V}$ with $\mathsf{T}^{k-1}(\mathbf{E}) \neq \mathbf{0}$. Since

$$\mathsf{T}(\mathsf{T}^{k-1}(\mathbf{E})) = \mathsf{T}^k(\mathbf{E}) = \mathbf{0} = 0\mathbf{E}$$

it follows that $\mathbf{0} \neq \mathsf{T}^{k-1}(\mathbf{E}) \in \ker \mathsf{T}$, so 0 is an eigenvalue of T, $\mathscr{V}_0 = \ker \mathsf{T}$. By definition, if e is any eigenvalue of T then there is a nonzero vector $\mathbf{F} \in \mathscr{V}$ such that

$$\mathsf{T}(\mathbf{F}) = e\mathbf{F}.$$

Then

$$\mathsf{T}^2(\mathbf{F}) = \mathsf{T}(\mathsf{T}(\mathbf{F})) = \mathsf{T}(e\mathbf{F})) = e\mathsf{T}(\mathbf{F}) = ee\mathbf{F} = e^2\mathbf{F},$$
$$\mathsf{T}^3(\mathbf{F}) = \mathsf{T}(\mathsf{T}^2(\mathbf{F})) = e^2\mathsf{T}(\mathbf{F}) = e^3\mathbf{F},$$

and continuing in this way we see

$$\mathsf{T}^k(\mathbf{F}) = e^k\mathbf{F}.$$

But $\mathsf{T}^k = 0$ so

$$\mathsf{T}^k(\mathbf{F}) = \mathbf{0}$$

and therefore

$$e^k\mathbf{F} = \mathbf{0}.$$

Since $\mathbf{F}$ is not the zero vector this implies $e^k = 0$ so $e = 0$ as required.

The only diagonalizable transformation with 0 as its only eigenvalue is the zero transformation. Therefore the only nilpotent transformation that is diagonalizable is the zero transformation! In a very strong sense this is the reason why not all transformations can be diagonalized. In fact if complex scalars are used then any transformation $\mathsf{T} : \mathscr{V} \to \mathscr{V}$ can be written as a sum $\mathsf{D} + \mathsf{N}$ where D is diagonalizable with the same eigenvalues (and multiplicities) as T and N is nilpotent. (This is the Jordan canonical form and the subject of Chapter 17.)

Proposition 14.4. *Let* $\mathsf{T} : \mathscr{V} \to \mathscr{V}$ *be an endomorphism of the vector space* $\mathscr{V}$. *If* e *is an eigenvalue of* T *then* $\mathscr{V}_e = \ker(\mathsf{T} - e\mathsf{I})$. *Hence* $\mathscr{V}_e$ *is a subspace of* $\mathscr{V}$.

PROOF. Immediate from the definitions. □

In view of these results it will be convenient to have a name for endomorphisms $\mathsf{T} : \mathscr{V} \to \mathscr{V}$ that are *not* isomorphisms.

Definition. An endomorphism $\mathsf{T} : \mathscr{V} \to \mathscr{V}$ is said to be *singular* iff T is *not* an isomorphism.

Note that in the course of proving (14.3) we also proved (look at 8.10 again):

Proposition 14.5. *Let* $\mathscr{V}$ *be a finite-dimensional vector space and* $\mathsf{T} : \mathscr{V} \to \mathscr{V}$ *an endomorphism. Then* T *is singular iff* $\ker \mathsf{T} \neq \{\mathbf{0}\}$.

EXAMPLE 4. Let $\mathsf{S} : \mathscr{P}(\mathbb{R}) \to \mathscr{P}(\mathbb{R})$ be the linear transformation given by

$$\mathsf{S}(p(x)) = xp(x).$$

Then S is an endomorphism of $\mathscr{P}(\mathbb{R})$. Clearly

$$\ker \mathsf{S} = \{\mathbf{0}\}.$$

On the other hand S is singular since the vector 1 is not in Im S. Thus (14.5) becomes false if we do not assume $\mathscr{V}$ to be finite dimensional.

EXAMPLE 5. Find the eigenvalues and associated eigenspaces for the linear transformation $\mathsf{T} : \mathbb{R}^2 \to \mathbb{R}^2$ with matrix

$$\begin{pmatrix} 1 & 2 \\ 4 & 3 \end{pmatrix}$$

relative to the standard basis.

Solution. The eigenvalues occur when the matrix

$$\begin{pmatrix} 1 & 2 \\ 4 & 3 \end{pmatrix} - e\begin{pmatrix} 1 & 0 \\ 0 & 1 \end{pmatrix}$$

is singular. (*Important*: note that since we have insisted on using the same basis in domain and range the matrix of I is

$$\begin{pmatrix} 1 & 0 \\ 0 & 1 \end{pmatrix}.$$

You should go back and look at Examples 1 and 2 of Chapter 12 again.) Now

$$\begin{pmatrix} 1 & 2 \\ 4 & 3 \end{pmatrix} - e\begin{pmatrix} 1 & 0 \\ 0 & 1 \end{pmatrix} = \begin{pmatrix} 1-e & 2 \\ 4 & 3-e \end{pmatrix}$$

is singular iff its determinant vanishes. So we must have

$$0 = \det\begin{pmatrix} 1-e & 2 \\ 4 & 3-e \end{pmatrix} = (1-e)(3-e) - 8$$

so

$$\begin{aligned} 0 &= (1-e)(3-e) - 8 = 3 - 4e + e^2 - 8 \\ &= e^2 - 4e - 5 = (e-5)(e+1). \end{aligned}$$

So the eigenvalues of T must be $e = 5$, $e = -1$. To find the eigenspace of T we must solve the equations

$$e = 5: \quad \begin{cases} -4x + 2y = 0 \\ 4x - 2y = 0 \end{cases}$$

$$e = -1: \quad \begin{cases} 2x + 2y = 0 \\ 4x + 4y = 0 \end{cases}$$

so

$$\begin{aligned} \mathscr{V}_5 &= \mathscr{L}(1, 2) = \{(x, y) | 2x - y = 0\} \\ \mathscr{V}_{-1} &= \mathscr{L}(1, -1) = \{(x, x) | x + y = 0\}. \end{aligned}$$

Example 6. Find the eigenvalues and eigenvectors of the linear transformation with matrix

$$\begin{pmatrix} 0 & 1 \\ -1 & 0 \end{pmatrix}$$

relative to the standard basis of $\mathbb{R}^2$.

Solution. We must find when

$$\det\begin{pmatrix} -e & 1 \\ -1 & -e \end{pmatrix} = 0.$$

But

$$\det\begin{pmatrix} -e & 1 \\ -1 & -e \end{pmatrix} = e^2 + 1$$

and hence is **never** zero. Therefore this linear transformation

$$\mathsf{T}: R^2 \to \mathbb{R}^2$$

has *no* eigenvalues.

It should be clear from the preceding examples that the problem of calculating the eigenvalues and eigenvectors of a linear transformation

$$\mathsf{T}: \mathbb{R}^n \to \mathbb{R}^n$$

for $n > 2$ depends on having a simple method to test when $\mathsf{T} - e\mathsf{I}$ is singular. Such a method exists, although for large n it is not really simple, as it is enormously tedious. It involves extending the determinant to square matrices of size greater than 2. We will do this by induction.

Definition. Let $\mathbf{A}$ be an $n \times n$ matrix. The square matrix obtained from $\mathbf{A}$ by deleting the ith row and jth column of $\mathbf{A}$ is called the *minor* of the element a_{ij} of $\mathbf{A}$, and is denoted by $\mathbf{M}_{ij}$.

EXAMPLE 7. Let

$$\mathbf{A} = \begin{pmatrix} 1 & 0 & 1 \\ 2 & 1 & 2 \\ 0 & 4 & 6 \end{pmatrix}$$

find the minors $\mathbf{M}_{12}$, $\mathbf{M}_{23}$, $\mathbf{M}_{33}$ of $\mathbf{A}$.

Solution. We have

$$\mathbf{M}_{12} = \begin{pmatrix} 2 & 2 \\ 0 & 6 \end{pmatrix}$$

$$\mathbf{M}_{23} = \begin{pmatrix} 1 & 0 \\ 0 & 4 \end{pmatrix}$$

$$\mathbf{M}_{33} = \begin{pmatrix} 1 & 0 \\ 2 & 1 \end{pmatrix}.$$

Definition. Let $\mathbf{A}$ be a square matrix of size n. Then the *determinant* of $\mathbf{A}$, denoted by $\det \mathbf{A}$ is defined by the inductive formula

$$\det(a) = a$$

$$\det \mathbf{A} = a_{11} \det \mathbf{M}_{11} - a_{12} \det \mathbf{M}_{12} + \cdots + (-1)^{n+1} a_{1n} \det \mathbf{M}_{1n}$$

$$= \sum_{j=1}^{n} (-1)^{1+j} a_{1j} \det \mathbf{M}_{1j}.$$

EXAMPLE 8. Calculate the determinant of

$$\mathbf{A}=\begin{pmatrix}1&0&1\\2&1&2\\0&4&6\end{pmatrix}.$$

Solution. We have

$$\det \mathbf{A} = 1\det\begin{pmatrix}1&2\\4&6\end{pmatrix} - 0\det\begin{pmatrix}2&2\\0&6\end{pmatrix} + 1\det\begin{pmatrix}2&1\\0&4\end{pmatrix}$$
$$= 6 - 8 + 0 + 8 = 6.$$

EXAMPLE 9. Calculate the determinant of

$$\mathbf{A}=\begin{pmatrix}1&2&-1&3\\0&1&4&2\\0&1&0&4\\1&0&2&1\end{pmatrix}.$$

Solution. We have

$$\det \mathbf{A} = 1\det\begin{pmatrix}1&4&2\\1&0&4\\0&2&1\end{pmatrix} - 2\det\begin{pmatrix}0&4&2\\0&0&4\\1&2&1\end{pmatrix}$$
$$-1\det\begin{pmatrix}0&1&2\\0&1&4\\1&0&1\end{pmatrix} - 3\det\begin{pmatrix}0&1&4\\0&1&0\\1&0&2\end{pmatrix}$$
$$= 1\left(\det\begin{pmatrix}0&4\\2&1\end{pmatrix} - 4\det\begin{pmatrix}1&4\\0&1\end{pmatrix} + 2\det\begin{pmatrix}1&0\\0&2\end{pmatrix}\right)$$
$$-2\left(0\det\begin{pmatrix}0&4\\2&1\end{pmatrix} - 4\det\begin{pmatrix}0&4\\1&1\end{pmatrix} + 2\det\begin{pmatrix}0&0\\1&2\end{pmatrix}\right)$$
$$-\left(0\det\begin{pmatrix}1&4\\0&1\end{pmatrix} - 1\det\begin{pmatrix}0&4\\1&1\end{pmatrix} + 2\det\begin{pmatrix}0&1\\1&0\end{pmatrix}\right)$$
$$-3\left(0\det\begin{pmatrix}1&0\\0&2\end{pmatrix} - \det\begin{pmatrix}0&0\\1&2\end{pmatrix} + 4\det\begin{pmatrix}0&1\\1&0\end{pmatrix}\right)$$
$$= (-8 - 4 + 4) - 2(0 - 4(-4) + 0) - (0 + 4 - 2) - 3(0 - 0 - 4)$$
$$= -30.$$

Hopefully the preceding example will convince you that the evaluation of determinants of large matrices is a most painful process. However for small values of n the utility of determinants cannot be denied.

There are a number of properties of determinants that make easier their computation. The verification of these properties requires an excursion into multilinear algebra which is undertaken in Chapter 14bis. The basic properties of determinants are summarized in properties (1)–(8) below, of which we make free use in the sequel.

Definition. The *transpose* of a matrix $\mathbf{A} = (a_{ij})$ is the matrix $\mathbf{A}^t = (a^t_{ij})$ where $a^t_{ij} = a_{ji}$.

For example when $n = 4$

$$\begin{pmatrix} 1 & 5 & 9 & 13 \\ 2 & 6 & 10 & 14 \\ 3 & 7 & 11 & 15 \\ 4 & 8 & 12 & 16 \end{pmatrix}^t = \begin{pmatrix} 1 & 2 & 3 & 4 \\ 5 & 6 & 7 & 8 \\ 9 & 10 & 11 & 12 \\ 13 & 14 & 15 & 16 \end{pmatrix}.$$

The transpose of $\mathbf{A}$ is the matrix $\mathbf{A}^t$ whose rows are the columns of $\mathbf{A}$ (see Chapter 11 Exercises 6–10).

Properties of determinants

Let $\mathbf{A}$ be a matrix.

(1) If $\mathbf{A}$ has a row of zeros then $\det \mathbf{A} = 0$.
(2) If $\mathbf{A}$ has two rows equal then $\det \mathbf{A} = 0$.
(3) If $\mathbf{B}$ is obtained from $\mathbf{A}$ by interchanging two rows then $\det \mathbf{B} = -\det \mathbf{A}$.
(4) If $\mathbf{B}$ is obtained from $\mathbf{A}$ by adding a multiple of one row to a *different* row then $\det \mathbf{B} = \det \mathbf{A}$.
(5) If $\mathbf{B}$ is obtained from $\mathbf{A}$ by multiplying a row of $\mathbf{A}$ by a number k then $\det \mathbf{B} = k \det \mathbf{A}$.
(6) $\det \mathbf{A}^t = \det \mathbf{A}$, hence everything we said about rows in (1)–(5) works also for columns.
(7) If $\mathbf{A}$ is upper triangular $\det \mathbf{A}$ is the product of the diagonal entries.
(8) $\det(\mathbf{AB}) = \det \mathbf{A} \det \mathbf{B}$.

Using these properties we may do Example 9 a little less painfully.

EXAMPLE 9 REVISITED. We have

$$\det\begin{pmatrix} 1 & 2 & -1 & 3 \\ 0 & 1 & 4 & 2 \\ 0 & 1 & 0 & 4 \\ 1 & 0 & 2 & 1 \end{pmatrix}$$

$$= \det\begin{pmatrix} 0 & 2 & -3 & 2 \\ 0 & 1 & 4 & 2 \\ 0 & 1 & 0 & 4 \\ 1 & 0 & 2 & 1 \end{pmatrix} = \det\begin{pmatrix} 0 & 0 & 0 & 1 \\ 2 & 1 & 1 & 0 \\ -3 & 4 & 0 & 2 \\ 2 & 2 & 4 & 1 \end{pmatrix}$$

$$= (-1)^{1+4} \det\begin{pmatrix} 2 & 1 & 1 \\ -3 & 4 & 0 \\ 2 & 2 & 4 \end{pmatrix} = -\det\begin{pmatrix} 0 & 0 & 1 \\ -3 & 4 & 0 \\ -6 & -2 & 4 \end{pmatrix}$$

$$= -(-1)^{1+3} \det\begin{pmatrix} -3 & 4 \\ -6 & -2 \end{pmatrix} = -(6 + 24) = -30.$$

Proposition 14.6. *Let* **A** *be an* $n \times n$ *matrix. Then* **A** *is invertible iff* det **A** $\neq 0$.

PROOF. Suppose that **A** is invertible. Then

$$\mathbf{A}(\mathbf{A}^{-1}) = \mathbf{I}.$$

Taking determinants of both sides gives

$$\det(\mathbf{A}(\mathbf{A}^{-1})) = \det \mathbf{I}.$$

Since **I** is upper triangular it follows from (7) that

$$\det \mathbf{I} = 1.$$

From (8) we find

$$\det(\mathbf{A}(\mathbf{A}^{-1})) = \det \mathbf{A} \det(\mathbf{A}^{-1}).$$

Equating gives

$$\det \mathbf{A} \det(\mathbf{A}^{-1}) = 1.$$

Hence det $\mathbf{A} \neq 0$ and in fact $\det(\mathbf{A}^{-1}) = (\det \mathbf{A})^{-1}$.

To prove the converse we will actually give a method for constructing $\mathbf{A}^{-1}$. Let us introduce the numbers

$$A_{ij} = (-1)^{i+j} \det \mathbf{M}_{ij}.$$

The number A_{ij} is called the *cofactor* of a_{ij}. In this notation we have

Lemma 14.7. det $\mathbf{A} = \sum_j a_{ij} A_{ij}$ *for any* $i = 1, 2, \ldots$

PROOF. Let **B** be the matrix obtained from **A** by interchanging the first and ith rows. Then

$$\det \mathbf{A} = -\det \mathbf{B} = -\sum b_{1j} B_{1j}.$$

Next note that the minors $\mathbf{M}_{ij}(B)$ of the ith row of B differ from the minors $\mathbf{M}_{1j}(\mathbf{A})$ of the first row of **A** in that the first row of $\mathbf{M}_{ij}(\mathbf{B})$ is the $(i-1)$th row of $\mathbf{M}_{1j}(\mathbf{A})$. By interchanging this row with the one below it in $\mathbf{M}_{1j}(\mathbf{B})$ $i-2$ times we can put it in the $(i-1)$th row. Therefore $B_{1j} = -A_{ij}$ and hence $\det \mathbf{A} = -\det \mathbf{B} = -\sum_j b_{1j} B_{1j} = \sum_j a_{ij} A_{ij}$. □

Remark. Note in view of the fact that $\det \mathbf{A} = \det \mathbf{A}^t$ it follows from this lemma that det **A** may be computed from the minors of any column also.

Lemma 14.8. $\sum a_{ij} A_{kj} = 0$ *if* $k \neq i$.

PROOF. Let **B** be the matrix obtained from **A** by replacing the kth row of **A** by the ith row. Then **B** has two rows equal, so det **B** $= 0$. On the other hand

$$B_{kj} = A_{kj}$$

and using (14.6) to evaluate det **B** by minors of the kth row we find

$$0 = \det \mathbf{B} = \sum b_{kj} B_{kj} = \sum a_{ij} A_{kj}$$

as required. □

Let us define now a matrix $\mathbf{A}^{\text{cof}} = (a_{ij}^*)$ called the *cofactor matrix of* **A** by $a_{ij}^* = A_{ji}$. (Note carefully the switch in index.) We then have

Lemma 14.9. $\mathbf{AA}^{\text{cof}} = (\det \mathbf{A})\mathbf{I}$.

PROOF. Let $\mathbf{AA}^{\text{cof}} = \mathbf{C} = (c_{ij})$. Then

$$c_{ik} = \sum_j a_{ij} a^*_{jk} = \sum_j a_{ij} A_{kj}$$

so by (14.7)

$$c_{ik} = \begin{cases} 0 & i \neq k \\ \det \mathbf{A} & i = k \end{cases}$$

so that $\mathbf{C} = (\det \mathbf{A})\mathbf{I}$ as required. □

Returning now to the proof of (14.6) we suppose that $\det \mathbf{A} \neq 0$. Let

$$\mathbf{B} = \frac{1}{\det \mathbf{A}} \mathbf{A}^{\text{cof}}.$$

Then

$$\mathbf{AB} = \frac{1}{\det \mathbf{A}} \mathbf{AA}^{\text{cof}} = \frac{1}{\det \mathbf{A}} (\det \mathbf{A})\mathbf{I} = \mathbf{I}$$

as required. □

EXAMPLE 10. Calculate the inverse of the matrix of Example 8.

Solution. We must calculate the cofactors of $\mathbf{A}$. We find

$$A_{11} = (-1)^{1+1} \det \begin{pmatrix} 1 & 2 \\ 4 & 6 \end{pmatrix} = 6 - 8 = -2$$

$$A_{12} = (-1)^{1+2} \det \begin{pmatrix} 2 & 2 \\ 0 & 6 \end{pmatrix} = -12$$

$$A_{13} = (-1)^{1+3} \det \begin{pmatrix} 2 & 1 \\ 0 & 4 \end{pmatrix} = 8$$

$$A_{21} = (-1)^{2+1} \det \begin{pmatrix} 0 & 1 \\ 4 & 6 \end{pmatrix} = -(0 - 4) = 4$$

$$A_{22} = (-1)^{2+2} \det \begin{pmatrix} 1 & 1 \\ 0 & 6 \end{pmatrix} = 6$$

$$A_{23} = (-1)^{2+3} \det \begin{pmatrix} 1 & 0 \\ 0 & 4 \end{pmatrix} = -4$$

$$A_{31} = (-1)^{3+1} \det \begin{pmatrix} 0 & 1 \\ 1 & 2 \end{pmatrix} = -1$$

$$A_{32} = (-1)^{3+2} \det \begin{pmatrix} 1 & 1 \\ 2 & 2 \end{pmatrix} = 0$$

$$A_{33} = (-1)^{3+3} \det \begin{pmatrix} 1 & 0 \\ 2 & 1 \end{pmatrix} = 1.$$

Hence

$$\mathbf{A}^{\text{cof}} = \begin{pmatrix} 2 & 4 & -1 \\ -12 & 6 & 0 \\ 8 & -4 & 1 \end{pmatrix}$$

and therefore

$$\mathbf{A}^{-1} = \frac{1}{6}\begin{pmatrix} -2 & 4 & -1 \\ -12 & 6 & 0 \\ 8 & -4 & 1 \end{pmatrix}.$$

EXAMPLE 11. (Compare Chapter 6 Example 17.) Determine when the vectors $(r, 1, 1), (1, r, 1), (1, 1, r)$ are a basis for $\mathbb{R}^3$.

Solution. Consider the linear transformation

$$\mathsf{T} : \mathbb{R}^3 \to \mathbb{R}^3$$

that is the linear extension of

$$\begin{aligned} \mathsf{T}(1, 0, 0) &= (r, 1, 1) \\ \mathsf{T}(0, 1, 0) &= (1, r, 1) \\ \mathsf{T}(0, 0, 1) &= (1, 1, r). \end{aligned}$$

Then T is an isomorphism iff $\{(r, 1, 1), (1, r, 1), (1, 1, r)\}$ is a basis for $\mathbb{R}^3$. The matrix of T relative to the standard basis of $\mathbb{R}^3$ is

$$\begin{pmatrix} r & 1 & 1 \\ 1 & r & 1 \\ 1 & 1 & r \end{pmatrix}$$

and so T is invertible iff

$$\begin{aligned} 0 \neq \det\begin{pmatrix} r & 1 & 1 \\ 1 & r & 1 \\ 1 & 1 & r \end{pmatrix} &= r \det\begin{pmatrix} r & 1 \\ 1 & r \end{pmatrix} + (-1)\det\begin{pmatrix} 1 & 1 \\ 1 & r \end{pmatrix} + 1 \det\begin{pmatrix} 1 & r \\ 1 & 1 \end{pmatrix} \\ &= r(r^2 - 1) - (r - 1) + (1 - r) \\ &= (r - 1)[r(r + 1) - 2] = (r - 1)(r^2 + r - 2) \\ &= (r - 1)^2(r + 2). \end{aligned}$$

Therefore T is invertible iff $r \neq 1, -2$, so $\{(r, 1, 1), (1, r, 1), (1, 1, r)\}$ is a basis for $\mathbb{R}^3$ iff $r \neq 1, -2$.

As a method for computing the above procedure is quite systematic but fraught with the perils of arithmetic error.

Let us return now to the problem of calculating the eigenvalues of an endomorphism $\mathsf{T} : \mathscr{V} \to \mathscr{V}$ of the finite-dimensional vector space $\mathscr{V}$. Choose a basis $\{\mathbf{A}_1, \ldots, \mathbf{A}_n\}$ for $\mathscr{V}$ and let T be represented by the matrix $\mathbf{A} = (a_{ij})$ in this basis.

Definition. The *characteristic polynomial* of T is the polynomial of degree n, $\Delta(t) = \det(\mathbf{A} - t\mathbf{I})$, where t is a variable.

Remark. It may seem as though the characteristic polynomial of T depends on the choice of basis $\{\mathbf{A}_1, \ldots, \mathbf{A}_n\}$. To see that this is not so suppose that $\{\mathbf{B}_1, \ldots, \mathbf{B}_n\}$ is another basis for $\mathscr{V}$ and that the matrix of T is $\mathbf{B}$ relative to this basis. Then by (12.6)

$$\mathbf{B} = \mathbf{PA}(\mathbf{P}^{-1})$$

(**since we are using the same bases in domain and range**). Thus

$$\begin{aligned} \mathbf{B} - t\mathbf{I} &= \mathbf{PAP}^{-1} - t\mathbf{I} = \mathbf{PAP}^{-1} - t\mathbf{PIP}^{-1} \\ &= \mathbf{P}(\mathbf{A} - t\mathbf{I})\mathbf{P}^{-1}. \end{aligned}$$

Therefore

$$\begin{aligned} \det(\mathbf{B} - t\mathbf{I}) &= \det(\mathbf{P}(\mathbf{A} - t\mathbf{I})\mathbf{P}^{-1}) \\ &= (\det \mathbf{P})(\det(\mathbf{A} - t\mathbf{I}))(\det \mathbf{P}^{-1}) \\ &= (\det \mathbf{P})(\det \mathbf{P}^{-1})(\det(\mathbf{A} - t\mathbf{I})) \\ &= (\det \mathbf{P})(\det \mathbf{P})^{-1}(\det(\mathbf{A} - t\mathbf{I})) \\ &= \det(\mathbf{A} - t\mathbf{I}). \end{aligned}$$

Therefore it does not matter what basis we use to represent T by a matrix to compute its characteristic polynomial.

By combining (14.3) and (14.6) we obtain the important:

Theorem 14.10 (Criteria for Eigenvalues.) *Let $\mathsf{T} : \mathscr{V} \to \mathscr{V}$ be an endomorphism with characteristic polynomial $\Delta(t)$. Then a number e is an eigenvalue of T iff e is a root of $\Delta(t)$.*

PROOF. Choose a basis for $\mathscr{V}$, $\{\mathbf{A}_1, \ldots, \mathbf{A}_n\}$ and let $\mathbf{A} = (a_{ij})$ be the matrix of T relative to this basis.

Suppose that e is an eigenvalue of T. Then by (14.3) the transformation

$$\mathsf{T} - e\mathsf{I} : \mathscr{V} \to \mathscr{V}$$

is singular. Its matrix relative to $\{\mathbf{A}_1, \ldots, \mathbf{A}_n\}$ is the matrix $\mathbf{A} - e\mathbf{I}$. Since $\mathsf{T} - e\mathsf{I}$ is not an isomorphism the matrix $\mathbf{A} - e\mathbf{I}$ has no inverse, (12.4). So,

$$\det(\mathbf{A} - e\mathbf{I}) = 0$$

which says that the number e is a root of the polynomial in t

$$\Delta(t) = \det(\mathbf{A} - t\mathbf{I}).$$

To prove the converse suppose that e is a root of $\Delta(t)$. Then

$$0 = \Delta(e) = \det(\mathbf{A} - e\mathbf{I}).$$

So by (14.5) $\mathbf{A} - e\mathbf{I}$ is not invertible and hence

$$\mathsf{T} - e\mathsf{I} : \mathscr{V} \to \mathscr{V}$$

is singular, so again by (14.3) e is an eigenvalue of T. □

EXAMPLE 12. Let $\mathsf{T} : \mathbb{R}^3 \to \mathbb{R}^3$ be the linear transformation given by the formula

$$\mathsf{T}(x, y, z) = (0, x, y).$$

Find the characteristic polynomial of T, T^2, and T^3.

Solution. The matrix of T relative to the standard basis is

$$\mathbf{A} = \begin{pmatrix} 0 & 0 & 0 \\ 1 & 0 & 0 \\ 0 & 1 & 0 \end{pmatrix}$$

so those of T^2 and T^3 are

$$\mathbf{A}^2 = \begin{pmatrix} 0 & 0 & 0 \\ 0 & 0 & 0 \\ 1 & 0 & 0 \end{pmatrix}, \qquad \mathbf{A}^3 = \mathbf{0}.$$

Therefore, (since we are dealing with more than one endomorphism we will employ subscripts to indicate the transformation's characteristic polynomial),

$$\begin{aligned} \Delta_{\mathsf{T}}(t) &= \det(\mathbf{A} - t\mathbf{I}) \\ &= \det\begin{pmatrix} -t & 0 & 0 \\ 1 & -t & 0 \\ 0 & 1 & -t \end{pmatrix} = -t^3, \\ \Delta_{\mathsf{T}^2} &= \det(\mathbf{A}^2 - t\mathbf{I}) \\ &= \det\begin{pmatrix} -t & 0 & 0 \\ 0 & -t & 0 \\ 1 & 0 & -t \end{pmatrix} = -t^3, \\ \Delta_{\mathsf{T}^3}(t) &= \det(\mathbf{A}^3 - t\mathbf{I}) \\ &= \det\begin{pmatrix} -t & 0 & 0 \\ 0 & -t & 0 \\ 0 & 0 & -t \end{pmatrix} = -t^3. \end{aligned}$$

Thus T, T^2 and $\mathsf{T}^3 = \mathsf{0}$ have the same characteristic polynomial.

EXAMPLE 13. Find the characteristic polynomial and the eigenvalues of the linear transformation $\mathsf{T}: \mathbb{R}^3 \to \mathbb{R}^3$ given by

$$\mathsf{T}(x, y, z) = (x - 2z, 0, -2x + 4z).$$

Solution. The matrix of T relative to the standard basis is

$$\mathbf{A} = \begin{pmatrix} 1 & 0 & -2 \\ 0 & 0 & 0 \\ -2 & 0 & 4 \end{pmatrix}$$

and thus the characteristic polynomial of T is

$$\begin{aligned} \det(\mathbf{A} - t\mathbf{I}) &= \det\begin{pmatrix} 1-t & 0 & -2 \\ 0 & -t & 0 \\ -2 & 0 & 4-t \end{pmatrix} \\ &= (1-t)(-t)(4-t) - (-2)(-t)(-2) = -t^3 + 5t^2. \end{aligned}$$

So the characteristic polynomial is $\Delta(t) = -t^3 + 5t^2$. The eigenvalues are $t = 0$, $t = 0$, $t = 5$. The corresponding eigenspaces are obtained by solving the systems of equations

$$t = 0 \quad \begin{cases} x \qquad\quad - 2z = 0 \\ -2x \qquad + 4z = 0 \end{cases}$$

$$t = 5 \quad \begin{cases} -4x \qquad\quad - 2z = 0 \\ \qquad - 5y \qquad = 0 \\ -2x \qquad\quad - \ z = 0. \end{cases}$$

We find

$$\begin{aligned} &\dim \mathscr{V}_0 = 2 \quad \text{spanned by } (0, 1, 0), (2, 0, 1) \\ &\dim \mathscr{V}_5 = 1 \quad \text{spanned by } (1, 0, -2). \end{aligned}$$

Thus $\mathbb{R}^3$ has the basis

$$\mathbf{E}_1 = (1, 0, -2) \qquad \mathbf{E}_2 = (0, 1, 0) \qquad \mathbf{E}_3 = (2, 0, 1)$$

composed of eigenvectors of T. Since

$$\begin{aligned} \mathsf{T}(\mathbf{E}_1) &= 5\mathbf{E}_1 \\ \mathsf{T}(\mathbf{E}_2) &= \mathbf{0} \\ \mathsf{T}(\mathbf{E}_3) &= \mathbf{0} \end{aligned}$$

the matrix of T relative to this basis is

$$\mathbf{B} = \begin{pmatrix} 5 & 0 & 0 \\ 0 & 0 & 0 \\ 0 & 0 & 0 \end{pmatrix}.$$

Notice that if we were to use the matrix **B** to compute the characteristic polynomial of T we would get

$$\Delta(t) = \det(\mathbf{B} - t\mathbf{I}) = \det\begin{pmatrix} 5-t & 0 & 0 \\ 0 & -t & 0 \\ 0 & 0 & -t \end{pmatrix}$$

$$= (5-t)(-t)(-t) = -t^3 + 5t^2$$

as before.

The procedure we have gone through is called **diagonalizing** the linear transformation T. *Warning*: Not every linear transformation can be diagonalized! Go back and stare at Example 6.

EXAMPLE 14. Diagonalize, if possible, the linear transformation:

$$\mathsf{T} : \mathbb{R}^4 \to \mathbb{R}^4$$

given by

$$\mathsf{T}(x, y, z, w) = (x, 2x + 5y + 6z + 7w, 3x + 8z + 9w, 4x + 10w)$$

Solution. The matrix of T is computed first. We have

$$\begin{aligned} \mathsf{T}(1, 0, 0, 0) &= (1, 2, 3, 4) \\ \mathsf{T}(0, 1, 0, 0) &= (0, 5, 0, 0) \\ \mathsf{T}(0, 0, 1, 0) &= (0, 6, 8, 0) \\ \mathsf{T}(0, 0, 0, 1) &= (0, 7, 9, 10). \end{aligned}$$

So the matrix of T relative to the standard base is

$$\mathbf{A} = \begin{pmatrix} 1 & 0 & 0 & 0 \\ 2 & 5 & 6 & 7 \\ 3 & 0 & 8 & 9 \\ 4 & 0 & 0 & 10 \end{pmatrix}.$$

The characteristic polynomial of T is thus

$$\Delta(t) = \det(\mathbf{A} - t\mathbf{I}) = \det\begin{pmatrix} 1-t & 0 & 0 & 0 \\ 2 & 5-t & 6 & 7 \\ 3 & 0 & 8-t & 9 \\ 4 & 0 & 0 & 10-t \end{pmatrix}$$

$$= (1-t)\det\begin{pmatrix} 5-t & 6 & 7 \\ 0 & 8-t & 9 \\ 0 & 0 & 10-t \end{pmatrix}$$

(by property 7 of determinants)

$$= (1-t)(5-t)(8-t)(10-t).$$

Thus the eigenvalues of T are

$$t = 1, 5, 8, 10.$$

The eigenspaces are computed by solving the systems

$$t = 1 \quad \begin{cases} 2x + 4y + 6z + 7w = 0 \\ 3x \qquad\quad + 7z + 9w = 0 \\ 4x \qquad\qquad\quad + 9w = 0 \end{cases}$$

$$-4x = 9w, \qquad 3x + 7z - 4x = 0, \qquad 7z = x$$

$$2x + 4y + \tfrac{6}{7}x - \tfrac{28}{9}x = 0 \qquad 4y = (\tfrac{28}{9} - \tfrac{6}{7} - 2)x$$

$$w = -\tfrac{4}{9}x \qquad 4y = \tfrac{16}{63}x$$

$$z = \tfrac{1}{7}x \qquad y = \tfrac{4}{63}x.$$

So

$$\mathscr{V}_1 = \mathscr{L}(63, 4, 9, -28).$$

$$t = 5 \quad \begin{cases} -4x \qquad\qquad\quad = 0 \\ \quad 2x + 6z + 7w = 0 \\ \quad 3x + 3z + 9w = 0 \\ \quad 4x \qquad\quad 5w = 0 \end{cases}$$

$$x = 0, \qquad z = 0, \qquad w = 0, \qquad y = \text{anything}.$$

So

$$\mathscr{V}_5 = \mathscr{L}(0, 1, 0, 0).$$

Next we have

$$t = 8 \quad \begin{cases} -7x \qquad\qquad\qquad\quad = 0 \\ \quad 2x - 3y + 6z + 7w = 0 \\ \quad 3x \qquad\qquad\quad + 9w = 0 \\ \quad 4x \qquad\qquad\quad + 2w = 0 \end{cases}$$

$$x = 0, \qquad w = 0, \qquad y = 2z.$$

Therefore

$$\mathscr{V}_8 = \mathscr{L}(0, 2, 1, 0).$$

Finally

$$t = 10 \quad \begin{cases} -9x \qquad\qquad\qquad\quad = 0 \\ \quad 2x - 5y + 6z + 7w = 0 \\ \quad 3x \qquad\quad - 2z + 9w = 0 \\ \quad 4x \qquad\qquad\qquad\quad = 0 \end{cases}$$

so

$$x = 0$$

$$z = \frac{9}{2} w$$

$$5y = 7w + 6z = 7w + \frac{6(9)}{2} w = 34w, \qquad y = \frac{34}{5}.$$

Thus

$$\mathscr{V}_{10} = \mathscr{L}(0, 68, 45, 10).$$

The vectors

$$\begin{aligned} \mathbf{E}_1 &= (63, 4, 9, -28) \\ \mathbf{E}_2 &= (0, 1, 0, 0) \\ \mathbf{E}_3 &= (0, 2, 1, 0) \\ \mathbf{E}_4 &= (0, 68, 45, 10) \end{aligned}$$

may be shown to be linearly independent. (See, for example, Proposition 14.11 below.) Thus $\mathbb{R}^4$ has as a basis the eigenvectors $\{\mathbf{E}_1, \mathbf{E}_2, \mathbf{E}_3, \mathbf{E}_4\}$ of T. The matrix of T relative to these vectors is

$$\mathbf{B} = \begin{pmatrix} 1 & 0 & 0 & 0 \\ 0 & 5 & 0 & 0 \\ 0 & 0 & 8 & 0 \\ 0 & 0 & 0 & 10 \end{pmatrix}.$$

Proposition 14.11. *Let* $\mathsf{T}: \mathscr{V} \to \mathscr{V}$ *be an endomorphism of* $\mathscr{V}$. *Suppose* $e_1, \ldots, e_m$ *are* distinct *eigenvalues of* T *and* $\mathbf{E}_1, \ldots, \mathbf{E}_m$ *are corresponding eigenvectors. Then* $\mathbf{E}_1, \ldots, \mathbf{E}_m$ *are linearly independent.*

Proof. Suppose to the contrary that $\mathbf{E}_1, \ldots, \mathbf{E}_m$ are linearly dependent. Applying Proposition 6.2 we may find an integer k such that $\mathbf{E}_k$ is linearly dependent on $\mathbf{E}_1, \ldots, \mathbf{E}_{k-1}$. We may also suppose that k is the smallest integer with this property and so by (6.2) again we may suppose $\mathbf{E}_1, \ldots, \mathbf{E}_{k-1}$ are linearly independent. Now let

(A) $$\mathbf{E}_k = a_1\mathbf{E}_1 + a_2\mathbf{E}_2 + \cdots + a_{k-1}\mathbf{E}_{k-1}.$$

Apply T to this equation and get

(B) $$\mathsf{T}(\mathbf{E}_k) = a_1\mathsf{T}(\mathbf{E}_1) + a_2\mathsf{T}(\mathbf{E}_2) + \cdots + a_{k-1}\mathsf{T}(\mathbf{E}_{k-1}).$$

Since the $\mathbf{E}_1, \ldots, \mathbf{E}_k$ are eigenvectors of T associated to the eigenvalues $e_1, \ldots, e_k$ this equation yields

(C) $$e_k\mathbf{E}_k = e_1a_1\mathbf{E}_1 + e_2a_2\mathbf{E}_2 + \cdots + e_{k-1}a_{k-1}\mathbf{E}_{k-1}.$$

Now multiply Equation (A) by e_k to obtain

(D) $$e_k\mathbf{E}_k = e_k a_1\mathbf{E}_1 + e_k a_2\mathbf{E}_2 + \cdots + e_k a_{k-1}\mathbf{E}_{k-1}.$$

and subtract Equation (D) from Equation (C) giving

(E) $$\mathbf{0} = a_1(e_1 - e_k)\mathbf{E}_1 + a_2(e_2 - e_k)\mathbf{E}_2 + \cdots + a_{k-1}(e_{k-1} - e_k)\mathbf{E}_{k-1}.$$

Since the vectors $\mathbf{E}_1, \ldots, \mathbf{E}_{k-1}$ are linearly independent it follows that

(F) $$\begin{aligned} a_1(e_1 - e_k) &= 0 \\ a_2(e_2 - e_k) &= 0 \\ &\vdots \\ a_{k-1}(e_{k-1} - e_k) &= 0. \end{aligned}$$

Next we recall that the eigenvalues $e_1, \ldots, e_k$ are distinct, so that

(G) $$\begin{aligned} (e_1 - e_k) &\neq 0 \\ &\vdots \\ (e_{k-1} - e_k) &\neq 0 \end{aligned}$$

so combining (F) and (G) we find

(H) $$\begin{aligned} a_1 &= 0 \\ &\vdots \\ a_{k-1} &= 0. \end{aligned}$$

Going back to Equation (A) this says

(I) $$\mathbf{E}_k = \mathbf{0}.$$

But an eigenvector is by definition **nonzero**, and hence our assumption that $\{\mathbf{E}_1, \ldots. \mathbf{E}_m\}$ is linearly dependent has led to a contradiction. Therefore $\{\mathbf{E}_1, \ldots, \mathbf{E}_m\}$ is linearly independent. □

If you think that the proof of (14.11) is hard compare it with the pain of proving directly that the vectors $\mathbf{E}_1, \mathbf{E}_2, \mathbf{E}_3, \mathbf{E}_4$ of Example 14 are independent.

Since the characteristic polynomial of an endomorphism $\mathsf{T}: \mathscr{V} \to \mathscr{V}$ of a vector space $\mathscr{V}$ of dimension n is of degree n we find the following useful diagonalization result.

Corollary 14.12. *Let $\mathsf{T}: \mathscr{V} \to \mathscr{V}$ be an endomorphism of the n-dimensional vector space $\mathscr{V}$. Suppose that the characteristic polynomial of T has n distinct real roots. Then there is a basis of $\mathscr{V}$ composed of eigenvectors of T. Relative to this basis the matrix of T is diagonal.*

PROOF. Let $e_1, \ldots, e_n$ be the eigenvalues of T. There are n of them because the characteristic polynomial has degree n. By assumption they are distinct. Choose eigenvectors $\mathbf{E}_1, \ldots, \mathbf{E}_n$ corresponding to $e_1, \ldots, e_n$. By (14.11) the vectors $\mathbf{E}_1, \ldots, \mathbf{E}_n$ are linearly independent. Therefore by (6.9) $\{\mathbf{E}_1, \ldots, \mathbf{E}_n\}$ is a basis for $\mathscr{V}$. □

It is quite natural to inquire what the situation is when the characteristic polynomial does not factor into distinct linear factors. In general this can mean a quite complicated structure for the endomorphism $\mathsf{T}: \mathscr{V} \to \mathscr{V}$. There are however a few general results we can deduce. We begin with a definition.

Definition. Let $\mathsf{T}: \mathscr{V} \to \mathscr{V}$ be an endomorphism of the finite-dimensional vector space $\mathscr{V}$ and e an eigenvalue of T. The *geometric multiplicity* of e is the dimension of the eigenspace $\mathscr{V}_e$. The *algebraic multiplicity* of e is the multiplicity of the root e of the characteristic polynomial $\Delta(t)$ of T.

Proposition 14.13. *Let $\mathsf{T}: \mathscr{V} \to \mathscr{V}$ be an endomorphism of the finite-dimensional vector space $\mathscr{V}$ and e an eigenvalue of T. Then the geometric multiplicity of e is less than or equal to the algebraic multiplicity of e.*

PROOF. Recall from the remark preceding Theorem 14.10 that we may calculate the characteristic polynomial of T by representing T as a matrix relative to any basis of $\mathscr{V}$. Let us now cleverly choose such a basis for $\mathscr{V}$. Let us suppose $\{\mathbf{A}_1, \ldots, \mathbf{A}_r\}$ a basis for $\mathscr{V}_e$, so that the geometric multiplicity of e is equal to r. Let n be the dimension of $\mathscr{V}$ and extend $\{\mathbf{A}_1, \ldots, \mathbf{A}_r\}$ to a basis $\{\mathbf{A}_1, \ldots, \mathbf{A}_r, \mathbf{A}_{r+1}, \ldots, \mathbf{A}_n\}$ of $\mathscr{V}$. The matrix of T relative to this basis is easily seen to be

$$\mathbf{A} = \begin{pmatrix} e & & & \\ & \ddots & & \mathbf{B} \\ & & e & \\ & \mathbf{0} & & \mathbf{C} \end{pmatrix}$$

where $\mathbf{B}$ is an $r \times n - r$ matrix, $\mathbf{C}$ is an $n - r \times n - r$ matrix and $\mathbf{0}$ an $n - r \times r$ matrix of zeros.

The characteristic polynomial of T is therefore

$$\Delta(t) = \det(\mathbf{A} - t\mathbf{I}) = \det \begin{pmatrix} e - t & & & \\ & e - t & & \mathbf{B} \\ & & \ddots & \\ & & & e - t \\ & \mathbf{0} & & \mathbf{C} - t\mathbf{I} \end{pmatrix}$$

and since $e - t$ is a factor of each of the first r columns we find

$$\Delta(t) = (e - t)^r \det \begin{pmatrix} \mathbf{I} & \mathbf{B} \\ \mathbf{0} & \mathbf{C} - t\mathbf{I} \end{pmatrix} = (e - t)^r q(t)$$

and therefore e is a root of $\Delta(t) = 0$ of multiplicity at least r which shows that the algebraic multiplicity of e is at least r and the result follows. □

Proposition 14.14. *If* $\mathsf{T} : \mathscr{V} \to \mathscr{V}$ *is a diagonalizable linear transformation then*

$$\Delta(t) = (t - e_1)^{m_1} \dots (t - e_k)^{m_k}$$

and

$$\dim \mathscr{V}_{e_i} = m_i \qquad i = 1, 2, \dots, k$$

where $e_1, \dots, e_k$ *are the distinct eigenvalues of* T *and* $m_1, \dots, m_k$ *their multiplicities.*

PROOF. Since $\mathbf{T}$ is diagonalizable it has a matrix representation

$$\mathbf{A} = \begin{pmatrix} \overbrace{e_1 \ \ddots \ e_1}^{n_1} & & & 0 \\ & \overbrace{e_2 \ \ddots \ e_2}^{n_2} & & \\ & & \ddots & \\ 0 & & & \overbrace{e_k \ \ddots \ e_k}^{n_k} \end{pmatrix}$$

where $n_i = \dim \mathscr{V}_{e_i}$ for $i = 1, \dots, k$. *Using this matrix representation to compute the characteristic polynomial of* T we get

$$\Delta(t) = \det \begin{pmatrix} \overbrace{e_1 - t \quad e_1 - t \ \ddots \ e_1 - t}^{n_1} & & 0 \\ & \ddots & \\ 0 & & \overbrace{e_k - t \quad e_k - t \ \ddots \ e_k - t}^{n_k} \end{pmatrix}$$

$$= (e_1 - t)^{n_1} \cdots (e_k - t)^{n_k}$$
$$= (-1)^n[t - e_1)^{n_1} \cdots (t - e_k)^{n_k}]$$

and therefore $n_1 = m_1, \dots, n_k = m_k$ as required. □

We may summarize all this in the following:

Theorem 14.15. *A linear transformation* $\mathsf{T} : \mathscr{V} \to \mathscr{V}$ *of the finite-dimensional vector space* $\mathscr{V}$ *is diagonalizable iff the following two conditions hold:*

(1) *The characteristic polynomial* $\Delta(t)$ *is a product of linear factors.*
(2) *For each eigenvalues of* T *the geometric and algebraic multiplicities are the same.*

At this point it is worthwhile to pause to reflect on the difference between the case of real or complex scalars. If we look at $\mathsf{T} : \mathbb{R}^2 \to \mathbb{R}^2$ given by

$$\mathsf{T}(x, y) = (y, -x)$$

which is a rotation through 90° we find that the matrix of T relative to the usual basis of $\mathbb{R}^2$ is

$$\begin{pmatrix} 0 & 1 \\ -1 & 0 \end{pmatrix}$$

so that

$$\Delta(t) = \det\begin{pmatrix} -t & 1 \\ -1 & -t \end{pmatrix} = t^2 + 1,$$

and of course as we all know

$$t^2 + 1 = 0$$

has no real roots. So this simple transformation of $\mathbb{R}^2$ has no real eigenvalues. On the other hand it does have two distinct complex roots, namely $\pm i = \pm\sqrt{-1} \in \mathbb{C}$. This means that the transformation

$$\mathsf{S} : \mathbb{C}^2 \to \mathbb{C}^2$$

given by

$$\mathsf{S}(u, v) = (v, -u)$$

which has characteristic polynomial

$$\Delta(t) = t^2 + 1$$

can be diagonalized, while the transformation T with the same (matrix and) characteristic polynomial does not have even a single eigenvalue. In fact an endomorphism $\mathsf{S} : \mathscr{V} \to \mathscr{V}$ of a (finite-dimensional) vector space over the complex number field $\mathbb{C}$ must *always* have an eigenvalue. Thus it has at least one eigenvector. To see this recall the following:

Fundamental Theorem of Algebra: *Suppose that* $p(t) = a_n t^n + \cdots + a_1 t + a_0$ *is a polynomial of degree n (that is* $a_n \neq 0$*) with complex coefficients. Then there are complex numbers* $r_1, \ldots, r_n$ *(unique up to change of order) such that* $p(t) = a_n(t - r_1)(t - r_2)\cdots(t - r_n)$.

Therefore the first condition of (14.15) for a linear transformation to be diagonalizable is always satisfied in the complex case.

EXAMPLE 15. Find the characteristic polynomial, eigenvalues, and eigenvectors of the endomorphism

$$\mathsf{T}: \mathbb{R}^3 \to \mathbb{R}^3$$

given by

$$\mathsf{T}(x, y, z) = (2x + y, y - z, 2y + 4z).$$

Solution. We begin by calculating the matrix of T relative to the standard basis of $\mathbb{R}^3$. We find

$$\begin{aligned} \mathsf{T}(1, 0, 0) &= (2, 0, 0) \\ \mathsf{T}(0, 1, 0) &= (1, 1, 2) \\ \mathsf{T}(0, 0, 1) &= (0, -1, 4) \end{aligned}$$

and so the matrix we seek is

$$\mathbf{A} = \begin{pmatrix} 2 & 1 & 0 \\ 0 & 1 & -1 \\ 0 & 2 & 4 \end{pmatrix}.$$

The characteristic polynomial of T is therefore

$$\begin{aligned} \Delta(t) &= \det\begin{pmatrix} 2-t & 1 & 0 \\ 0 & 1-t & -1 \\ 0 & 2 & 4-t \end{pmatrix} \\ &= (2-t)\det\begin{pmatrix} 1-t & -1 \\ 2 & 4-t \end{pmatrix} \\ &= (2-t)[(1-t)(4-t)+2] = (2-t)[4-5t+t^2+2] \\ &= (2-t)[t^2-5t+6] = (2-t)(t-2)(t-3) \\ &= -(t-2)^2(t-3). \end{aligned}$$

So the eigenvalues of T are 2, 3 with 2 having algebraic multiplicity 2.

To calculate the eigenspaces we must solve the linear systems below.

$$e = 2 \quad \begin{cases} 0x + y + 0z = 0 \\ 0 - y - z = 0 \\ 0 + 2y + 2z = 0 \end{cases}$$

which yields

$$y = 0 \qquad z = 0$$

and therefore $\mathscr{V}_2 = \mathscr{L}(1, 0, 0)$. That is the eigenspace associated to the eigenvalue 2 is 1-dimensional and is spanned by the vector (1, 0, 0).

$$e = 3 \quad \begin{cases} -x + y + 0z = 0 \\ 0x - 2y - z = 0 \\ 0x + 2y + z = 0 \end{cases}$$

which yields

$$x - y = 0 \qquad 2y + z = 0$$

so

$$x = y \quad z = -2y$$

and therefore the eigenspace corresponding to 3 is spanned by the vector $(1, 1, -2)$, that is $\mathscr{V}_3 = \mathscr{L}(1, 1, -2)$. Notice that while the eigenvalue 2 has algebraic multiplicity 2 it has geometric multiplicity only 1. It is therefore *not* possible to find a basis fo $\mathbb{R}^3$ composed of eigenvectors of T and hence T is not diagonalizable.

EXAMPLE 16. Let $\mathsf{T}: \mathbb{R}^3 \to \mathbb{R}^3$ be defined by

$$\mathsf{T}(x, y, z) = (0, x, y).$$

Determine if T can be diagonalized.

Solution. The matrix of T relative to the standard basis of $\mathbb{R}^3$ is

$$\begin{pmatrix} 0 & 0 & 0 \\ 1 & 0 & 0 \\ 0 & 1 & 0 \end{pmatrix}$$

so the characteristic polynomial of T is

$$\Delta(t) = \det \begin{pmatrix} -t & 0 & 0 \\ 1 & -t & 0 \\ 0 & 1 & -t \end{pmatrix} = -t^3$$

and the only eigenvalue of T is 0 with algebraic multiplicity 3. The geometric multiplicity cannot also be 3 because $\mathscr{V}_0 = \ker \mathsf{T}$ and $3 = \dim \mathscr{V}_0 = \dim \ker \mathsf{T}$ implies $\mathsf{T} = \mathsf{0}$. But certainly $\mathsf{T} \neq \mathsf{0}$. In fact the kernel of T is the subspace spanned by (0, 0, 1) so the geometric multiplicity of the eigenvalue 0 is 1.

EXAMPLE 17. Let $\mathsf{T} : \mathbb{R}^3 \to \mathbb{R}^3$ be defined by

$$\mathsf{T}(x, y, z) = (-x, -z, y)$$

and let $\mathsf{S} : \mathbb{C}^3 \to \mathbb{C}^3$ be defined by (the complex formula)

$$\mathsf{S}(u, v, w) = (-u, -w, v).$$

Determine if T or S can be diagonalized, and if so find their diagonal forms.

Solution. At least part of the problem can be worked simultaneously for both T and S because relative to the standard bases of $\mathbb{R}^3$ and $\mathbb{C}^3$ respectively both T and S have the same matrix:

$$\begin{pmatrix} -1 & 0 & 0 \\ 0 & 0 & -1 \\ 0 & 1 & 0 \end{pmatrix}$$

so their characteristic polynomials are the same, namely,

$$\Delta(t) = \det\begin{pmatrix} -1-t & 0 & 0 \\ 0 & -t & -1 \\ 0 & 1 & -t \end{pmatrix} = -(t+1)(t^2+1),$$

and now the difference between the real and complex case appears!

$$\Delta(t) = -(t+1)(t^2+1)$$

has only one real root so T does not diagonalize (though geometrically it is quite simple: reflection across the yz-plane followed by a 90° rotation around the x-axis) and S does since

$$\Delta(t) = -(t+1)(t-i)(t+i)$$

has three distinct eigenvalues, $-1, i, -i$. The diagonal form of S is

$$\begin{pmatrix} 1 & 0 & 0 \\ 0 & i & 0 \\ 0 & 0 & -i \end{pmatrix}.$$

The corresponding basis may be computed as follows.

For $e_1 = -1$:

$$\mathsf{S}(u, v, w) = -(u, v, w) = (-u, -v, -w)$$

so

$$(-u, -w, v) = (-u, -v, -w)$$

and passing to components gives

$$-w = -v \quad \text{and} \quad v = -w$$

so

$$w = v = 0$$

and therefore a basis for $\mathscr{V}_{-1}$ is $(1, 0, 0)$.

For $e_2 = i$:

$$\mathsf{S}(u, v, w) = i(u, v, w)$$

so

$$(-u, -w, v) = (iu, iv, iw),$$

taking components gives

$$-u = iu \qquad -w = iv \qquad v = iw$$

so

$$u = 0, \qquad v = i, \qquad w = 1.$$

Thus a basis for $\mathscr{V}_i$ is $(0, i, 1)$.

For $e_3 = -i$: From $\mathbf{S}(u, v, w) = -i(u, v, w)$, as before passing to components gives

$$-u = -iu \qquad -w = -iv \qquad v = -iw$$

so

$$u = 0, \qquad v = i, \qquad w = -1.$$

Thus a basis for $\mathscr{V}_{-i}$ is $(0, i, -1)$.

Exercises

1. Find the rank of each of the following linear transformations

$$\mathsf{T}: \mathbb{R}^3 \to \mathbb{R}^3, \qquad \mathsf{T}(x, y, z) = (x + y, y, z + x)$$

$$\mathsf{T}: \mathbb{R}^4 \to \mathbb{R}^4, \qquad \mathsf{T}(x, y, z, w) = (x - y, z - w, w + x, x - w)$$

$$\mathsf{T}: \mathscr{P}_3(\mathbb{R}) \to \mathscr{P}_3(\mathbb{R}), \qquad \mathsf{T}(p(x)) = x\frac{d}{dx}p(x)$$

$$\mathsf{T}: \mathscr{P}_3(\mathbb{R}) \to \mathscr{P}_3(\mathbb{R}), \qquad \mathsf{T}(p(x)) = \frac{d}{dx}(xp(x)).$$

2. Find the cofactor of a_{24} and a_{42} in the matrix

$$\begin{pmatrix} 1 & 7 & 7 & 1 & 7 & 7 \\ 2 & 8 & 6 & 2 & 8 & 6 \\ 3 & 9 & 5 & 3 & 9 & 5 \\ 4 & 10 & 4 & 4 & 10 & 4 \\ 5 & 9 & 3 & 5 & 9 & 3 \\ 6 & 8 & 2 & 6 & 8 & 2 \end{pmatrix}$$

3. Find the determinant of each of the following matrices

$$\begin{pmatrix} 1 & 2 \\ 3 & 4 \end{pmatrix}, \qquad \begin{pmatrix} 1 & 4 & 7 \\ 2 & 5 & 8 \\ 3 & 6 & 9 \end{pmatrix}$$

$$\begin{pmatrix} 1 & 2 & 3 & 4 \\ 1 & 2 & 3 & 4 \\ 1 & 2 & 3 & 4 \\ 1 & 2 & 3 & 4 \end{pmatrix}, \qquad \begin{pmatrix} 1 & 0 & 3 & 4 & 0 \\ 0 & 2 & 0 & 4 & 0 \\ 1 & 0 & 0 & 0 & 5 \\ 0 & 2 & 0 & 4 & 0 \\ 1 & 0 & 3 & 4 & 0 \end{pmatrix}.$$

4. Determine which of the following matrices are invertible. For those that are, compute the inverses.

$$\begin{pmatrix} 1 & 3 \\ 2 & 4 \end{pmatrix}, \qquad \begin{pmatrix} 1 & 0 & 0 & 0 \\ 0 & 2 & 0 & 0 \\ 0 & 0 & 3 & 0 \\ 0 & 0 & 0 & 4 \end{pmatrix}$$

$$\begin{pmatrix} 2 & 1 \\ 1 & -1 \\ 2 & 2 \end{pmatrix}, \qquad \begin{pmatrix} 1 & 1 & 2 & 3 \\ 1 & 0 & 1 & 0 \\ 3 & 0 & 4 & 5 \end{pmatrix}$$

$$\begin{pmatrix} 1 & 1 & -1 & 2 \\ 0 & 1 & 0 & -1 \\ 0 & 0 & 1 & 0 \\ 0 & 0 & 0 & 1 \end{pmatrix}$$

5. Can you find a 2×2 matrix $\mathbf{A}$ such that $\mathbf{A} = \mathbf{A}^{\text{cof}}$?

6. Find the characteristic polynomial, eigenvalues, and eigenvectors and diagonalize *if possible* each of the following linear tansformations:

$$\begin{array}{ll} \mathsf{T} : \mathbb{R}^3 \to \mathbb{R}^3 & \mathsf{T}(x, y, z) = (x + 2y, 2y + x, z) \\ \mathsf{T} : \mathbb{R}^3 \to \mathbb{R}^3 & \mathsf{T}(x, y, z) = (3x - z, y - x + 2z, 4z) \\ \mathsf{T} : \mathbb{R}^4 \to \mathbb{R}^4 & \mathsf{T}(x, y, z, w) = (x + y, x, z + 2w, 2z + w) \\ \mathsf{T} : \mathbb{R}^4 \to \mathbb{R}^4 & \mathsf{T}(x, y, z, w) = (x + y, y, 2z, y + 2z + 2w) \\ \mathsf{T} : \mathbb{R}^2 \to \mathbb{R}^2 & \mathsf{T}(x, y) = (y, -x) \\ \mathsf{T} : \mathbb{R}^2 \to \mathbb{R}^3 & \mathsf{T}(x, y) = (y, -x, z). \end{array}$$

7. Prove that if a transformation $\mathsf{T}: \mathscr{V} \to \mathscr{V}$ is nilpotent then all its eigenvalues are zero.

8. Let $\mathsf{T} : \mathbb{R}^3 \to \mathbb{R}^3$. Prove that T has at least one eigenvalue.

9. Suppose that $\mathsf{T}: \mathscr{V} \to \mathscr{V}$ is nonsingular. What relation is there between the eigenvalues of T and T^{-1}?

10. Give an example of a linear transformation $\mathsf{T} : \mathbb{R}^2 \to \mathbb{R}^2$ without any eigenvalues. Repeat the problem for $\mathsf{T} : \mathbb{R}^4 \to \mathbb{R}^4$.

11. Suppose that e is an eigenvalue of $\mathsf{T} : \mathscr{V} \to \mathscr{V}$ with corresponding eigenvector $\mathbf{E}$. Prove that $\mathbf{E}$ is an eigenvector of T^k corresponding to the eigenvalue e^k.

12. Suppose that $\mathsf{T} : \mathbb{R}^n \to \mathbb{R}^n$ is a linear transformation with distinct eigenvalues $e_1, \ldots, e_n$. Show that $\det \mathsf{T} = e_1 \cdots e_n$.

13. Let $\mathsf{T} : \mathscr{V} \to \mathscr{V}$ be a diagonalizable endomorphism. Show that T^2 is also diagonalizable. If the diagonal form of T is

$$\begin{pmatrix} e_1 & & & \\ & \ddots & & 0 \\ & & e_2 & \\ 0 & & & \ddots \\ & & & & e_n \end{pmatrix}$$

what is the diagonal form of T^2?

14. Let $\mathbf{A} = (a_{ij})$, $i, j = 1, 2, 3$. Then $\det \mathbf{A} = a_{11} \det \mathbf{M}_{11} - a_{12} \det \mathbf{M}_{12} + a_{12} \det \mathbf{M}_{13}$, where $\mathbf{M}$, where $\mathbf{M}_{1j}$, $j = 1, 2, 3$. are minors of the element a_{1j} of $\mathbf{A}$. Show that

$$\det \mathbf{A} = a_{11}a_{22}a_{33} + a_{12}a_{23}a_{31} + a_{13}a_{21}a_{32} - a_{13}a_{22}a_{31} - a_{12}a_{21}a_{33} - a_{11}a_{23}a_{32}.$$

15. Using the above expression of $\det \mathbf{A}$ for 3×3 matrix, show properties (1)–(8) of $\det \mathbf{A}$ hold for 3×3 matrices.

16. Let

$$\mathbf{A} = \begin{pmatrix} a_1 & b_1 & c_1 \\ a_2 & b_2 & c_2 \\ a_3 & b_3 & c_3 \end{pmatrix}.$$

Show that $\det \mathbf{A} \neq 0$ iff column vectors form a set of linearly independent vectors. (*Hint*: consider $\mathbf{A}$ to be the matrix of a linear transformation with respect to the standard basis.)

17. Let (a, b, c), (a', b', c') be two independent vectors. Suppose (x, y, z) is another vector which lies in the linear span of (a, b, c) and (a', b', c'). Then

$$\det \begin{pmatrix} x & y & z \\ a & b & c \\ a' & b' & c' \end{pmatrix} = 0,$$

i.e., $(bc' - cb')x + (ca' - ac')y + (ab' - ba')z = 0$. (Compare to (**) preceding Example 5 of Chapter 1.)

18. Given vectors $(a, b, c), (a', b', c') \in \mathbb{R}^3$ the vector

$$(bc' - b'c, ca' - ac', ab' - ba')$$

is called the outer-product (or cross-product) of vectors (a, b, c) and (a', b', c'), and denoted by $(a, b, c) \times (a', b', c')$. Show

(a) $(a, b, c) \times (a', b', c') = -(a', b', c') \times (a, b, c)$.
(b) $(a, b, c) \times (a, b, c) = 0$.

19. (Cramer's rule). Let

$$a_{i1}x_1 + a_{i2}x_2 + \cdots + a_{in}x_n = b_i, \qquad i = 1, 2, \ldots, n,$$

and suppose that the matrix $\mathbf{A} = (a_{ij})$, gotten from coefficients of the above system of linear equations, has the inverse $\mathbf{A}^{-1}$. Then

$$\mathbf{X} = \begin{pmatrix} x_1 \\ x_2 \\ \vdots \\ x_n \end{pmatrix} = \mathbf{A}^{-1}\mathbf{B}, \quad \text{where } \mathbf{B} = \begin{pmatrix} b_1 \\ b_2 \\ \vdots \\ b_n \end{pmatrix}.$$

Also, $x_i = \det \mathbf{A}_i / \det \mathbf{A}$, $i = 1, 2, \ldots, n$, where $\mathbf{A}_i$ is the matrix $\mathbf{A}$ whose ith column is replaced by $\mathbf{B}$, $i = 1, 2, \ldots, n$. Verify the above statement for $n = 3$. The statement holds for any n, (finite!). But for all practical purpose it is useless for a large n, in computing x_i, since you have to compute $n + 1$ determinants of size n matrices (such a tedious task!).

20. Solve the following system of linear equations:

$$\begin{aligned} x_1 + x_2 + x_3 + x_4 &= 5 \\ x_2 + 2x_3 - x_4 &= -3 \\ x_1 + 3x_2 + 4x_3 - x_4 &= 0 \\ 2x_1 - x_2 - x_3 + x_4 &= 4 \end{aligned}$$

using Cramer's rule.

21. Using Cramer's rule solve the following systems of linear equations:

(a) $$\begin{aligned} x + y + z &= 0 \\ 2x - y + z &= 6 \\ 4x + 5y - z &= 2 \end{aligned}$$

(b) $$\begin{aligned} 3x + y - z &= 1 \\ -x - y + 4z &= 7 \\ 2x + y - 5z &= -8 \end{aligned}$$

22. Compute det **H**, where

$$\mathbf{H} = \begin{pmatrix} 1 & x & x^2 & x^3 \\ 1 & y & y^2 & y^3 \\ 1 & z & z^2 & z^3 \\ 1 & w & w^2 & w^3 \end{pmatrix}.$$

23. Find the inverse of each of the following matrices

$$\begin{pmatrix} 0 & 1 & 1 \\ 1 & 0 & 1 \\ 1 & 1 & 0 \end{pmatrix}, \quad \begin{pmatrix} 0 & 1 & 1 & 1 \\ 1 & 0 & 1 & 1 \\ 1 & 1 & 0 & 1 \\ 1 & 1 & 1 & 0 \end{pmatrix}, \quad \begin{pmatrix} 0 & 1 & 1 & 1 & 1 \\ 1 & 0 & 1 & 1 & 1 \\ 1 & 1 & 0 & 1 & 1 \\ 1 & 1 & 1 & 0 & 1 \\ 1 & 1 & 1 & 1 & 0 \end{pmatrix}$$

Can you find a general pattern?

24. Diagonalize each of the matrices in Exercise 23.

25. Let $\mathsf{T}: \mathscr{M}_i \to \mathscr{M}_i$ be the linear transformation given by

$$\mathsf{T}(\mathbf{M}) = \mathbf{M}^t.$$

(a) Find the characteristic polynomial of T.
(b) Determine the eigenvalues and eigenspaces of T.
(c) Is T diagonalizable? If so, diagonalize it; if not, prove it is not.

26. Recall (Chapter 12, Exercise 15) that a linear transformation $\mathsf{T}: \mathscr{V} \to \mathscr{V}$ is an n-dimensional vector space $\mathscr{V}$ is called a reflection if there is an $(n - 1)$-dimensional subspace $\mathscr{H}$ in $\mathscr{V}$ and a nonzero vector $\mathbf{v}$ in $\mathscr{V}$, not in $\mathscr{H}$ so that

$$\mathsf{T}(\mathbf{h}) = \mathbf{h} \quad \text{for all } \mathbf{h} \in \mathscr{H},$$

$$\mathsf{T}(\mathbf{v}) = -\mathbf{v}.$$

(a) Show that T is diagonalizable with diagonal form

$$\begin{pmatrix} -1 & & & 0 \\ & 1 & & \\ & & \ddots & \\ 0 & & & 1 \end{pmatrix}.$$

(b) Show $\det \mathsf{T} = -1$.

(c) Let T and S be reflections. Is T, S also a reflection?

27. Let $\mathsf{T}: \mathscr{V} \to \mathscr{V}$ be a cyclic transformation in the n-dimensional vector space $\mathscr{V}$ with cyclic vector **A**. Then there exist unique numbers $a_0, \ldots, a_{n-1}$ such that

$$\mathsf{T}^n(\mathbf{A}) = a_0\mathbf{A} + a_1\mathsf{T}(\mathbf{A}) + \cdots + a_{n-1}\mathsf{T}^{n-1}(\mathbf{A}).$$

Show that the characteristic polynomial of T is:

$$(-1)^{n+1}(a_0 + a_1 t + \cdots + a_{n-1}t^{n-1} - t^n)$$

28. Show that the only eigenvalues of an involution are ± 1.

29. Is $\det: \mathscr{M}_n \to \mathbb{R}$ a linear transformation?

30. Is the trace $tr: \mathscr{M}_n \to \mathbb{R}$ a linear transformation? (See p. 248 for the definition of the trace of a square matrix.)

Multilinear algebra: determinants 14 bis

Let n be a positive integer, and let $\mathscr{V}_1, \ldots, \mathscr{V}_n$, $\mathscr{W}$ be vector spaces. A function

$$f: \mathscr{V}_1 \times \cdots \times \mathscr{V}_n \to \mathscr{W}$$

is called a *multilinear form* iff for each integer i, $1 \leq i \leq n$, and each $(n-1)$-tuple $(v_1, \ldots, v_{i-1}, v_{i+1}, \ldots, v_n)$ the function

$$F: \mathscr{V}_i \to \mathscr{W}$$

defined by

$$F(v) = f(v_1, \ldots, v_{i-1}, v, v_{i+1}, \ldots, v_n)$$

is a linear transformation. For $n = 1$ a multilinear form is simply a linear transformation. For $n = 2$ we speak of a bilinear form. Often we just say *form* for a multilinear form.

EXAMPLE 1. The determinant of a 2×2 matrix defines a form

$$\mathbb{R}^2 \times \mathbb{R}^2 \to \mathbb{R}$$

EXAMPLE 2. Let $\mathscr{V}_1, \ldots, \mathscr{V}_n$ be vectorspaces then the map

$$f: \mathscr{V}_1 \times \cdots \times \mathscr{V}_n \to \mathscr{V}_2 \oplus \cdots \oplus \mathscr{V}_n$$

defined by

$$f(v_1, \ldots, v_n) = (v_1, v_2, \ldots, v_n)$$

is a multilinear form.

EXAMPLE 3. Let a, b be positive integers. Then multiplication of polynomials defines a bilinear form

$$\mu: \mathscr{P}_a(\mathbb{R}) \times \mathscr{P}_b(\mathbb{R}) \to \mathscr{P}_{a+b}(\mathbb{R}).$$

EXAMPLE 4. Let a, b be positive integers and define

$$f: \mathbb{R}^{a+1} \times \mathbb{R}^{b+1} \to \mathbb{R}^{a+b-1}$$

by (note the indexing starts with a zero)

$$f(x, y) = (z_0, \ldots, z_{a+b}),$$

where

$$x = (x_0, \ldots, x_a), \qquad y = (y_0, \ldots, y_b),$$

and

$$z_c = \sum_{i+j=c} x_i y_j.$$

Then f is a multilinear form.

Definition. Two multilinear forms

$$f': \mathscr{V}'_1 \times \cdots \times \mathscr{V}'_n \to \mathscr{W}',$$

$$f'': \mathscr{V}''_1 \times \cdots \times \mathscr{V}''_n \to \mathscr{W}'',$$

are said to be *equivalent* iff there exist vector space isomorphisms

$$\varphi_i: \mathscr{V}'_i \to \mathscr{V}''_{\sigma(i)}, \qquad i = 1, \ldots, n,$$

$$\psi: \mathscr{W}' \to \mathscr{W}'',$$

where $(\sigma(1), \ldots, \sigma(n))$ is a rearrangement (i.e., *permutation*) of the integers $(1, \ldots, n)$ such that

$$f'(v'_1, \ldots, v'_n) = f''(v''_1, \ldots, v''_n) \quad \text{for all } v'_i \in \mathscr{V}'_i, \quad i = 1, \ldots, n,$$

where

$$v''_j = \varphi_j(v'_{\tau(j)}), \qquad j = 1, \ldots, n,$$

and τ is the inverse rearrangement of σ, that is,

$$\tau(\sigma(j)) = j, \qquad j = 1, \ldots, n.$$

Theorem 14bis.1 (Linear Extension Construction for Forms). *Let n be a positive integer and let $\mathscr{V}_1, \ldots, \mathscr{V}_n$, $\mathscr{W}$ be vector spaces. Assume*

$$\dim \mathscr{V}_i = d_i$$

is finite, and let $\{e_{i,1}, \ldots, e_{i,d_i}\}$ be a basis for $\mathscr{V}_i$, $i = 1, \ldots, n$. For each index $(j_1, \ldots, j_n)$ with $1 \le j_s \le d_s$ let $w_{(j_1 \ldots j_n)} \in \mathscr{W}$. Then there is one and only one form

$$f: \mathscr{V}_1 \times \cdots \times \mathscr{V}_n \to \mathscr{W}$$

such that

$$f(e_{1,j_1},\ldots,e_{n,j_n}) = w_{(j,\ldots,j_n)}.$$

PROOF. This is analogous to (8.13), namely, the linear extension construction. To construct f we suppose $v_i \in \mathscr{V}_i$, $i = 1, \ldots, n$. We can then write

$$v_i = a_{i,1}e_{i,1} + \cdots + a_{i,d_i}e_{i,d_i}$$

for unique scalars $a_{i,1}, \ldots, a_{i,d_i}$. We now set (this definition is forced by the desired multilinearity)

$$f(v_1, \ldots, v_n) = \sum_{\substack{(j_1,\ldots,j_n)\\ 1 \le j_i \le d_i}} a_{(1,j_1)}a_{(2,j_2)}\cdots a_{(n,j_n)}w_{(j_1,\ldots,j_n)}.$$

A simple check of the definitions shows that f is indeed a multilinear form. Furthermore, if

$$g: \mathscr{V}_1 \times \cdots \times \mathscr{V}_n \to \mathscr{W}$$

is any multilinear form such that

$$g(e_{1,j_1}, \ldots, e_{n,j_n}) = w_{(j_1,\ldots,j_n)}, \qquad 1 \le j_i \le d_i,$$

then by multilinearity

$$g(v_1, \ldots, v_n) = \sum a_{(1,j_1)}\cdots a_{(n,j_n)}w_{(j_1,\ldots,j_n)},$$

that is, $f = g$. □

If a is a scalar and

$$f, g: \mathscr{V}_1 \times \cdots \times \mathscr{V}_n \to \mathscr{W}$$

are multilinear forms, then so are

$$f + g: \mathscr{V}_1 \times \cdots \times \mathscr{V}_n \to \mathscr{W}, \to (f+g)(v_1, \ldots, v_n) = f(v_1, \ldots, v_n) + g(v_1, \ldots, v_n),$$

$$af: \mathscr{V}_1 \times \cdots \times \mathscr{V}_n \to \mathscr{W}, \to (af(v_1, \ldots, v_n) = af(v_1, \ldots, v_n).$$

Here is a partial verification of the first for $n = 2$.

$$\begin{aligned}(f+g)(x'+x'', y) &= f(x'+x'', y) + g(x'+x'', y) \quad \text{Def}^n.\\ &= f(x', y) + f(x'', y) + g(x', y) + g(x'', y) \quad \text{bilinearity}\\ &= f(x', y) + g(x', y) + f(x'', y) + g(x'', y)\\ &= (f+g)(x', y) + (f+g)(x'', y).\end{aligned}$$

The verification in general is completely analogous. We have thus proved:

Proposition 14bis.2. *The set of all multilinear forms*

$$f: \mathscr{V}_1 \times \cdots \times \mathscr{V}_n \to \mathscr{W}$$

is a vector space with respect to the addition of forms and multiplication of a form by a scalar as defined above.

Proposition 14bis.3. *The dimension of the vector space of multilinear forms*

$$f: \mathscr{V}_1 \times \cdots \times \mathscr{V}_n \to \mathscr{W}$$

is $d \cdot d_1 \cdots d_n$ where $d_i = \dim \mathscr{V}_i$, $i = 1, \ldots, n$ and $d = \dim \mathscr{W}$.

PROOF. Follow from (14bis.1). □

The study of multilinear forms (multilinear algebra) leads to many interesting, and in some cases, totally unsolved problems. As an illustration of what is possible we first introduce one new concept.

Definition. A form

$$f: \mathscr{V}_1 \times \cdots \times \mathscr{V}_n \to \mathscr{W}$$

is called *nonsingular* iff whenever

$$f(v_1, v_2, \ldots, v_n) = 0$$

at least one of $v_1, \ldots, v_n$ is the zero vector.

EXAMPLE 5. $\mu: \mathbb{R}^2 \times \mathbb{R}^2 \to \mathbb{R}^2$ defined by

$$\mu: (x, y) = (x_1 y_1 - x_2 y_2, x_1 y_2 + x_2 y_1)$$

is a nonsingular bilinear form.

It is possible to construct nonsingular bilinear forms

$$\mathbb{R}^4 \times \mathbb{R}^4 \to \mathbb{R}^4,$$

$$\mathbb{R}^8 \times \mathbb{R}^8 \to \mathbb{R}^8.$$

Just as μ devolves from multiplication of complex numbers these forms derive from multiplication of quaternions and Cayley numbers. These facts were known in the nineteenth century. Only in 1960 was it proved: if there is a nonsingular form

$$\mathbb{R}^n \times \mathbb{R}^n \to \mathbb{R}^n,$$

then $n = 1, 2, 4$ or 8.

Open Problem. Let n, k be positive integers. Find the largest integer $b(k, n)$ such that there exists a nonsingular bilinear form

$$\mathbb{R}^n \times \mathbb{R}^{b(k,n)} \to \mathbb{R}^k.$$

Our purpose for introducing multilinear forms is to define and establish the basic properties of the determinant of $n \times n$ matrices. We turn to this now. We will be concerned with multilinear forms

$$f: \mathscr{V}_1 \times \cdots \times \mathscr{V}_n \to \mathscr{W},$$

where $\mathscr{V}_1 = \cdots = \mathscr{V}_n$, that is, multilinear forms of the type

$$f: \mathscr{V} \times \cdots \times \mathscr{V} \to \mathscr{W}.$$

Definition. A form

$$f: \mathscr{V} \times \cdots \times \mathscr{V} \to \mathscr{W}$$

is called *sign-symmetric* iff for every pair of integers i, j, $1 \leq i \neq j \leq n$ and $v_1, \ldots, v_n \in \mathscr{V}$

$$f(v_1, \ldots, v_i, \ldots, v_j, \ldots, v_n) = -f(v_1, \ldots, v_j, \ldots, v_i, \ldots, v_n).$$

A sign-symmetric form is also referred to as **alternating**. Note that whenever $v_i = v_j$ for some $1 \leq i \neq j \leq n$, then interchanging v_i and v_j does not change the value of $f(v_1, \ldots, v_n)$. If f is sign-symmetric on the other hand the sign of the value must change. This is only possible if

$$f(v_1, \ldots, v_n) = 0,$$

whenever there is a pair of integers i, j such that $1 \leq i \neq j \leq n$ and $v_i = v_j$.

Example 6. Let

$$\Delta: \mathbb{R}^2 \times \mathbb{R}^2 \to \mathbb{R}$$

be defined by

$$\Delta(a, b), (c, d)) = ad - bc.$$

Then Δ is bilinear and sign-symmetric. The verification, while tedious, is straightforward and left to the reader.

Theorem 14bis.4. *Let $\mathscr{V}$ be an n-dimensional vector space. Then the sign-symmetric forms (Note: $n = \dim \mathscr{V}$)*

$$f: \underbrace{\mathscr{V} \times \cdots \times \mathscr{V}}_{n} \to \mathbb{R}$$

form a one-dimensional subspace in the space of all forms.

Proof. Let $\{e_1, \ldots, e_n\}$ be a basis for $\mathscr{V}$. By (14bis.1) a form

$$f: \mathscr{V} \times \cdots \times \mathscr{V} \to \mathbb{R}$$

is completely determined when we know the values $f(e_{i_1}, \ldots, e_{i_n}) \in \mathbb{R}$, $1 \leq i_j \leq n$. If the form is alternating then

$$f(e_{i_1}, \ldots, e_{i_n}) = 0,$$

whenever $e_{i_a} = e_{i_b}$ for a pair indexes $i_a \neq i_b$. This means that an alternating form

$$f : \underbrace{\mathscr{V} \times \cdots \times \mathscr{V}}_{n} \to \mathbb{R}, \qquad n = \dim \mathscr{V}$$

is determined by the values

$$f(e_{i_1}, \ldots, e_{i_n}),$$

where the indices are all distinct. If the form is sign-symmetric and $(e_{i_1}, \ldots, e_{i_n})$ has distinct indices, then

$$f(e_{i_1}, \ldots, e_{i_n}) = (-1)^s f(e_1, e_2, \ldots, e_n),$$

where s is the number of pairs of indices that have to be switched to rearrange the n-tuple $(i_1, \ldots, i_n)$ to give $(1, \ldots, n)$. Thus a sign-symmetric/alternating from

$$f : \underbrace{\mathscr{V} \times \cdots \times \mathscr{V}}_{n} \to \mathbb{R}, \qquad n = \dim \mathscr{V}$$

is completely determined by

sign-symmetry

and

$$f(e_1, e_2, \ldots, e_n) \in \mathbb{R}.$$

Let

$$g : \underbrace{\mathscr{V} \times \cdots \times \mathscr{V}}_{n} \to \mathbb{R}, \qquad n = \dim \mathscr{V}$$

be the sign-*symmetric/alternating* form defined by

$$g(e_1, \ldots, e_n) = 1.$$

Let

$$f : \mathscr{V} \times \cdots \times \mathscr{V} \to \mathbb{R}, \qquad n = \dim \mathscr{V}$$

be a sign-*symmetric/alternating* form. Then

$$h := f - f(e_1, \ldots, e_n) - g, \qquad \mathscr{V} \times \cdots \times \mathscr{V} \to \mathbb{R}$$

is also sign-symmetric and alternating. Moreover

$$h(e_1, \ldots, e_n) = 0.$$

Therefore h is identically 0. Thus $f = f(e_1, \ldots, e_n) \cdot g$ and g spans the space of sign-symmetric forms

$$f: \mathscr{V} \times \cdots \times \mathscr{V} \to \mathbb{R}, \qquad n = \dim \mathscr{V}$$

so this space is one-dimensional. □

Suppose $\mathsf{T}: \mathscr{V} \to \mathscr{V}$ is a linear transformation and

$$g: \underbrace{\mathscr{V} \times \cdots \times \mathscr{V}}_{n} \to \mathbb{R}, \qquad n = \dim \mathscr{V}$$

is a nonzero sign-symmetric/alternating form. Then

$$g_{\mathsf{T}}: \mathscr{V} \times \cdots \times \mathscr{V} \to \mathbb{R}$$

defined by

$$g_{\mathsf{T}}(v_1, \ldots, v_n) = g(\mathsf{T}v_1, \ldots, \mathsf{T}v_n) \in \mathbb{R}$$

is also a sign-symmetric/alternating form. Since the space of such forms is one-dimensional, and $g \neq 0$, it follows that

$$g_{\mathsf{T}} = d_g(\mathsf{T}) \cdot g$$

for a unique scalar $d_g(\mathsf{T}) \in \mathbb{R}$. Thus we are led to:

Definition. Let $\mathbf{M} \in \mathscr{M}_{nn}$ be an $n \times n$ matrix. Let

$$g: \mathbb{R}^n \times \cdots \times \mathbb{R}^n \to \mathbb{R}$$

be the sign-symmetric/alternating form defined by

$$g(\mathbf{E}_1, \ldots, \mathbf{E}_n) = 1,$$

where $\{\mathbf{E}_1, \ldots, \mathbf{E}_n\}$ is the standard basis for $\mathbb{R}^n$. Let

$$\mathsf{M}: \mathbb{R}^n \to \mathbb{R}^n$$

be the linear transformation represented by the matrix $\mathbf{M}$ with respect to the standard basis. The *determinant* of M is the unique real number $\det(\mathbf{M})$ such that $g_{\mathsf{M}} = \det(\mathbf{M}) \cdot g$.

Theorem 14bis.5. *The function*

$$\text{Det}: \mathscr{M}_{nn} \to \mathbb{R}$$

has the following properties:

(1) *If* $\mathbf{A}$ *has a column of zero then* $\det \mathbf{A} = 0$.
(2) *If* $\mathbf{A}$ *has two columns equal then* $\det \mathbf{A} = 0$.
(3) *If* $\mathbf{B}$ *is obtained from* $\mathbf{A}$ *by interchanging two columns then* $\det \mathbf{B} = -\det \mathbf{A}$.

(4) *If* **B** *is obtained from* **A** *by adding a multiple of one column to a different column then* det **B** = det **A**.
(5) *If* **B** *is obtained by multiplying a column of* **A** *by a number* k *then* det **B** = k det **A**.
(6) det $\mathbf{A}^t$ = det **A**; *thus everything said about columns in* (1)–(5) *holds also for rows.*
(7) *If* **A** *is upper triangular then* det **A** *is the product of the diagonal entries.*
(8) det(**A** · **B**) = det **A** det **B**.
(9) **A** *is invertible* = det **A** ≠ 0.

Before taking up the proof proper we require some preliminary manouvering. The proof of (14bis.4) shows that a sign-symmetric form

$$f:\underbrace{\mathscr{V}\times\cdots\times\mathscr{V}}_{n}\to\mathbb{R},\qquad n=\dim,$$

is completely determined by the value

$$f(e_1,\ldots,e_n)\in\mathbb{R},$$

where $\{e_1,\ldots,e_n\}$ is a basis for $\mathscr{V}$. In fact, if

$$g:\underbrace{\mathscr{V}\times\cdots\times\mathscr{V}}_{n}\to\mathbb{R},\qquad n=\dim\mathscr{V},$$

is the unique sign-symmetric/alternating form such that

$$g(e_1,\ldots,e_n)=1,$$

then $f = f(e_1,\ldots,e_n)\cdot g$ as forms. Thus we have shown:

Corollary 14bis.6. *Two sign-symmetric/alternating forms*

$$f',f'':\underbrace{\mathscr{V}\times\cdots\times\mathscr{V}}_{n}\to\mathbb{R},\qquad n=\dim\mathscr{V}$$

are equal ⇔ *for some basis* $\{e_1,\ldots,e_n\}$ *of* $\mathscr{V}$

$$f'(e_1,e_2,\ldots,e_n)=f''(e_1,\ldots,e_n).$$

Let $\mathbf{A}\in\mathscr{M}_{nn}$ and write

$$\mathsf{L}(\mathbf{A}):\mathbb{R}^n\to\mathbb{R}^n$$

for the linear transformation corresponding to the matrix **A** with respect to the standard basis $\{\mathbf{E}_1,\ldots,\mathbf{E}_n\}$ for $\mathbb{R}^n$ (see (12.1)). Let

$$\mathbf{A}=(\mathbf{A}_1,\ldots,\mathbf{A}_n),$$

where

$$\mathbf{A}_i=\begin{pmatrix}a_{i1}\\ a_{in}\end{pmatrix},\qquad i=1,\ldots,n,$$

is the ith column of $\mathbf{A}$. Then

$$\mathsf{L}(\mathbf{A})(\mathbf{E}_i) = \mathbf{A}_i^t, \qquad i = 1, \ldots, n,$$

where $\mathbf{A}_i^t$ denotes the transpose of $\mathbf{A}_i$; that is, the column $\mathbf{A}_i$ rewritten as a row.

Let

$$g\colon \underbrace{\mathbb{R}^n \times \cdots \times \mathbb{R}^n}_{n} \to \mathbb{R}$$

be the unique symmetric form defined by

$$g(\mathbf{E}_1, \ldots, \mathbf{E}_n) = 1.$$

To compute the determinant of $\mathbf{A}$ we introduce the form

$$g_{\mathsf{L}(\mathbf{A})}\colon \mathbb{R}^n \times \cdots \times \mathbb{R}^n \to \mathbb{R}^n$$

defined by

$$g_{\mathsf{L}(\mathbf{A})}(v_1, \ldots, v_n) = g(\mathbf{A}v_1, \ldots, \mathbf{A}v_n).$$

Then $\det \mathbf{A}$ is the unique number such that

$$g_{\mathbf{A}} = (\det \mathbf{A}) \cdot g\colon \mathbb{R}^n \times \cdots \times \mathbb{R}^n \to \mathbb{R}.$$

The remarks preceding (14bis.6) thus show

$$\begin{aligned} g_{\mathbf{A}} &= g_{\mathbf{A}}(\mathbf{E}_1, \ldots, \mathbf{E}_n) \cdot g = g(\mathsf{L}(\mathbf{A})\mathbf{E}_1, \ldots, \mathsf{L}(\mathbf{A})(\mathbf{E}_n)) \\ &= g(\mathbf{A}_1^t, \ldots, \mathbf{A}_n^t), \end{aligned}$$

and therefore we conclude:

Lemma 14bis.7. *In the preceding notations we have*

$$\det(\mathbf{A}) = g(\mathbf{A}_1^t, \ldots, \mathbf{A}_n^t)$$

for $\mathbf{A} \in \mathcal{M}_{nn}$. □

Proof of 14bis.5. In permuted order here are the proofs.

(2) Suppose $\mathbf{A}$ has two columns equal, say $\mathbf{A}_i = \mathbf{A}_j$, $i \neq j$. Then $\mathbf{A}_i^t = \mathbf{A}_j^t$, so $0 = g(\mathbf{A}_1^t, \ldots, \mathbf{A}_i^t, \ldots, \mathbf{A}_j^t, \ldots, \mathbf{A}_n) = \det(\mathbf{A})$ because g is alternating.

(1) Suppose $\mathbf{A}$ has a column of zeros, say $\mathbf{A}_i = (0)$. Then $\mathbf{A}_i^t = (0, \ldots, 0)$ so $0 = g(\mathbf{A}_1^t, \ldots, \mathbf{A}_i^t, \ldots, \mathbf{A}_n^t) = \det \mathbf{A}$ by multilinearity.

(3) If $\mathbf{B}$ is obtained from $\mathbf{A}$ by interchanging the two columns $\mathbf{A}_i$ and $\mathbf{A}_j$, then if we arrange the notations so that $i < j$ we see

$$\begin{aligned} \det(\mathbf{B}) &= g(\mathbf{A}_1, \ldots, \mathbf{A}_j, \ldots, \mathbf{A}_i, \ldots, \mathbf{A}_n) \\ &= -g(\mathbf{A}_1, \ldots, \mathbf{A}_i, \ldots, \mathbf{A}_j, \ldots, \mathbf{A}_n) = -\det(\mathbf{A}), \end{aligned}$$

because g is sign-symmetric.

(4) Suppose $\mathbf{B}$ is obtained from $\mathbf{A}$ by adding $\mathbf{A}_i$ to $\mathbf{A}_j$, $i \neq j$. Then

$$\begin{aligned}
\det(\mathbf{B}) &= g(\mathbf{A}_1^t, \ldots, \mathbf{A}_i^t, \ldots, \mathbf{A}_j^t + a\mathbf{A}_i^t, \ldots, \mathbf{A}_n^t) \\
&= g(\mathbf{A}_1^t, \ldots, \mathbf{A}_i^t, \ldots, \mathbf{A}_j^t, \ldots, \mathbf{A}_n^t) \\
&\quad + g(\mathbf{A}_1^t, \ldots, \mathbf{A}_i^t, \ldots, a\mathbf{A}_i^t, \ldots, \mathbf{A}_n^t) \\
&= \det \mathbf{A} + ag(\mathbf{A}_1^t, \ldots, \mathbf{A}_i^t, \ldots, \mathbf{A}_i^t, \ldots, \mathbf{A}_n^t) \\
&= \det \mathbf{A} + 0 = \det \mathbf{A},
\end{aligned}$$

by property (2).

(5) Suppose $\mathbf{B}$ is obtained from $\mathbf{A}$ by multiplying $\mathbf{A}_i$ by a. Then

$$\begin{aligned}
\det \mathbf{B} &= g(\mathbf{A}_1^t, \ldots, a\mathbf{A}_i^t, \ldots, \mathbf{A}_n^t) \\
&= ag(\mathbf{A}_1^t, \ldots, \mathbf{A}_n^t) = a \det(\mathbf{A}),
\end{aligned}$$

by multilinearity.

(7) Suppose that $\mathbf{A}$ is an upper triangular matrix, that is

$$\mathbf{A} = \begin{bmatrix} a_{11} & a_{12} & & \\ & a_{22} & & \\ & & \ddots & \\ 0 & & & a_{nn} \end{bmatrix}$$

Then

$$\begin{aligned}
\mathbf{A}_1^t &= a_{11}\mathbf{E}_1, \\
\mathbf{A}_2^t &= a_{12}\mathbf{E}_1 + a_{22}\mathbf{E}_2, \\
&\vdots \\
\mathbf{A}_n^t &= a_{1n}\mathbf{E} + \cdots + a_{nn}\mathbf{E}_n,
\end{aligned}$$

as vectors in $\mathbb{R}^n$. Therefore

$$\begin{aligned}
\det \mathbf{A} &= g(\mathbf{A}_1^t, \ldots, \mathbf{A}_n^t) \\
&= g(a_{11}\mathbf{E}_1, \mathbf{A}_2^t, \ldots, \mathbf{A}_n^t),
\end{aligned}$$

$$\begin{aligned}
&\text{(by linearity in the first variable)} = a_{11}g(\mathbf{E}_1, a_{12}\mathbf{E}_1 + a_{22}\mathbf{E}_2, \mathbf{A}_3^t, \ldots, \mathbf{A}_n^t), \\
&\text{(by linearity in the second variable)} = a_{11}a_{12}g(\mathbf{E}_1, \mathbf{E}_1, \mathbf{A}_3^t, \ldots, \mathbf{A}_n^t) \\
&\qquad + a_{11}a_{22}g(\mathbf{E}_1, \mathbf{E}_2, \mathbf{A}_3^t, \ldots, \mathbf{A}_n^t), \\
&\text{(since } g \text{ is alternating)} = a_{11}a_{12}\cdot 0 + a_{11}a_{22}g(\mathbf{E}_1, \mathbf{E}_2, \mathbf{A}_3^t, \ldots, \mathbf{A}_n^t), \\
&\text{(by linearity in the third variable)} = a_{11}a_{22}a_{13}g(\mathbf{E}_1, \mathbf{E}_2, \mathbf{E}_1, \ldots, \mathbf{A}_n^t) \\
&\qquad + a_{11}a_{22}a_{23}g(\mathbf{E}_1, \mathbf{E}_2, \mathbf{E}_2, \mathbf{A}_4^t, \ldots, \mathbf{A}_n^t) \\
&\qquad + a_{11}a_{22}a_{33}g(\mathbf{E}_1, \mathbf{E}_2, \mathbf{E}_3, \mathbf{A}_4, \ldots, \mathbf{A}_n^t),
\end{aligned}$$

$$
\begin{aligned}
\text{(since } g \text{ is alternating)} &= a_{11}a_{22}a_{13}\cdot 0 + a_{11}a_{22}a_{23}\cdot 0 \\
&\quad + a_{11}a_{22}a_{23}g(\mathbf{E}_1, \mathbf{E}_2, \mathbf{E}_3, \mathbf{A}_4^t, \ldots, \mathbf{A}_n^t) \\
&\ \vdots \\
&= a_{11}a_{22}, \ldots, a_{1n}\cdot 0 + \cdots + a_{11}a_{22}, \ldots, a_{1,n-i}0 \\
&\quad + a_{11}, \ldots, a_{nn}g(\mathbf{E}_1, \ldots, \mathbf{E}_n) \\
&= a_{11}, \ldots, a_{nn},
\end{aligned}
$$

as required.

It is convenient, before proving (8) to prove (9).

(9) Let $\mathbf{A} \in \mathscr{M}_{nn}$. Suppose that $\mathbf{A}$ is invertible. Then $\{\mathbf{AE}_1, \ldots, \mathbf{AE}_n\}$ is a basis for $\mathbb{R}^n$. Suppose

$$g(\mathbf{AE}_1, \ldots, \mathbf{AE}_n) = 0.$$

By (14bis.4) this implies that

$$g = 0: \underbrace{\mathbb{R}^n \times \cdots \times \mathbb{R}^n}_{n} \to \mathbb{R}$$

as multilinear form, since, of course

$$0(\mathbf{AE}_1, \ldots, \mathbf{AE}_n) = 0$$

and a symmetric/alternating form is completely determined by its value on *any* ordered basis, in particular, on the ordered basis $\{\mathbf{AE}_1, \ldots, \mathbf{AE}_n\}$. Therefore

$$0 \neq g_{\mathbf{A}} = \det \mathbf{A} \cdot g = \det \mathbf{A} \neq 0,$$

as required.

(8) Suppose $\mathbf{B}$ is singular. Then so is $\mathbf{A} \cdot \mathbf{B}$ so by (9)

$$\det(\mathbf{AB}) = 0 = \det \mathbf{A} \cdot 0 = \det \mathbf{A} \cdot \det \mathbf{B}.$$

If $\mathbf{B}$ is nonsingular, then the form

$$g_{\mathbf{B}}: \mathbb{R}^n \times \cdots \times \mathbb{R}^n \to \mathbb{R}$$

is nonzero, since by (9)

$$g_{\mathbf{B}} = \det \mathbf{B} \cdot g \neq 0.$$

Therefore $g_{\mathbf{B}}$ is a basis for the symmetric/alternating forms since the space of such forms is one-dimensional. Thus as forms there exists $d \in \mathbb{R}$ such that

$$g_{\mathbf{AB}} = dg_{\mathbf{B}}: \underbrace{\mathbb{R}^n \times \cdots \times \mathbb{R}^n}_{n} \to \mathbb{R}.$$

Therefore we obtain from the definitions

$$g_{\mathbf{AB}} = dg_{\mathbf{B}} = d\det(\mathbf{B})g,$$

and

$$g_{\mathbf{AB}} = \det(\mathbf{A} \cdot \mathbf{B})g.$$

So it remains to show that $d = \det \mathbf{A}$. To this end note that since $\mathbf{B}$ is nonsingular can consider the point

$$(\mathbf{B}^{-1}\mathbf{E}_1, \ldots, \mathbf{B}^{-1}\mathbf{E}_n) \in \underbrace{\mathbb{R}^n \times \cdots \times \mathbb{R}^n}_{n}.$$

So evaluating the two sides of the equation $g_{\mathbf{AB}} = dg_{\mathbf{B}}$ on

$$(\mathbf{B}^{-1}\mathbf{E}_1, \ldots, \mathbf{B}^{-1}\mathbf{E}_n) \in \mathbb{R}^n \times \cdots \times \mathbb{R}^n$$

gives, on the one hand,

$$\begin{aligned} g_{\mathbf{AB}}(\mathbf{B}^{-1}\mathbf{E}_1, \ldots, \mathbf{B}^{-1}\mathbf{E}_n) &= g(\mathbf{ABB}^{-1}\mathbf{E}_1, \ldots, \mathbf{ABB}^{-1}\mathbf{E}_n) \\ &= g(\mathbf{AE}_1, \ldots, \mathbf{AE}_n) \\ &= \det \mathbf{A} g(\mathbf{E}_1, \ldots, \mathbf{E}_n) = \det \mathbf{A}, \end{aligned}$$

and, on the other hand,

$$\begin{aligned} dg_{\mathbf{B}}(\mathbf{B}^{-1}\mathbf{E}_1, \ldots, \mathbf{B}^{-1}\mathbf{E}_n) &= dg(\mathbf{BB}^{-1}\mathbf{E}_1, \ldots, \mathbf{BB}^{-1}\mathbf{E}_n) \\ &= dg(\mathbf{E}_1, \ldots, \mathbf{E}_n) = d, \end{aligned}$$

and so equating yields $d = \det \mathbf{A}$ as required.

(6) We require a lemma.

Lemma. *Let*

$$\mathbf{H}_i \colon \mathbb{R}^{n-1} \to \mathbb{R}^n \mid H_i(x_1, \ldots, x_{n-1}) = (x_1, \ldots, x_{i-1}, 0, x_1, \ldots, x_n),$$

and let

$$g_n \colon \underbrace{\mathbb{R}^n \times \cdots \times \mathbb{R}^n}_{n} \to \mathbb{R},$$

and

$$g_{n-1} \colon \overbrace{\mathbb{R}^{n-1} \times \cdots \times \mathbb{R}^{n-1}}^{n-1} \to \mathbb{R}$$

be the fundamental sign-symmetric/alternating forms. Define

$$g_{n-1}^i \colon \underbrace{\mathbb{R}^{n-1} \times \cdots \times \mathbb{R}^{n-1}}_{n-1} \to \mathbb{R}$$

by

$$g_{n-1}^i(\mathbf{A}_1, \ldots, \mathbf{A}_{n-1}) = g_n(\mathbf{H}_i \mathbf{A}_1, \ldots, \mathbf{H}_i, \mathbf{A}_{n-1}, \mathbf{E}_i).$$

Then

$$g_{n-1}^{i} = (-1)^{n-i} g_{n-1}.$$

Proof. It is enough to show

$$g_{n-1}^{i}(\mathbf{E}_1, \ldots, \mathbf{E}_{n-1}) = (-1)^{i-1}.$$

Note that

$$H_i(\mathbf{E}_j) = \begin{cases} \mathbf{E}_j, & j \leq i-1, \\ \mathbf{E}_{j+1}, & j \geq i, \end{cases}$$

so by definition

$$\begin{aligned} g_{n-1}^{i}(\mathbf{E}_1, \ldots, \mathbf{E}_{n-1}) &= g_n(\mathbf{E}_1, \ldots, \mathbf{E}_{i-1}, \mathbf{E}_{i+1}, \ldots, \mathbf{E}_n, \mathbf{E}_i) \\ n - \text{interchanges} &\begin{cases} = -g_n(\mathbf{E}_1, \ldots, \mathbf{E}_{i-1}, \mathbf{E}_{i+1}, \ldots, \mathbf{E}_i, E_n) \\ = (-1)^{i-1} g_n(\mathbf{E}_1, \ldots, \mathbf{E}_n) = (-1)^{n-i}, \end{cases} \end{aligned}$$

as required. □

Let $\mathbf{A} = (\mathbf{A}_1, \ldots, \mathbf{A}_n) \in \mathscr{M}_{nn}$ where

$$\mathbf{A}_i = \begin{pmatrix} a_{1i} \\ a_{2i} \\ a_{ni} \end{pmatrix}, \qquad i = 1, \ldots, n,$$

are the column of $\mathbf{A}$. Then

$$\mathbf{A}_i^t = \sum_{k=1}^{n} a_{ki} \mathbf{E}_k = \mathbf{B}_i + a_{ni} \mathbf{E}_n$$

which defines $\mathbf{B}_1, \ldots, \mathbf{B}_n \in \operatorname{Span}\{\mathbf{E}_1, \ldots, \mathbf{E}_{n-1}\}$. By definition we have

$$\begin{aligned} \det \mathbf{A} &= g(\mathbf{A}_1^t, \ldots, \mathbf{A}_n^t) \\ &= g(\mathbf{B}_1 + a_{n1}\mathbf{E}_n, \ldots, \mathbf{B}_n + a_{nn}\mathbf{E}_n). \end{aligned}$$

We are going to expand this last expression using the multilinearity of $g(\ ,\ldots,\ )$. Among the terms of this expansion will be

$$g(\mathbf{B}_1, \ldots, \mathbf{B}_n).$$

This term is however zero. To see this notice that

$$g(\mathbf{B}_1, \ldots, \mathbf{B}_n) = \det \begin{pmatrix} a_{11} & \cdots & a_{1n} \\ a_{n-11} & \cdots & a_{n,n} \\ 0 & \cdots & 0 \end{pmatrix}.$$

The matrix on the right is obviously singular because it has a row of zeros. By property (8) we conclude that the determinant of this matrix is zero, so

$$g(\mathbf{B}_1, \ldots, \mathbf{B}_n) = 0.$$

Another typical term of the expansion is

$$g(\mathbf{B}_1, \ldots, \mathbf{B}_{i-1}, a_{ni}\mathbf{E}_n, \mathbf{B}_{i+1}, \ldots, a_{nj}\mathbf{E}_n, \mathbf{B}_{j+1}, \ldots, B_n).$$

This term also vanishes because by multilinearity and sign-symmetry it is equal to

$$a_{ni}a_{nj}g(\mathbf{B}_1, \ldots, \mathbf{B}_{i-1}, \mathbf{E}_n, \mathbf{B}_{i+1}, \ldots, \mathbf{B}_{j-1}, \mathbf{E}_n, \mathbf{B}_{j+1}, \ldots, \mathbf{B}_n) = a_{n1}a_{nj} \cdot 0 = 0.$$

These considerations then lead to the following formula

$$\det \mathbf{A} = g(a_{n1}\mathbf{E}_n, \mathbf{B}_2, \ldots, \mathbf{B}_n) + g(\mathbf{B}_1, a_{n2}, \mathbf{E}_n, \mathbf{B}_3, \ldots, \mathbf{B}_n) + \cdots + g(\mathbf{B}_1, \ldots, \mathbf{B}_{n-1} + a_{nn}\mathbf{E}_n).$$

So by linearity and the preceding lemma we get

$$\begin{aligned}\det \mathbf{A} &= a_{n1}g(\mathbf{E}_n, \mathbf{B}_2, \ldots, \mathbf{B}_n) + \cdots + a_{nn}g(\mathbf{B}_1, \ldots, \mathbf{B}_{n-1}, \mathbf{E}_n)\\ &= \sum_{i=1}^{n}(-1)^{n-1}a_{ni}g(\mathbf{B}_1, \ldots, \mathbf{B}_{i-1}, \mathbf{B}_{i+1}, \ldots, \mathbf{B}_n)\\ &= \sum_{i=1}^{n}(-1)^{n-i}a_{ni} \det \mathbf{M}_{ni},\end{aligned}$$

by the definition of the (n, i) minor of $\mathbf{A}$ and the sign-symmetry of g. This formula is the *Lagrange expansion formula with respect to the* nth *row*. There is a similar formula for det $\mathbf{A}$ that employs the nth *column* of $\mathbf{A}$. To obtain this we write

$$\begin{aligned}\det \mathbf{A} = g(\mathbf{A}_1^t, \ldots, \mathbf{A}_n^t) &= g\left(a_1^t, \ldots, \mathbf{A}_{n-1}^t, \sum_{k=1}^{n} a_{kn}\mathbf{E}_k\right)\\ &= \sum_{k=1}^{n} a_{kn}g(\mathbf{A}_1^t, \ldots, \mathbf{A}_{n-1}^t, \mathbf{E}_k).\end{aligned}$$

Let us now analyze the individual terms of this sum. For $1 \le k \le n$ we write

$$\mathbf{A}_i^t = \mathbf{C}_{k,i} + a_{ki}\mathbf{E}_k.$$

Then reasoning as above, and using the fact that

$$g(\mathbf{C}_{k,1}, \ldots, \mathbf{C}_{k,n}) = 0,$$

since it may be interpreted as the determinant of a matrix with a column of zeros, and

$$g(\mathbf{C}_{k,1}, \ldots, a_{ki}\mathbf{E}_k, \mathbf{C}_{k+i+1}, \ldots, a_{k,j}\mathbf{E}_k, \mathbf{C}_{k,j+1}, \ldots, C_{k,n})$$

vanishes by sign-symmetry, we get the formula

$$\begin{aligned}\det \mathbf{A} &= \sum_{k=1}^{n} a_{k,n}g(\mathbf{C}_{k,1}, \ldots, \mathbf{C}_{k,n-1}, \mathbf{E}_k)\\ &= \sum_{k=1}^{n}(-1)^{n-k}a_{k,n}g(\mathbf{C}_{k,1}, \ldots, \mathbf{C}_{k,n-1})\\ &= \sum_{k=1}^{n}(-1)^{n-k}a_{kn} \det(\mathbf{M}_{kn}),\end{aligned}$$

which is the *Lagrange expansion formula with respect to the* nth *column*.

We may now prove (6) by observing that the Lagrange expansion with respect to the nth column of $\mathbf{A}^t$ is the same as the Lagrange expansion of $\mathbf{A}$ with respect to the nth row. But the Lagrange expansions compute det, so

$$\begin{aligned}\det \mathbf{A}^t &= \text{Lagrange expansion of } \mathbf{A}^t \text{ with respect to the } n\text{th column} \\ &= \text{Lagrange expansion of } \mathbf{A} \text{ with respect to the } n\text{th row} \\ &= \det \mathbf{A},\end{aligned}$$

as required. □

Let us derive an explicit form for the determinant of a 2×2 matrix. Let

$$\mathbf{A} = (\mathbf{A}_1, \mathbf{A}_2),$$

where

$$\mathbf{A}_i = \begin{pmatrix} a_{1i} \\ a_{2i} \end{pmatrix}, \qquad i = 1, 2,$$

are the columns of $\mathbf{A}$. As in the discussion preceding the proof of (14bis.5)(7) we see

$$\begin{aligned}\det \mathbf{A} &= g(\mathbf{A}_1^t, \mathbf{A}_2^t) \\ &= g(a_{11}\mathbf{E}_1 + a_{21}\mathbf{E}_2, a_{12}\mathbf{E}_1 + a_{22}\mathbf{E}_2) \\ &= a_{11}g(\mathbf{E}_1, a_{12}\mathbf{E}_1 + a_{22}\mathbf{E}_2) + a_{21}g(\mathbf{E}_2, a_{12}\mathbf{E}_1 + a_{22}\mathbf{E}_2) \\ &= a_{11}(a_{12}g(\mathbf{E}_1, \mathbf{E}_1) + a_{22}g(\mathbf{E}_1, \mathbf{E}_2)) \\ &\quad + a_{21}(a_{12}g(\mathbf{E}_2, \mathbf{E}_1) + a_{22}g(\mathbf{E}_2, \mathbf{E}_2)) \\ &= a_{11}(0 + a_{22}) + a_{21}(-a_{12} + 0) \\ &= a_{11}a_{22} - a_{21}a_{12},\end{aligned}$$

because

$$\begin{aligned}g(\mathbf{E}_1, \mathbf{E}_1) &= 0 = g(\mathbf{E}_2, \mathbf{E}_2), \\ g(\mathbf{E}_1, \mathbf{E}_2) &= 1 = -g(\mathbf{E}_2, \mathbf{E}_1).\end{aligned}$$

Thus

$$\det\begin{pmatrix} a & b \\ c & d \end{pmatrix} = ad - cb$$

exactly as we had already defined the determinant in Chapter 13 (immediately preceding Example 11) for 2×2 matrices.

It should be apparent that one can employ the preceding procedure to arrive at a formula for det $\mathbf{A}$, where $\mathbf{A}$ is an $n \times n$ matrix. Such a formula would look like

$$(*) \qquad \det \mathbf{A} = \sum (-1)^{\sigma(j_1, \ldots, j_n)} a_{j_1 1} \cdots a_{j_n n},$$

where each term in the sum

$$a_{j_1 1} \cdots a_{j_n n}$$

which is a product of matrix elements, one element chosen out of each column, so that no two elements lie in the same row, and $\sigma(j_1, \ldots, j_n)$ is the number of pairs of indices that must be interchanged to bring $(\mathbf{E}_{j_1}, \ldots, \mathbf{E}_{j_n})$ to the form $(\mathbf{E}_1, \ldots, \mathbf{E}_n)$.

Note carefully. There are as many terms in the sum $(*)$ as there are possible rearrangements of $(\mathbf{E}_1, \ldots, \mathbf{E}_n)$; or, what is the same thing, $(1, \ldots, n)$. An elementary counting argument shows that the number of such arrangements is

$$n! = n(n - 1)(\cdots)(2)(1).$$

So for a 5×5 matrix the preceding sum contains

$$5! = 5 \cdot 4 \cdot 3 \cdot 2 \cdot 1 = 120$$

terms. Thus the formula $(*)$ is not particularly efficacious for computations.

We close our discussion of determinants with the Lagrange expansion formula.

Proposition 14bis.8. *If* $\mathbf{A}$ *is a square matrix of size n, then*

$$\det \mathbf{A} = \sum_{j=1}^{n} (-1)^{i+j} a_{j_i} \det \mathbf{M}_{j_i}$$

(see the definition preceding Example 7, *Chapter* 14 *for the definition of the minor* M_{ij}*).*

PROOF. This follows from the Lagrange expansion formula with respect to the nth column, which we proved in $(14^{\text{bis}}.5)(6)$, by interchanging columns. □

EXERCISES

1. Find a nonsingular bilinear form

$$f: \mathbb{R}^4 \times \mathbb{R}^4 \to \mathbb{R}^4.$$

2. Let $\mathscr{V}_1, \ldots, \mathscr{V}_n$ be subspaces of $\mathscr{W}$ and

$$\sigma: \mathscr{V}_1 \times \cdots \times \mathscr{V}_n \to \mathscr{W}$$

the multilinear form

$$\sigma(v_1, \ldots, v_n) = v_1 + \cdots + v_n.$$

Show that σ is nonsingular iff

$$\mathscr{V}_i \cap \mathscr{V}_j = \{0\}, \qquad i = j.$$

3. Show that the bilinear forms of Examples 3 and 4 are isomorphic.

4. (a) Show that $b(k, n) \geq 1$.
(b) If k and n are even show $b(k, n) \geq 2$.

5. (a) Compute the determinant of each of the following matrices

$$\begin{pmatrix} 0 & 1 & 1 & 1 \\ 1 & 0 & 1 & 1 \\ 1 & 1 & 0 & 1 \\ 1 & 1 & 1 & 0 \end{pmatrix}, \quad \begin{pmatrix} 1 & -1 & -1 & -1 \\ -1 & 1 & -1 & -1 \\ -1 & -1 & 1 & -1 \\ -1 & -1 & -1 & 1 \end{pmatrix},$$

$$\begin{pmatrix} 1 & 2 & 3 \\ 4 & 5 & 6 \\ 7 & 8 & 9 \end{pmatrix}, \quad \begin{pmatrix} 1 & -1 & 1 \\ -1 & 1 & -1 \\ 1 & -1 & 1 \end{pmatrix},$$

$$\begin{pmatrix} -1 & 1 & 1 \\ 1 & -1 & 1 \\ -1 & 1 & 1 \end{pmatrix}, \quad \begin{pmatrix} 4 & -1 & -1 \\ -1 & 4 & -1 \\ 4 & -1 & -1 \end{pmatrix},$$

(b) which of the preceding matrices are invertible?

6. (a) Suppose $\mathbf{A}, \mathbf{D} \in \mathscr{M}_{rr}$, $\mathbf{B}, \mathbf{C} \in \mathscr{M}_{ss}$ and let $\mathbf{M}$ be the matrix

$$\begin{pmatrix} \mathbf{A} & \mathbf{B} \\ \mathbf{C} & \mathbf{D} \end{pmatrix}.$$

Is it true in general that

$$(*) \qquad \det \mathbf{M} = \det \mathbf{A} \det \mathbf{D} - \det \mathbf{B} \det \mathbf{C}?$$

(Review some of the examples in Exercise 5(a).)

(b) Suppose $\mathbf{A}, \mathbf{D}$ and $\mathbf{B}, \mathbf{C}$ commute with each other. Show that $(*)$ holds

7. (a) Let

$$\mathbf{A} = \begin{pmatrix} a_{11} & a_{12} & a_{13} \\ a_{21} & a_{22} & a_{23} \\ a_{31} & a_{32} & a_{33} \end{pmatrix}.$$

The *diagonals* of $\mathbf{A}$ are

$$\begin{pmatrix} a_{11} & & & a_{12} & & & a_{13} & & \\ & a_{22} & & & a_{23} & & & a_{21} & \\ & & a_{33} & & & a_{31} & & & a_{32} \end{pmatrix},$$

and the *antidiagonals* are

$$\begin{pmatrix} & & {}_{13} & & & a_{12} & & & {}_{11} \\ & a_{22} & & & a_{21} & & & a_{23} & \\ a_{31} & & & a_{33} & & & a_{32} & & \end{pmatrix}.$$

Show that the determinant of $\mathbf{A}$ is the sum of the products of the diagonal elements minus the sum of the products of the antidiagonal elements.

(b) Show by an example that the analogous rule is false for 4×4 matrices.

8. Let $\mathbf{A} = (a_1, a_2)$, $\mathbf{B} = (a_1, b_2) \in \mathbb{R}^2$, $\mathbf{A} \neq \mathbf{B}$. Show that the equation of the line through $\mathbf{A}$ and $\mathbf{B}$ is given by

$$\det\begin{pmatrix} x & y & 1 \\ a_1 & b_1 & 1 \\ a_2 & b_2 & 1 \end{pmatrix} = 0.$$

9. Show that the area of the triangle in $\mathbb{R}^2$ with vertices (a_1, b_1), (a_2, b_2), (a_3, b_3) is given by

$$\det\begin{pmatrix} a_1 & b_1 & 1 \\ a_2 & b_2 & 1 \\ a_3 & b_3 & 1 \end{pmatrix}.$$

10. If $\mathbf{A} \in \mathscr{M}_{nn}$ is skew-symmetric and n is odd show $\det \mathbf{A} = 0$.

11. Let σ be a *permutation* of $(1, \ldots, n)$, that is a rearrangement $(\sigma(1), \ldots, \sigma(n))$ of $(1, \ldots, n)$ in perhaps some other order. Define $\operatorname{sgn}(\sigma)$ by

$$\prod_{i<j} (x_{\sigma(i)} - x_{(j)}) = \operatorname{sgn}(\sigma)\Delta(x),$$

where

$$\Delta(x) = \prod_{i<j} (x_i - x_j).$$

(a) Show that $\operatorname{sgn}(\sigma) = -1$ if σ is the permutation where

$$\sigma(k) = \begin{cases} k, & k \neq i, j, \\ j, & k = i, \\ i, & k = j, \end{cases}$$

such a permutation is simply the interchange of i and j.

(b) Show any permutation can be accomplished by a succession of interchanges.

(c) Show $\operatorname{sgn}(\sigma) = (-1)^s$ where s is the number of interchanges required to accomplish σ.

(d) Let σ be a permutation and $\mathsf{L}(\sigma)\colon \mathbb{R}^n \to \mathbb{R}^n$ by taking the linear extension of

$$\mathsf{L}(\sigma)(\mathbf{E}_i) = \mathbf{E}_{\sigma(i)}, \qquad i = 1, \ldots, n.$$

The matrix of $\mathsf{L}(\sigma)$ with respect to the standard basis of $\mathbb{R}^n$ is called a *permutation matrix*. Write down all the permutation matrices of size 3 and compute their determinants.

(e) Show $\det \mathsf{L}(\sigma) = \operatorname{sgn}(\sigma)$.

12. Let $\mathbf{A} \in \mathscr{M}_{nn}$, Show

$$\det \mathbf{A} = \sum (-1)^{\operatorname{sgn}(\sigma)} a_{\sigma(1)} \cdots a_{\sigma(n)},$$

where the sum extends over all permutations of $(1, \ldots, n)$.

13. (a) Evaluate

$$\det\begin{pmatrix} 1 & x_1 & x_1^2 \\ 1 & x_2 & x_2^2 \\ 1 & x_3 & x_3^2 \end{pmatrix}.$$

(b) Show

$$\det\begin{pmatrix} 1 & x_1 & x_1^2 & \cdots & x_1^{n-1} \\ \vdots & \vdots & \vdots & & \vdots \\ 1 & x_n & x_n^2 & & x_n^{n-1} \end{pmatrix} = \prod_{i<j} (x_i - x_j).$$

14. Let $\mathbf{A} \in \mathcal{M}_{mn}$, $\mathbf{B} \in \mathcal{M}_{nm}$ where $m > n$. Show $\det(\mathbf{AB}) = 0$. (*Note*: $\mathbf{A} \cdot \mathbf{B}$ is square of size $m \times m$.)

15 Inner product spaces

So far in our study of vector spaces and linear transformations we have made no use of the notions of length and angle, although these concepts play an important role in our intuition for the vector algebra of $\mathbb{R}^2$ and $\mathbb{R}^3$. In fact the length of a vector and the angle between two vectors play very important parts in the further development of linear algebra and it is now time to introduce these ingredients into our study. There are many ways to do this and in the approach that we will follow both length and angle will be derived from a more fundamental concept called a *scalar* or *inner* product of two vectors. No doubt the student has encountered the scalar product in the guise of the *dot* product of two vectors in $\mathbb{R}^3$ which is usually defined by the equation

$$\mathbf{A} \cdot \mathbf{B} = |\mathbf{A}||\mathbf{B}| \cos \theta$$

where $|\mathbf{A}|$ is the length of the vector $\mathbf{A}$, similarly for $\mathbf{B}$, and θ is the *angle between* $\mathbf{A}$ and $\mathbf{B}$. In the study of vectors in $\mathbb{R}^3$ this is a reasonable way to introduce the scalar product, because lengths and angles are already defined and well-studied concepts of geometry. In a more abstract study of linear algebra, such as we are undertaking, such an approach is not possible, for what is the length of a polynomial (vector) in $\mathscr{P}_4(\mathbb{R})$? This problem can be overcome by the use of the axiomatic method. Having introduced vector spaces by axioms it is not at all unreasonable to employ additional axioms to impose further structure on them.

Definition. A *scalar* (or *inner*) *product* on a vector space $\mathscr{V}$ is a function which assigns to each pair of vectors $\mathbf{A}$, $\mathbf{B}$ in $\mathscr{V}$ a real number, denoted by $\langle \mathbf{A}, \mathbf{B} \rangle$ having the following properties:

(1) $\langle \mathbf{A}, \mathbf{B} \rangle = \langle \mathbf{B}, \mathbf{A} \rangle$ for all $\mathbf{A}$, $\mathbf{B}$ in $\mathscr{V}$.
(2) For all vectors $\mathbf{A}$, $\mathbf{B}$ in $\mathscr{V}$ and numbers r in $\mathbb{R}$ we have $\langle r\mathbf{A}, \mathbf{B} \rangle = r\langle \mathbf{A}, \mathbf{B} \rangle = \langle \mathbf{A}, r\mathbf{B} \rangle$.
(3) For all vectors $\mathbf{A}$, $\mathbf{B}$, and $\mathbf{C}$ in $\mathscr{V}$ we have
   (a) $\langle \mathbf{A} + \mathbf{B}, \mathbf{C} \rangle = \langle \mathbf{A}, \mathbf{C} \rangle + \langle \mathbf{B}, \mathbf{C} \rangle$
   (b) $\langle \mathbf{A}, \mathbf{B} + \mathbf{C} \rangle = \langle \mathbf{A}, \mathbf{B} \rangle + \langle \mathbf{A}, \mathbf{C} \rangle$.
(4) For all $\mathbf{A}$ in $\mathscr{V}$, $\langle \mathbf{A}, \mathbf{A} \rangle \geq 0$ and $\langle \mathbf{A}, \mathbf{A} \rangle = 0$ iff $\mathbf{A} = \mathbf{0}$.

Notice in reading the above axioms that we have used the notations $+$, 0, etc. for two different ideas, namely $+$ for vectors and $+$ for numbers, the vector "zero" and the number "zero," etc. This should be taken into account in reading Axioms (2), (3), and (4).

It is not difficult to verify that for the dot product

$$\mathbf{A} \cdot \mathbf{B} = |\mathbf{A}||\mathbf{B}| \cos \theta$$

defined in $\mathbb{R}^3$ the above axioms are satisfied. The only condition required of $\cdot$ that is not immediately clear is (3). To verify (3, *a*) consider Figure 15.1 where θ is the angle between $\mathbf{A}$ and $\mathbf{B}$ and the angles with subscripts denote the angle between the vector subscript and $\mathbf{C}$. Now use the law of cosines. Thus the scalar product extends to general vector spaces the dot product of everyday experience.

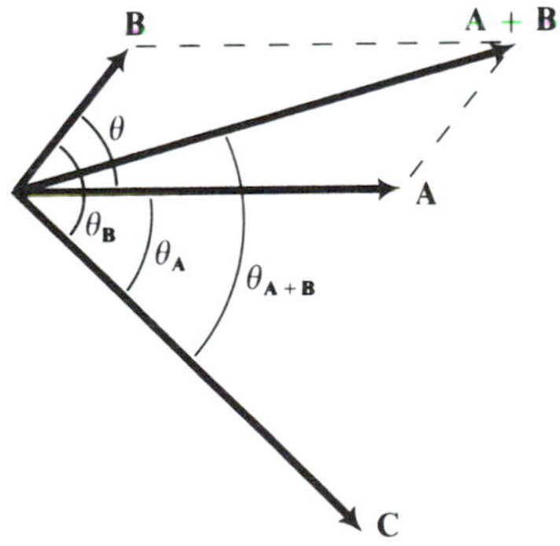

Figure 15.1

Why *scalar* product? That is, why is the concept introduced in the definition above called a *scalar* product? The answer should be apparent: because it associates to a pair of vectors a scalar (=real number). (Compare to Exercise 18 of Chapter 14.)

The notion of inner product can also be introduced into a complex vector space (see Exercise 17 at the end of this chapter where *a priori* we have not a single example where length and angle have any intuitive content.

EXAMPLE 1. *The standard scalar product on* $\mathbb{R}^n$.

For two vectors **A**, **B** in $\mathbb{R}^n$ we *define*

$$\langle \mathbf{A}, \mathbf{B} \rangle = a_1 b_1 + a_2 b_2 + \cdots + a_n b_n$$

where $\mathbf{A} = (a_1, \ldots, a_n)$, $\mathbf{B} = (b_1, \ldots, b_n)$. We assert that $\langle \mathbf{A}, \mathbf{B} \rangle$ is a scalar product. To verify this assertion we must check that Axioms (1)–(4) of the definition of a scalar product $\langle\ ,\ \rangle$ are satisfied for the function $\langle \mathbf{A}, \mathbf{B} \rangle = a_1 b_1 + a_2 b_2 + \cdots + a_n b_n$.

Let's see, we have

$$\begin{aligned} \langle \mathbf{A}, \mathbf{B} \rangle &= a_1 b_1 + a_2 b_2 + \cdots + a_n b_n = b_1 a_1 + b_2 a_2 + \cdots + b_n a_n \\ &= \langle \mathbf{B}, \mathbf{A} \rangle \end{aligned}$$

so that (1) holds. Next note that

$$\begin{aligned} \langle r\mathbf{A}, \mathbf{B} \rangle &= ra_1 b_1 + ra_2 b_2 + \cdots + ra_n b_n \\ &= r(a_1 b_1 + a_2 b_2 + \cdots + a_n b_n) = r\langle \mathbf{A}, \mathbf{B} \rangle \\ &= a_1 r b_1 + a_2 r b_2 + \cdots + a_n r b_n = \langle \mathbf{A}, r\mathbf{B} \rangle \end{aligned}$$

so that Axiom (2) also holds. To verify (3a) we compute as follows

$$\begin{aligned} \langle \mathbf{A} + \mathbf{B}, \mathbf{C} \rangle &= (a_1 + b_1)c_1 + (a_2 + b_2)c_2 + \cdots + (a_n + b_n)c_n \\ &= a_1 c_1 + b_1 c_1 + a_2 c_2 + b_2 c_2 + \cdots + a_n c_n + b_n c_n \\ &= a_1 c_1 + a_2 c_2 + \cdots + a_n c_n + b_1 c_1 + b_2 c_2 + \cdots + b_n c_n \\ &= \langle \mathbf{A}, \mathbf{C} \rangle + \langle \mathbf{B}, \mathbf{C} \rangle. \end{aligned}$$

The verification of (3b) is similar. Finally note that

$$\langle \mathbf{A}, \mathbf{A} \rangle = a_1 a_1 + a_2 a_2 + \cdots + a_n a_n = a_1^2 + a_2^2 + \cdots + a_n^2$$

and since a sum of squares is always nonnegative, and is zero iff all the terms are zero we see that (4) is also satisfied.

Therefore

$$\langle \mathbf{A}, \mathbf{B} \rangle = a_1 b_1 + a_2 b_2 + \cdots + a_n b_n$$

defines a scalar product on $\mathbb{R}^n$ called the standard scalar product of $\mathbb{R}^n$.

EXAMPLE 2. *The standard scalar product in* $\mathscr{V}$ *with basis.*

The preceding example may be generalized to any finite-dimensional vector space $\mathscr{V}$ with a **fixed chosen** basis $\{\mathbf{F}_1, \ldots, \mathbf{F}_n\}$. Namely if **A** and **B** belong to $\mathscr{V}$ then we know that

$$\begin{aligned} \mathbf{A} &= a_1 \mathbf{F}_1 + \cdots + a_n \mathbf{F}_n \\ \mathbf{B} &= b_1 \mathbf{F}_1 + \cdots + b_n \mathbf{F}_n \end{aligned}$$

for unique n-tuples $(a_1, \ldots, a_n)$, $(b_1, \ldots, b_n)$; the components of **A** and **B** relative to the basis $\{\mathbf{F}_1, \ldots, \mathbf{F}_n\}$ (see (6.6)). Now define

$$\langle \mathbf{A}, \mathbf{B} \rangle = a_1 b_1 + a_2 b_2 + \cdots + a_n b_n.$$

The same computations made in Example 1 show that $\langle \mathbf{A}, \mathbf{B} \rangle$ is a scalar product on $\mathscr{V}$. Notice that this scalar product depends on the choice of basis $\{\mathbf{F}_1, \ldots, \mathbf{F}_n\}$. A different choice of basis will give a different scalar product. For example, if in $\mathbb{R}^2$ we choose the standard basis and use the above definition then we obtain the standard scalar product of $\mathbb{R}^2$. On the other hand we could use the basis $\mathbf{F}_1 = (1, 0), \mathbf{F}_2 = (1, 1)$ in which case we would obtain an inner product $\langle \mathbf{A}, \mathbf{B} \rangle_{\mathbf{F}}$ where

$$\langle \mathbf{A}, \mathbf{B} \rangle_{\mathbf{F}} = a_1 b_1 - a_1 b_2 - b_2 a_1 + 2a_2 b_2$$

where $\mathbf{A} = (a_1, a_2) = a_1 \mathbf{E}_1 + a_2 \mathbf{E}_2$ and $\mathbf{B} = (b_1, b_2) = b_1 \mathbf{E}_1 + b_2 \mathbf{E}_2$. Thus many "different" scalar products seem possible on $\mathscr{V}$, one for each choice of bases. This difference is more apparent than real as we shall presently see.

EXAMPLE 3. *A scalar product on* $\mathscr{P}_k(\mathbb{R})$.

For the space $\mathscr{P}_k(\mathbb{R})$ of polynomials of degree at most k we may define the scalar product

$$\langle p(x), q(x) \rangle = \int_0^1 p(x)q(x)dx.$$

Certainly this definition is not at all like those of Example 1 or 2. In fact it is much more natural than those because it does not refer to a fixed basis. To verify the scalar product properties we freely use facts about integration from the calculus:

(1) $$\langle p(x), q(x) \rangle = \int_0^1 p(x)q(x)dx = \int_0^1 q(x)p(x)dx = \langle q(x), p(x) \rangle$$

(2) $$\langle rp(x), q(x) \rangle = \int_0^1 rp(x)q(x)dx = r \int_0^1 p(x)q(x)dx = r\langle p(x), q(x) \rangle$$
$$= \int_0^1 p(x)rq(x)dx = \langle p(x), rq(x) \rangle.$$

(3a) $$\langle p(x) + q(x), r(x) \rangle = \int_0^1 (p(x) + q(x))r(x)dx$$
$$= \int_0^1 (p(x)r(x) + q(x)r(x))dx$$
$$= \int_0^1 p(x)r(x)dx + \int_0^1 q(x)r(x)dx$$
$$= \langle p(x), r(x) \rangle + \langle q(x), r(x) \rangle,$$

and similarly for (3b).

(4) Recall that for a nonnegative continuous function $f(x)$

$$\int_0^1 f(x)dx \geq 0$$

and

$$\int_0^1 f(x)dx = 0 \quad \text{iff } f(x) = 0 \text{ for all } x \in [0, 1].$$

Note

$$\langle p(x), p(x)\rangle = \int_0^1 (p(x))^2 \, dx$$

and that $p(x)^2$ is nonnegative. Hence

$$\langle p(x), p(x)\rangle = \int_0^1 (p(x))^2 \, dx \geq 0.$$

Finally suppose that $\langle p(x), p(x)\rangle = 0$. Then

$$\int_0^1 (p(x))^2 \, dx = 0 \quad \text{implies} \quad p(x) = 0 \qquad x \in [0, 1].$$

A nonzero polynomial of degree k has at most k roots, but $p(x) = 0$ for all $x \in [0, 1]$ so has lots more than k roots. Therefore $p(x) = 0$.

Thus the axioms for a scalar product on $\mathscr{P}_k(\mathbb{R})$ are satisfied by the formula

$$\langle p(x), q(x)\rangle = \int_0^1 p(x)q(x)dx.$$

EXAMPLE 4. *The trace scalar product on* $\mathscr{M}_{nn}$.

Let $\mathscr{M}_{nn}$ denote the vector space of $n \times n$ matrices. Recall that $\mathbf{E}_{rs}$ is the $n \times n$ matrix defined by

$$\mathbf{E}_{rs} = (e_{ij}), \qquad e_{ij} = \begin{cases} 1 & \text{if } i = r, j = s, \\ 0 & \text{otherwise.} \end{cases}$$

The vectors $\{\mathbf{E}_{rs}\}$ form a basis for $\mathscr{M}_{nn}$ (see (11.2)). Therefore

$$\langle \mathbf{A}, \mathbf{B}\rangle = \sum_{i,j} a_{ij}b_{ij} \qquad \mathbf{A} = (a_{ij}), \mathbf{B} = (b_{ij})$$

defines a scalar product on $\mathscr{M}_{nn}$ corresponding to the basis $\{\mathbf{E}_{rs}\}$. There is another way to write this scalar product that is sort of interesting.

Definition. If $\mathbf{A} = (a_{ij})$ is an $n \times n$ matrix, the *trace of* $\mathbf{A}$ is the number $\text{tr}(\mathbf{A})$ defined by

$$\text{tr}(\mathbf{A}) = a_{11} + a_{22} + \cdots + a_{nn}.$$

Thus $\text{tr}(\mathbf{A})$ is the sum of the diagonal entries of $\mathbf{A}$. For example ($n = 3$)

$$\text{tr}\begin{pmatrix} 1 & 4 & 7 \\ 2 & 5 & 8 \\ 3 & 6 & 9 \end{pmatrix} = 15,$$

and

$$\operatorname{tr}\begin{pmatrix} 1 & & 0 \\ & \ddots & \\ 0 & & 1 \end{pmatrix} = n = \operatorname{tr}(\mathbf{I})$$

where $\mathbf{I} \in \mathscr{M}_{nn}$ is the identity matrix.

One way in which the trace and the inner product of Example 4 are connected is by the following formula

$$\langle \mathbf{A}, \mathbf{B} \rangle = \operatorname{tr}(\mathbf{AB}^t)$$

To verify this we need only compute the right-hand side and compare it to the definition of $\langle \mathbf{A}, \mathbf{B} \rangle$. We have

$$\operatorname{tr}(\mathbf{AB}^t) = \sum_i \sum_j a_{ij} b^t{}_{ji}$$

because the entry in the (i, i) position of the product matrix $\mathbf{AB}^t$ is $\sum_j a_{ij} b^t{}_{ji}$

$$= \sum_i \sum_j a_{ij} b_{ij}$$

$$= \sum_{i,j} a_{ij} b_{ij} = \langle \mathbf{A}, \mathbf{B} \rangle$$

as we claimed.

The trace of a square matrix is quite a handy concept and although we will not make use of it, it does play a rather important role in further developments of linear algebra.

Definition. A vector space $\mathscr{V}$ equipped with a fixed scalar product $\langle\ ,\ \rangle$ is called an *inner product space*. (Don't ask me why it is called an inner product space and not a scalar product space, I don't know. Because of this strange turn of the terminology, we will often call a scalar product an *inner product*.)

Let us turn now to the concepts of length and angle in an inner product space.

Definition. Let $(\mathscr{V}, \langle\ ,\ \rangle)$ be an inner product space and $\mathbf{A}$ a vector of $\mathscr{V}$. The *length of* $\mathbf{A}$, denoted by $|\mathbf{A}|$, is the nonnegative number

$$|\mathbf{A}| = \sqrt{\langle \mathbf{A}, \mathbf{A} \rangle}.$$

(Notice by Axiom (4) of the definition of scalar product that $\langle \mathbf{A}, \mathbf{A} \rangle \geq 0$ for all vectors $\mathbf{A}$ in $\mathscr{V}$ and hence $\langle \mathbf{A}, \mathbf{A} \rangle$ has a unique nonnegative square root $\sqrt{\langle \mathbf{A}, \mathbf{A} \rangle}$ which we have defined to be $|\mathbf{A}|$.)

Proposition 15.1. *Let $(\mathscr{V}, \langle\ ,\ \rangle)$ be an inner product space. Then the length function $|\ \ |$ has the following properties:*

(1) *For each* $\mathbf{A}$ *in* $\mathscr{V}$, $|\mathbf{A}| \geq 0$ *and* $|\mathbf{A}| = 0$ *iff* $\mathbf{A} = \mathbf{0}$.
(2) *For each* $\mathbf{A}$ *in* $\mathscr{V}$ *and number* r, $|rA| = |r||\mathbf{A}|$.

(As is becoming usual in our use of notations we have written $|\ \ |$ for two different concepts above. In (2) $|r\mathbf{A}|$ is the length of the vector $r\mathbf{A}$ while $|r|$ is the absolute value of the number r. Actually no confusion should arise because if we recall that $\mathbb{R} = \mathbb{R}^1$ is a vector space, then the absolute value $|r|$ is just the length of r when $\mathbb{R}^1$ is given the standard scalar product.)

Proof. Condition (1) is just a translation of Axiom (4) for scalar products while the computation

$$\begin{aligned} |r\mathbf{A}| &= \sqrt{\langle r\mathbf{A}, r\mathbf{A}\rangle} = \sqrt{r\langle \mathbf{A}, r\mathbf{A}\rangle} \\ &= \sqrt{r^2\langle \mathbf{A}, \mathbf{A}\rangle} = \sqrt{r^2}\sqrt{\langle \mathbf{A}, \mathbf{A}\rangle} \\ &= |r||\mathbf{A}| \end{aligned}$$

verifies condition (2). □

Proposition 15.2 (Schwarz inequality). *Let $(\mathscr{V}, \langle\ ,\ \rangle)$ be an inner product space. Then for any pair of vectors* $\mathbf{A}$ *and* $\mathbf{B}$ *in* $\mathscr{V}$,

$$|\langle \mathbf{A}, \mathbf{B}\rangle| \leq |\mathbf{A}||\mathbf{B}|.$$

Proof. We begin by observing that for $\mathbf{A} = \mathbf{0}$ we have

$$|\langle \mathbf{A}, \mathbf{B}\rangle| = \mathbf{0} \quad \text{and} \quad |\mathbf{A}||\mathbf{B}| = \mathbf{0}$$

so that the inequality is true in this case. We may therefore assume that $\mathbf{A} \neq \mathbf{0}$ and hence that $|\mathbf{A}| \neq \mathbf{0}$. Let

$$\mathbf{C} = \mathbf{B} - \frac{\langle \mathbf{A}, \mathbf{B}\rangle}{\langle \mathbf{A}, \mathbf{A}\rangle}\mathbf{A}.$$

Then

$$\begin{aligned} 0 \leq \langle \mathbf{C}, \mathbf{C}\rangle &= \left\langle \mathbf{B} - \frac{\langle \mathbf{A}, \mathbf{B}\rangle}{\langle \mathbf{A}, \mathbf{A}\rangle}\mathbf{A}, \mathbf{B} - \frac{\langle \mathbf{A}, \mathbf{B}\rangle}{\langle \mathbf{A}, \mathbf{A}\rangle}\mathbf{A}\right\rangle \\ &= \left\langle \mathbf{B} - \frac{\langle \mathbf{A}, \mathbf{B}\rangle}{\langle \mathbf{A}, \mathbf{A}\rangle}\mathbf{A}, \mathbf{B}\right\rangle - \left\langle \mathbf{B} - \frac{\langle \mathbf{A}, \mathbf{B}\rangle}{\langle \mathbf{A}, \mathbf{A}\rangle}\mathbf{A}, \frac{\langle \mathbf{A}, \mathbf{B}\rangle}{\langle \mathbf{A}, \mathbf{A}\rangle}\mathbf{A}\right\rangle \\ &= \langle \mathbf{B}, \mathbf{B}\rangle - \frac{\langle \mathbf{A}, \mathbf{B}\rangle}{\langle \mathbf{A}, \mathbf{A}\rangle}\langle \mathbf{A}, \mathbf{B}\rangle - \frac{\langle \mathbf{A}, \mathbf{B}\rangle}{\langle \mathbf{A}, \mathbf{A}\rangle}\langle \mathbf{B}, \mathbf{A}\rangle \\ &\quad + \frac{\langle \mathbf{A}, \mathbf{B}\rangle^2}{\langle \mathbf{A}, \mathbf{A}\rangle^2}\langle \mathbf{A}, \mathbf{A}\rangle \\ &= \langle \mathbf{B}, \mathbf{B}\rangle - \frac{\langle \mathbf{A}, \mathbf{B}\rangle^2}{\langle \mathbf{A}, \mathbf{A}\rangle} - \frac{\langle \mathbf{A}, \mathbf{B}\rangle\langle \mathbf{A}, \mathbf{B}\rangle}{\langle \mathbf{A}, \mathbf{A}\rangle} + \frac{\langle \mathbf{A}, \mathbf{B}\rangle^2}{\langle \mathbf{A}, \mathbf{A}\rangle} \\ &= \langle \mathbf{B}, \mathbf{B}\rangle - \frac{\langle \mathbf{A}, \mathbf{B}\rangle^2}{\langle \mathbf{A}, \mathbf{A}\rangle}. \end{aligned}$$

That is

$$\langle \mathbf{B}, \mathbf{B} \rangle - \frac{\langle \mathbf{A}, \mathbf{B} \rangle^2}{\langle \mathbf{A}, \mathbf{A} \rangle} \geq 0$$

Multiply both sides by $\langle \mathbf{A}, \mathbf{A} \rangle$, which is nonnegative and hence does not change the sense of the inequality, so we get

$$\langle \mathbf{A}, \mathbf{A} \rangle \langle \mathbf{B}, \mathbf{B} \rangle - \langle \mathbf{A}, \mathbf{B} \rangle^2 \geq 0$$

which may be rewritten as

$$|\mathbf{A}|^2 |\mathbf{B}|^2 \geq \langle \mathbf{A}, \mathbf{B} \rangle^2$$

taking positive square roots then gives

$$|\mathbf{A}||\mathbf{B}| \geq |\langle \mathbf{A}, \mathbf{B} \rangle|$$

as required. □

Proposition 15.3 (The triangle inequality). *Let* $(\mathscr{V}, \langle\ ,\ \rangle)$ *be an inner product space. Then for any vectors* $\mathbf{A}$ *and* $\mathbf{B}$ *in* $\mathscr{V}$ *we have*

$$|\mathbf{A} + \mathbf{B}| \leq |\mathbf{A}| + |\mathbf{B}|.$$

PROOF. We compute as follows

$$\begin{aligned} |\mathbf{A} + \mathbf{B}|^2 &= \langle \mathbf{A} + \mathbf{B}, \mathbf{A} + \mathbf{B} \rangle \\ &= \langle \mathbf{A}, \mathbf{A} + \mathbf{B} \rangle + \langle \mathbf{B}, \mathbf{A} + \mathbf{B} \rangle \\ &= \langle \mathbf{A}, \mathbf{A} \rangle + \langle \mathbf{A}, \mathbf{B} \rangle + \langle \mathbf{B}, \mathbf{A} \rangle + \langle \mathbf{B}, \mathbf{B} \rangle \\ &= |\mathbf{A}|^2 + \langle \mathbf{A}, \mathbf{B} \rangle + \langle \mathbf{A}, \mathbf{B} \rangle + |\mathbf{B}|^2 \\ &= |\mathbf{A}|^2 + 2\langle \mathbf{A}, \mathbf{B} \rangle + |\mathbf{B}|^2 \\ &\leq |\mathbf{A}|^2 + 2|\langle \mathbf{A}, \mathbf{B} \rangle| + |\mathbf{B}|^2 \\ &\leq |\mathbf{A}|^2 + 2|\mathbf{A}||\mathbf{B}| + |\mathbf{B}|^2 = (|\mathbf{A}| + |\mathbf{B}|)^2. \end{aligned}$$

That is,

$$|\mathbf{A} + \mathbf{B}|^2 \leq (|\mathbf{A}| + |\mathbf{B}|)^2$$

and taking positive square roots gives the desired conclusion. □

To see why (15.3) is called the *triangle inequality* let us define the *distance between* two vectors $\mathbf{A}$ and $\mathbf{B}$ in an inner product space $(\mathscr{V}, \langle\ ,\ \rangle)$ by the formula

$$d(\mathbf{A}, \mathbf{B}) = |\mathbf{A} - \mathbf{B}|.$$

The distance function d is easily seen to satisfy:

$$d(\mathbf{A}, \mathbf{B}) \geq 0$$

and

$$d(\mathbf{A}, \mathbf{B}) = 0 \quad \text{iff } \mathbf{A} = \mathbf{B}.$$

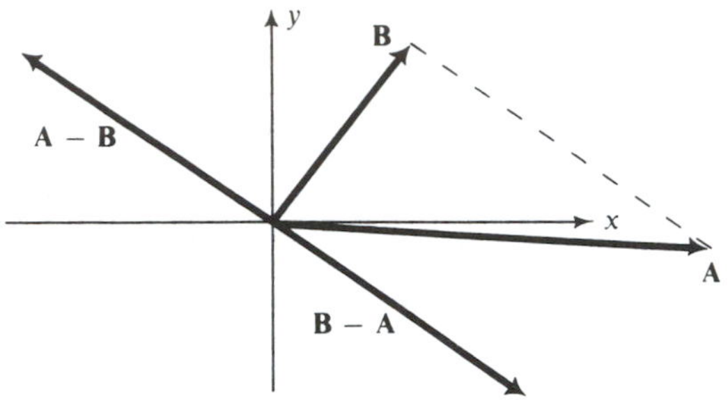

Figure 15.2

In $\mathbb{R}^2$ one has the picture shown in Figure 15.2. Here the length of $\mathbf{A} - \mathbf{B}$ is equal to the length of $\mathbf{B} - \mathbf{A}$ which is the length of the dotted segment since they are opposite sides of a parallelogram. This shows that $d(\mathbf{A}, \mathbf{B})$ is what you would think it was in this case.

The inequality of (15.3) now reads

$$d(\mathbf{A}, -\mathbf{B}) \leq |\mathbf{A}| + |-\mathbf{B}|,$$

namely the sum of the lengths of two sides of a triangle is greater than or equal to the length of the third side.

We have already used the inner product to define the length of a vector, we now use it to define angles.

Definition. Let $(\mathscr{V}, \langle\ ,\ \rangle)$ be an inner product space. If $\mathbf{A}$ and $\mathbf{B}$ are *non-zero* vectors of $\mathscr{V}$ the *angle between* $\mathbf{A}$ *and* $\mathbf{B}$ is defined by

$$\cos\theta = \frac{\langle \mathbf{A}, \mathbf{B} \rangle}{|\mathbf{A}||\mathbf{B}|}$$

and $0 \leq \theta \leq \pi$.

(Note that by the Schwarz inequality

$$-1 \leq \frac{\langle \mathbf{A}, \mathbf{B} \rangle}{|\mathbf{A}||\mathbf{B}|} \leq 1$$

so that θ is well defined by the above formula.)

This definition gives the formula

$$\langle \mathbf{A}, \mathbf{B} \rangle = |\mathbf{A}||\mathbf{B}| \cos\theta$$

that we are familiar with in $\mathbb{R}^3$ so we may safely assume that the definition of θ is a reasonable extension of the concept of angle to an arbitrary inner product space.

Proposition 15.4. *Two vectors* $\mathbf{A}$ *and* $\mathbf{B}$ *in an inner product space* $\mathscr{V}$ *are perpendicular* (*or orthogonal*) iff $\langle \mathbf{A}, \mathbf{B} \rangle = 0$.

PROOF. By definition $\mathbf{A}$ and $\mathbf{B}$ are perpendicular iff the angle θ between them is $\pi/2$ radians $= 90°$. For such θ we have $\cos\theta = 0$. Therefore

$$\frac{\langle \mathbf{A}, \mathbf{B}\rangle}{|\mathbf{A}||\mathbf{B}|} = \cos\theta = 0$$

so $\langle \mathbf{A}, \mathbf{B}\rangle = 0$. The reverse implication, namely $\langle \mathbf{A}, \mathbf{B}\rangle = 0$ implies $\theta = \pi/2$ is immediate from the definition of θ. □

Thus the inner product $\langle\ ,\ \rangle$ provides a very handy check for perpendicularity of vectors.

EXAMPLE 5. Are the vectors $(1, -1, 2), (2, 4, 1)$ in $\mathbb{R}^3$ perpendicular? (We assume tht $\mathbb{R}^3$ has its standard inner product).

Solution. One checks

$$\langle (1, -1, 2), (2, 4, 1)\rangle = 2 - 4 + 2 = 0$$

so they are perpendicular. □

Definition. A set of vectors S in an inner product space $(\mathscr{V}, \langle\ ,\ \rangle)$ is said to be an *orthogonal set of vectors* iff for any two distinct vectors $\mathbf{A}, \mathbf{B}$ in S, $\langle \mathbf{A}, \mathbf{B}\rangle = 0$. An orthogonal set of vectors is said to be *orthonormal* iff $|\mathbf{A}| = 1$ for every $\mathbf{A}$ in S.

Proposition 15.5. *An orthogonal set of* nonzero vectors *S in an inner product space $(\mathscr{V}, \langle\ ,\ \rangle)$ is linearly independent.*

PROOF. Suppose that

$$a_1\mathbf{A}_1 + \cdots + a_n\mathbf{A}_n = \mathbf{0}$$

is a linear relation between vectors $\mathbf{A}_1, \ldots, \mathbf{A}_n$ of $\mathscr{V}$. Then we have

$$\begin{aligned} 0 = \langle \mathbf{A}_i, \mathbf{0}\rangle &= \langle \mathbf{A}_i, a_1\mathbf{A}_1 + \cdots + a_n\mathbf{A}_n\rangle \\ &= a_1\langle \mathbf{A}_i, \mathbf{A}_1\rangle + a_2\langle \mathbf{A}_i, \mathbf{A}_2\rangle + \cdots + a_n\langle \mathbf{A}_i, \mathbf{A}_n\rangle \\ &= a_i\langle \mathbf{A}_i, \mathbf{A}_i\rangle \end{aligned}$$

since $\langle \mathbf{A}_i, \mathbf{A}_j\rangle = 0$ for $i \neq j$ because the set S is orthogonal. But $\mathbf{A}_i$ is not the zero vector and therefore $\langle \mathbf{A}_i, \mathbf{A}_i\rangle \neq 0$, so the equation

$$0 = a_i\langle \mathbf{A}_i, \mathbf{A}_i\rangle$$

implies $a_i = 0$. Since this is true for $i = 1, 2, \ldots, n$ there can be no nontrivial linear relation between the vectors of S, so S is a linearly independent set. □

EXAMPLE 6. Which of the following sets of vectors in $\mathbb{R}$ are orthogonal (in the standard inner product)

(a) $\{(0, 0, 0), (1, 1, 1)\}$?
(b) $\{(1, 0, 0), (0, 1, 0), (0, 0, 1)\}$?
(c) $\{(1, 1, 1), (1, 1, 0), (1, 0, 0)\}$?
(d) $\{(1, 0, 0), (0, 1, 0), (0, 0, 1), (1, 1, 1)\}$?

Solution. The set (a) is orthogonal because it consists of exactly two vectors, one of which is $\mathbf{0}$. The set (b) is orthogonal but the set (c) is not, because

$$\langle (1, 1, 1), (1, 0, 0) \rangle = 1 \neq 0.$$

The set (d) is not orthogonal because it contains 4 *nonzero* vectors which if it were orthogonal would have to be linearly independent by (15.5) contrary to the fact that $\mathbb{R}^3$ is only 3-dimensional.

EXAMPLE 7. Is the set of vectors

$$\{1, x - \tfrac{1}{2}, x^2 - x + \tfrac{1}{6}\}$$

orthogonal in $\mathscr{P}_2(\mathbb{R})$?

Solution. The answer is yes, and you might well ask how do I know. The "real answer" lies in (15.7) and Example 10. For the moment all we can do is compute. We have

$$\left\langle 1, x - \frac{1}{2} \right\rangle = \int_0^1 \left(x - \frac{1}{2}\right) dx = \frac{x^2}{2} - \frac{1}{2}x \Big|_0^1 = 0$$

$$\left\langle 1, x^2 - x + \frac{1}{6} \right\rangle = \int_0^1 \left(x^2 - x + \frac{1}{6}\right) dx = \frac{x^3}{3} - \frac{x^2}{2} + \frac{1}{6}\Big|_0^1$$

$$= \frac{1}{3} - \frac{1}{2} + \frac{1}{6} = 0$$

$$\left\langle x - \frac{1}{2}, x^2 - x + \frac{1}{6} \right\rangle = \int_0^1 \left(x - \frac{1}{2}\right)\left(x^2 - x + \frac{1}{6}\right) dx$$

$$= \int_0^1 \left(x^3 - \frac{3}{2}x^2 + \frac{2}{3}x - \frac{1}{12}\right) dx$$

$$= \frac{x^4}{4} - \frac{1}{2}x^3 + \frac{1}{3}x^2 - \frac{1}{12}x \Big|_0^1$$

$$= \frac{1}{4} - \frac{1}{2} + \frac{1}{3} - \frac{1}{12} = \frac{3 - 6 + 4 - 1}{12}$$

$$= 0.$$

The remaining scalar products are seen to be zero by symmetry of $\langle \ , \ \rangle$ in the two variables.

Note that by (15.4) the vectors $\{1, x - \frac{1}{2}, x^2 - x + \frac{1}{6}\}$ are linearly independent. Since $\dim \mathscr{P}_2(\mathbb{R}) = 3$ it follows that these vectors are actually an orthogonal *basis* for $\mathscr{P}_2(\mathbb{R})$.

The construction of orthonormal sets in $\mathscr{P}_k(\mathbb{R})$ for large k is a fascinating business involving many areas of mathematics including analytic number theory and differential equations. The polynomials that most often occur are named after the mathematical giants of preceding centuries.

Proposition 15.6. *Suppose that* $\{\mathbf{A}_1, \ldots, \mathbf{A}_n\}$ *is an orthogonal set of nonzero vectors in the inner product space* $\mathscr{V}$. *If*

$$\mathbf{A} \in \mathscr{L}(\mathbf{A}_1, \ldots, \mathbf{A}_n)$$

then[1]

$$\mathbf{A} = \frac{\langle \mathbf{A}, \mathbf{A}_1 \rangle}{\langle \mathbf{A}_1, \mathbf{A}_1 \rangle} \mathbf{A}_1 + \cdots + \frac{\langle \mathbf{A}, \mathbf{A}_n \rangle}{\langle \mathbf{A}_n, \mathbf{A}_n \rangle} \mathbf{A}_n.$$

Proof. The proof is very similar to that of (15.5). Suppose that $\mathbf{A}$ belongs to the linear span of $\mathbf{A}_1, \ldots, \mathbf{A}_k$. Then there are numbers $a_1, \ldots, a_n$ such that

$$\mathbf{A} = a_1\mathbf{A}_1 + \cdots + a_n\mathbf{A}_n.$$

What we are trying to prove is that the number a_i must be the particular number $\langle \mathbf{A}, \mathbf{A}_i \rangle / \langle \mathbf{A}_i, \mathbf{A}_i \rangle$. This is easy to prove. Take the inner product of both sides of the preceding equation with $\mathbf{A}_i$ giving

$$\begin{aligned} \langle \mathbf{A}_i, \mathbf{A} \rangle &= \langle \mathbf{A}_i, a_1\mathbf{A}_1 + \cdots + a_n\mathbf{A}_n \rangle \\ &= a_1\langle \mathbf{A}_i, \mathbf{A}_1 \rangle + \cdots + a_n\langle \mathbf{A}_i, \mathbf{A}_n \rangle \\ &= a_i\langle \mathbf{A}_i, \mathbf{A}_i \rangle \end{aligned}$$

and since $\mathbf{A}_i$ is nonzero so is $\langle \mathbf{A}_i, \mathbf{A}_i \rangle$ so

$$a_i = \frac{\langle \mathbf{A}_i, \mathbf{A} \rangle}{\langle \mathbf{A}_i, \mathbf{A}_i \rangle} = \frac{\langle \mathbf{A}, \mathbf{A}_i \rangle}{\langle \mathbf{A}_i, \mathbf{A}_i \rangle}$$

as required. □

Remarks. The numbers $\langle \mathbf{A}, \mathbf{A}_i \rangle / \langle \mathbf{A}_i, \mathbf{A}_i \rangle$ are usually called the *Fourier coefficients* of $\mathbf{A}$ with respect to $\{\mathbf{A}_1, \ldots, \mathbf{A}_n\}$. The vector

$$\frac{\langle \mathbf{A}, \mathbf{A}_i \rangle}{\langle \mathbf{A}_i, \mathbf{A}_i \rangle} \mathbf{A}_i$$

is called the *component of* $\mathbf{A}$ *along* $\mathbf{A}_i$.

Since the vector

$$\frac{\langle \mathbf{A}, \mathbf{A}_1 \rangle}{\langle \mathbf{A}_1, \mathbf{A}_1 \rangle} \mathbf{A}_1 + \cdots + \frac{\langle \mathbf{A}, \mathbf{A}_n \rangle}{\langle \mathbf{A}_n, \mathbf{A}_n \rangle} \mathbf{A}_n$$

[1] Note that since $\mathbf{A}_i \neq \mathbf{0}$, $\langle \mathbf{A}_i, \mathbf{A}_i \rangle \neq 0$ also. Thus the right-hand side is well defined.

belongs to $\mathscr{L}(\mathbf{A}_1, \ldots, \mathbf{A}_n)$ for any vector $\mathbf{A}$ in $\mathscr{V}$ we have actually proven more than is claimed in (15.6), namely

$$\mathbf{A} \in \mathscr{L}(\mathbf{A}_1, \ldots, \mathbf{A}_n)$$

iff

$$\mathbf{A} = \frac{\langle \mathbf{A}, \mathbf{A}_1 \rangle}{\langle \mathbf{A}_1, \mathbf{A}_1 \rangle} \mathbf{A}_1 + \cdots + \frac{\langle \mathbf{A}, \mathbf{A}_n \rangle}{\langle \mathbf{A}_n, \mathbf{A}_n \rangle} \mathbf{A}_n.$$

Here are some examples to illustrate this point:

EXAMPLE 8. Find the Fourier coefficients of the vector

$$x^2 + x + 1$$

with respect to the orthogonal set of vectors

$$\{1, x - \tfrac{1}{2}, x^2 - x + \tfrac{1}{6}\}$$

in $\mathscr{P}_2(\mathbb{R})$.

Solution. We must compute

$$\langle 1, x^2 + x + 1 \rangle = \int_0^1 (x^2 + x + 1)dx = \frac{x^3}{3} + \frac{x^2}{2} + x \Big|_0^1 = \frac{11}{6}$$

$$\begin{aligned}
\left\langle x - \frac{1}{2}, x^2 + x + 1 \right\rangle &= \int_0^1 \left(x - \frac{1}{2}\right)(x^2 + x + 1)dx \\
&= \int_0^1 \left(x^3 + \frac{1}{2}x^2 + \frac{1}{2}x - \frac{1}{2}\right)dx \\
&= \frac{x^4}{4} + \frac{x^3}{6} + \frac{x^2}{4} - \frac{1}{2}x \Big|_0^1 = \frac{1}{4} + \frac{1}{6} + \frac{1}{4} - \frac{1}{2} \\
&= \frac{1}{6}
\end{aligned}$$

$$\begin{aligned}
\left\langle x^2 - x + \frac{1}{6}, x^2 + x + 1 \right\rangle &= \int_0^1 \left(x^2 - x + \frac{1}{6}\right)(x^2 + x + 1)dx \\
&= \int_0^1 \left(x^4 + \frac{1}{6}x^2 - \frac{5}{6}x + \frac{1}{6}\right)dx \\
&= \frac{x^5}{5} + \frac{x^3}{18} - \frac{5x^2}{12} + \frac{x}{6}\Big|_0^1 = \frac{1}{5} + \frac{1}{18} - \frac{5}{12} + \frac{1}{6} \\
&= \frac{36 + 10 - 75 + 30}{180} = \frac{1}{180}
\end{aligned}$$

$$\langle 1, 1 \rangle = \int_0^1 1\, dx = 1$$

$$\left\langle x - \frac{1}{2}, x - \frac{1}{2} \right\rangle = \int_0^1 \left(x - \frac{1}{2}\right)^2 dx = \frac{1}{3}\left(x - \frac{1}{2}\right)^3\Big|_0^1 = \frac{1}{12},$$

$$\begin{aligned}\left\langle x^2 - x + \frac{1}{6}, x^2 - x + \frac{1}{6} \right\rangle &= \int_0^1 \left(x^2 - x + \frac{1}{6}\right)^2 dx \\ &= \int_0^1 \left(x^4 - 2x^3 + \frac{4}{3}x^2 - \frac{x}{3} + \frac{1}{36}\right) dx \\ &= \frac{x^5}{5} - \frac{x^4}{2} + \frac{4}{9}x^3 - \frac{x^2}{6} + \frac{1}{36}x\Big|_0^1 \\ &= \frac{1}{5} - \frac{1}{2} + \frac{4}{9} - \frac{1}{6} + \frac{1}{36} \\ &= \frac{36 - 90 + 80 - 30 + 5}{180} \\ &= \frac{1}{180}.\end{aligned}$$

Thus the Fourier coefficients of $1 + x + x^2$ with respect to $\{1, x - \frac{1}{2}, x^2 - x + \frac{1}{6}\}$ are $\{\frac{11}{6}, 2, 1\}$. (Note order counts!).

Example 9. Write the vector $\mathbf{A} = (1, -1, 1)$ as a linear combination of the vectors

$$\begin{aligned}\mathbf{V}_1 &= (1, 1, 1)\\ \mathbf{V}_2 &= (0, 1, -1)\\ \mathbf{V}_3 &= (-2, 1, 1).\end{aligned}$$

Solution. Before plunging ahead and solving lists of linear equations it pays to note that $\{\mathbf{V}_1, \mathbf{V}_2, \mathbf{V}_3\}$ are orthogonal and hence independent by (15.5) which means, since $\dim \mathbb{R}^3 = 3$, that they are a basis for $\mathbb{R}^3$. Accordingly by (15.6) we must have

$$\mathbf{A} = \frac{\langle \mathbf{A}, \mathbf{V}_1 \rangle}{\langle \mathbf{V}_1, \mathbf{V}_1 \rangle}\mathbf{V}_1 + \frac{\langle \mathbf{A}, \mathbf{V}_2 \rangle}{\langle \mathbf{V}_2, \mathbf{V}_2 \rangle}\mathbf{V}_2 + \frac{\langle \mathbf{A}, \mathbf{V}_3 \rangle}{\langle \mathbf{V}_3, \mathbf{V}_3 \rangle}\mathbf{V}_3$$

and rather than solve systems of equations we need only compute inner products. We find

$$\begin{aligned}\langle \mathbf{A}, \mathbf{V}_1 \rangle &= 1 \qquad & \langle \mathbf{V}_1, \mathbf{V}_1 \rangle &= 3\\ \langle \mathbf{A}, \mathbf{V}_2 \rangle &= -2 \qquad & \langle \mathbf{V}_2, \mathbf{V}_2 \rangle &= 2\\ \langle \mathbf{A}, \mathbf{V}_3 \rangle &= -2 \qquad & \langle \mathbf{V}_3, \mathbf{V}_3 \rangle &= 6\end{aligned}$$

and therefore

$$(1, -1, 1) = \tfrac{1}{3}(1, 1, 1) - 1(0, 1, -1) - \tfrac{1}{3}(-2, 1, 1),$$

which is all rather easy. □

Proposition 15.7. *Let $(\mathscr{V}, \langle\ ,\ \rangle)$ be an inner product space and $\mathscr{W}$ a finite-dimensional subspace of $\mathscr{V}$. Then there exists in $\mathscr{V}$ an orthonormal set of vectors $\{\mathbf{A}_1, \ldots, \mathbf{A}_n\}$ that is a basis for $\mathscr{W}$.*

PROOF. The method of proof that we will employ is called the *Gram–Schmidt orthonormalization* process. It is a manufacturing process that starts with any old basis for $\mathscr{W}$ as raw material and manufactures one of the required type. We therefore begin by choosing a basis $\{\mathbf{B}_1, \ldots, \mathbf{B}_n\}$ for $\mathscr{W}$. This we know we can do by (6.1). We now define in succession vectors $\mathbf{A}_1, \ldots, \mathbf{A}_n$ by the equations

$$\mathbf{A}_1 = \mathbf{B}_1$$

$$\mathbf{A}_2 = \mathbf{B}_2 - \frac{\langle \mathbf{B}_2, \mathbf{A}_1 \rangle}{\langle \mathbf{A}_1, \mathbf{A}_1 \rangle} \mathbf{A}_1$$

$$\mathbf{A}_n = \mathbf{B}_n - \frac{\langle \mathbf{B}_n, \mathbf{A}_{n-1} \rangle}{\langle \mathbf{A}_{n-1}, \mathbf{A}_{n-1} \rangle} \mathbf{A}_{n-1} - \cdots - \frac{\langle \mathbf{B}_n, \mathbf{A}_1 \rangle}{\langle \mathbf{A}_1, \mathbf{A}_1 \rangle} \mathbf{A}_1.$$

Before proceeding further we must check that we have not committed the cardinal sin of dividing by 0 in one of the above formulas. To do so note that $\mathbf{A}_i$ is actually a linear combination of $\mathbf{B}_1, \ldots, \mathbf{B}_i$ in disguise, that is $\mathbf{A}_i \in \mathscr{L}(\mathbf{B}_1, \ldots, \mathbf{B}_i)$. If for some j, $\langle \mathbf{A}_j, \mathbf{A}_j \rangle = 0$, then $\mathbf{A}_j = 0$ and therefore we get from the equation defining $\mathbf{A}_j$

$$\mathbf{B}_j = \frac{\langle \mathbf{B}_j, \mathbf{A}_{j-1} \rangle}{\langle \mathbf{A}_{j-1}, \mathbf{A}_{j-1} \rangle} \mathbf{A}_{j-1} + \cdots + \frac{\langle \mathbf{B}_j, \mathbf{A}_1 \rangle}{\langle \mathbf{A}_1, \mathbf{A}_1 \rangle} \mathbf{A}_1 \in \mathscr{L}(\mathbf{A}_1, \ldots, \mathbf{A}_{j-1})$$

and therefore $\mathbf{B}_j \in \mathscr{L}(\mathbf{B}_1, \ldots, \mathbf{B}_{j-1})$ contrary to the linear independence of $\{\mathbf{B}_1, \ldots, \mathbf{B}_n\}$. Therefore $\langle \mathbf{A}_j, \mathbf{A}_j \rangle$ is never zero.

Next we claim that $\mathscr{L}(\mathbf{A}_1, \ldots, \mathbf{A}_i) = \mathscr{L}(\mathbf{B}_1, \ldots, \mathbf{B}_i)$. We have already seen that $\mathbf{A}_1, \ldots, \mathbf{A}_i \in \mathscr{L}(\mathbf{B}_1, \ldots, \mathbf{B}_i)$ so that $\mathscr{L}(\mathbf{A}_1, \ldots, \mathbf{A}_i) \subset \mathscr{L}(\mathbf{B}_1, \ldots, \mathbf{B}_i)$. It will therefore suffice to show $\mathscr{L}(\mathbf{B}_1, \ldots, \mathbf{B}_i) \subset \mathscr{L}(\mathbf{A}_1, \ldots, \mathbf{A}_i)$ which is easy.

Our next task is to show that the vectors $\{\mathbf{A}_1, \ldots, \mathbf{A}_n\}$ are orthogonal. We will do this by showing successively that the sets $\{\mathbf{A}_1\}$, $\{\mathbf{A}_1, \mathbf{A}_2\}, \ldots,$ $\{\mathbf{A}_1, \ldots, \mathbf{A}_n\}$ are orthogonal. The first set is orthogonal because $\mathbf{A}_1 = \mathbf{B}_1 \neq \mathbf{0}$. Suppose as an induction hypothesis that we have already shown the set $\{\mathbf{A}_1, \ldots, \mathbf{A}_i\}$ to be orthogonal. In order to prove that $\{\mathbf{A}_1, \ldots, \mathbf{A}_{i+1}\}$ is orthogonal it is only necessary to show that $\mathbf{A}_{i+1}$ is orthogonal to $\mathbf{A}_1, \ldots, \mathbf{A}_i$. Now we compute for $1 \leq j \leq i$

$$\begin{aligned}
\langle \mathbf{A}_{i+1}, \mathbf{A}_j \rangle &= \left\langle \mathbf{B}_{i+1} - \frac{\langle \mathbf{B}_{i+1}, \mathbf{A}_i \rangle}{\langle \mathbf{A}_i, \mathbf{A}_i \rangle} \mathbf{A}_i - \cdots - \frac{\langle \mathbf{B}_{i+1}, \mathbf{A}_1 \rangle}{\langle \mathbf{A}_1, \mathbf{A}_1 \rangle} \mathbf{A}_1, \mathbf{A}_j \right\rangle \\
&= \langle \mathbf{B}_{i+1}, \mathbf{A}_j \rangle - \frac{\langle \mathbf{B}_{i+1}, \mathbf{A}_i \rangle}{\langle \mathbf{A}_i, \mathbf{A}_i \rangle} \langle \mathbf{A}_i, \mathbf{A}_j \rangle - \cdots \\
&\quad - \frac{\langle \mathbf{B}_{i+1}, \mathbf{A}_1 \rangle}{\langle \mathbf{A}_1, \mathbf{B}_1 \rangle} \langle \mathbf{A}_1, \mathbf{A}_j \rangle \\
&= \langle \mathbf{B}_{i+1}, \mathbf{A}_j \rangle - \frac{\langle \mathbf{B}_{i+1}, \mathbf{A}_i \rangle}{\langle \mathbf{A}_i, \mathbf{A}_i \rangle} 0 - \cdots - \frac{\langle \mathbf{B}_{i+1}, \mathbf{A}_j \rangle}{\langle \mathbf{A}_j, \mathbf{A}_j \rangle} \langle \mathbf{A}_j, \mathbf{A}_j \rangle - \cdots \\
&\quad - \frac{\langle \mathbf{B}_{i+1}, \mathbf{A}_1 \rangle}{\langle \mathbf{A}_1, \mathbf{A}_1 \rangle} 0 \\
&= \langle \mathbf{B}_{i+1}, \mathbf{A}_j \rangle - \langle \mathbf{B}_{i+1}, \mathbf{A}_j \rangle = 0,
\end{aligned}$$

where the key step was to use the fact $\langle \mathbf{A}_k, \mathbf{A}_j \rangle = 0$ if $1 \leq k \leq i$ and $k \neq j$. This computation shows that $\{\mathbf{A}_1, \ldots, \mathbf{A}_{i+1}\}$ is an orthogonal set of vectors. By induction we may conclude that $\{\mathbf{A}_1, \ldots, \mathbf{A}_n\}$ is an orthogonal set of nonzero vectors. By (15.5) it follows that the set $\{\mathbf{A}_1, \ldots, \mathbf{A}_n\}$ is linearly independent. Since $\mathscr{L}(\mathbf{A}_1, \ldots, \mathbf{A}_n) \subset \mathscr{W}$ and $n = \dim \mathscr{W}$ it follows from (6.9) that $\{\mathbf{A}_1, \ldots, \mathbf{A}_n\}$ is a basis for $\mathscr{W}$.

But we are not yet done! We have only constructed an orthogonal set of vectors $\{\mathbf{A}_1, \ldots, \mathbf{A}_n\}$ that is a basis for $\mathscr{V}$. What we want is an ortho*normal* set. However this is easy. We just **normalize** the set $\{\mathbf{A}_1, \ldots, \mathbf{A}_n\}$ by defining $\overline{\mathbf{A}}_i = (1/|\mathbf{A}_i|)\mathbf{A}_i$, which we may do since $|\mathbf{A}_i| \neq 0$. We then have

$$|\overline{\mathbf{A}}_i| = \left| \frac{1}{|\mathbf{A}_i|} \mathbf{A}_i \right| = \frac{1}{|\mathbf{A}_i|} |\mathbf{A}_i| = 1$$

and

$$\langle \overline{\mathbf{A}}_r, \overline{\mathbf{A}}_s \rangle = \frac{1}{|\mathbf{A}_r|} \frac{1}{|\mathbf{A}_s|} \langle \mathbf{A}_r, \mathbf{A}_s \rangle = 0, \qquad r \neq s$$

so that $\{\overline{\mathbf{A}}_1, \ldots, \overline{\mathbf{A}}_n\}$ is both a basis and an orthonormal set. □

The kind of basis that appears in the above proposition is very important and therefore comes in for a special name.

Definition. Let $(\mathscr{V}, \langle\ ,\ \rangle)$ be a finite-dimensional inner product space. A basis $\{\mathbf{A}_1, \ldots, \mathbf{A}_n\}$ for $\mathscr{V}$ is called an *orthonormal basis* for $\mathscr{V}$ iff $\{\mathbf{A}_1, \ldots, \mathbf{A}_n\}$ is a basis for $\mathscr{V}$ and an orthonormal set, that is

$$\langle \mathbf{A}_i, \mathbf{A}_j \rangle = \begin{cases} 1 & \text{if } i = j, \\ 0 & \text{if } i \neq j, \end{cases}$$

or what is the same thing, the vectors are of unit length and pairwise orthogonal.

While the Gram–Schmidt process may seem quite formidable it really isn't. For once one has established that it works, one can apply it without worrying about the inductive check that the vectors it manufactures are orthogonal. However one should not forget to normalize at the end if required to do so.

EXAMPLE 10. Apply the Gram–Schmidt orthogonalization process to find an orthogonal basis for $\mathscr{P}_2(\mathbb{R})$.

Solution. We may start with any old basis for $\mathscr{P}_2(\mathbb{R})$ to apply the process to, so let us choose the obvious basis $\{1, x, x^2\}$. We then define vectors

$$\mathbf{A}_1 = 1$$

$$\mathbf{A}_2 = x - \frac{\langle x, \mathbf{A}_1\rangle}{\langle \mathbf{A}_1, \mathbf{A}_1\rangle}\mathbf{A}_1$$

$$\mathbf{A}_3 = x^2 - \frac{\langle x^2, \mathbf{A}_1\rangle}{\langle \mathbf{A}_1, \mathbf{A}_1\rangle}\mathbf{A}_1 - \frac{\langle x^2, \mathbf{A}_2\rangle}{\langle \mathbf{A}_2, \mathbf{A}_2\rangle}\mathbf{A}_2$$

and so we have some integrations to carry out.

Since $\mathbf{A}_1 = 1$ we get

$$\langle x, \mathbf{A}_1\rangle = \langle x, 1\rangle = \int_0^1 x\,dx = \tfrac{1}{2}$$

$$\langle \mathbf{A}_1, \mathbf{A}_1\rangle = \langle 1, 1\rangle = \int_0^1 1\,dx = 1$$

so that $\mathbf{A}_2 = x - \frac{1}{2}$ and we get

$$\langle x^2, \mathbf{A}_1\rangle = \langle x^2, 1\rangle = \int_0^1 x^2\,dx = \tfrac{1}{3}$$

$$\langle x^2, A_2\rangle = \int_0^1 x^2\left(x - \frac{1}{2}\right)dx = \int_0^1 \left(x^3 - \frac{1}{2}x^2\right)dx$$

$$= \frac{x^4}{4} - \frac{x^3}{6}\bigg|_0^1 = \frac{1}{4} - \frac{1}{6} = \frac{1}{12}$$

$$\langle \mathbf{A}_2, \mathbf{A}_2\rangle = \int_0^1 (x - \tfrac{1}{2})^2\,dx = \tfrac{1}{3}(x - \tfrac{1}{2})^3\big|_0^1$$

$$= \tfrac{1}{3}[(1 - \tfrac{1}{2})^3 - (0 - \tfrac{1}{2})^3]$$

$$= \tfrac{1}{3}[\tfrac{1}{8} - (-\tfrac{1}{8})] = \tfrac{1}{3}(\tfrac{1}{4}) = \tfrac{1}{12}$$

so that we find

$$\begin{aligned}\mathbf{A}_1 &= 1\\ \mathbf{A}_2 &= x - \tfrac{1}{2}\\ \mathbf{A}_3 &= x^2 - \tfrac{1}{3}\cdot 1 - (x - \tfrac{1}{2}) = x^2 - x + \tfrac{1}{6}\end{aligned}$$

is an orthogonal basis for $\mathscr{P}_2(\mathbb{R})$. (Go back and look at Example 7 again.)

EXAMPLE 11. Find an orthonormal basis for the subspace $\mathscr{W}$ given by

$$x + 2y + 3z = 0$$

of $\mathbb{R}^2$.

Solution. A basis for $\mathscr{W}$ is easily seen to be

$$\mathbf{B}_1 = (-3, 0, 1), \qquad \mathbf{B}_2 = (-2, 1, 0).$$

Applying the Gram–Schmidt process we get the orthogonal basis

$$\begin{aligned}\mathbf{A}_1 &= (-3, 0, 1)\\ \mathbf{A}_2 &= (-2, 1, 0) - \frac{\langle(-3, 0, 1), (-2, 1, 0)\rangle}{\langle(-3, 0, 1), (-3, 0, 1)\rangle}(-3, 0, 1)\\ &= (-2, 1, 0) - \tfrac{6}{10}(-3, 0, 1)\\ &= (-\tfrac{1}{5}, 1, -\tfrac{3}{5}).\end{aligned}$$

But we are not done yet! The vectors $\{\mathbf{A}_1, \mathbf{A}_2\}$ are only orthogonal, not ortho*normal*. We must therefore normalize them getting

$$\mathbf{C}_1 = \frac{\mathbf{A}_1}{|\mathbf{A}_1|} = \frac{1}{\sqrt{10}}(-3, 0, 1) = \left(-\frac{3}{\sqrt{10}}, 0, \frac{1}{\sqrt{10}}\right)$$

$$\mathbf{C}_2 = \frac{\mathbf{A}_2}{|\mathbf{A}_2|} = \frac{\sqrt{5}}{\sqrt{7}}\left(-\frac{1}{5}, 1, -\frac{3}{5}\right) = \left(-\frac{1}{\sqrt{35}}, \frac{\sqrt{5}}{\sqrt{7}}, -\frac{3}{\sqrt{35}}\right)$$

which is an orthonormal basis for $\mathscr{W}$.

There is a very important consequence of (15.7), namely that a finite-dimensional inner product space has an orthonormal basis. We record this for future reference.

Corollary 15.8. *Let $(\mathscr{V}, \langle\ ,\ \rangle)$ be a finite-dimensional inner product space. Then there is an orthonormal basis for $\mathscr{V}$.* □

One reason why orthonormal bases are so useful in an inner product space is the following

Theorem 15.9. *Let* $\{\mathbf{A}_1, \ldots, \mathbf{A}_n\}$ *be an orthonormal basis for the inner product space* $(\mathscr{V}, \langle\ ,\ \rangle)$. *If* $\mathbf{A}$ *belongs to* $\mathscr{V}$ *then*

$$\mathbf{A} = \langle \mathbf{A}, \mathbf{A}_1 \rangle \mathbf{A}_1 + \langle \mathbf{A}, \mathbf{A}_2 \rangle \mathbf{A}_2 + \cdots + \langle \mathbf{A}, \mathbf{A}_n \rangle \mathbf{A}_n.$$

If

$$\begin{aligned} \mathbf{B} &= b_1 \mathbf{A}_1 + b_2 \mathbf{A}_2 + \cdots + b_n \mathbf{A}_n \\ \mathbf{C} &= c_1 \mathbf{A}_1 + c_2 \mathbf{A}_2 + \cdots + c_n \mathbf{A}_n \end{aligned}$$

then

$$\langle \mathbf{B}, \mathbf{C} \rangle = b_1 c_1 + b_2 c_2 + \cdots + b_n c_n.$$

PROOF. The first assertion follows directly from (15.6) because $\langle \mathbf{A}_1, \mathbf{A}_1 \rangle = \langle \mathbf{A}_2, \mathbf{A}_2 \rangle = \cdots = \langle \mathbf{A}_n, \mathbf{A}_n \rangle = 1$. To prove the second assertion we just compute:

$$\begin{aligned} \langle \mathbf{B}, \mathbf{C} \rangle &= \langle b_1 \mathbf{A}_1 + \cdots + b_n \mathbf{A}_n, c_1 \mathbf{A}_1 + \cdots + c_n \mathbf{A}_n \rangle \\ &= \sum_{i,j} b_i c_j \langle \mathbf{A}_i, \mathbf{A}_j \rangle \\ &= b_1 a_1 + b_2 a_2 + \cdots + b_n a_n \end{aligned}$$

because

$$\langle \mathbf{A}_i, \mathbf{A}_j \rangle = \begin{cases} 1 & \text{if } i = j, \\ 0 & \text{otherwise,} \end{cases}$$

since $\{\mathbf{A}_1, \ldots, \mathbf{A}_n\}$ is orthonormal. □

Thus by choosing an orthonormal basis $\{\mathbf{A}_1, \ldots, \mathbf{A}_n\}$ in a finite-dimensional inner product space and expressing vectors in terms of their components with respect to $\{\mathbf{A}_1, \ldots, \mathbf{A}_n\}$ we obtain formulas for $\langle \mathbf{A}, \mathbf{B} \rangle$ that look exactly like the formulas for $\mathbb{R}^n$ with its standard inner product. There is a good reason for this as we will now explain.

In studying vector spaces we found that it was important to also consider linear transformations, that is, functions which preserved the basic vector operations. For vector spaces equipped with an inner product it seems reasonable to demand that the inner product also be preserved.

Definition. Let $\mathscr{V}$ and $\mathscr{W}$ be inner product spaces. A linear transformation $\mathsf{T}: \mathscr{V} \to \mathscr{W}$ is called an *isometric embedding* iff

$$\langle \mathsf{T}(\mathbf{A}), \mathsf{T}(\mathbf{B}) \rangle = \langle \mathbf{A}, \mathbf{B} \rangle$$

for all vectors $\mathbf{A}, \mathbf{B}$ in $\mathscr{V}$.

Isometric embeddings are a very restricted class of linear transformations. We have:

Proposition 15.10. *Let $\mathscr{V}$ and $\mathscr{W}$ be inner product spaces and $\mathsf{T}: \mathscr{V} \to \mathscr{W}$ an isometric embedding. Then* $\ker T = \{\mathbf{0}\}$.

PROOF. Suppose that $\mathsf{T}(\mathbf{A}) = \mathbf{0}$. Then since T is an isometric embedding

$$\langle \mathbf{A}, \mathbf{A} \rangle = \langle \mathsf{T}(\mathbf{A}), \mathsf{T}(\mathbf{A}) \rangle = \langle \mathbf{0}, \mathbf{0} \rangle = 0$$

and therefore $\mathbf{A} = \mathbf{0}$. □

Corollary 15.11. *Let $\mathscr{V}$ and $\mathscr{W}$ be finite-dimensional inner product spaces of the same dimension and $\mathsf{T}: \mathscr{V} \to \mathscr{W}$ an isometric embedding. Then T is an isomorphism.*

PROOF. By (8.11) it suffices to show that $\operatorname{Im} \mathsf{T} = \mathscr{W}$. By (8.10) we obtain $\dim \mathscr{V} = \dim \operatorname{Im} \mathsf{T}$. Since $\dim \mathscr{V} = \dim \mathscr{W}$ by hypothesis, $\dim \mathscr{W} = \dim \operatorname{Im} \mathsf{T}$. Therefore since $\operatorname{Im} \mathsf{T}$ is a linear subspace of $\mathscr{W}$ (8.9) it follows from (6.9) that $\operatorname{Im} \mathsf{T} = \mathscr{W}$. □

EXAMPLE 12. An example of an isometric embedding of $\mathbb{R}^2$ in $\mathbb{R}^4$ is given by

$$\mathsf{T}: \mathbb{R}^2 \to \mathbb{R}^4$$

such that

$$\mathsf{T}(x, y) = \frac{1}{\sqrt{2}}(x, y, y, x).$$

To see this we note that

$$\begin{aligned}
\langle \mathsf{T}(\mathbf{A}), \mathsf{T}(\mathbf{B}) \rangle &= \left\langle \frac{1}{\sqrt{2}}(a_1, a_2, a_2, a_1), \frac{1}{\sqrt{2}}(b_1, b_2, b_2, b_1) \right\rangle \\
&= \tfrac{1}{2}(a_1 b_1 + a_2 b_2 + a_2 b_2 + a_1 b_1) \\
&= \tfrac{1}{2}(2a_1 b_1 + 2a_2 b_2) \\
&= a_1 b_1 + a_2 b_2 = \langle \mathbf{A}, \mathbf{B} \rangle
\end{aligned}$$

as required. □

Definition. Let $\mathscr{V}$ and $\mathscr{W}$ be inner product spaces. A linear transformation $\mathsf{T}: \mathscr{V} \to \mathscr{W}$ is called an *isometric isomorphism*, an *isometry* for short, iff T is an isometric embedding, and there exists an isometric embedding $\mathsf{S}: \mathscr{W} \to \mathscr{V}$ such that

$$\mathsf{S} \cdot \mathsf{T} = \mathsf{I}: \mathscr{V} \to \mathscr{V}$$

and

$$\mathsf{T} \cdot \mathsf{S} = \mathsf{I}: \mathscr{W} \to \mathscr{W}.$$

If there is an isometry $\mathsf{T}: \mathscr{V} \to \mathscr{W}$ we say that $\mathscr{V}$ and $\mathscr{W}$ are *isometric*.

Proposition 15.12. *Let $\mathscr{V}$ and $\mathscr{W}$ be finite-dimensional inner product spaces of the same dimension. If $\mathsf{T}: \mathscr{V} \to \mathscr{W}$ is an isometric embedding then T is an isometric isomorphism.*

PROOF. By (15.11) T is an isomorphism. Let $\mathsf{S} = \mathsf{T}^{-1}$. We need only show that $\mathsf{S}: \mathscr{W} \to \mathscr{V}$ is an isometric embedding. Let $\mathbf{A}, \mathbf{B} \in \mathscr{W}$. Write $\mathbf{A} = \mathsf{T}(\mathbf{C})$, $\mathbf{B} = \mathsf{T}(\mathbf{D})$. Then

$$\langle \mathsf{S}(\mathbf{A}), \mathsf{S}(\mathbf{B}) \rangle = \langle \mathbf{C}, \mathbf{D} \rangle = \langle \mathsf{T}(\mathbf{C}), \mathsf{T}(\mathbf{D}) \rangle = \langle \mathbf{A}, \mathbf{B} \rangle$$

as required. □

Proposition 15.13. *Let $\mathscr{V}$ and $\mathscr{W}$ be finite-dimensional inner product spaces with* $\dim \mathscr{V} \leq \dim \mathscr{W}$. *Then there exists an isometric embedding $\mathsf{T}: \mathscr{V} \to \mathscr{W}$.*

PROOF. Choose orthonormal bases $\{\mathbf{V}_1, \ldots, \mathbf{V}_m\}$ and $\{\mathbf{W}_1, \ldots, \mathbf{W}_n\}$ for $\mathscr{V}$ and $\mathscr{W}$ respectively and let $\mathsf{T}: \mathscr{V} \to \mathscr{W}$ be the linear extension of

$$\mathsf{T}(\mathbf{V}_i) = \mathbf{W}_i \qquad i = 1, \ldots, m.$$

Thus if $\mathbf{A}, \mathbf{B} \in \mathscr{V}$, say

$$\mathbf{A} = \sum a_i \mathbf{V}_i, \qquad \mathbf{B} = \sum b_i \mathbf{V}_i$$

we obtain from (15.9) the formula

$$\langle \mathbf{A}, \mathbf{B} \rangle = \sum a_i b_i,$$

while since

$$\mathsf{T}(\mathbf{A}) = \sum a_i \mathbf{W}_i, \qquad \mathsf{T}(\mathbf{B}) = \sum b_i \mathbf{W}_i$$

we get

$$\langle \mathsf{T}(\mathbf{A}), \mathsf{T}(\mathbf{B}) \rangle = \sum a_i b_i$$

so T is an isometric embedding. □

Summarizing we obtain:

Theorem 15.14. *Let $\mathscr{V}$ and $\mathscr{W}$ be finite-dimensional inner product spaces. Then $\mathscr{V}$ and $\mathscr{W}$ are isometric* iff $\dim \mathscr{V} = \dim \mathscr{W}$.

PROOF. If $\mathscr{V}$ and $\mathscr{W}$ are isometric then they are certainly isomorphic, so $\dim \mathscr{V} = \dim \mathscr{W}$. On the other hand if $\dim \mathscr{V} = \dim \mathscr{W}$ then we may apply (15.13) to construct an isometric embedding $\mathsf{T}: \mathscr{V} \to \mathscr{W}$ which by (15.12) will be an isometric isomorphism. □

Corollary 15.15. *If $\mathscr{V}$ is an inner product space of finite dimension n then $\mathscr{V}$ is isometric to $\mathbb{R}^n$.*

EXAMPLE 13. According to (15.15) the spaces $\mathbb{R}^3$ and $\mathscr{P}_2(\mathbb{R})$ are isometric. An explicit isometry is given by

$$\mathsf{T}(u, v, w) = 6\sqrt{5}wx^2 + (2\sqrt{3}v - 6\sqrt{5}w)x + (u - \sqrt{3}v + \sqrt{5}w).$$

This isometry is constructed as follows. Note that $\{(1, 0, 0), (0, 1, 0), (0, 0, 1)\}$ is an orthonormal basis for $\mathbb{R}^3$ while $\{1, \sqrt{12}(x - \frac{1}{2}), \sqrt{180}(x^2 - x + \frac{1}{6})\}$ is an orthonormal basis for $\mathscr{P}_2(\mathbb{R})$. Therefore the linear extension of

$$\begin{aligned} \mathsf{T}(1, 0, 0) &= 1 \\ \mathsf{T}(0, 1, 0) &= \sqrt{12}(x - \tfrac{1}{2}) \\ \mathsf{T}(0, 0, 1) &= \sqrt{180}(x^2 - x + \tfrac{1}{6}) \end{aligned}$$

will be an isometry $\mathsf{T} : \mathbb{R}^2 \to \mathscr{P}_2(\mathbb{R})$. A little algebraic manipulation brings forth the above formula.

Thus, exactly as with (8.15), we see that the added abstraction of the axiomatics has added nothing new, that is, up to isometry there is only one inner product space of finite dimension n, namely $\mathbb{R}^n$. Still, the axiomatic approach is quite useful in that it provides a natural setting for rather exotic inner products on $\mathbb{R}^n$ such as that of Example 3. These inner products occur in nature and should not be denied.

For the last topic in this chapter we are going to show how a linear transformation

$$\mathsf{T} : \mathscr{V} \to \mathbb{R}$$

of an inner product space $\mathscr{V}$ to the 1-dimensional space $\mathbb{R}$ may be expressed in terms of the inner product $\langle\ ,\ \rangle$. This will prove quite useful in the next chapter.

Definition. Let $\mathscr{V}$ be an inner product space and $\mathscr{W}$ a linear subspace. The *orthogonal complement* $\mathscr{W}^\perp$ of $\mathscr{W}$ in $\mathscr{V}$ is defined by

$$\mathscr{W}^\perp = \{\mathbf{A} \in \mathscr{V} \mid \langle \mathbf{A}, \mathbf{B} \rangle = 0 \text{ for every } \mathbf{B} \in \mathscr{W}\}.$$

Proposition 15.16. *Let $\mathscr{V}$ be an inner product space and $\mathscr{W}$ a linear subspace. Then $\mathscr{W}^\perp$ is a linear subspace and $\mathscr{W} \cap \mathscr{W}^\perp = \{\mathbf{0}\}$.*

PROOF. Let $\mathbf{B}$, $\mathbf{C}$ belong to $\mathscr{W}^\perp$ then

$$\langle \mathbf{A}, \mathbf{B} + \mathbf{C} \rangle = \langle \mathbf{A}, \mathbf{B} \rangle + \langle \mathbf{A}, \mathbf{C} \rangle = 0 + 0 = 0$$

so $\mathbf{B} + \mathbf{C}$ belongs to $\mathscr{W}^\perp$. If r is a number

$$\langle \mathbf{A}, r\mathbf{B} \rangle = r\langle \mathbf{A}, \mathbf{B} \rangle = r \cdot 0 = 0$$

so $r\mathbf{B} \in \mathscr{W}^\perp$. Therefore $\mathscr{W}^\perp$ is closed under scalar products and vector addition so is a linear subspace of $\mathscr{V}$.

If $\mathbf{A}$ belongs to $\mathscr{W} \cap \mathscr{W}^\perp$ then $\langle \mathbf{A}, \mathbf{A} \rangle = 0$ since $\mathbf{A} \in \mathscr{W}$ and $\mathbf{A} \in \mathscr{W}^\perp$. Therefore $\mathbf{A} = \mathbf{0}$. □

EXAMPLE 14. What is the orthogonal complement of the subspace

$$\mathscr{W} = \{(x, y, z) | x + 2y + 3z = 0\}$$

in $\mathbb{R}^3$?

Solution. The orthogonal complement of the plane $\mathscr{W}$ will be a line $\mathscr{L}$. In this case the line is the subspace spanned by (1, 2, 3). (These coordinates are the coefficients of the equation used to define $\mathscr{W}$. This always works). For if $(x, y, z) \in \mathscr{W}$ then

$$\langle(1, 2, 3), (x, y, z)\rangle = x + 2y + 3z = 0$$

because $(x, y, z) \in \mathscr{W}$.

EXAMPLE 15. What is the orthogonal complement of the subspace

$$\mathscr{W} = \{(x, y, z) | x + y + z = 0 \quad \text{and} \quad x - y + z = 0\}?$$

Solution. The subspace $\mathscr{W}$ is a line. Its orthogonal complement will therefore be a plane. A basis for $\mathscr{W}$ is the vector $(1, 0, -1)$, since it satisfies the equations of both the planes

$$x + y + z = 0 \quad \text{and} \quad x - y + z = 0$$

and must therefore be on their line of intersection.

The orthogonal complement of $\mathscr{W}$ must therefore be the plane $\mathscr{S}$ whose normal vector at the origin is $(1, 0, -1)$. Therefore

$$\mathscr{S} = \{(x, y, z) | x - z = 0\}.$$

as is required.

Proposition 15.17. *Let $\mathscr{V}$ be a finite-dimensional inner product space and $\mathscr{W}$ a linear subspace of $\mathscr{V}$. Then (see* (13.4))

$$\mathscr{W} + \mathscr{W}^{\perp} = \mathscr{V}.$$

PROOF. What we must do is show that any vector $\mathbf{A}$ in $\mathscr{V}$ mav be written as a sum $\mathbf{B} + \mathbf{C}$ with $\mathbf{B}$ in $\mathscr{W}$ and $\mathbf{C}$ in $\mathscr{W}^{\perp}$. To this end apply (15.7) to choose an orthonormal basis $\mathbf{W}_1, \ldots, \mathbf{W}_k$ for $\mathscr{W}$. Let $\mathbf{A} \in \mathscr{V}$ and define

$$\begin{aligned} \mathbf{B} &= \langle \mathbf{W}_1, \mathbf{A}\rangle \mathbf{W}_1 + \cdots + \langle \mathbf{W}_k, \mathbf{A}\rangle \mathbf{W}_k \\ \mathbf{C} &= \mathbf{A} - \mathbf{B}. \end{aligned}$$

Clearly $\mathbf{A} = \mathbf{B} + \mathbf{C}$ and $\mathbf{B}$ belongs to $\mathscr{W}$ So what it remains to show is that $\mathbf{C}$ belongs to $\mathscr{W}^{\perp}$. First of all note

$$\mathbf{C} = \mathbf{A} - \langle \mathbf{W}_1, \mathbf{A}\rangle \mathbf{W}_1 + \cdots + \langle \mathbf{W}_k, \mathbf{A}\rangle \mathbf{W}_k.$$

Hence

$$\begin{aligned} \langle \mathbf{C}, \mathbf{W}_i\rangle &= \langle \mathbf{A}, \mathbf{W}_i\rangle - \langle \mathbf{W}_1, \mathbf{A}\rangle\langle \mathbf{W}_1, \mathbf{W}_i\rangle + \cdots + \langle \mathbf{W}_k \mathbf{A}\rangle\langle \mathbf{W}_k, \mathbf{W}_i\rangle \\ &= \langle \mathbf{A}, \mathbf{W}_i\rangle - \langle \mathbf{W}_i, \mathbf{A}\rangle = 0 \end{aligned}$$

since

$$\langle \mathbf{W}_j, \mathbf{W}_i \rangle = \begin{cases} 1 & \text{if } j = i \\ 0 & \text{otherwise.} \end{cases}$$

Next suppose that $\mathbf{W}$ belongs to $\mathscr{W}$. Then

$$\mathbf{W} = w_1\mathbf{W}_1 + \cdots + w_k\mathbf{W}_k$$

so

$$\langle \mathbf{C}, \mathbf{W} \rangle = \sum w_i \langle \mathbf{C}, \mathbf{W}_i \rangle = \sum w_i 0 = 0$$

and hence $\mathbf{C}$ belongs to $\mathscr{W}^\perp$. □

Proposition 15.18. *Let $\mathscr{V}$ be a finite-dimensional inner product space and $\{\mathbf{A}_1, \ldots, \mathbf{A}_k\}$ an orthonormal set of vectors in $\mathscr{V}$. Then there are vectors $\{\mathbf{A}_{k+1}, \ldots, \mathbf{A}_n\}$ such that $\{\mathbf{A}_1, \ldots, \mathbf{A}_k, \mathbf{A}_{k+1}, \ldots, \mathbf{A}_n\}$ is an orthonormal basis for $\mathscr{V}$.*

PROOF. Let $\mathscr{W} = \mathscr{L}(\mathbf{A}_1, \ldots, \mathbf{A}_k)$. By (15.6) $\mathscr{W}^\perp$ is a linear subspace of $\mathscr{V}$ and since $\mathscr{V}$ is finite dimensional so is $\mathscr{W}^\perp$. Let $\{\mathbf{B}_1, \ldots, \mathbf{B}_l\}$ be an orthonormal basis for $\mathscr{W}^\perp$. One readily checks that the set $\{\mathbf{A}_1, \ldots, \mathbf{A}_k, \mathbf{B}_1, \ldots, \mathbf{B}_l\}$ is an orthonormal set, and since none of the vectors in the set is $\mathbf{0}$, this set is linearly independent by (15.5). By (15.17) $\mathscr{L}(\mathbf{A}_1, \ldots, \mathbf{A}_k, \mathbf{B}_1, \ldots, \mathbf{B}_l) = \mathscr{V}$ and so setting

$$\mathbf{A}_{k+1} = \mathbf{B}_1, \ldots, \mathbf{A}_n = \mathbf{B}_l$$

we have the required orthonormal basis. □

Notice the following important consequence of the proof of (15.18).

Corollary 15.19. *Let $\mathscr{V}$ be a finite-dimensional inner product space of dimension n and $\mathscr{W}$ a subspace of $\mathscr{V}$ of dimension m. Then* $\dim \mathscr{W}^\perp = n - m$.

Theorem 15.20. *Let $\mathscr{V}$ be a finite-dimensional inner product space and $\mathsf{T}: \mathscr{V} \to \mathbb{R}$ a linear transformation. Then there exists a unique vector $\mathbf{A}$ in $\mathscr{V}$ (depending on T) such that*

$$\mathsf{T}(\mathbf{B}) = \langle \mathbf{B}, \mathbf{A} \rangle$$

for all $\mathbf{B}$ in $\mathscr{V}$.

PROOF. Let the dimension of $\mathscr{V}$ be n. If $\operatorname{Im} \mathsf{T} = \{\mathbf{0}\}$ set $\mathbf{A} = \mathbf{0}$. Then

$$0 = \mathsf{T}(\mathbf{B})$$
$$0 = \langle \mathbf{B}, \mathbf{0} \rangle$$

so the theorem is true in this case. If $\operatorname{Im} \mathsf{T} \neq \{\mathbf{0}\}$ then we must have $\operatorname{Im} \mathsf{T} = \mathbb{R}$ since $\{\mathbf{0}\}$ and $\mathbb{R}$ are the only subspaces of $\mathbb{R}$. Therefore by (8.10) the kernel of T has dimension $n - 1$. Let $\{\mathbf{A}_1, \ldots, \mathbf{A}_{n-1}\}$ be an orthonormal basis for $\ker \mathsf{T}$.

By (15.18) there is a vector $\mathbf{A}_n$ such that $\{\mathbf{A}_1, \ldots, \mathbf{A}_{n-1}, \mathbf{A}_n\}$ is a basis for $\mathscr{V}$. Let $\mathbf{B} \in \mathscr{V}$. By (15.6)

$$\mathbf{B} = \langle \mathbf{B}, \mathbf{A}_1 \rangle \mathbf{A}_1 + \cdots + \langle \mathbf{B}, \mathbf{A}_n \rangle \mathbf{A}_n$$

so

$$\begin{aligned} \mathsf{T}(\mathbf{B}) &= \langle \mathbf{B}, \mathbf{A}_1 \rangle \mathsf{T}(\mathbf{A}_1) + \cdots + \langle \mathbf{B}, \mathbf{A}_n \rangle \mathsf{T}(\mathbf{A}_n) \\ &= \langle \mathbf{B}, \mathbf{A}_n \rangle \mathsf{T}(\mathbf{A}_n) \\ &= \langle \mathbf{B}, \mathsf{T}(\mathbf{A}_n)\mathbf{A}_n \rangle \end{aligned}$$

(remember $\mathsf{T}(\mathbf{A}_n)$ is only a number). Thus $\mathbf{A} = \mathsf{T}(\mathbf{A}_n)\mathbf{A}_n$ is the required vector.

To see that the vector $\mathbf{A}$ is unique suppose that $\mathbf{C}$ is another vector such that $\mathsf{T}(\mathbf{B}) = \langle \mathbf{B}, \mathbf{C} \rangle$ for all $\mathbf{B}$ in. Therefore

$$\begin{aligned} \langle \mathbf{B}, \mathbf{A} - \mathbf{C} \rangle &= \langle \mathbf{B}, \mathbf{A} \rangle - \langle \mathbf{B}, \mathbf{C} \rangle \\ &= \mathsf{T}(\mathbf{B}) - \mathsf{T}(\mathbf{B}) \\ &= 0 \end{aligned}$$

for all $\mathbf{B}$ in $\in \mathscr{V}$. Put $\mathbf{B} = \mathbf{A} - \mathbf{C}$ to get

$$\langle \mathbf{A} - \mathbf{C}, \mathbf{A} - \mathbf{C} \rangle = 0$$

but this implies by Axiom (4) of an inner product that $\mathbf{A} - \mathbf{C} = \mathbf{0}$ and hence $\mathbf{A} = \mathbf{C}$. □

EXAMPLE 16. Consider the linear transformation

$$\mathsf{T} : \mathbb{R}^3 \to \mathbb{R}$$

given by

$$\mathsf{T}(x, y, z) = x + y + z.$$

The kernel of T is thus the plane

$$\ker \mathsf{T} = \{(x, y, z) \mid x + y + z = 0\}$$

and a vector orthogonal to $\ker \mathsf{T}$ is $(1, 1, 1)$, which normalizes to $(1/\sqrt{3}, 1/\sqrt{3}, 1/\sqrt{3})$. Since

$$\mathsf{T}\left(\frac{1}{\sqrt{3}}(1, 1, 1)\right) = \frac{3}{\sqrt{3}} = \sqrt{3}$$

we find that

$$\mathsf{T}(x, y, z) = \langle (x, y, z), (1, 1, 1) \rangle$$

which readily checks.

In fact more generally if

$$\mathsf{T} : \mathbb{R}^3 \to \mathbb{R}$$

is given by

$$\mathsf{T}(x, y, z) = ax + by + cz$$

then the vector $\mathbf{V} = (a, b, c)$ satisfies

$$\mathsf{T}(x, y, z) = \langle (x, y, z), (a, b, c) \rangle$$

as one easily checks.

EXAMPLE 17. Consider the liner transformation

$$\mathsf{T} : \mathscr{P}_2(\mathbb{R}) \to \mathbb{R}$$

given by

$$\mathsf{T}(p(x)) = p(1).$$

The kernel of T is the subspace

$$\begin{aligned} \ker \mathsf{T} &= \{p(x) \mid p(1) = 0\} \\ &= \{a_0 + a_1 x + a_2 x^2 \mid a_0 + a_1 + a_2 = 0\}. \end{aligned}$$

A basis for the kernel of T consists of $1 - x$, $x - x^2$. The three vectors $\{1 - x, x - x^2, 1\}$ are a basis for $\mathscr{P}_2(\mathbb{R})$. Apply the Gram–Schmidt process to orthogonalize them. (Note the order!) We get

$$\begin{aligned} \mathbf{A}_1 &= 1 - x \\ \mathbf{A}_2 &= x - x^2 - \frac{\langle x - x^2, 1 - x \rangle}{\langle 1 - x, 1 - x \rangle} \cdot (1 - x) \\ &= -\tfrac{1}{4} + \tfrac{5}{4}x - x^2 \\ \mathbf{A}_3 &= 1 - \frac{\langle 1, -\frac{1}{4} + \frac{5}{4}x - x^2 \rangle}{\langle -\frac{1}{4} + \frac{5}{4}x - x^2, -\frac{1}{4} + \frac{5}{4} - x^2 \rangle} \cdot (-\tfrac{1}{4} + \tfrac{5}{4}x - x^2) \\ &\quad - \frac{\langle 1, 1 - x \rangle}{\langle 1 - x, 1 - x \rangle} \cdot (1 - x) \\ &= \frac{1}{2046} [1028 + 3044x + 20x^2] \end{aligned}$$

so if $\alpha = \dfrac{1}{\langle \mathbf{A}_3, \mathbf{A}_3 \rangle}$ a *unit* vector orthogonal to $\ker \mathsf{T}$ is

$$\alpha \mathbf{A}_3 = \frac{\alpha}{2046} [1028 + 3044x + 20x^2]$$

finally since

$$\begin{aligned} \mathsf{T}(\alpha \mathbf{A}_3) &= \frac{\alpha}{2046} [1028 + 3044 + 20] \\ &= \alpha \frac{4092}{2046} = 2\alpha \end{aligned}$$

we find that

$$\mathsf{T}(p(x)) = \frac{\alpha}{1023} \int_0^1 p(x) [1028 + 3044x + 20x^2] \, dx$$

Exercises

1. Compute the following inner products in $\mathbb{R}^3$ relative to the standard inner product. Find the cosine of the angle between the indicated pairs of vectors:

$$\langle(1, 2, 3), (4, 5, 6)\rangle \qquad \langle(1, 0, 1), (0, 4, 0)\rangle$$
$$\langle(1, 2, 3), (3, 2, 1)\rangle \qquad \langle(1, 1, 1), (1, -2, 1)\rangle$$

2. Compute the following inner products in $\mathscr{P}_2(\mathbb{R})$. Find the cosine of the angle between the indicated pair of vectors:

$$\langle 1 + 2x + 3x^2, 4 + 5x + 6x^2\rangle$$
$$\langle 1 + 2x + 3x^2, 3 + 2x + x^2\rangle$$
$$\langle 1 + x^2, 4x\rangle$$
$$\langle 1 + x + x^2, 1 - 2x + x^2\rangle$$

3. Compute the following inner products in $\mathscr{M}_{33}$ using the trace inner product:

$$\left\langle \begin{pmatrix} 1 & 2 & 3 \\ 0 & 4 & 0 \\ 0 & 1 & 3 \end{pmatrix}, \begin{pmatrix} 4 & 0 & 1 \\ 2 & 1 & 0 \\ 1 & 1 & 4 \end{pmatrix} \right\rangle$$
$$\left\langle \begin{pmatrix} 1 & 1 & 1 \\ 0 & 1 & 0 \\ 1 & 0 & 1 \end{pmatrix}, \begin{pmatrix} 1 & 0 & 1 \\ 0 & 0 & 0 \\ 1 & 0 & 1 \end{pmatrix} \right\rangle$$

4. Let $(\mathscr{V}, \langle\ ,\ \rangle)$ be an inner product space and $\{\mathbf{A}_1, \ldots, \mathbf{A}_n\}$ an orthogonal set of vectors in $\mathscr{V}$. Establish the following:

$$\textit{Bessel's inequality}: \quad \sum_{i=1}^{n} \frac{|\langle \mathbf{B}, \mathbf{A}_i\rangle|^2}{|\mathbf{A}_i|^2} \leq |\mathbf{B}|^2$$

$$\textit{Parseval's identity}: \quad \langle \mathbf{B}, \mathbf{C}\rangle = \sum \langle \mathbf{B}, \mathbf{A}_i\rangle\langle \mathbf{A}_i, \mathbf{C}\rangle$$

5. Let $\mathscr{V}$ be an inner product space. Prove the following: (Draw pictures in $\mathbb{R}^2$ to see what is going on!)

$$\textit{Law of cosines}: \quad |\mathbf{A} - \mathbf{B}|^2 = |\mathbf{A}|^2 + |\mathbf{B}|^2 - 2|\mathbf{A}||\mathbf{B}| \cos\theta$$

where θ is the angle between **A** and **B**.

6. Which of the following sets of vectors in $\mathbb{R}^3$ are orthogonal?

$$\{(0, 0, 0), (1, 1, 1), (0, 1, -1)\}$$
$$\{(1, 1, 1), (1, -2, 1), (-2, 1, -2)\}$$
$$\{(1, 0, 0), (0, 1, 0), (0, 0, 1), (1, 1, 1)\}$$
$$\{(1, -1, 2), (2, 0, -1), (0, 1, 1), (0, 0, 1)\}$$

7. Which of the following sets of vectors in $\mathscr{P}_3(\mathbb{R})$ are orthogonal?

$$\{1, x - \tfrac{1}{2}\}$$
$$\{1, x - 1\}$$
$$\{x, x^2 - 1\}$$
$$\{x, 3x^2 - 2, 2x^3\}$$

8. Which of the following sets of vectors in $\mathcal{M}_{33}$ are orthogonal?

$$\left\{\begin{pmatrix}1 & 2 & 3\\ 0 & 1 & 2\\ 0 & 0 & 1\end{pmatrix}, \begin{pmatrix}1 & 0 & 0\\ 0 & -2 & 0\\ 0 & 0 & 1\end{pmatrix}, \begin{pmatrix}2 & 3 & 4\\ 0 & -1 & 1\\ 0 & 0 & -1\end{pmatrix}\right\}$$

$$\left\{\begin{pmatrix}1 & 2 & 3\\ 0 & 2 & 3\\ 0 & 0 & 3\end{pmatrix}, \begin{pmatrix}-3 & 0 & 0\\ 0 & 0 & 0\\ 0 & 0 & 1\end{pmatrix}, \begin{pmatrix}1 & 0 & 0\\ 1 & -2 & 0\\ 0 & 1 & 1\end{pmatrix}\right\}$$

9. Find an orthonormal basis for each of the indicated subspaces:

$$\{(x, y, z) \in \mathbb{R}^3 \,|\, x + y - z = 0\}$$

$$\left\{p(x) \in \mathcal{P}_3(\mathbb{R}) \,\middle|\, x\frac{d}{dx}p(x) = p(x)\right\}$$

$$\{\mathbf{A} \in \mathcal{M}_{33} \,|\, \mathrm{tr}(\mathbf{A}) = 0\}$$

10. Which of the following linear transformations are isometric embeddings?

$$\mathsf{T}: \mathbb{R}^3 \to \mathbb{R}^4 \qquad \mathsf{T}(x, y, z) = (x + y, y + z, z + x, x + y + z)$$

$$\mathsf{T}: \mathbb{R}^2 \to \mathbb{R}^3 \qquad \mathsf{T}(x, y) = (x, 0, y)$$

$$\mathsf{T}: \mathcal{P}_2(\mathbb{R}) \to \mathcal{P}_3(\mathbb{R}) \qquad \mathsf{T}(p(x)) = xp(x)$$

$$\mathsf{T}: \mathbb{R}^3 \to \mathcal{M}_{33} \qquad \mathsf{T}(x, y, z) = \begin{pmatrix}x & 0 & 0\\ 0 & y & 0\\ 0 & 0 & z\end{pmatrix}$$

$$\mathsf{T}: \mathbb{R}^3 \to \mathcal{M}_{33} \qquad \mathsf{T}(x, y, z) = \begin{pmatrix}x & 0 & 0\\ y & 0 & 0\\ z & 0 & 0\end{pmatrix}.$$

11. Find the orthogonal complement of each of the subspaces in Exercise 9.

12. A linear transformation $\mathsf{T}: \mathbb{R}^2 \to \mathbb{R}^2$ is called a *rotation* iff T is an isometry and $\det \mathsf{T} > 0$. Show that the matrix of T relative to the standard basis must be

$$\begin{pmatrix}\cos\theta & -\sin\theta\\ \sin\theta & \cos\theta\end{pmatrix}$$

for a unique angle θ. In fact, $\cos\theta = \langle \mathbf{E}_1, \mathsf{T}(\mathbf{E}_1)\rangle$. Draw a picture.

13. If $\mathsf{T}_1, \mathsf{T}_2: \mathbb{R}^2 \to \mathbb{R}^2$ are rotations show that $\mathsf{T}_1\mathsf{T}_2$ is also a rotation. Show $\mathsf{T}_1\mathsf{T}_2 = \mathsf{T}_2\mathsf{T}_1$.

14. Suppose $\{\mathbf{A}_1, \ldots, \mathbf{A}_n\}$ is an orthonormal basis for $\mathscr{V}$. If $\mathbf{A}$ is a vector in $\mathscr{V}$ such that

$$\langle \mathbf{A}, \mathbf{A}_i\rangle = 0, \qquad i = 1, 2, \ldots, n$$

show that $\mathbf{A} = \mathbf{0}$.

15. Verify that the trace of a square matrix has the following properties

(a) $\mathrm{tr}(\mathbf{A}) = \mathrm{tr}(\mathbf{A}^t)$

(b) $\mathrm{tr}(\mathbf{AB}) = \mathrm{tr}(\mathbf{BA})$

(c) $\mathrm{tr}(\mathbf{A} + \mathbf{B}) = \mathrm{tr}(\mathbf{A}) + \mathrm{tr}(\mathbf{B})$.

(d) $\mathrm{tr}(a\mathbf{A}) = a\,\mathrm{tr}\mathbf{A}$

(e) If $\mathbf{A}$ is invertible then $\mathrm{tr}(\mathbf{ABA}^{-1}) = \mathrm{tr}(\mathbf{B})$ for any two square matrices $\mathbf{A}$ and $\mathbf{B}$ and numbers a.

16. Show that there do not exist square matrices $\mathbf{A}$ and $\mathbf{B}$ such that $\mathbf{AB} - \mathbf{BA} = \mathbf{I}$. (*Hint*: see Exercise 15).

The notion of inner product can be introduced in complex vector spaces also, by a slight modification (remember: definitions are useful or useless *not* true or false) of the axioms we have used in the real case. To this end recall that for a complex number $z = a + bi$ (a, b real numbers) the *complex conjugate* of z is the complex number

$$\bar{z} = a - bi.$$

It is very important to note that

$$z\bar{z} = (a + bi)(a - bi) = a^2 - b^2i^2 = a^2 + b^2$$

is a positive real number.

17. Let $\mathsf{T} : \mathbb{C} \to \mathbb{R}^2$ be given by $\mathsf{T}(a + bi) = (a, b)$. Show that T is an isomorphism of real vector spaces and $|\mathsf{T}(z)|^2 = z\bar{z}$.

Definition. If $\mathscr{V}$ is a complex vector space, an inner-product on $\mathscr{V}$ is a function that assigns to each pair $\mathbf{A}$, $\mathbf{B}$ of vectors in $\mathscr{V}$ a complex number $\langle \mathbf{A}, \mathbf{B} \rangle$ such that

(1) $\langle \mathbf{A}, \mathbf{B} \rangle = \overline{\langle \mathbf{B}, \mathbf{A} \rangle}$ for all $\mathbf{A}, \mathbf{B} \in \mathscr{V}$.

(2) For all $\mathbf{A}, \mathbf{B} \in \mathscr{V}$ and $z \in \mathbb{C}$ we have

$$\langle z\mathbf{A}, \mathbf{B} \rangle = z\langle \mathbf{A}, \mathbf{B} \rangle,$$
$$\langle \mathbf{A}, z\mathbf{B} \rangle = \bar{z}\langle \mathbf{A}, \mathbf{B} \rangle.$$

(3) For all vectors $\mathbf{A}, \mathbf{B}, \mathbf{C} \in \mathscr{V}$ we have:

$$\langle \mathbf{A}, \mathbf{B} + \mathbf{C} \rangle = \langle \mathbf{A}, \mathbf{B} \rangle + \langle \mathbf{A}, \mathbf{C} \rangle,$$
$$\langle \mathbf{A} + \mathbf{B}, \mathbf{C} \rangle = \langle \mathbf{A}, \mathbf{C} \rangle + \langle \mathbf{B}, \mathbf{C} \rangle.$$

(4) For all $\mathbf{A}$ in $\mathscr{V}$, $\langle \mathbf{A}, \mathbf{A} \rangle \geq 0$ and $\langle \mathbf{A}, \mathbf{A} \rangle = 0$ iff $\mathbf{A} = \mathbf{0}$.

18. On $\mathbb{C}^n$ define

$$\langle \mathbf{Z}, \mathbf{W} \rangle = \sum_{i=1}^{n} z_i \overline{w}_i$$

where $\mathbf{Z} = (z_1, z_2, \ldots, z_n)$, $\mathbf{W} = (w_1, \ldots, w_n)$. Verify that $\langle\ ,\ \rangle$ is an inner product on $\mathbb{C}^n$.

19. Compute the following inner products in $\mathbb{C}^3$:

$$\langle(3, i, 2 - i), (5, -i, 1)\rangle$$
$$\langle(i, i, i,), (1, 1, 1)\rangle$$
$$\langle(i, 1, i), (1, i, 1)\rangle$$

20. The *length* of a vector $\mathbf{Z}$ in a complex inner product space is by definition

$$|\mathbf{Z}| = \langle \mathbf{Z}, \mathbf{Z}\rangle^{1/2}.$$

Verify the triangle inequality in the complex case.

21. A basis for a complex inner product space is orthonormal if it consists of orthogonal vectors of length 1. Which of the following are orthonormal bases for $\mathbb{C}^3$?

(a) $(i, 0, 0), (0, i, 0), (0, 0, i)$
(b) $(1, 0, 0), (0, i, 0), (0, 0, 1)$
(c) $\sqrt{\tfrac{1}{2}}(1 + i, 1 - i, 0), \sqrt{\tfrac{1}{2}}(0, 1, 1), (0, 0, 1)$
(d) $(1, i, 0), (-1, i, 0), (0, 0, 1)$

22. What changes, if any, are needed to prove the theorems of this section in the complex case. In particular show that two complex inner product spaces of finite dimensions are isometric iff they have the same dimension. In particular a complex inner product space of dimension n is isometric to $\mathbb{C}^n$.

23. Let $\mathsf{T}: \mathscr{V} \to \mathscr{V}$ be a linear transformation in an n-dimensional inner product space $\mathscr{V}$. T is called an *orthogonal* reflection iff there exists an $(n - 1)$-dimensional subspace $\mathscr{H}$ in $\mathscr{V}$ and a nonzero vector $\mathbf{v}$ perpendicular to $\mathscr{H}$ such that

$$\mathsf{T}(\mathbf{h}) = \mathbf{h} \quad \text{for all } \mathbf{h} \text{ in } \mathscr{H},$$
$$\mathsf{T}(\mathbf{v}) = -\mathbf{v}.$$

show

$$\mathsf{T}(\mathbf{w}) = \mathbf{w} - 2\langle \mathbf{w}, \mathbf{v}\rangle \cdot \mathbf{v}\rangle \quad \text{for all } \mathbf{w}.$$

24. Let $\mathsf{T}: \mathscr{V} \to \mathscr{V}$ be a reflection. Show that with respect to a properly chosen inner product T is orthogonal.

25. Let S be a finite set, $\mathscr{F}(S)$ the vector space of real valued functions on S, and $\varphi: S \to S$ a function. Define

$$\varphi^*: \mathscr{F}(S) \to \mathscr{F}(S)$$

by

$$\varphi^*(\mathbf{f})(s) = \mathbf{f}\,(\varphi(s)).$$

By Chapter 8, Exercise 22, φ^* is a linear transformation. Show that the trace of φ^* is the number of fixed points of φ. (A point $x \in S$ is called a fixed point of φ iff $\varphi(x) = x$.)

16 The spectral theorem and quadratic forms

So far in our study of linear transformations we have concentrated on trying to find conditions that assure the matrix of the transformation has a particular form. We have not asked the related question of studying properties of those transformations whose matrix is *assumed* to have a particularly simple form. There is in fact a good reason for this and it is tied up with our work of the last chapter. For example we might propose to study those linear transformations whose matrix is symmetric. We would therefore like to introduce the following:

Proposed Definition. Let $\mathsf{T} : \mathscr{V} \to \mathscr{V}$ be a linear transformation. We say that T is *symmetric* iff there is a basis $\{\mathbf{A}_1, \ldots, \mathbf{A}_n\}$ for $\mathscr{V}$ such that the matrix of T relative to this basis is symmetric.

Now the reason we have indicated that this is only a proposed definition, and not one we intend to work with, is because it is not really a useful concept. Remember, a definition is not right or wrong, only useful or useless. The preceding one is pretty much useless. The following example will help illustrate why.

EXAMPLE 1. Let $\mathsf{T} : \mathbb{R}^2 \to \mathbb{R}^2$ be the linear transformation given by

$$\mathsf{T}(x, y) = (x + 2y, 2x + y).$$

Let us calculate the matrix of T relative to the standard basis of $\mathbb{R}^2$. We find

$$\mathsf{T}(1, 0) = (1, 2), \qquad \mathsf{T}(0, 1) = (2, 1)$$

so the matrix we seek is:

$$\mathbf{A} = \begin{pmatrix} 1 & 2 \\ 2 & 1 \end{pmatrix}$$

which is a symmetric matrix. On the other hand we may compute the matrix of T relative to the basis $\{\mathbf{F}_1 = (1, 1), \mathbf{F}_2 = (1, 0)\}$. We get

$$\begin{aligned} \mathsf{T}(1, 1) &= (3, 3) = 3\mathbf{F}_1 + 0\mathbf{F}_2 \\ \mathsf{T}(1, 0) &= (1, 2) = 2\mathbf{F}_1 - \mathbf{F}_2 \end{aligned}$$

so the required matrix is

$$\mathbf{B} = \begin{pmatrix} 3 & 2 \\ 0 & -1 \end{pmatrix}$$

which is not symmetric.

The preceding example illustrates a crucial drawback to using the proposed definition of symmetric transformation. The concept is not a property of the linear transformation that is invariant under change of basis. This has several consequences. For one, as defined, symmetry is more a property of the coordinate system in which the matrix of the transformation is computed. Thus we will have to specify this basis and work only with it if we wish to take advantage of the symmetry property. Secondly, and more importantly, a linear transformation T might be symmetric in terms of the proposed definition and we would never know it because we are unable to discover a basis in which T has a symmetric matrix!

The aforementioned difficulties can be overcome by introducing a slightly different concept. One which depends only on the linear transformation and not on the choice of basis. *However* we must assume that the ambient vector space is equipped with a fixed inner product. The particular definition we are going to give was arrived at through a process of trial, error, and experimentation; the basis of all modern science.

We begin with an important construction.

Lemma 16.1. *Let $\mathscr{V}$ be a finite-dimensional inner product space and $\mathsf{T} : \mathscr{V} \to \mathscr{V}$ a linear endomorphism. For each vector* $\mathbf{A}$ *in $\mathscr{V}$ define the function*

$$\mathsf{T}_{\mathbf{A}} : \mathscr{V} \to \mathbb{R} \quad \text{by} \quad \mathsf{T}_{\mathbf{A}}(\mathbf{B}) = \langle \mathsf{T}(\mathbf{B}), \mathbf{A} \rangle$$

where $\langle\ ,\ \rangle$ is the inner product in $\mathscr{V}$. Then the function $\mathsf{T}_{\mathbf{A}} : \mathscr{V} \to \mathbb{R}$ is a linear transformation.

PROOF. Suppose $\mathbf{B}_1$, $\mathbf{B}_2$ belong to $\mathscr{V}$. Then

$$\begin{aligned} \mathsf{T}_{\mathbf{A}}(\mathbf{B}_1 + \mathbf{B}_2) &= \langle \mathsf{T}(\mathbf{B}_1 + \mathbf{B}_2), \mathbf{A} \rangle \\ &= \langle \mathsf{T}(\mathbf{B}_1) + \mathsf{T}(\mathbf{B}_2), \mathbf{A} \rangle \\ &= \langle \mathsf{T}(\mathbf{B}_1), \mathbf{A} \rangle + \langle \mathsf{T}(\mathbf{B}_2), \mathbf{A} \rangle \\ &= \mathsf{T}_{\mathbf{A}}(\mathbf{B}_1) + \mathsf{T}_{\mathbf{A}}(\mathbf{B}_2). \end{aligned}$$

A similar computation shows that $\mathsf{T}_\mathbf{A}$ preserves scalar multiples, and hence $\mathsf{T}_\mathbf{A}$ is a linear transformation. □

Suppose again that $\mathsf{T}: \mathscr{V} \to \mathscr{V}$ is a linear transformation in the finite-dimensional inner product space $\mathscr{V}$. For each **A** in $\mathscr{V}$ there is also the linear transformation

$$\mathsf{T}_\mathbf{A}: \mathscr{V} \to \mathbb{R} \quad \text{given by} \quad \mathsf{T}_\mathbf{A}(\mathbf{B}) = \langle \mathsf{T}(\mathbf{B}), \mathbf{A} \rangle.$$

According to (15.20) there exists a unique vector $\mathbf{A}^*$ in $\mathscr{V}$ such that

$$\mathsf{T}_\mathbf{A}(\mathbf{B}) = \langle \mathbf{B}, \mathbf{A}^* \rangle$$

for all **B** in $\mathscr{V}$. Define a function

$$\mathsf{T}^*: \mathscr{V} \to \mathscr{V} \quad \text{by} \quad \mathsf{T}^*(\mathbf{A}) = \mathbf{A}^*$$

The function T^* is called the *adjoint* of T.

Proposition 16.2. *Let* $\mathsf{T}: \mathscr{V} \to \mathscr{V}$ *be a linear transformation in the finite-dimensional inner product space* $\mathscr{V}$. *Then the adjoint of* T,

$$\mathsf{T}^*: \mathscr{V} \to \mathscr{V}$$

is also a linear transformation.

PROOF. We are going to beat to death the word unique that occurs in the statement of (15.20).

The adjoint of T is characterized by the equations

$$\langle \mathsf{T}(\mathbf{B}), \mathbf{A} \rangle = \mathsf{T}_\mathbf{A}(\mathbf{B}) = \langle \mathbf{B}, \mathsf{T}^*(\mathbf{A}) \rangle$$

for all vectors **A**, **B** in $\mathscr{V}$. So suppose that $\mathbf{A}_1, \mathbf{A}_2$ belong to $\mathscr{V}$. Then for any **B** in $\mathscr{V}$

$$\begin{aligned}\langle \mathbf{B}, \mathsf{T}^*(\mathbf{A}_1 + \mathbf{A}_2) \rangle &= \langle \mathsf{T}(\mathbf{B}), \mathbf{A}_1 + \mathbf{A}_2 \rangle \\ &= \langle \mathsf{T}(\mathbf{B}), \mathbf{A}_1 \rangle + \langle \mathsf{T}(\mathbf{B}), \mathbf{A}_2 \rangle \\ &= \langle \mathbf{B}, \mathsf{T}^*(\mathbf{A}_1) \rangle + \langle \mathbf{B}, \mathsf{T}^*(\mathbf{A}_2) \rangle \\ &= \langle \mathbf{B}, \mathsf{T}^*(\mathbf{A}_1) + \mathsf{T}^*(\mathbf{A}_2) \rangle.\end{aligned}$$

Now set $\mathbf{A} = \mathbf{A}_1 + \mathbf{A}_2$. Then the above equations show

$$\langle \mathbf{B}, \mathsf{T}^*(\mathbf{A}_1 + \mathbf{A}_2) \rangle = \mathsf{T}_\mathbf{A}(\mathbf{B}) = \langle \mathbf{B}, \mathsf{T}^*(\mathbf{A}_1) + \mathsf{T}^*(\mathbf{A}_2) \rangle$$

for all **B** in $\mathscr{V}$. By the **uniqueness** part of (15.20) we therefore find

$$\mathsf{T}^*(\mathbf{A}_1 + \mathbf{A}_2) = \mathsf{T}^*(\mathbf{A}_1) + \mathsf{T}^*(\mathbf{A}_2)$$

so that T^* preserves vector sums. To show that T^* preserves scalar products we suppose **A** in $\mathscr{V}$ and r a number. Then for all **B** in $\mathscr{V}$

$$\begin{aligned}\langle \mathbf{B}, \mathsf{T}^*(r\mathbf{A}) \rangle &= \langle \mathsf{T}(\mathbf{B}), r\mathbf{A} \rangle \\ &= r\langle \mathsf{T}(\mathbf{B}), \mathbf{A} \rangle \\ &= r\langle \mathbf{B}, \mathsf{T}^*(\mathbf{A}) \rangle \\ &= \langle \mathbf{B}, r\mathsf{T}^*(\mathbf{A}) \rangle\end{aligned}$$

and therefore

$$\langle \mathbf{B}, \mathsf{T}^*(r\mathbf{A})\rangle = \mathsf{T}_{r\mathbf{A}}(\mathbf{B}) = \langle \mathbf{B}, r\mathsf{T}^*(\mathbf{A})\rangle$$

and the uniqueness part of (15.20) yields

$$\mathsf{T}^*(r\mathbf{A}) = r\mathsf{T}^*(\mathbf{A})$$

so that T^* preserves scalar products. □

Proposition 16.3. *Let $\mathsf{T} : \mathscr{V} \to \mathscr{V}$ be a linear transformation in the finite-dimensional inner product space $\mathscr{V}$. Then*

$$\langle \mathsf{T}(\mathbf{B}), \mathbf{A}\rangle = \langle \mathbf{B}, \mathsf{T}^*(\mathbf{A})\rangle$$

for all $\mathbf{A}, \mathbf{B} \in \mathscr{V}$.

PROOF. This follows instantly from the formulas

$$\mathsf{T}_{\mathbf{A}}(\mathbf{B}) = \langle \mathsf{T}(\mathbf{B}), \mathbf{A}\rangle$$
$$\mathsf{T}_{\mathbf{A}}(\mathbf{B}) = \langle \mathbf{B}, \mathsf{T}^*(\mathbf{A})\rangle$$

which have already found application in (16.2). □

Let us work a few numerical examples.

EXAMPLE 2. Let $\mathsf{T} : \mathbb{R}^2 \to \mathbb{R}^2$ be the linear transformation given by

$$\mathsf{T}(x, y) = (x + 2y, 2x + y).$$

Calculate $\mathsf{T}^* : \mathbb{R}^2 \to \mathbb{R}^2$.

Solution. To compute $\mathsf{T}^*(u, v)$ we will use the equation

$$\langle (x, y), \mathsf{T}^*(u, v)\rangle = \langle \mathsf{T}(x, y), (u, v)\rangle$$

for all (x, y), (u, v) in $\mathbb{R}^2$. We find

$$\begin{aligned}\langle (x, y), \mathsf{T}^*(u, v)\rangle &= \langle \mathsf{T}(x, y), (u, v)\rangle\\ &= \langle (x + 2y, y + 2x), (u, v)\rangle\\ &= xu + 2yu + yv + 2xv\\ &= xu + 2xv + yv + 2yu\\ &= x(u + 2v) + y(v + 2u)\\ &= \langle (x, y), (u + 2v, v + 2u)\rangle\end{aligned}$$

and therefore

$$\mathsf{T}^*(u, v) = (u + 2v, v + 2u)$$

or if you prefer x and y

$$\mathsf{T}^*(x, y) = (x + 2y, y + 2x).$$

Therefore $\mathsf{T}^* = \mathsf{T}$.

EXAMPLE 3. Let $\mathsf{T} : \mathbb{R}^2 \to \mathbb{R}^2$ be the linear transformation given by

$$\mathsf{T}(x, y) = (x + 3y, 2x + y).$$

Calculate $\mathsf{T}^* : \mathbb{R}^2 \to \mathbb{R}^2$.

Solution. We have

$$\begin{aligned}\langle (x, y), \mathsf{T}^*(u, v)\rangle &= \langle \mathsf{T}(x, y), (u, v)\rangle \\ &= \langle (x + 3y, 2x + y), (u, v)\rangle \\ &= xu + 3yu + 2xv + yv \\ &= xu + 2xv + 3yu + yv \\ &= x(u + 2v) + y(3u + v) \\ &= \langle (x, y), (u + 2v, 3u + v)\rangle\end{aligned}$$

and therefore

$$\mathsf{T}^*(u, v) = (u + 2v, 3u + v)$$

and $\mathsf{T} \neq \mathsf{T}^*$.

Definition. Let $\mathscr{V}$ be a finite-dimensional inner product space. A linear transformation $\mathsf{T} : \mathscr{V} \to \mathscr{V}$ is called *self-adjoint* iff $\mathsf{T} = \mathsf{T}^*$.

Notice that if $\mathsf{T} : \mathscr{V} \to \mathscr{V}$ is self-adjoint then since $\mathsf{T} = \mathsf{T}^*$

$$\langle \mathsf{T}(\mathbf{B}), \mathbf{A}\rangle = \langle \mathbf{B}, \mathsf{T}(\mathbf{A})\rangle$$

for all vectors $\mathbf{A}$, $\mathbf{B}$ in $\mathscr{V}$. On the other hand if $\mathsf{T} : \mathscr{V} \to \mathscr{V}$ is a linear transformation such that

$$\langle \mathsf{T}(\mathbf{B}), \mathbf{A}\rangle = \langle \mathbf{B}, \mathsf{T}(\mathbf{A})\rangle$$

for all vectors $\mathbf{A}$, $\mathbf{B}$ in $\mathscr{V}$ then T is self-adjoint. To see this recall we always have the equation

$$\langle \mathsf{T}(\mathbf{B}), \mathbf{A}\rangle = \langle \mathbf{B}, \mathsf{T}^*(\mathbf{A})\rangle$$

for all vectors $\mathbf{A}$, $\mathbf{B}$ in $\mathscr{V}$. Hence for our hypothesized T we will have

$$\langle \mathbf{B}, \mathsf{T}(\mathbf{A})\rangle = \langle \mathbf{B}, \mathsf{T}^*(\mathbf{A})\rangle$$

for all vectors $\mathbf{A}$, $\mathbf{B}$ in $\mathscr{V}$. This may be rewritten as

$$\langle \mathbf{B}, \mathsf{T}(\mathbf{A}) - \mathsf{T}^*(\mathbf{A})\rangle = 0$$

for all vectors $\mathbf{A}$, $\mathbf{B}$ in $\mathscr{V}$. In particular putting $\mathbf{B} = \mathsf{T}(\mathbf{A}) - \mathsf{T}^*(\mathbf{A})$ we find

$$\langle \mathsf{T}(\mathbf{A}) - \mathsf{T}^*(\mathbf{A}), \mathsf{T}(\mathbf{A}) - \mathsf{T}^*(\mathbf{A})\rangle = 0$$

for all vectors $\mathbf{A}$ in $\mathscr{V}$. Therefore $\mathsf{T}(\mathbf{A}) - \mathsf{T}^*(\mathbf{A}) = \mathbf{0}$ for all vectors $\mathbf{A}$ in $\mathscr{V}$ or what is the same thing

$$\mathsf{T}(\mathbf{A}) = \mathsf{T}^*(\mathbf{A})$$

for all vectors $\mathbf{A}$ in $\mathscr{V}$. Hence $\mathsf{T} = \mathsf{T}^*$ so that T is self-adjoint as we claimed.

As an example of how we may apply the above fact to check if a transformation is self-adjoint consider the linear transformation

$$\mathsf{T} : \mathcal{M}_{nn} \to \mathcal{M}_{nn}$$

given by

$$\mathsf{T}(\mathbf{A}) = \mathbf{A}^t$$

where $\mathbf{A}^t$ is the transpose of the $n \times n$ matrix $\mathbf{A}$. We have

$$\begin{aligned} \langle \mathsf{T}(\mathbf{B}^t), \mathbf{A} \rangle &= \operatorname{tr}(\mathsf{T}(\mathbf{B}^t)\mathbf{A}^t) \\ &= \operatorname{tr}(\mathbf{B}^{tt}\mathbf{A}^t) \\ &= \operatorname{tr}(\mathbf{B}\mathbf{A}^t) \end{aligned}$$

and

$$\begin{aligned} \langle \mathbf{B}^t, \mathsf{T}(\mathbf{A}) \rangle &= \operatorname{tr}(\mathbf{B}^t(\mathsf{T}(\mathbf{A}))^t) \\ &= \operatorname{tr}(\mathbf{B}^t\mathbf{A}^{tt}) \\ &= \operatorname{tr}(\mathbf{B}^t\mathbf{A}). \end{aligned}$$

Using the definition of the trace it is very easy to verify (see Chapter 15, Exercise 15)

$$\begin{aligned} \operatorname{tr}(\mathbf{C}^t) &= \operatorname{tr}(\mathbf{C}) \\ \operatorname{tr}(\mathbf{C}\mathbf{D}) &= \operatorname{tr}(\mathbf{D}\mathbf{C}). \end{aligned}$$

Recall also that

$$(\mathbf{C}\mathbf{D})^t = \mathbf{D}^t\mathbf{C}^t.$$

We therefore may compute as follows:

$$\operatorname{tr}(\mathbf{B}^t\mathbf{A}) = \operatorname{tr}(\mathbf{A}\mathbf{B}^t) = \operatorname{tr}((\mathbf{A}\mathbf{B}^t)^t) = \operatorname{tr}(\mathbf{B}^{tt}\mathbf{A}^t) = \operatorname{tr}(\mathbf{B}\mathbf{A}^t).$$

Therefore

$$\langle \mathsf{T}(\mathbf{B}^t), \mathbf{A} \rangle = \operatorname{tr}(\mathbf{B}\mathbf{A}^t) = \operatorname{tr}(\mathbf{B}^t\mathbf{A}) = \langle \mathbf{B}^t, \mathsf{T}(\mathbf{A}) \rangle$$

and hence $\mathsf{T} : \mathcal{M}_{nn} \to \mathcal{M}_{nn}$ must be self-adjoint.

Of course there are other methods to check if a linear transformation is self-adjoint. One such method is to examine the matrix relative to an orthonormal basis.

Theorem 16.4. *Let $\mathcal{V}$ be a finite-dimensional inner product space, $\mathsf{T} : \mathcal{V} \to \mathcal{V}$ a linear transformation and $\{\mathbf{A}_1, \ldots, \mathbf{A}_n\}$ an* orthonormal *basis for $\mathcal{V}$. Let $\mathbf{M}$ be the matrix of T relative to the basis $\{\mathbf{A}_1, \ldots, \mathbf{A}_n\}$ and $\mathbf{N}$ the matrix of T^* relative to the basis $\{\mathbf{A}_1, \ldots, \mathbf{A}_n\}$. Then $\mathbf{N} = \mathbf{M}^t$.*

PROOF. What we need is (15.9) and the definitions. Let $(\mathbf{M}) = (m_{ij})$ and $\mathbf{N} = (n_{ij})$. Then

$$\begin{aligned} \mathsf{T}(\mathbf{A}_j) &= \sum m_{ij}\mathbf{A}_i \\ \mathsf{T}^*(\mathbf{A}_s) &= \sum n_{rs}\mathbf{A}_r. \end{aligned}$$

According to (15.9)

$$m_{ij} = \langle \mathsf{T}(\mathbf{A}_j), \mathbf{A}_i \rangle$$

and

$$n_{rs} = \langle \mathsf{T}^*(\mathbf{A}_s), \mathbf{A}_r \rangle.$$

But

$$\langle \mathsf{T}(\mathbf{A}_j), \mathbf{A}_i \rangle = \langle \mathbf{A}_j, \mathsf{T}^*(\mathbf{A}_i) \rangle = \langle \mathsf{T}^*(\mathbf{A}_i), \mathbf{A}_j \rangle$$

so that

$$m_{ij} = n_{ji}.$$

But this just says $\mathbf{N} = \mathbf{M}^t$ as required. □

Theorem 16.5. *Let $\mathscr{V}$ be a finite-dimensional inner produce space and $\mathsf{T} : \mathscr{V} \to \mathscr{V}$ a self-adjoint linear transformation. If $\{\mathbf{A}_1, \ldots, \mathbf{A}_n\}$ is an orthonormal basis for $\mathscr{V}$ then the matrix of T relative to this basis is symmetric.*

For this reason self-adjoint transformations are often called symmetric.

Proof. Let $\mathbf{M}$ be the matrix of T and $\mathbf{N}$ the matrix of T^*, both relative to the basis $\{\mathbf{A}_1, \ldots, \mathbf{A}_n\}$. Then $\mathbf{N} = \mathbf{M}^t$ by (16.4). But since T is self-adjoint $\mathsf{T} = \mathsf{T}^*$ so $\mathbf{N} = \mathbf{M}$. Thus $\mathbf{M} = \mathbf{N} = \mathbf{M}^t$. That is, $\mathbf{M} = \mathbf{M}^t$ so $\mathbf{M}$ is symmetric. □

Example 4. Let $\mathsf{T} : \mathbb{R}^3 \to \mathbb{R}^3$ be the linear transformation given by

$$\mathsf{T}(x, y, z) = (x - y + z, x + y, z - x).$$

Decide if T is self-adjoint.

Solution. The matrix of T relative to the standard orthonormal base may be easily computed. We get

$$\begin{aligned} \mathsf{T}(1, 0, 0) &= (1, 1, -1) \\ \mathsf{T}(0, 1, 0) &= (-1, 1, 0) \\ \mathsf{T}(0, 0, 1) &= (1, 0, 1), \end{aligned}$$

so the required matrix is

$$\mathbf{M} = \begin{pmatrix} 1 & -1 & 1 \\ 1 & 1 & 0 \\ -1 & 0 & 1 \end{pmatrix}.$$

Then

$$\mathbf{M}^t = \begin{pmatrix} 1 & 1 & -1 \\ -1 & 1 & 0 \\ 1 & 0 & 1 \end{pmatrix}$$

so that $\mathbf{M} \neq \mathbf{M}^t$, that is $\mathbf{M}$ is not symmetric. By (16.5) it follows that T is not self-adjoint.

Actually we have enough information to compute the adjoint $\mathsf{T}^* : \mathbb{R}^3 \to \mathbb{R}^3$. In view of (16.4) we need only calculate

$$\mathbf{M}^t\begin{pmatrix} x \\ y \\ z \end{pmatrix} = \begin{pmatrix} 1 & 1 & -1 \\ -1 & 1 & 0 \\ 1 & 0 & 1 \end{pmatrix}\begin{pmatrix} x \\ y \\ z \end{pmatrix} = \begin{pmatrix} x + y - z \\ y - x \\ x + z \end{pmatrix}$$

from which we get the formula

$$\mathsf{T}^*(x, y, z) = (x + y - z, y - x, x + z).$$

Of course we could also calculate T* by using the method of Examples 2 and 3. You ought to try this.

Theorem 16.6. *Let* $\mathsf{T} : \mathscr{V} \to \mathscr{V}$ *be an endomorphism of the finite-dimensional inner product space* $\mathscr{V}$. *Let* $\{\mathbf{A}_1, \ldots, \mathbf{A}_n\}$ *be an orthonormal basis for* $\mathscr{V}$ *and suppose that the matrix of* T *is symmetric with respect to this basis. Then* T *is self-adjoint.*

PROOF. Let $\mathbf{M}$ be the matrix of T relative to the basis $\{\mathbf{A}_1, \ldots, \mathbf{A}_n\}$. By (16.4) the matrix of T* relative to this basis is also $\mathbf{M}$ since $\mathbf{M}$ is symmetric. Therefore by (12.1) $\mathsf{T} = \mathsf{T}^*$. □

ALTERNATE PROOF. Let $\mathbf{M}$ be the matrix of T relative to the basis $\{\mathbf{A}_1, \ldots, \mathbf{A}_n\}$. Let $\mathbf{A}$ and $\mathbf{B}$ be vectors in $\mathscr{V}$ with

$$\mathbf{A} = a_1\mathbf{A}_1 + a_2\mathbf{A}_2 + \cdots + a_n\mathbf{A}_n$$
$$\mathbf{B} = b_1\mathbf{A}_1 + b_2\mathbf{B}_2 + \cdots + b_n\mathbf{A}_n.$$

Then if $\mathbf{M} = (m_{ij})$

$$\mathsf{T}(\mathbf{A}) = \sum_{i,j} m_{ij}a_j\mathbf{A}_i$$

$$\mathsf{T}(\mathbf{B}) = \sum_{i,j} m_{ij}b_j\mathbf{A}_i.$$

Applying (15.9) we get

$$\langle \mathsf{T}(\mathbf{A}), \mathbf{B} \rangle = \sum_{i,j} m_{ij}a_jb_i = \sum_{r,s} a_rb_sm_{sr}$$

$$\langle \mathbf{A}, \mathsf{T}(\mathbf{B}) \rangle = \sum_{i,j} a_im_{ij}b_j = \sum_{r,s} a_rb_sm_{rs}.$$

Since $\mathbf{M}$ is symmetric $m_{rs} = m_{sr}$ so we get

$$\langle \mathsf{T}(\mathbf{A}), \mathbf{B} \rangle = \sum_{r,s} a_rb_sm_{sr} = \sum_{r,s} a_rb_sm_{rs} = \langle \mathbf{A}, \mathsf{T}(\mathbf{B}) \rangle$$

for all $\mathbf{A}, \mathbf{B}$ in $\mathscr{V}$. But according to the definition of T*,

$$\langle \mathsf{T}(\mathbf{B}), \mathbf{A} \rangle = \langle \mathbf{B}, \mathsf{T}^*(\mathbf{A}) \rangle,$$

for all **A**, **B** in $\mathscr{V}$. Therefore we get

$$\langle \mathbf{B}, \mathsf{T}(\mathbf{A})\rangle = \langle \mathsf{T}(\mathbf{A}), \mathbf{B}\rangle = \langle \mathbf{A}, \mathsf{T}(\mathbf{B})\rangle = \langle \mathsf{T}(\mathbf{B}), \mathbf{A}\rangle = \langle \mathbf{B}, \mathsf{T}^*(\mathbf{A})\rangle$$

so that

$$\langle \mathbf{B}, \mathsf{T}(\mathbf{A})\rangle = \langle \mathbf{B}, \mathsf{T}^*(\mathbf{A})\rangle$$

for all **A**, **B** in $\mathscr{V}$. The definition of the adjoint now gives $\mathsf{T}(\mathbf{A}) = \mathsf{T}^*(\mathbf{A})$ for all **A** in $\mathscr{V}$. Thus $\mathsf{T} = \mathsf{T}^*$. □

Notice that (16.5) provides us with a painless method for constructing loads of self-adjoint linear transformations. All we have to do is take a symmetric matrix S of size n, an n-dimensional inner product space $\mathscr{V}$ and an orthonormal basis $\{\mathbf{A}_1, \ldots, \mathbf{A}_n\}$ for $\mathscr{V}$. The linear transformation $\mathsf{T}: \mathscr{V} \to \mathscr{V}$ whose matrix is S in the basis $\{\mathbf{A}_1, \ldots, \mathbf{A}_n\}$ will then be self-adjoint.

Notice also that (16.5) suggests ways to modify the proposed definition we made of a symmetric linear transformation so that we get a useful concept. We will not pursue this idea here, but relegate it to a flock of exercises.

We are ready now to state the main result of this chapter (and indeed of this book so far).

Theorem 16.7 (Spectral theorem). *Let $\mathsf{T}: \mathscr{V} \to \mathscr{V}$ be a self-adjoint linear transformation in the finite-dimensional vector space $\mathscr{V}$. Then there exists an orthonormal basis $\{\mathbf{A}_1, \ldots, \mathbf{A}_n\}$ and numbers $\lambda_1, \ldots, \lambda_n$ such that*

$$\mathsf{T}(\mathbf{A}_i) = \lambda_i \mathbf{A}_i \qquad i = 1, 2, \ldots, n.$$

Therefore the matrix of T relative to this basis is

$$\begin{pmatrix} \lambda_1 & & & 0 \\ & \lambda_2 & & \\ & & \ddots & \\ 0 & & & \lambda_n \end{pmatrix}.$$

Note that it is not asserted that the numbers $\lambda_1, \ldots, \lambda_n$ are distinct; they need not be. What Theorem (16.7) says is that the characteristic polynomial of T factors completely and the geometric multiplicity of any eigenvalue coincides with its algebraic multiplicity.

The proof of (16.7) is an induction argument. We will give the proof for $n = 2$ and 3. The general case is sketched out in the exercises. We need two preliminary results.

Proposition 16.8. *Let $\mathsf{T}: \mathscr{V} \to \mathscr{V}$ be a self-adjoint linear transformation in the finite-dimensional inner product space $\mathscr{V}$. Suppose that $\mathscr{W}$ is a linear subspace of $\mathscr{V}$ with the property that $\mathsf{T}(\mathscr{W}) \subset \mathscr{W}$. Then $\mathscr{W}^\perp$ has the same property, namely, $\mathsf{T}(\mathscr{W}^\perp) \subset \mathscr{W}^\perp$.*

PROOF. Let $\mathbf{A} \in \mathscr{W}^{\perp}$. We must show that

$$\langle \mathsf{T}(\mathbf{A}), \mathbf{B} \rangle = 0$$

for all $\mathbf{B}$ in $\mathscr{W}$. We have

$$\begin{aligned}\langle \mathsf{T}(\mathbf{A}), \mathbf{B} \rangle &= \langle \mathbf{A}, \mathsf{T}^*(\mathbf{B}) \rangle \\ &= \langle \mathbf{A}, \mathsf{T}(\mathbf{B}) \rangle \quad \text{since } \mathsf{T} = \mathsf{T}^* \\ &= 0 \quad \text{since } \mathsf{T}(\mathbf{B}) \in \mathscr{W} \text{ and } \mathbf{A} \in \mathscr{W}^{\perp}\end{aligned}$$

as required. □

Proposition 16.9. *Let $\mathsf{T} : \mathscr{V} \to \mathscr{V}$ be a self-adjoint linear transformation in the finite-dimensional inner product space $\mathscr{V}$. Suppose that e' and e'' are distinct eigenvalues of T with corresponding eigenvectors $\mathbf{E}'$ and $\mathbf{E}''$. Then $\langle \mathbf{E}', \mathbf{E}'' \rangle = 0$, that is, $\mathbf{E}'$ and $\mathbf{E}''$ are perpendicular.*

PROOF. By hypothesis we have

$$\mathsf{T}(\mathbf{E}') = e'\mathbf{E}', \qquad \mathsf{T}(\mathbf{E}'') = e''\mathbf{E}''$$

so since T is self-adjoint we get

$$\begin{aligned}e'\langle \mathbf{E}', \mathbf{E}'' \rangle &= \langle e'\mathbf{E}', \mathbf{E}'' \rangle = \langle \mathsf{T}(\mathbf{E}'), \mathbf{E}'' \rangle \\ &= \langle \mathbf{E}', \mathsf{T}^*(\mathbf{E}'') \rangle = \langle \mathbf{E}', \mathsf{T}(\mathbf{E}'') \rangle \\ &= \langle \mathbf{E}', e''\mathbf{E}'' \rangle = e''\langle \mathbf{E}', \mathbf{E}'' \rangle\end{aligned}$$

which means

$$e'\langle \mathbf{E}', \mathbf{E}'' \rangle = e''\langle \mathbf{E}', \mathbf{E}'' \rangle.$$

Since $e' \neq e''$ by hypothesis we must have $\langle \mathbf{E}', \mathbf{E}'' \rangle = 0$ as required. □

PROOF OF (16.7) FOR $n = 2$. Let $\{\mathbf{E}_1, \mathbf{E}_2\}$ be an orthonormal base for $\mathscr{V}$. By (16.4) the matrix $\mathbf{M}$ of T relative to this basis is symmetric, say

$$\mathbf{M} = \begin{pmatrix} a & b \\ b & c \end{pmatrix}.$$

If $b = 0$ there is nothing to prove, so we will assume that $b \neq 0$. The characteristic polynomial of T is given by

$$\begin{aligned}\Delta(t) = \det(\mathbf{M} - t\mathbf{I}) &= \det \begin{pmatrix} a - t & b \\ b & c - t \end{pmatrix} \\ &= (a - t)(c - t) - b^2 \\ &= t^2 - (a + c)t + ac - b^2.\end{aligned}$$

By the quadratic formula the roots of $\Delta(t)$ are given by

$$\begin{aligned} r_{\pm} &= \frac{a + c \pm \sqrt{(a + c)^2 - 4(ac - b^2)}}{2} \\ &= \frac{a + c \pm \sqrt{a^2 + 2ac + c^2 - 4ac + 4ab^2}}{2} \\ &= \frac{a + c \pm \sqrt{4b^2 + (a - c)^2}}{2}. \end{aligned}$$

Notice that since $b \neq 0$

$$4b^2 + (a - c)^2 > 0$$

so that the characteristic roots are distinct. Let $\{\mathbf{F}_+, \mathbf{F}_-\}$ be the corresponding eigenvectors. By (16.8) $\mathbf{F}_+$ and $\mathbf{F}_-$ are orthogonal. Since we may normalize $\mathbf{F}_+$ and $\mathbf{F}_-$ they constitute an orthonormal basis of eigenvectors completing the proof. □

PROOF OF (16.7) FOR $n = 3$. Let $\Delta(t)$ be the characteristic polynomial of T. Then $\Delta(t)$ is of degree 3 and hence must have at least one *real* root, say r. Then r is an eigenvalue of T. Let $\mathbf{E}$ be a corresponding eigenvector of unit length and $\mathscr{W} = \mathscr{L}(\mathbf{E}) \subset \mathscr{V}$. Of course $\mathsf{T}(\mathscr{W}) \subset \mathscr{W}$ and therefore by (16.7) $\mathsf{T}(\mathscr{W}^{\perp}) \subset \mathscr{W}^{\perp}$. Let $\mathbb{F}_1, \mathbb{F}_2$ be an orthonormal basis for $\mathscr{W}^{\perp}$. Then $\{\mathbf{E}, \mathbb{F}_1, \mathbb{F}_2\}$ is an orthonormal basis for $\mathscr{V}$ (see the proof of (15.18)). By (16.4) the matrix $\mathbf{M}$ of T relative to the basis $\{\mathbf{E}, \mathbb{F}_1, \mathbb{F}_2\}$ will be symmetric. Since e is an eigenvalue it must look like

$$\mathbf{M} = \begin{pmatrix} e & 0 & 0 \\ 0 & a & b \\ 0 & b & c \end{pmatrix}$$

The characteristic polynomial of T is therefore

$$\Delta(t) = (e - t)(t^2 - (a + c)t + ac - b^2).$$

If $b = 0$ there is nothing to prove. If $b \neq 0$ then as in the case $n = 2$ there are two distinct eigenvalues, e_1, e_2 the zeros of the quadratic

$$t^2 - (a + c)t + ac - b^2.$$

If $\mathbf{E}_1$ and $\mathbf{E}_2$ are corresponding eigenvectors then $\mathbf{E}_1, \mathbf{E}_2$ are orthogonal by (16.8) while they both are orthogonal to $\mathbf{E}$ since they belong to $\mathscr{W}^{\perp}$. Therefore $\{\mathbf{E}, \mathbf{E}_1, \mathbf{E}_2\}$ is an orthonormal basis for $\mathscr{V}$ and the matrix of T relative to this basis is

$$\mathbf{M} = \begin{pmatrix} e & 0 & 0 \\ 0 & e_1 & 0 \\ 0 & 0 & e_2 \end{pmatrix}.$$

(Note that it is possible that $e = e_1$ or $e = e_2$ even if $b \neq 0$. If $b = 0$ it is even possible for $e_1 = e = e_2$.) □

EXAMPLE 5. Let $\mathsf{T} : \mathbb{R}^2 \to \mathbb{R}^2$ be the linear transformation given by

$$\mathsf{T}(x, y) = (x + 2y, 2y + x).$$

Find an orthonormal basis of eigenvectors of T for $\mathbb{R}^2$.

Solution. We already have seen that T is self-adjoint. Therefore by applying (16.7) we know that it is possible to find an orthonormal basis for $\mathbb{R}^2$ composed of eigenvectors of T. As a matter of fact the orthogonality will take care of itself because of (16.1). This will even provide a numerical check on our work.

The matrix of T relative to the standard orthonormal basis of $\mathbb{R}^2$ was computed to be

$$\mathbf{M} = \begin{pmatrix} 1 & 2 \\ 2 & 1 \end{pmatrix}$$

in Example 1. The characteristic polynomial of T is therefore

$$\begin{aligned} \Delta(t) = \det(\mathbf{M} - t\mathbf{I}) &= \det\begin{pmatrix} 1 - t & 2 \\ 2 & 1 - t \end{pmatrix} \\ &= (1 - t)^2 - 4 = t^2 - 2t + 1 - 4 \\ &= (t^2 - 2t - 3) = (t - 3)(t + 1). \end{aligned}$$

The eigenvalues of T are therefore 3 and -1. Corresponding eigenvectors are easily computed

$$e = 3 \quad \begin{cases} -2x + 2y = 0 \\ -2x + 2y = 0 \end{cases}$$

$x = y$ so $(1, 1)$ is an eigenvector and $(\sqrt{2}/2, \sqrt{2}/2)$ an eigenvector of unit length.

$$e = -1 \quad \begin{cases} 2x + 2y = 0 \\ 2x + 2y = 0 \end{cases}$$

$x = -y$ so $(1, -1)$ is an eigenvector and $(\sqrt{2}/2, -\sqrt{2}/2)$ an eigenvector of unit length. Therefore $\mathbf{F}_1 = (\sqrt{2}/2, \sqrt{2}/2)$, $\mathbf{F}_2 = (\sqrt{2}/2, -\sqrt{2}/2)$ is, by (16.6), an orthonormal basis of eigenvectors of T for $\mathbb{R}^2$. We can check the orthogonality easily enough

$$\langle \mathbf{F}_1, \mathbf{F}_2 \rangle = \tfrac{2}{4} - \tfrac{2}{4} = 0$$

as required.

EXAMPLE 6. Let $\mathsf{T} : \mathbb{R}^3 \to \mathbb{R}^3$ be given by

$$\mathsf{T}(x, y, z) = (3x - z, 2y, -x + 3z).$$

Verify that T is self-adjoint and put it into diagonal form.

Solution. By direct calculation

$$\begin{aligned} \mathsf{T}(1,0,0) &= (3,0,-1)\\ \mathsf{T}(0,1,0) &= (0,2,0)\\ \mathsf{T}(0,0,1) &= (-1,0,3). \end{aligned}$$

Therefore the matrix of T relative to the standard basis of $\mathbb{R}^3$ is

$$\begin{pmatrix} 3 & 0 & -1\\ 0 & 2 & 0\\ -1 & 0 & 3 \end{pmatrix}$$

which is symmetric, so T is self-adjoint. The characteristic polynomial of T is

$$\begin{aligned} \Delta(t) &= \det\begin{pmatrix} 3-t & 0 & -1\\ 0 & 2-t & 0\\ -1 & 0 & 3-t \end{pmatrix}\\ &= (2-t)\det\begin{pmatrix} 3-t & -1\\ -1 & 3-t \end{pmatrix}\\ &= (2-t)((3-t)^2-1) = (2-t)(t^2-6t+8)\\ &= (2-t)(t-4)(t-2) \end{aligned}$$

Therefore the eigenvalues of T are $t = 2$, $t = 4$ with multiplicities 2 and 1 respectively. A basis for the eigenspace corresponding to the eigenvalue 2 is obtained by solving the linear system

$$\mathsf{T}(x, y, z) = 2(x, y, z),$$

that is,

$$\begin{aligned} 3x - z &= 2x\\ 2y &= 2y\\ -x + 3z &= 2z \end{aligned}$$

or

$$\begin{aligned} x - z &= 0\\ -x + z &= 0 \end{aligned}$$

so a basis for the eigenspace corresponding to eigenvalue 2 consists of the vectors (1, 0, 1), (0, 1, 0).

To find an eigenvector corresponding to eigenvalue 4 we solve

$$\mathsf{T}(x, y, z) = 4(x, y, z)$$

that is,

$$\begin{aligned} 3x - z &= 4x\\ 2y &= 4y\\ -x + 3z &= 4z \end{aligned}$$

or

$$\begin{aligned} -z &= x \\ -2y &= 0 \\ -x &= z \end{aligned}$$

which has $(-1, 0, 1)$ as a solution. Therefore the matrix of T relative to the ordered basis $\mathbf{f}_1 = (1, 0, 1)$, $\mathbf{f}_2 = (0, 1, 0)$, $\mathbf{f}_3 = (-1, 0, 1)$ is

$$\begin{pmatrix} 2 & 0 & 0 \\ 0 & 2 & 0 \\ 0 & 0 & 4 \end{pmatrix}.$$

Notice that $\{\mathbf{f}_1, \mathbf{f}_2, \mathbf{f}_3\}$ is an orthogonal basis.

EXAMPLE 7. Put $\mathsf{T}: \mathcal{M}_{2,2} \to \mathcal{M}_{2,2}$ given by $\mathsf{T}(\mathbf{A}) = \mathbf{A}^t$ into diagonal form.

Solution. We showed that T is self adjoint, so by the spectral theorem T is diagonalizable. As an orthonormal basis for $\mathcal{M}_{2,2}$ we choose:

$$\mathbf{M}_1 = \begin{pmatrix} 1 & 0 \\ 0 & 0 \end{pmatrix}, \qquad \mathbf{M}_2 = \begin{pmatrix} 0 & 1 \\ 0 & 0 \end{pmatrix},$$

$$\mathbf{M}_3 = \begin{pmatrix} 0 & 0 \\ 1 & 0 \end{pmatrix}, \qquad \mathbf{M}_4 = \begin{pmatrix} 0 & 0 \\ 0 & 1 \end{pmatrix}.$$

So

$$\mathsf{T}(\mathbf{M}_1) = \mathbf{M}_1, \qquad \mathsf{T}(\mathbf{M}_2) = \mathbf{M}_3, \qquad \mathsf{T}(\mathbf{M}_3) = \mathbf{M}_2, \qquad \mathsf{T}(\mathbf{M}_4) = \mathbf{M}_4,$$

and the matrix of T relative to this basis is

$$\begin{pmatrix} 1 & 0 & 0 & 0 \\ 0 & 0 & 1 & 0 \\ 0 & 1 & 0 & 0 \\ 0 & 0 & 0 & 1 \end{pmatrix}.$$

Note. From $\mathsf{T}(\mathbf{M}_1) = \mathbf{M}_1$, and $\mathsf{T}(\mathbf{M}_4) = \mathbf{M}_4$, we see immediately that $\mathbf{M}_1$ and $\mathbf{M}_4$ are eigenvectors of T corresponding to eigenvalues 1.

The characteristic polynomial of T is

$$\begin{aligned} \Delta(t) &= \det\begin{pmatrix} 1-t & 0 & 0 & 0 \\ 0 & -t & 1 & 0 \\ 0 & 1 & -t & 0 \\ 0 & 0 & 0 & 1-t \end{pmatrix} \\ &= (1-t)\det\begin{pmatrix} -t & 1 & 0 \\ 1 & -t & 0 \\ 0 & 0 & 1-t \end{pmatrix} \\ &= (1-t)(-1)^{3+3}(1-t)\det\begin{pmatrix} -t & 1 \\ 1 & -t \end{pmatrix} \\ &= (1-t)^2(t^2-1) = (t-1)^3(t+1), \end{aligned}$$

so the eigenvalues of T are $+1$, and -1 with multiplicity 3 and 1, respectively. By the spectral theorem $\dim \mathscr{V}_1 = 3$, $\dim \mathscr{V}_{-1} = 1$. We can find three orthonormal eigenvectors of T corresponding to eigenvalue $+1$ by solving the system

$$\begin{pmatrix} x \\ y \\ z \\ w \end{pmatrix} = \begin{pmatrix} 1 & 0 & 0 & 0 \\ 0 & 0 & 1 & 0 \\ 0 & 1 & 0 & 0 \\ 0 & 0 & 0 & 1 \end{pmatrix} \begin{pmatrix} x \\ y \\ z \\ w \end{pmatrix} = \begin{pmatrix} x \\ z \\ y \\ w \end{pmatrix}$$

which is

$$x = x, \qquad y = z, \qquad z = y, \qquad w = w.$$

So

$$(1, 0, 0, 0), \qquad \sqrt{\tfrac{1}{2}}(0, 1, 1, 0), \qquad (0, 0, 0, 1)$$

are the components relative to $\{\mathbf{M}_1, \mathbf{M}_2, \mathbf{M}_3, \mathbf{M}_4\}$ of an orthonormal basis for $\mathscr{V}_1$. The corresponding vectors of $\mathscr{M}_{2.2}$ are

$$\mathbf{N}_1 = \begin{pmatrix} 1 & 0 \\ 0 & 1 \end{pmatrix}, \qquad \mathbf{N}_2 = \sqrt{\frac{1}{2}}\begin{pmatrix} 0 & 1 \\ 1 & 0 \end{pmatrix}, \qquad \mathbf{N}_3 = \begin{pmatrix} 0 & 0 \\ 0 & 1 \end{pmatrix}.$$

To find a basis for the eigenspace $\mathscr{V}_{-1}$ requires we solve the system

$$-\begin{pmatrix} x \\ y \\ z \\ w \end{pmatrix} = \begin{pmatrix} 1 & 0 & 0 & 0 \\ 0 & 0 & 1 & 0 \\ 0 & 1 & 0 & 0 \\ 0 & 0 & 0 & 1 \end{pmatrix} \begin{pmatrix} x \\ y \\ z \\ w \end{pmatrix} = \begin{pmatrix} x \\ z \\ y \\ w \end{pmatrix},$$

which is

$$-x = x, \qquad -y = z, \qquad -z = y, \qquad -w = w.$$

Thus $x = 0$, $w = 0$, and $y = 1$, $z = -1$. Normalizing this we have

$$\sqrt{\tfrac{1}{2}}(0, 1, -1, 0)$$

as a basis for the solution space relative to the basis $\{\mathbf{M}_1, \mathbf{M}_2, \mathbf{M}_3, \mathbf{M}_4\}$ so that the vector

$$\mathbf{N}_4 = \sqrt{\frac{1}{2}}\begin{pmatrix} 0 & 1 \\ -1 & 0 \end{pmatrix}$$

is an orthonormal base for $\mathscr{V}_{-1}$. The matrix of T relative to the basis $\{\mathbf{N}_1, \mathbf{N}_2, \mathbf{N}_3, \mathbf{N}_4\}$ is

$$\begin{pmatrix} 1 & 0 & 0 & 0 \\ 0 & 1 & 0 & 0 \\ 0 & 0 & 1 & 0 \\ 0 & 0 & 0 & -1 \end{pmatrix}.$$

The final topic we will take up is the so-called *principal axis theorem* for quadratic forms. In the calculus one spends considerable time in studying the graphs of quadratic equations; these graphs being the conic sections, ellipse, hyperbola, parabola, and various degenerate cases. There is a foolproof procedure for drawing the graph of say

$$2x^2 + 3y^2 = 7,$$

but for

$$x^2 + 4xy + y^2 = 7$$

there is a difficulty caused by the cross-product term "$4xy$." Of course there is a formula for rotating the coordinate system so as to eliminate the cross-product term, and it is just this formula that we wish to discuss in the light of our work with self-adjoint linear transformations! Rather than introduce the final, sophisticated version of the theory let us consider an example first.

EXAMPLE 8. Sketch the graph of

$$x^2 + 4xy + y^2 = 7.$$

Solution. Make the following observation

$$x^2 + 4xy + y^2 = (x, y)\begin{pmatrix} 1 & 2 \\ 2 & 1 \end{pmatrix}\begin{pmatrix} x \\ y \end{pmatrix}.$$

What we wish to discover is a new coordinate system for $\mathbb{R}^2$ in which the quadratic form has no cross-product term. Let

$$\mathsf{T} : \mathbb{R}^2 \to \mathbb{R}^2$$

be the linear transformation whose matrix in the standard coordinate system is

$$\begin{pmatrix} 1 & 2 \\ 2 & 1 \end{pmatrix}.$$

That is,

$$\mathsf{T}(x, y) = (x + 2y, y + 2x).$$

Then we see that

$$(x, y)\begin{pmatrix} 1 & 2 \\ 2 & 1 \end{pmatrix}\begin{pmatrix} x \\ y \end{pmatrix} = \langle (x, y), \mathsf{T}(x, y) \rangle.$$

That is, for a vector $\mathbf{A}$ in $\mathbb{R}^2$ with coordinates (x, y) we have

$$x^2 + 4xy + y^2 = \langle \mathbf{A}, \mathsf{T}(\mathbf{A}) \rangle.$$

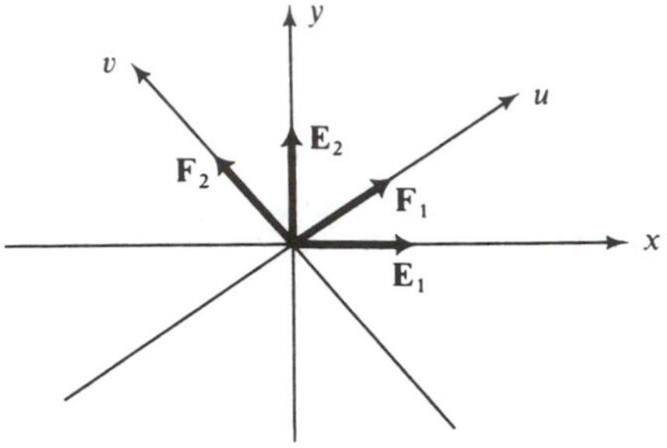

Figure 16.1

The number $\langle \mathbf{A}, \mathsf{T}(\mathbf{A})\rangle$ depends, not on the coordinate system in which we express $\mathbf{A}$, but only on T. Thus we could equally well use the coordinate system determined by the unit eigenvectors of T, namely

$$\mathbf{F}_1 = \left(\frac{\sqrt{2}}{2}, \frac{\sqrt{2}}{2}\right), \qquad \mathbf{F}_2 = \left(-\frac{\sqrt{2}}{2}, \frac{\sqrt{2}}{2}\right).$$

See Figure 16.1. Let us write (u, v) for the coordinates of a vector in the coordinate system determined by the vectors $\mathbf{F}_1, \mathbf{F}_2$. Then

$$\langle \mathbf{A}, \mathsf{T}(\mathbf{A})\rangle = (u, v)\begin{pmatrix} 3 & 0 \\ 0 & -1 \end{pmatrix}\begin{pmatrix} u \\ v \end{pmatrix} = 3u^2 - v^2$$

since the matrix of T relative to the $\{\mathbf{F}_1, \mathbf{F}_2\}$ base is

$$\begin{pmatrix} 3 & 0 \\ 0 & -1 \end{pmatrix}.$$

Thus our original quadratic form

$$7 = x^2 + 4xy + y^2$$

may be written (changing coordinate systems has not changed the value of $\langle \mathbf{A}, \mathsf{T}(\mathbf{A})\rangle$)

$$7 = 3u^2 - v^2$$

in the (u, v) coordinate system. One readily checks that this is a hyperbola and the graph is easily sketched as in Figure 16.2. The vertices of the hyperbola are at the points $(u, v) = (\pm\sqrt{7/3}, 0)$. The (x, y) coordinates of these points are easily computed

$$(x, y) = \pm\sqrt{\frac{7}{3}}\,\mathbf{F}_1 + 0\mathbf{F}_2 = \pm\sqrt{\frac{7}{3}}\left(\frac{\sqrt{2}}{2}, \frac{\sqrt{2}}{2}\right).$$

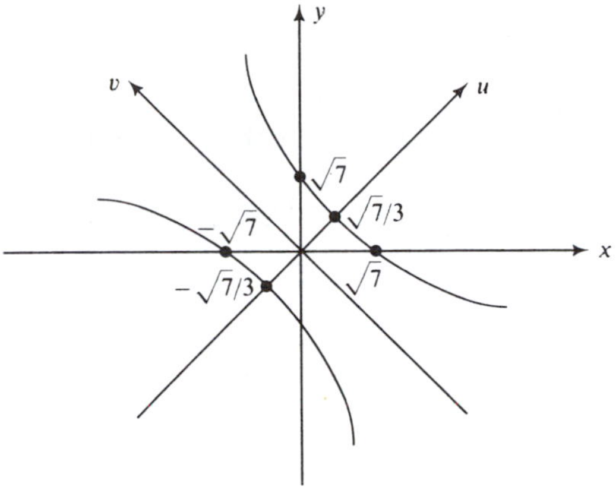

Figure 16.2

The relation between the (x, y) and (u, v) coordinates of a point $\mathbf{A}$ in $\mathbb{R}^2$ is computed thusly:

$$\begin{aligned} x\mathbf{E}_1 + y\mathbf{E}_2 = \mathbf{A} &= u\mathbf{F}_1 + v\mathbf{F}_2 \\ &= u\left(\frac{\sqrt{2}}{2}\mathbf{E}_1 + \frac{\sqrt{2}}{2}\mathbf{E}_2\right) + v\left(-\frac{\sqrt{2}}{2}\mathbf{E}_1 + \frac{\sqrt{2}}{2}\mathbf{E}_2\right) \\ &= \frac{\sqrt{2}}{2}(u - v)\mathbf{E}_1 + \frac{\sqrt{2}}{2}(u + v)\mathbf{E}_2 \end{aligned}$$

whence,

$$x = \frac{\sqrt{2}}{2}u - \frac{\sqrt{2}}{2}v$$

$$y = \frac{\sqrt{2}}{2}u + \frac{\sqrt{2}}{2}v$$

or equivalently

$$\begin{aligned} u\mathbf{F}_1 + v\mathbf{F}_2 = \mathbf{A} &= x\mathbf{E}_1 + y\mathbf{E}_2 \\ &= x\left(\frac{\sqrt{2}}{2}\mathbf{F}_1 - \frac{\sqrt{2}}{2}\mathbf{F}_2\right) + y\left(\frac{\sqrt{2}}{2}\mathbf{F}_1 + \frac{\sqrt{2}}{2}\mathbf{F}_2\right) \\ &= \frac{\sqrt{2}}{2}(x + y)\mathbf{F}_1 + \frac{\sqrt{2}}{2}(y - x)\mathbf{F}_2 \end{aligned}$$

whence

$$u = \frac{\sqrt{2}}{2}x + \frac{\sqrt{2}}{2}y$$

$$v = -\frac{\sqrt{2}}{2}x + \frac{\sqrt{2}}{2}y.$$

The equations

$$\sqrt{3}u \pm v = \sqrt{7}$$

for the asymptotes of the hyperbola may thus be rewritten

$$\sqrt{3}\left(\frac{\sqrt{2}}{2}x + \frac{\sqrt{2}}{2}y\right) \pm \left(-\frac{\sqrt{2}}{2}x + \frac{\sqrt{2}}{2}y\right) = \sqrt{7}$$

or

$$\frac{\sqrt{6} \mp \sqrt{2}}{2}x + \frac{\sqrt{6} \pm \sqrt{2}}{2}y = \sqrt{7}.$$

Notice that the angle θ between the (u, v) and (x, y) coordinate systems (that is, the angle between $\mathbf{E}_1$ and $\mathbf{F}_1$) may be computed from the formula

$$\cos\theta = \langle \mathbf{E}_1, \mathbf{F}_1 \rangle = \frac{\sqrt{2}}{2}.$$

It is now time to reveal the basic theory behind the preceding example. This will require a definition.

Definition. Let $\mathscr{V}$ be a finite-dimensional inner product space. A function $\varphi : \mathscr{V} \to \mathbb{R}$ is called *quadratic* iff there exists a self-adjoint linear transformation $\mathsf{T} : \mathscr{V} \to \mathscr{V}$ such that

$$\varphi(\mathbf{A}) = \langle \mathbf{A}, \mathsf{T}(\mathbf{A}) \rangle$$

for all vectors $\mathbf{A}$ in $\mathscr{V}$.

EXAMPLE 9. The function

$$\varphi(x, y) = x^2 + 4xy - y^2$$

is quadratic. For if

$$\mathsf{T} : \mathbb{R}^2 \to \mathbb{R}^2$$

is the linear transformation with matrix

$$\begin{pmatrix} 1 & 2 \\ 2 & -1 \end{pmatrix}$$

in the standard orthonormal coordinate system then T is self-adjoint by (16.5). Moreover

$$\langle (x, y), \mathsf{T}(x, y) \rangle = (x, y)\begin{pmatrix} 1 & 2 \\ 2 & -1 \end{pmatrix}\begin{pmatrix} x \\ y \end{pmatrix} = x^2 + 4xy - y^2 = \varphi(x, y)$$

and so φ is quadratic. More generally

$$\varphi(x, y) = ax^2 + 2bxy + cy^2$$

is quadratic. We let

$$\mathsf{T} : \mathbb{R}^2 \to \mathbb{R}^2$$

have matrix

$$\begin{pmatrix} a & b \\ b & c \end{pmatrix}$$

in the standard orthonormal basis, so that T is self adjoint and one checks

$$\varphi(x, y) = \langle (x, y), \mathsf{T}(x, y) \rangle.$$

Thus φ is quadratic.

Strictly speaking we should call a function $\varphi : \mathscr{V} \to \mathbb{R}$ given by

$$\varphi(\mathbf{A}) = \langle \mathbf{A}, \mathsf{T}(\mathbf{A}) \rangle$$

a *homogeneous* quadratic function, because it has no linear terms.

Theorem 16.10 (Principal Axis Theorem). *Suppose that*

$$\varphi : \mathscr{V} \to \mathbb{R}$$

is a homogeneous quadratic function on the finite-dimensional inner product space $\mathscr{V}$. *Then there exists a basis* $\{\mathbf{A}_1, \ldots, \mathbf{A}_n\}$ *for* $\mathscr{V}$ *and numbers* $\lambda_1, \ldots, \lambda_n$ *such that for a vector* $\mathbf{X} = x_1\mathbf{A}_1 + \cdots + x_n\mathbf{A}_n$ *in* $\mathscr{V}$ *the value of* φ *is given by the formula*

$$\varphi(\mathbf{X}) = \lambda_1 x_1^2 + \lambda_2 x_2^2 + \cdots + \lambda_n x_n^2.$$

Proof. Let $\mathsf{T} : \mathscr{V} \to \mathscr{V}$ be a self-adjoint transformation such that

$$\varphi(\mathbf{X}) = \langle \mathbf{X}, \mathsf{T}(\mathbf{X}) \rangle$$

for all $\mathbf{X}$ in $\mathscr{V}$. By the spectral theorem we may find a basis $\mathbf{A}_1, \ldots, \mathbf{A}_n$ for $\mathscr{V}$ composed of eigenvectors of T. Let $\lambda_1, \ldots, \lambda_n$ be the corresponding eigenvalues. If $\mathbf{X} = x_1\mathbf{A}_1 + \cdots + x_n\mathbf{A}_n$ then by (15.9) we get

$$\begin{aligned}
\varphi(\mathbf{X}) &= \langle \mathbf{X}, \mathsf{T}(\mathbf{X}) \rangle \\
&= \langle x_1\mathbf{A}_1 + \cdots + x_n\mathbf{A}_n, \mathsf{T}(x_1\mathbf{A}_1 + \cdots + x_n\mathbf{A}_n) \rangle \\
&= \langle x_1\mathbf{A}_1 + \cdots + x_n\mathbf{A}_n, x_1\mathsf{T}(\mathbf{A}_1) + \cdots + x_n\mathsf{T}(\mathbf{A}_n) \rangle \\
&= \langle x_1\mathbf{A}_1 + \cdots + x_n\mathbf{A}_n, x_1\lambda_1\mathbf{A}_1 + \cdots + x_n\lambda_n\mathbf{A}_n \rangle \\
&= x_1\lambda_1 x_1 + x_2\lambda_2 x_2 + \cdots + x_n\lambda_n x_n \\
&= \lambda_1 x_1^2 + \lambda_2 x_2^2 + \cdots + \lambda_n x_n^2,
\end{aligned}$$

as required. □

Note that the proof of (16.10) is itself important as it provides a method for finding the basis $\mathbf{A}_1, \ldots, \mathbf{A}_n$ whose elements are called the *principal axes* of the quadratic form φ.

Example 10. Find the principal axes of the quadratic function

$$\varphi : \mathbb{R}^3 \to \mathbb{R}$$

given by

$$\varphi(x, y, z) = 3x^2 + 2y^2 + 3z^2 - 2xy - 2yz.$$

Solution. A little experimentation will show that

$$\varphi(x, y, z) = \langle (x, y, z), \mathsf{T}(x, y, z) \rangle$$

where $\mathsf{T} : \mathbb{R}^3 \to \mathbb{R}^3$ is the self-adjoint linear transformation whose matrix is (do you see the gimmick?)

$$\begin{pmatrix} 3 & -1 & 0 \\ -1 & 2 & -1 \\ 0 & -1 & 3 \end{pmatrix}$$

relative to the standard orthonormal basis of $\mathbb{R}^3$. The characteristic polynomial of T is

$$\begin{aligned}
\Delta(t) &= \det \begin{pmatrix} 3-t & -1 & 0 \\ -1 & 2-t & -1 \\ 0 & -1 & 3-t \end{pmatrix} \\
&= (3-t)\det\begin{pmatrix} 2-t & -1 \\ -1 & 3-t \end{pmatrix} + \det\begin{pmatrix} -1 & -1 \\ 0 & 3-t \end{pmatrix} \\
&= (3-t)[(2-t)(3-t) - 1] - (3-t) \\
&= (3-t)[6 - 5t + t^2 - 1 - 1] \\
&= (3-t)(t^2 - 5t + 4) \\
&= -(t-3)(t-4)(t-1),
\end{aligned}$$

so the eigenvalues of T are 1, 3, and 4. Corresponding eigenvectors may be readily found. We get:

$$t = 1 \quad \begin{cases} 2x - y = 0 \\ -x + y - z = 0 \\ -y + 2z = 0 \end{cases}$$

which yields

$$\begin{aligned} y &= 2x \\ x - z &= 0 \\ -2x + 2z &= 0 \end{aligned}$$

so

$$\begin{aligned} y &= 2x \\ z &= x \end{aligned}$$

which has a nontrivial solution

$$x = 1, y = 2, z = 1$$

and a unit eigenvector is thus $\mathbf{A}_1 = (1/\sqrt{6}, 2/\sqrt{6}, 1/\sqrt{6})$.

Likewise we find $\mathbf{A}_2 = (1/\sqrt{2}, 0, -1/\sqrt{2})$ is a unit eigenvector corresponding to $t = 3$ and $\mathbf{A}_3 = (1/\sqrt{3}, -1/\sqrt{3}, 1/\sqrt{3})$ is an eigenvector corresponding to $t = 4$.

Thus the principal axes of φ are

$$\mathbf{A}_1 = \left(\frac{\sqrt{6}}{6}, \frac{\sqrt{6}}{3}, \frac{\sqrt{6}}{6}\right)$$

$$\mathbf{A}_2 = \left(\frac{\sqrt{2}}{2}, 0, -\frac{\sqrt{2}}{2}\right)$$

$$\mathbf{A}_3 = \left(\frac{\sqrt{3}}{3}, -\frac{\sqrt{3}}{3}, \frac{\sqrt{3}}{3}\right)$$

and if (u, v, w) denotes the coordinates of a vector $\mathbf{X}$ with respect to the principal axes

$$\varphi(\mathbf{X}) = u^2 + 3v^2 + 4w^2.$$

Example 11. Sketch the graph of the quadratic equation

$$2x^2 + 3xy - 2y^2 = 10.$$

Solution. Consider the function

$$\varphi(x, y) = 2x^2 + 3xy - 2y^2$$

Of course $\varphi : \mathbb{R}^2 \to \mathbb{R}$ is a quadratic function and

$$\varphi(x, y) = \langle (x, y), \mathsf{T}(x, y) \rangle$$

where $\mathsf{T} : \mathbb{R}^2 \to \mathbb{R}^2$ is the self-adjoint transformation whose matrix in the standard orthonormal basis of $\mathbb{R}^2$ is

$$\begin{pmatrix} 2 & \frac{3}{2} \\ \frac{3}{2} & -2 \end{pmatrix}.$$

The characteristic polynomial of T is

$$\begin{aligned}\Delta(t) &= \det\begin{pmatrix} 2 - t & \frac{3}{2} \\ \frac{3}{2} & -2 - t \end{pmatrix} = (2 - t)(-2 - t) - \tfrac{9}{4} \\ &= t^2 - 4 - \tfrac{9}{4} \\ &= t^2 - \tfrac{25}{4}.\end{aligned}$$

The eigenvalues of T are therefore $\pm 5/2$. Corresponding unit eigenvectors are easily found to be

$$\left(\frac{3\sqrt{10}}{10}, \frac{\sqrt{10}}{10}\right) = \mathbf{F}_+ \quad \text{corresponding to } \tfrac{5}{2}$$

$$\left(\frac{-\sqrt{10}}{10}, \frac{3\sqrt{10}}{10}\right) = \mathbf{F}_- \quad \text{corresponding to } -\tfrac{5}{2}.$$

If we write (u, v) for the coordinates of a vector $\mathbf{A}$ in $\mathbb{R}^2$ with respect to the orthonormal basis $\{\mathbf{F}_+, \mathbf{F}_-\}$ we find

$$\varphi(\mathbf{A}) = \frac{5u^2}{2} - \frac{5v^2}{2}.$$

We seek the graph of the equation

$$\varphi(\mathbf{A}) = 10.$$

But if we employ the (u, v) coordinates determined by the principal axes of T then this is just the graph of the equation

$$\frac{5u^2}{2} - \frac{5v^2}{2} = 10$$

or

$$u^2 - v^2 = 4.$$

One recognizes this straight off as a hyperbola (see Figure 16.3). The angle θ between the principal axes of θ and the usual axes is given by

$$\cos\theta = \langle \mathbf{E}_1, \mathbf{F}_+ \rangle = \frac{3\sqrt{10}}{10}$$

which is about 18.5°.

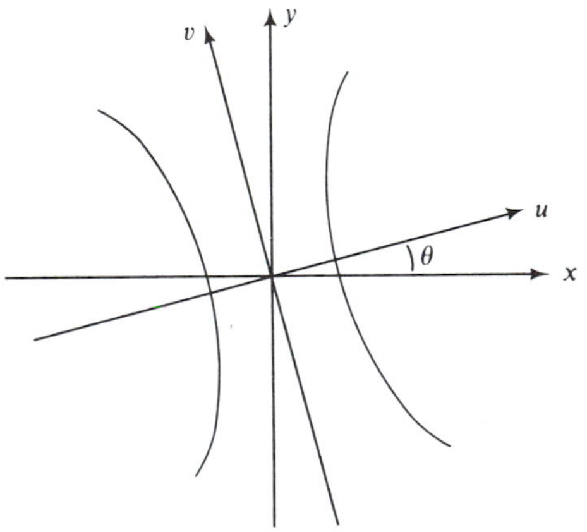

Figure 16.3

The coordinates of the vertices of the hyperbola are

$$(u, v) = (\pm 2, 0)$$

or

$$(x, y) = \left(\frac{3\sqrt{10}}{5}, \frac{\sqrt{10}}{5}\right).$$

Etc.

As a final example we wish to indicate that the application of the principal axis theorem to the study of conic sections is not limited to the homogeneous case.

EXAMPLE 12. Sketch the graph of the equation

$$16x^2 - 24xy + 9y^2 - 30x - 40y = 0.$$

Solution. Let us first find the principal axes of the homogeneous quadratic function $\varphi : \mathbb{R}^2 \to \mathbb{R}^2$ given by

$$\varphi(x, y) = 16x^2 - 24xy + 9y^2.$$

We begin by noting that

$$\varphi(x, y) = \langle (x, y), \mathsf{T}(x, y) \rangle$$

where $\mathsf{T} : \mathbb{R}^2 \to \mathbb{R}^2$ is the self-adjoint transformation with matrix

$$\begin{pmatrix} 16 & -12 \\ -12 & 9 \end{pmatrix}$$

relative to the standard orthonormal basis of $\mathbb{R}^2$. The characteristic polynomial of T is

$$\begin{aligned} \Delta(t) &= \det \begin{pmatrix} 16 - t & -12 \\ -12 & 9 - t \end{pmatrix} = (9 - t)(16 - t) - 144 \\ &= t^2 - 25t + 144 - 144 \\ &= t^2 - 25t. \end{aligned}$$

The eigenvalues of T are therefore $t = 0, t = 25$. Corresponding eigenvectors are $\mathbf{F}_1 = (\frac{3}{5}, \frac{4}{5})$ and $\mathbf{F}_2 = (-\frac{4}{5}, \frac{3}{5})$. If we let (u, v) denote the coordinates of a vector in the $\{\mathbf{F}_1, \mathbf{F}_2\}$ basis then

$$\begin{aligned} x\mathbf{E}_1 + y\mathbf{E}_2 = \mathbf{A} = u\mathbf{F}_1 + v\mathbf{F}_2 &= u\left(\frac{3}{5}\mathbf{E}_1 + \frac{4}{5}\mathbf{E}_2\right) + v\left(\frac{-4}{5}\mathbf{E}_1 + \frac{3}{5}\mathbf{E}_2\right) \\ &= \left(\frac{3u}{5} - \frac{4v}{5}\right)\mathbf{E}_1 + \left(\frac{4u}{5} + \frac{3v}{5}\right)\mathbf{E}_2. \end{aligned}$$

That is, the relation between the (x, y) and (u, v) coordinate systems is

$$x = \frac{3u}{5} - \frac{4v}{5}$$

$$y = \frac{4u}{5} + \frac{3v}{5}.$$

To graph the equation

$$16x - 24xy - 9y - 30x - 40y = 0$$

is therefore the same as to graph the equation

$$\varphi(x, y) - 30x - 40y = 0$$

which is the same as

$$\varphi(u, v) - 30\left(\frac{3u}{5} - \frac{4v}{5}\right) - 40\left(\frac{4u}{5} + \frac{3v}{5}\right) = 0$$

which is the same as

$$0u^2 + 25v^2 - 18u + 24v - 32u - 24v = 0$$

or

$$25v^2 = 50u$$

or

$$v^2 = 2u$$

which is instantly seen to be a parabola (see Figure 16.4). The angle θ between the (u, v) and (x, y) coordinate systems is given by

$$\cos\theta = \langle \mathbf{E}_1, \mathbf{F}_1 \rangle = \tfrac{3}{5}$$

which is about 53.1°.

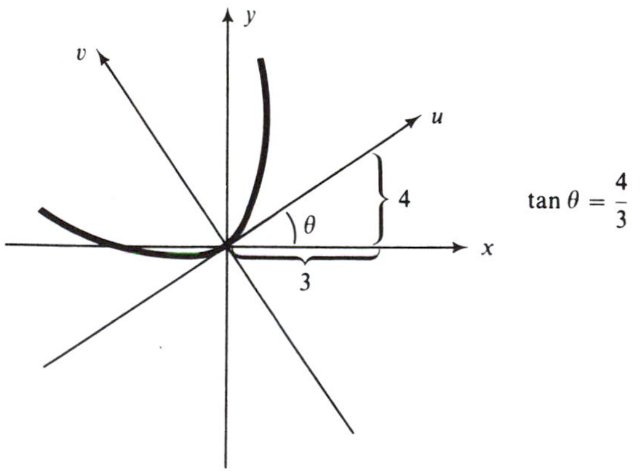

Figure 16.4

Exercises

1. Compute the adjoint of each of the following

(a) $\mathsf{T}:\mathbb{R}^3 \to \mathbb{R}^3$ $\quad \mathsf{T}(x, y, z) = (x + y + z, x + 2y + 2z, x + 2y + 3z)$

(b) $\mathsf{T}:\mathbb{R}^3 \to \mathbb{R}^3$ $\quad \mathsf{T}(x, y, z) = (x + y - z, -x + 2y + 2z, x + 2y + 3z)$

(c) $\mathsf{T}:\mathscr{P}_2(\mathbb{R}) \to \mathscr{P}_2(\mathbb{R})$ $\quad \mathsf{T}(p(x)) = x\dfrac{d}{dx}p(x) - \dfrac{d}{dx}(xp(x)).$

(d) $\mathsf{T}:\mathbb{R}^4 \to \mathbb{R}^4$ $\quad \mathsf{T}(x, y, z, w) = (x + y, y + z, z + w, w + x).$

(e) $\mathsf{T}:\mathscr{M}_{33} \to \mathscr{M}_{33}$ $\quad \mathsf{T}(\mathbf{A}) = \mathbf{A}^t + \mathbf{A}$

(f) $\mathsf{T}:\mathbb{R}^3 \to \mathbb{R}^3$ $\quad \mathsf{T}(x, y, z) = (-y + z, -x + 2z, x + 2y)$

2. Let $\mathsf{S}, \mathsf{T}:\mathscr{V} \to \mathscr{V}$. Prove:

(a) $(\mathsf{S} + \mathsf{T})^* = \mathsf{S}^* + \mathsf{T}^*$.
(b) $(a\mathsf{S})^* = a\mathsf{S}^*$.
(c) $(\mathsf{ST})^* = \mathsf{T}^*\mathsf{S}^*$.
(d) $\mathsf{T} + \mathsf{T}^*$ is self-adjoint.

3. Diagonalize each of the following self-adjoint transformations

(a) $\mathsf{T}:\mathbb{R}^2 \to \mathbb{R}^2$ $\quad \mathsf{T}(x, y) = (2x + y, 2y + x)$
(b) $\mathsf{T}:\mathbb{R}^3 \to \mathbb{R}^3$ $\quad \mathsf{T}(x, y, z) = (2x + y + z, x + 2y + z, x + y + 2z)$
(c) $\mathsf{T}:\mathbb{R}^5 \to \mathbb{R}^5$ $\quad \mathsf{T}(x, y, z, u, v) = (2x + y, 2y + x, 2z + u + v, z + 2u + v, z + u + 2v)$
(d) $\mathsf{T}:\mathbb{R}^3 \to \mathbb{R}^3$ $\quad \mathsf{T}(x, y, z) = (y + z, x + z, x + y)$
(e) $\mathsf{T}:\mathscr{M}_{33} \to \mathscr{M}_{33}$ $\quad \mathsf{T}(\mathbf{A}) = \mathbf{A}^t$
(f) $\mathsf{T}:\mathscr{M}_{33} \to \mathscr{M}_{33}$ $\quad \mathsf{T}(\mathbf{A}) = \mathbf{A} - \mathbf{A}^t$

4. Sketch the Graph of each of the following

(a) $8x^2 - 12xy + 17y^2 = 80$
(b) $3x^2 + 2xy + 3y^2 = 4$
(c) $5x^2 - 8xy + 5y^2 = 9$
(d) $11x^2 - 24xy + 4y^2 + 6x + 8y = -15$
(e) $16x^2 - 24xy + 9y^2 - 30x + 40y = 5$

5. Suppose that $\varphi:\mathscr{V} \to \mathbb{R}$ is a quadratic form. Show that for any two vectors $\mathbf{A}, \mathbf{B}$ in $\mathscr{V}$ that

$$\langle \mathbf{A}, \mathbf{B} \rangle = \tfrac{1}{2}(\varphi(\mathbf{A} + \mathbf{B}) - \varphi(\mathbf{A}) - \varphi(\mathbf{B}))$$

that is, the inner product in $\mathscr{V}$ may be computed from *any* quadratic form.

6. Suppose that $\varphi:\mathscr{V} \to \mathbb{R}$ is a quadratic form and

$$\varphi(A) = \langle \mathbf{A}, \mathsf{T}(\mathbf{A}) \rangle$$
$$\varphi(A) = \langle \mathbf{A}, \mathsf{S}(\mathbf{A}) \rangle$$

for two self-adjoint transformations $\mathsf{S}, \mathsf{T}: \mathscr{V} \to \mathscr{V}$. Show that $\mathsf{S} = \mathsf{T}$. (*Hint*: use (4) to show that $\langle \mathbf{B}, \mathsf{T}(\mathbf{A})\rangle = \langle \mathbf{B}, \mathsf{S}(\mathbf{A})\rangle$ for all $\mathbf{B}$ and $\mathbf{A}$ in $\mathscr{V}$. Then recall that $\langle \mathsf{S}(\mathbf{B}), \mathbf{A}\rangle = \langle \mathbf{B}, \mathsf{S}(\mathbf{A})\rangle$ for all $\mathbf{A}, \mathbf{B}$ in $\mathscr{V}$. Use this to show that $\mathsf{T} = \mathsf{S}^*$. Appeal to self-adjointness.)

7. Let $* : \mathscr{L}(\mathscr{V}) \to \mathscr{L}(\mathscr{V})$. Show that $*$ is a linear transformation and is self-adjoint.

The exercises that follow are intended to lead to a proof of the spectral theorem for arbitrary dimensions n. There are two basic steps in all the proofs that I know of. The first of these is to show that the transformation always has at least one eigenvector. Two different ways to obtain this are offered in Exercises 8 and 9. The second step is to show that the subspace orthogonal to an eigenvector is invariant. The proof can then be completed by induction. This is the purpose of Exercise 10.

8. Let $\mathsf{T}: \mathscr{V} \to \mathscr{V}$ be a self-adjoint linear transformation on the finite-dimensional inner product space $\mathscr{V}$. Let $\mathsf{S}(\mathscr{V}) = \{x \in \mathscr{V} \mid \|x\| = 1\}$ be the unit sphere of $\mathscr{V}$. **We assume from the multivariable calculus the fundamental theorem that a real-valued function on $\mathsf{S}(\mathscr{V})$ always has a minimum.** Define

$$f: \mathsf{S}(\mathscr{V}) \to \mathbb{R}$$

by the formula

$$f(x) = \langle \mathsf{T}(x), x\rangle$$

and let $v \in \mathsf{S}(\mathscr{V})$ be a vector at which f assumes its minimum value. Let $y \in \mathsf{S}(\mathscr{V})$ be a *fixed* vector orthogonal to v.

(a) Show that the vector $(v + ty)/\sqrt{1 + t^2}$ has unit length for all real numbers t. Define the function $g : \mathbb{R} \to \mathbb{R}$ by

$$g(t) = f\left(\frac{v + ty}{\sqrt{1 + t^2}}\right)$$

(b) Show that $g(0) = f(v)$ is the minimum value of $g(t)$.

(c) Show that

$$\frac{dg}{dt} = \frac{-2t}{(1 + t^2)^2}\langle \mathsf{T}(v + ty), v + ty\rangle.$$

(*Hint*: Write $g(t) = \langle \mathsf{T}((v + ty)/\sqrt{1 + t^2}), (v + ty)/\sqrt{1 + t^2}\rangle$ and expand the inner product remembering v and y are fixed, have unit length, and are perpendicular.)

(d) Evaluate

$$\left.\frac{dg}{dt}\right|_{t=0}$$

(Answer: $2\langle \mathsf{T}(y), v\rangle$).

(e) Show that $\mathsf{T}(v)$ is perpendicular to y. (*Hint*: Use (d), self-adjointness and the fact that $t = 0$ is a critical point of g.)

(f) Show $\mathsf{T}(v) = \lambda v$ for some number λ, that is, v is an eigenvector of T. (*Hint*: Use (e) to show $\mathsf{T}(v)$ is perpendicular to the orthogonal complement of $\mathscr{L}(v)$, and so $\mathsf{T}(v)$ belongs to $\mathscr{L}(v)$.)

9. Let $\mathsf{T} : \mathscr{V} \to \mathscr{V}$ be a self-adjoint linear transformation on the finite-dimensional inner product space $\mathscr{V}$. Let $\{e_1, \ldots, e_n\}$ be an orthonormal basis for $\mathscr{V}$ and $\mathbf{A}$ the matrix of T relative to $\{e_1, \ldots, e_n\}$. Introduce the linear transformation $\mathsf{T}_{\mathbb{C}} : \mathbb{C}^n \to \mathbb{C}^n$ whose matrix with respect to the standard basis is $\mathbf{A}$.

(a) Show the characteristic polynomial of $\mathsf{T}_{\mathbb{C}}$ is equal to the characteristic polynomial of T.

The fundamental theorem of algebra tells us that $\Delta_{\mathsf{T}_{\mathbb{C}}}(x)$ must have a root, which may, however, be complex. We want to show that it is real so that $\Delta_{\mathsf{T}}(x)$ has a real root, and hence T an eigenvalue. To do this let λ be a root of $\Delta_{\mathsf{T}_{\mathbb{C}}}(x)$, $w \in \mathbb{C}^n$ a corresponding eigenvector, and $\bar{w}$ the vector whose coordinates are the complex conjugates of those of w.
(b) Show that $\mathsf{T}_{\mathbb{C}}(\bar{w}) = \bar{\lambda}\bar{w}$ where $\bar{\lambda}$ is the complex conjugate of λ. (*Hint*: Compute the complex conjugate of the matrix product $\mathbf{A}w$ and use the fact that the entries of $\mathbf{A}$ are real.)
(c) Show that the number $\bar{w} \cdot w^t$ is nonzero ($\cdot$ denotes the matrix product of the row vector $\bar{w}$ with the column vector w^t).
(d) Show $\bar{w}\mathbf{A}w^t = \lambda(\bar{w} \cdot w^t)$, $w\mathbf{A}\bar{w}^t = \bar{\lambda}(w \cdot \bar{w}^t)$.
(e) Show that λ is real. (*Hint*: Use (d) and take transposes to show $\bar{w}\mathbf{A}w^t = w\mathbf{A}\bar{w}^t$ because $\mathbf{A}$ is symmetric.)
(f) Show that the characteristic polynomial of T has a real root.

10. Let $\mathsf{T} : \mathscr{V} \to \mathscr{V}$ be a self-adjoint linear transformation on the finite-dimensional inner product space $\mathscr{V}$. Using either (8) or (9) we know that T has at least one eigenvector. Let v be such an eigenvector with corresponding eigenvalue λ.

(a) If $y \in \mathscr{V}$ is orthogonal to v then $\mathsf{T}(y)$ is also orthogonal to v. (*Hint*: Use $\langle \mathsf{T}(y), v\rangle = \langle y, \mathsf{T}(v)\rangle$.)
(b) Let $\mathscr{W}$ be the subspace of $\mathscr{V}$ orthogonal to $\mathscr{L}(v)$. Show $\mathsf{T}(\mathscr{W}) \subset \mathscr{W}$.
(c) Let $\mathsf{T}|_{\mathscr{W}} : \mathscr{W} \to \mathscr{W}$ be the restriction of T to $\mathscr{W}$. Show $\mathsf{T}|_{\mathscr{W}}$ is self-adjoint.
(d) Show that there is an orthonormal basis $\{v, w_1, \ldots, w_m\}$ for $\mathscr{V}$ with respect to which the matrix of T is

$$\begin{pmatrix} \lambda & 0 & \cdots & 0 \\ 0 & & & \\ \vdots & & \mathbf{B} & \\ 0 & & & \end{pmatrix}$$

where $\mathbf{B}$ is symmetric.
(e) Use Exercise (8f) or (9f), (10d) and induction to prove the Spectral theorem (Theorem 16.7).

11. (a) Let $\mathsf{T} : \mathbb{R}^3 \to \mathbb{R}^3$ be the linear transformation whose matrix with respect to the standard orthonormal basis of $\mathbb{R}^3$ is

$$\begin{pmatrix} 0 & 1 & 1 \\ 1 & 0 & 1 \\ 1 & 1 & 0 \end{pmatrix}$$

Diagonalize T.

(b) Same problem for each of the following matrices

$$\begin{pmatrix} 1 & 1 & 1 \\ 1 & -1 & 1 \\ 1 & 1 & -1 \end{pmatrix}, \quad \begin{pmatrix} 1 & 0 & 1 \\ 0 & 1 & 0 \\ 1 & 0 & 1 \end{pmatrix}.$$

$$\begin{pmatrix} 0 & 1 & 1 \\ 1 & 0 & -1 \\ 1 & -1 & 0 \end{pmatrix}, \quad \begin{pmatrix} 0 & 1 & 0 \\ 1 & 0 & -1 \\ 0 & -1 & 0 \end{pmatrix}.$$

12. Let $\mathsf{T}: \mathbb{C}^2 \to \mathbb{C}^2$ be the linear transformation given by

$$\mathsf{T}(a, b) = (2ia + b, a),$$

where a, b are complex numbers.

(a) Find the matrix of T with respect to the standard basis of $\mathbb{C}^2$.
(b) Show the matrix of T is symmetric.
(c) Find the eigenvalues of T.
(d) Show that T is *not* diagonalizable.
(e) What do you conclude about the validity of the following statement? Symmetric matrices are diagonalizable. (*Hint*: Show this is true if the entries of the matrix are real, and can be false if the entries are allowed to be complex.)

13. Suppose $\mathbf{A} \in \mathscr{M}_{nn}$ and that the columns of $\mathbf{A}$ regarded as vectors in $\mathbb{R}^n$ are orthogonal. Show that regarded as vectors in $\mathbb{R}^n$ the rows are also orthogonal. (*Hint*: Think about the relation between matrices and linear transformations.)

Jordan canonical form 17

This chapter is of a distinctly more difficult nature than the preceding ones. In it we will treat the problem of finding a particular canonical, simple, matrix, representative of a linear transformation. In the preceding chapter we treated this problem for *self-adjoint* linear transformations in a finite-dimensional inner product space. For a self-adjoint linear transformation

$$\mathsf{T}: \mathscr{V} \to \mathscr{V}$$

in the finite-dimensional inner product space $\mathscr{V}$ we saw that we could always find an orthonormal basis for $\mathscr{V}$ so that the corresponding matrix of T was diagonal. The required basis would be composed of the normed eigenvectors of T and the diagonal entries of the corresponding matrix are the eigenvalues of T. The problem we now consider is: Can we achieve a similar *normal form* in general? An answer to this question leads to the Jordan normal form. Before we arrive at this goal we will require quite a few preparatory steps.

To begin with, let us consider where we might encounter difficulties in case we were to proceed by analogy with the case of self-adjoint linear transformations. What we achieved in the self-adjoint case is a decomposition

$$\mathscr{V} = \mathscr{U}_1 \oplus \cdots \oplus \mathscr{U}_q$$

of the vector space $\mathscr{V}$ into a direct sum of eigenspaces $\mathscr{U}_i$, $1 \leq i \leq q$, of T. For a general linear transformation

$$\mathsf{T}: \mathscr{V} \to \mathscr{V},$$

there are two problems that arise to prevent an analogous result.

(1) T may not have any eigenvalues at all.
(2) T has eigenvalues, but the sum of the eigenspaces $\mathscr{U}_1 + \cdots + \mathscr{U}_q$ is not all of $\mathscr{V}$.

The first problem is not so serious. We can avoid it by working with vector spaces over the *complex numbers*. (The advantage so gained has already been indicated in the exercises to Chapter 16.) On the other hand the second problem is more serious. It reflects more of a defect in the linear transformation T rather than in that of the scalars. Here are some examples to illustrate this point.

Note. A direct sum decomposition of a vector space with respect to subspaces will be defined, explained, and illustrated in what follows.

EXAMPLE 1. We will consider three linear transformations from $\mathbb{R}^3$ to itself, viz.

$$\mathsf{T}_i\colon \mathbb{R}^3 \to \mathbb{R}^3, \qquad i = 1, 2, 3;$$

each one of which has the real number e as eigenvalue with multiplicity 3, and furthermore, the corresponding eigenspace $\mathscr{V}_e$ will have dimensions 1, 2, and 3, respectively.

(1) $\mathsf{T}_1\colon \mathbb{R}^3 \to \mathbb{R}^3$ is the linear transformation represented by the matrix

$$\mathbf{A}_1 = \begin{pmatrix} e & 0 & 0 \\ 0 & e & 0 \\ 0 & 0 & e \end{pmatrix},$$

with respect to the standard basis of $\mathbb{R}^3$. Then e is an eigenvalue of T_1 with the geometric multiplicity 3 and the standard basis vectors are eigenvectors of T_1.

(2) $\mathsf{T}_2\colon \mathbb{R}^3 \to \mathbb{R}^3$ is the linear transformation represented by the matrix

$$\mathbf{A}_2 = \begin{pmatrix} e & 1 & 0 \\ 0 & e & 0 \\ 0 & 0 & e \end{pmatrix},$$

with respect tc the standard basis of $\mathbb{R}^3$. The characteristic polynomial of T_2 is

$$p(t) = (t - e)^3,$$

so e is an eigenvalue from algebraic multiplicity 3.

But there are only two linearly independent eigenvectors belonging to this eigenvalue. To see this remember the eigenvectors $\mathbf{E}$ are the nonzero solutions of

$$(*) \qquad (\mathsf{T}_2 - e\mathsf{I})(\mathbf{E}) = 0.$$

The vectors $\mathbf{E} = (x, y, z) \in \mathbb{R}^3$ that satisfy $(*)$ are the solutions of the linear system

$$y = 0.$$

So $(1, 0, 0), (0, 0, 1)$ is a basis for $\mathscr{V}_e$ and e is of geometric multiplicity 2.

(3) $\mathsf{T}_3: \mathbb{R}^3 \to \mathbb{R}^3$ is the linear transformation represented by the matrix

$$\mathbf{A}_3 = \begin{pmatrix} e & 1 & 0 \\ 0 & e & 1 \\ 0 & 0 & e \end{pmatrix},$$

with respect to the standard basis of $\mathbb{R}^3$. The characteristic polynomial is again

$$p(t) = (t - e)^3,$$

so e is again an eigenvalue with algebraic multiplicity 3. To find the geometric dimension we must find a basis for

$$\operatorname{Ker}(\mathsf{T}_3 - e\mathsf{I}).$$

This leads to the system of linear equations

$$\begin{pmatrix} 0 & 1 & 0 \\ 0 & 0 & 1 \\ 0 & 0 & 0 \end{pmatrix}\begin{pmatrix} x \\ y \\ z \end{pmatrix} = \begin{pmatrix} 0 \\ 0 \\ 0 \end{pmatrix},$$

namely

$$y = 0,$$

$$z = 0,$$

so that $\mathscr{V}_e = \mathscr{L}((1, 0, 0))$ and e has only geometric dimension 1.

From this example we learn that there are linear transformations that simply do not have enough eigenvectors to be representable by a diagonal matrix. In order to represent such a linear transformation by as simple a matrix as possible, we must attempt to cleverly enlarge the eigenspaces whose dimension is smaller than the algebraic multiplicity of the corresponding eigenvalue. The enlargement should be made in such a way that the enlarged space admits a basis with respect to which the matrix of T is "practically" diagonal.

To achieve this program we are going to need a number of new ideas.

Definition. Let $\mathsf{T}: \mathscr{V} \to \mathscr{V}$ be a linear transformation. A vector subspace $\mathscr{U}$ in $\mathscr{V}$ is called T-*invariant* iff $\mathsf{T}(\mathscr{U}) \subseteq \mathscr{U}$.

EXAMPLE 2. If $\mathsf{T}: \mathscr{V} \to \mathscr{V}$ has an eigenvalue e then the eigenspace $\mathscr{V}_e$ is T-invariant. To see this note that $\mathbf{E} \in \mathscr{V}_e$, then $\mathsf{T}(\mathbf{E}) = e\mathbf{E}$ is a multiple of $\mathbf{E}$ and so belongs to $\mathscr{V}_e$ since $\mathscr{V}_e$ is a subspace.

EXAMPLE 3. Let $\mathsf{T}: \mathbb{R}^3 \to \mathbb{R}^3$ be defined by the matrix (Example 1(2))

$$\mathbf{A} = \begin{pmatrix} e & 1 & 0 \\ 0 & e & 0 \\ 0 & 0 & e \end{pmatrix}.$$

Then $\mathscr{U} = \mathscr{L}((1, 0, 0), (0, 1, 0))$ is an invariant subspace. To see this note that

$$\mathsf{T}(1, 0, 0) = (e, 0, 0) = e(1, 0, 0) + 0(0, 1, 0),$$

$$\mathsf{T}(0, 1, 0) = (1, e, 0) = 1(1, 0, 0) + e(0, 1, 0),$$

so $\mathsf{T}(1, 0, 0), \mathsf{T}(0, 1, 0) \in \mathscr{U}$. By linearity of T this implies $\mathsf{T}(\mathscr{U}) \subseteq \mathscr{U}$, as $\mathscr{U}$ is invariant.

When there is an invariant subspace we can find a simple matrix representation.

Proposition 17.1. *Let* $\mathsf{T}: \mathscr{V} \to \mathscr{V}$ *be a linear transformation in the n-dimensional vector space* $\mathscr{V}$. *Suppose* $\mathscr{U}$ *is an m-dimensional* T*-invariant subspace of* $\mathscr{V}$. *Then with respect to a suitable basis for* $\mathscr{V}$, *T can be represented by a matrix in block form*

$$\underbrace{}_{}\left(\begin{array}{c|c} \mathbf{A} & \mathbf{C} \\ \hline 0 & \mathbf{B} \end{array}\right)\begin{array}{l} \}m \\ \}n - m \end{array} \qquad \text{columns: } \underbrace{\ }_{m} \ \underbrace{\ }_{n-m}$$

PROOF. Choose a basis $\{\mathbf{A}_1, \ldots, \mathbf{A}_m\}$ for $\mathscr{U}$ and extend this to a basis

$$\{\mathbf{A}_1, \ldots, \mathbf{A}_m, \mathbf{B}_1, \ldots, \mathbf{B}_{n-m}\}$$

for $\mathscr{V}$. Then because $\mathsf{T}\mathbf{A}_i \in \mathscr{U} = \mathscr{L}(\mathbf{A}_1, \ldots, \mathbf{A}_m)$ we have

$$\mathsf{T}(\mathbf{A}_i) = \sum_{j=1}^{m} a_{ji}\mathbf{A}_j,$$

so the first m columns of the matrix of T with respect to this basis have no entries in rows $m + 1$ through n. This says the matrix is of the above block form. □

Definition. Let $\mathscr{V}$ be a vector space and let $\mathscr{S}, \mathscr{T}$ be subspaces of $\mathscr{V}$. We say $\mathscr{V}$ is the *direct sum* of $\mathscr{S}$ and $\mathscr{T}$ iff

$$\mathscr{V} = \mathscr{S} + \mathscr{T} \quad \text{and} \quad \mathscr{S} \cap \mathscr{T} = \{0\}.$$

In this case we write $\mathscr{V} = \mathscr{S} \oplus \mathscr{T}$.

Proposition 17.2. *Let* $\mathscr{V}$ *be a vector space and let* $\mathscr{S}, \mathscr{T}$ *be subspaces of* $\mathscr{V}$. *Then* $\mathscr{V} = \mathscr{S} \oplus \mathscr{T}$ *iff every vector in* $\mathscr{V}$ *can be written* uniquely *in the form*

$$\mathbf{A} = \mathbf{S} + \mathbf{T}, \qquad \mathbf{S} \in \mathscr{S}, \quad \mathbf{T} \in \mathscr{T}.$$

PROOF. Suppose $\mathscr{V} = \mathscr{S} \oplus \mathscr{T}$. Let $\mathbf{A} \in \mathscr{V}$. Since $\mathscr{V} = \mathscr{S} + \mathscr{T}$. There exist vectors $\mathbf{S} \in \mathscr{S}, \mathbf{T} \in \mathscr{T}$ such that

$$(*) \qquad \mathbf{A} = \mathbf{S} + \mathbf{T}.$$

If there were vectors $\overline{\mathbf{S}} \in \mathscr{S}$, $\overline{\mathbf{T}} \in \mathscr{T}$ so that

$$\mathbf{A} = \overline{\mathbf{S}} + \overline{\mathbf{T}}$$

then

$$\mathbf{S} + \mathbf{T} = \mathbf{A} = \overline{\mathbf{S}} + \overline{\mathbf{T}}$$

implies

$$\mathbf{S} - \overline{\mathbf{S}} = \mathbf{T} - \overline{\mathbf{T}}.$$

So $\mathbf{S} - \overline{\mathbf{S}} \in \mathscr{T}$. But $\mathbf{S} - \overline{\mathbf{S}} \in \mathscr{S}$ since $\mathscr{S}$ is a subspace. Therefore

$$\mathbf{S} - \mathbf{S} \in \mathscr{S} \cap \mathscr{T} = \{0\},$$

so $\mathbf{S} - \overline{\mathbf{S}} = 0$ or $\mathbf{S} = \overline{\mathbf{S}}$. A similar argument shows $\mathbf{T} - \overline{\mathbf{T}} = 0$ or $\mathbf{T} = \overline{\mathbf{T}}$, so the representation of $\mathbf{A}$ in $(*)$ is unique.

The converse is immediate. □

Proposition 17.3. *Let $\mathscr{V}$ be a finite-dimensional vector space and let $\mathscr{S}$, $\mathscr{T}$ be subspaces of $\mathscr{V}$. If $\mathscr{V} = \mathscr{S} \oplus \mathscr{T}$ then*

$$\dim \mathscr{V} = \dim \mathscr{S} + \dim \mathscr{T}.$$

Moreover if $\{\mathbf{A}_1, \ldots, \mathbf{A}_s\}$ is a basis for $\mathscr{S}$ and $\{\mathbf{B}_1, \ldots, \mathbf{B}_t\}$ a basis for $\mathscr{T}$, then $\{\mathbf{A}_1, \ldots, \mathbf{A}_s, \mathbf{B}_1, \ldots, \mathbf{B}_t\}$ is a basis for $\mathscr{V}$.

PROOF. The proof is similar to (8.10). □

Proposition 17.4. *Let $\mathscr{V}$ be a finite-dimensional vector space, and let $\mathscr{S}$, $\mathscr{T}$ be subspaces of $\mathscr{V}$ such that $\mathscr{V} = \mathscr{S} + \mathscr{T}$. Then $\mathscr{V} = \mathscr{S} \oplus \mathscr{T}$ iff*

$$\dim \mathscr{V} = \dim \mathscr{S} + \dim \mathscr{T}$$

PROOF. Do Exercises 8 and 9 of Chapter 6. □

Corollary 17.5. *Let $\mathscr{V}\langle\ ,\ \rangle$ be an inner product space and let $\mathscr{W}$ be a subspace of $\mathscr{V}$. Then*

$$\mathscr{V} = \mathscr{W} \oplus \mathscr{W}^{\perp}.$$

PROOF. Follow from (15.16). □

EXAMPLE 4. In $\mathbb{R}^3$ consider the subspaces

$$\mathscr{S} = \mathscr{L}((1, -1, 0), (0, 1, -1)),$$
$$\mathscr{T} = \mathscr{L}((1, 1, 1)).$$

Then

$$\mathbb{R}^3 = \mathscr{S} \oplus \mathscr{T}.$$

If $\mathcal{R} = \mathcal{L}((1, 1, 1), (1, -2, 1))$. Then

$$\mathbb{R}^3 = \mathcal{R} + \mathcal{S}.$$

But $\mathbb{R}^3 \neq \mathcal{R} \oplus \mathcal{S}$ because

$$\mathcal{R} \cap \mathcal{S} \neq \{0\}.$$

In fact

$$1 \cdot (1, -1, 0) + (-1)(0, 1, -1) = (1, -2, 1)$$

shows

$$(1, -2, 1) \in \mathcal{R} \cap \mathcal{S}.$$

Definition. Let $\mathsf{T} : \mathcal{V} \to \mathcal{V}$ be a linear transformation. A T-invariant subspace $\mathcal{U}$ in $\mathcal{V}$ is called *completely invariant* or *fully invariant* iff there exists a T-invariant subspace in $\mathcal{V}$ such that $\mathcal{V} = \mathcal{U} \oplus \mathcal{W}$. The subspace $\mathcal{W}$ is called a T-*invariant complement* for $\mathcal{U}$.

Corollary 17.6. *Let* $\mathsf{T}: \mathcal{V} \to \mathcal{V}$ *be a linear transformation in the n-dimensional vector space* $\mathcal{V}$. *Suppose* $\mathcal{U}$ *is a* T*-completely invariant subspace of dimension m in* $\mathcal{V}$. *Then with respect to a suitable basis for* $\mathcal{V}$, T *can be represented in block matrix form*

$$\begin{matrix} m\{ \\ n\{ \end{matrix}\left(\begin{array}{c|c} \mathbf{A} & 0 \\ \hline 0 & \mathbf{B} \end{array}\right)$$
$$\qquad\underbrace{}_{m}\ \underbrace{}_{n-m}$$

PROOF. Let $\mathcal{W}$ be a T-invariant complement for $\mathcal{U}$. Let $\{\mathbf{A}_1, \ldots, \mathbf{A}_m\}$ be a basis for $\mathcal{U}$ and $\{\mathbf{B}_1, \ldots, \mathbf{B}_{n-m}\}$ a basis for $\mathcal{W}$. Then $\{\mathbf{A}_1, \ldots, \mathbf{A}_m, \mathbf{B}_1, \ldots, \mathbf{B}_{n-m}\}$ is a basis for $\mathcal{V}$. The matrix of T with respect to this basis is of the required form. □

This corollary shows that one way to simplify a matrix representative for T is to find a complementary pair of T-invariant subspaces. But, how does one find such pairs? The following results explain how.

Proposition 17.7. *Suppose given linear transformations*

$$\mathsf{S}, \mathsf{T} : \mathcal{V} \to \mathcal{V}$$

such that

$$\mathsf{S} \circ \mathsf{T} = \mathsf{T} \circ \mathsf{S}$$

(*that is,* S *and* T commute *with each other*). *Then* Ker S *and* Im S *are* T*-invariant subspaces.*

PROOF. If $\mathbf{A} \in \operatorname{Ker} \mathsf{S}$, then

$$\mathsf{S} \circ \mathsf{T}(\mathbf{A}) = \mathsf{T} \circ \mathsf{S}(\mathbf{A}) = \mathsf{T}(0) = 0$$

shows that $\mathsf{T}\mathbf{A} \in \operatorname{Ker} \mathsf{S}$, so $\operatorname{Ker} \mathsf{S}$ is a T-invariant subspace. Similarly, if $\mathbf{B} \in \operatorname{Im} \mathsf{S}$, then $\mathbf{B} = \mathsf{S}(\mathbf{C})$ for some $\mathbf{C} \in \mathscr{V}$ and, therefore

$$\mathsf{T}(\mathbf{B}) = \mathsf{T} \circ \mathsf{S}(\mathbf{C}) = \mathsf{S} \circ (\mathsf{T}(\mathbf{C}))$$

shows that $\mathsf{T}(\mathbf{B}) \in \operatorname{Im} \mathsf{S}$, so $\operatorname{Im} \mathsf{S}$ is a T-invariant subspace. □

Corollary 17.8. *Let* $\mathsf{T}: \mathscr{V} \to \mathscr{V}$ *be a linear transformation and*

$$\mathsf{S} = a_0\mathsf{I} + a_1\mathsf{T} + \cdots + a_m\mathsf{T}^m: \mathscr{V} \to \mathscr{V},$$

where $a_0, \ldots, a_m \in \mathbb{C}$. *Then* $\operatorname{Ker} \mathsf{S}$ *and* $\operatorname{Im} \mathsf{S}$ *are* T*-invariant.*

PROOF. Clearly S and T commute, so we may apply (17.7).

The next step in our program, the key that unlocks the door to the Jordan canonical form, is the generalization of (12bis.3) on the structure of nilpotent transformations with a cyclic vector to completely general nilpotent transformations.

Theorem 17.9. *Let* $\mathscr{V}$ *be a finite-dimensional vector space and suppose that* $\mathsf{T}: \mathscr{V} \to \mathscr{V}$ *is a nilpotent linear transformation. Then there exists a direct sum decomposition*

$$\mathscr{V} = \mathscr{U}_1 \oplus \cdots \oplus \mathscr{U}_t$$

such that $\mathscr{U}_i$ *is* T*-invariant and the induced map* $\mathsf{T}: \mathscr{U}_i \to \mathscr{U}_i$ *has a cyclic vector for* $i = 1, \ldots, t$.

PROOF. Let $p = \operatorname{Index}(\mathsf{T})$, that is, p is the smallest integer such that $\mathsf{T}^p = 0$. Define

$$H_i = \operatorname{Ker} \mathsf{T}^i, \qquad i = 0, 1, \ldots, p.$$

Then the H_i are vector subspaces of $\mathscr{V}$. Notice that $H_0 = \{0\}$ and $H_p = \mathscr{V}$. Moreover, if $0 \le i \le p$ then we have

$$\mathbf{A} \in H_i \quad \Rightarrow \quad \mathsf{T}^i\mathbf{A} = 0 \quad \Rightarrow \quad \mathsf{T}^{i+1}(\mathbf{A}) = \mathsf{T}(\mathsf{T}^i(\mathbf{A})) = \mathsf{T}(0) = 0,$$

so $\mathbf{A} \in H_{i+1}$. Thus we have a chain of subspaces

$$\{0\} = H_0 \subset H_1 \subset \cdots \subset H_p = \mathscr{V},$$

and $\mathsf{T}(H_i) \subset H_{i-1}$.

Remark. In the case that T has a cyclic vector $\mathbf{A}$ then

$$H_i = \mathscr{L}(\mathsf{T}^{p-i}(\mathbf{A}), \ldots, \mathsf{T}^{p-1}(\mathbf{A}))$$

for $i = 1, \ldots, p$ and $H_0 = \{0\}$.

We proceed now to construct the invariant subspaces $\mathscr{U}_1, \ldots, \mathscr{U}_t$ of $\mathscr{V}$. To do this we need a new idea.

Definition. Let $\mathscr{V}$ be a vector space and $\mathscr{W}$ a subspace. A set of vectors $\mathbf{E}$ is said to be *linearly dependent over* $\mathscr{W}$ or *modulo* $\mathscr{W}$ iff there are distinct vectors $\mathbf{A}_1, \ldots, \mathbf{A}_k$ in $\mathbf{E}$ and numbers $a_1, \ldots, a_k$, not all zero such that

$$(*) \qquad a_1\mathbf{A}_1 + \cdots + a_k\mathbf{A}_k \in \mathscr{W}.$$

A set of vectors $\mathbf{E}$ that is not linearly dependent over $\mathscr{W}$ is said to be *linearly independent over* $\mathscr{W}$.

Proposition A. *Let $\mathscr{V}$ be a vector space and $\mathscr{W}$ a finite-dimensional subspace of $\mathscr{V}$. A set of vectors $\mathbf{E}$ is linearly independent over $\mathscr{W}$ iff for every basis $\{\mathbf{B}_1, \ldots, \mathbf{B}_s\}$ of $\mathscr{W}$, the set of vectors $\mathbf{E} \cup \{\mathbf{B}_1, \ldots, \mathbf{B}_s\}$ is linearly independent in $\mathscr{V}$.*

PROOF. Suppose that $\mathbf{E}$ is linearly independent over $\mathscr{W}$ and $\{\mathbf{B}_1, \ldots, \mathbf{B}_s\}$ is a basis for $\mathscr{W}$. Let $\{\mathbf{A}_1, \ldots, \mathbf{A}_k\}$ be vectors of $\mathbf{E}$, $a_1, \ldots, a_k$, and $b_1, \ldots, b_s$ numbers, such that

$$a_1\mathbf{A}_1 + \cdots + a_k\mathbf{A}_k + b_1\mathbf{B}_1 + \cdots + b_s B_s = 0.$$

Then, of course

$$a_1\mathbf{A}_1 + \cdots + a_k\mathbf{A}_k = -(b_1\mathbf{B}_1 + \cdots + b_s\mathbf{B}_s) \in \mathscr{W},$$

so $a_1 = \cdots = a_k = 0$. But then

$$b_1\mathbf{B}_1 + \cdots + b_s\mathbf{B}_s = 0,$$

so $b_1 = \cdots = b_s = 0$ also. The converse is just as easy. □

Definition. Let $\mathscr{V}$ be a vector space and $\mathscr{W}$ a subspace of $\mathscr{V}$. A set of vectors $\mathbf{E}$ is said to *span* $\mathscr{V}$ *over* $\mathscr{W}$ or *modulo* $\mathscr{W}$ iff $\mathscr{V} = \mathscr{L}(\mathbf{E}) + \mathscr{W}$. If, in addition, $\mathbf{E}$ is linearly independent over $\mathscr{W}$, then we say that $\mathbf{E}$ is a *basis for* $\mathscr{V}$ *over* $\mathscr{W}$.

Proposition B. *If $\mathscr{V}$ is a finite-dimensional vector space and $\mathscr{W}$ is a subspace then there exists a basis for $\mathscr{V}$ over $\mathscr{W}$. The number of vectors in a basis for $\mathscr{V}$ over $\mathscr{W}$ is* $\dim \mathscr{V} - \dim \mathscr{W}$.

PROOF. To prove the existence of a basis for $\mathscr{V}$ over $\mathscr{W}$ is easy. Choose a basis $\{\mathbf{B}_1, \ldots, \mathbf{B}_s\}$ for $\mathscr{W}$. Extend this to a basis $\{\mathbf{A}_1, \ldots, \mathbf{A}_k, \mathbf{B}_1, \ldots, \mathbf{B}_s\}$ for $\mathscr{V}$. Then by Proposition A the set $\{\mathbf{A}_1, \ldots, \mathbf{A}_k\}$ is linearly independent over $\mathscr{W}$. Since

$$\begin{aligned}\mathscr{V} &= \mathscr{L}(\mathbf{A}_1, \ldots, \mathbf{A}_k, \mathbf{B}_1, \ldots, \mathbf{B}_s) = \mathscr{L}(\mathbf{A}_1, \ldots, \mathbf{A}_k) + \mathscr{L}(\mathbf{B}_1, \ldots, \mathbf{B}_s) \\ &= \mathscr{L}(\mathbf{A}_1, \ldots, \mathbf{A}_k) + \mathscr{W}\end{aligned}$$

it follows that $\{\mathbf{A}_1, \ldots, \mathbf{A}_k\}$ is a basis for $\mathscr{V}$ over $\mathscr{W}$. Since $s + k = \dim \mathscr{V}$ it follows that $k = \dim \mathscr{V} - \dim \mathscr{W}$. □

We return now to the proof of (17.7). We have the nested chain of subspaces

$$\{0\} = H_0 \subset H_1 \subset \cdots \subset H_p = \mathscr{V}.$$

Choose a basis $\{\mathbf{A}_1, \ldots, \mathbf{A}_r\}$ for $\mathscr{V} = H_p$ over H_{p-1}. Then

$$\mathsf{T}(\mathbf{A}_1), \ldots, \mathsf{T}(\mathbf{A}_r) \in H_{p-1}.$$

We claim that $\{\mathsf{T}(A_1), \ldots, \mathsf{T}(\mathbf{A}_r)\}$ are linearly independent over H_{p-2}. For suppose that there are numbers $a_1, \ldots, a_r$ such that

$$a_1\mathsf{T}(\mathbf{A}_1) + \cdots + a_r\mathsf{T}(\mathbf{A}_r) \in H_{p-2}.$$

Apply T^{p-2} to this equation. We get by definition of H_{p-2}

$$0 = \mathsf{T}^{p-2}(a_1\mathsf{T}(\mathbf{A}_1) + \cdots + a_r\mathsf{T}(\mathbf{A}_r)) = \mathsf{T}^{p-1}(a_1\mathbf{A}_1 + \cdots + a_r\mathbf{A}_r).$$

Therefore

$$a_1\mathbf{A}_1 + \cdots + a_r\mathbf{A}_r \in H_{p-1}.$$

But $\mathbf{A}_1, \ldots, \mathbf{A}_r$ is a basis for H_p over H_{p-1} and so $a_1 = \cdots = a_r = 0$ as required.

Extend $\{\mathsf{T}(\mathbf{A}_1, \ldots, \mathsf{T}(\mathbf{A}_r)\}$ to a basis $\{\mathsf{T}(\mathbf{A}_1), \ldots, \mathsf{T}(\mathbf{A}_r), \mathbf{A}_{r+1}, \ldots, \mathbf{A}_q\}$ for $\mathbf{H}_{p-1}$ over $\mathbf{H}_{p-2}$. Then

$$\mathsf{T}^2(\mathbf{A}_1), \ldots, \mathsf{T}^2(\mathbf{A}_r), \mathsf{T}(\mathbf{A}_{r+1}), \ldots, \mathsf{T}(\mathbf{A}_q) \in \mathbf{H}_{p-2},$$

and the argument we just gave may be applied again to show that $\{\mathsf{T}^2(\mathbf{A}_1), \ldots, \mathsf{T}^2(\mathbf{A}_r)\ \ \mathsf{T}(\mathbf{A}_{r+1}), \ldots, \mathsf{T}(\mathbf{A}_q)\}$ are linearly independent over H_{p-3}.

If we continue in this way we obtain after p steps a basis for $\mathscr{V}$ arranged as in the following scheme:

$$\begin{array}{l}
\mathbf{A}_1, \ldots, \mathbf{A}_r, \\
\mathsf{T}(\mathbf{A}_1), \ldots, \mathsf{T}(\mathbf{A}_r), \mathbf{A}_{r+1}, \ldots, \mathbf{A}_q, \\
\cdots\cdots\cdots\cdots\cdots\cdots\cdots\cdots\cdots\cdots \\
\mathsf{T}^{p-1}(\mathbf{A}_1), \ldots, \mathsf{T}^{p-1}(\mathbf{A}_r), \mathsf{T}^{p-2}(\mathbf{A}_{r+1}), \ldots, \mathbf{A}_m.
\end{array}$$

Notice that the vectors in each column span an invariant subspace for which the vector at the top of the column is a cyclic vector. □

Here is an example to help clarify what is going on in (17.9).

Example 5. Let $\mathsf{T}: \mathbb{R}^5 \to \mathbb{R}^5$ be the linear transformation given by

$$\mathsf{T}(x, y, u, v, w) = (0, u + v, 0, u, x + v).$$

Then

$$\begin{aligned}
\mathsf{T}^3(x, y, u, v, w) &= \mathsf{T}^2(0, u + v, 0, u, x + v) \\
&= \mathsf{T}(0, u, 0, 0, u) \\
&= (0, 0, 0, 0, 0)
\end{aligned}$$

so T is nilpotent of index 3. The matrix of T with respect to the standard basis of $\mathbb{R}^5$ is

$$\begin{pmatrix} 0 & 0 & 0 & 0 & 0 \\ 0 & 0 & 1 & 1 & 0 \\ 0 & 0 & 0 & 0 & 0 \\ 0 & 0 & 1 & 0 & 0 \\ 1 & 0 & 0 & 1 & 0 \end{pmatrix},$$

which despite the large number of zero entries is not very informative. The procedure of (17.9) provides a more suitable basis for $\mathbb{R}^5$. We first need the chain of subspaces

$$\{0\} = H_0 \subset H_1 \subset H_2 \subset H_3 = \mathbb{R}^5.$$

A basis for $H_2 = \operatorname{Ker} \mathsf{T}^2$ is given by

$$\mathbf{B}_1 = (1, 0, 0, 0, 0),$$
$$\mathbf{B}_2 = (0, 1, 0, 0, 0),$$
$$\mathbf{B}_3 = (0, 0, 0, 1, 0),$$
$$\mathbf{B}_4 = (0, 0, 0, 0, 1).$$

This can be extended to a basis for $\mathbb{R}^5$ by adjoining

$$\mathbf{A}_1 = (0, 0, 1, 0, 0).$$

Then $\mathsf{T}\mathbf{A}_1 \in H_2$, which together with the basis for $H_1 = \operatorname{Ker} \mathsf{t}$

$$\mathbf{C}_1 = (0, 1, 0, 0, 0),$$
$$\mathbf{C}_2 = (0, 0, 0, 0, 1),$$

fails by one vector to be a basis for H_2. So we add the vector

$$\mathbf{A}_2 = (1, 0, 0, 0, 0)$$

to obtain the basis

$$\mathbf{A}_1, \mathsf{T}\mathbf{A}_1, \mathbf{A}_2, \mathbf{C}_1, \mathbf{C}_2$$

for H_3. The vectors

$$\mathsf{T}^2\mathbf{A}_1, \mathsf{T}\mathbf{A}_2 \in H_1,$$

and are a basis for H_1. Here the process stops with the basis scheme

$$\begin{matrix} \mathbf{A}_1, & \\ \mathsf{T}\mathbf{A}_1, & \mathbf{A}_2, \\ \mathsf{T}^2\mathbf{A}_1, & \mathsf{T}\mathbf{A}_2, \end{matrix}$$

for $\mathbb{R}^5$. Set

$$\mathscr{U}_1 = \mathscr{L}(\mathbf{A}_1, \mathsf{T}\mathbf{A}_1, \mathsf{T}^2\mathbf{A}_1),$$

$$\mathscr{U}_2 = \mathscr{L}(\mathbf{A}_2, \mathsf{T}\mathbf{A}_2).$$

Then $\mathscr{U}_1$, $\mathscr{U}_2$ are complementary T-invariant subspaces of $\mathbb{R}^5$, so

$$\mathbb{R}^5 = \mathscr{U}_1 \oplus \mathscr{U}_2.$$

$\mathbf{A}_1 \in \mathscr{U}_1$ is a cyclic vector of index 3 and $\mathbf{A}_2 \in \mathscr{U}_2$ is a cyclic vector of index 2. The matrix of T with respect to the basis

$$\mathbf{A}_1, \mathsf{T}\mathbf{A}_1, \mathsf{T}^2\mathbf{A}_1, \mathbf{A}_2, \mathsf{T}\mathbf{A}_2$$

for $\mathbb{R}^5$ is

$$\left(\begin{array}{ccc:cc} 0 & 0 & 0 & 0 & 0 \\ 1 & 0 & 0 & 0 & 0 \\ 0 & 1 & 0 & 0 & 0 \\ \hdashline 0 & 0 & 0 & 0 & 0 \\ 0 & 0 & 0 & 1 & 0 \end{array}\right)$$

which is the block sum of two shift operators; one of index 3 and one of index 2.

The matrix interpretation of (17.9) is contained in the following corollary.

Corollary 17.10. *Let $\mathsf{T}: \mathscr{V} \to \mathscr{V}$ be a nilpotent linear transformation in the finite-dimensional vector space $\mathscr{V}$. Let p be the index of T. Then with respect to an appropriate basis for $\mathscr{V}$ the matrix of T is of the form*

$$\begin{pmatrix} \mathbf{J}_p & & & & & & \\ & \ddots & & & & \mathbf{0} & \\ & & \mathbf{J}_p & & & & \\ & & & \mathbf{J}_{p-1} & & & \\ & & & & \ddots & & \\ & & & & & \mathbf{J}_{p-1} & \\ & \mathbf{0} & & & & & \ddots \\ & & & & & & \mathbf{J}_1 \\ & & & & & & & \ddots \\ & & & & & & & & \mathbf{J}_1 \end{pmatrix},$$

where J_i occurs n_i times

$$\mathbf{J}_i = \underbrace{\begin{pmatrix} 0 & & & 0 \\ 1 & \ddots & & \\ & \ddots & \ddots & \\ 0 & \cdots & 1 & 0 \end{pmatrix}}_{i}$$

and $n_i \geq 0$, $i = 1, \ldots, p$. (*Note*: $n_p \neq 0$, *but it can happen that* $n_{p-1} = 0$, *for example.*)

Proof. By (17.9) we can find a direct sum decomposition

$$\mathscr{V} = \mathscr{U}_1 \oplus \cdots \oplus \mathscr{U}_h$$

of $\mathscr{V}$ where each $\mathscr{U}_i$ is T-invariant and the induced linear transformation

$$\mathsf{T}: \mathscr{U}_i \to \mathscr{U}_i$$

has a cyclic vector. If we choose a basis for $\mathscr{U}_i$ composed of $\mathbf{A}_i, \mathsf{T}\mathbf{A}_i, \ldots, \mathsf{T}^{p_i - 1}\mathbf{A}_i$ where $\mathsf{T}p^i(\mathbf{A}_i) = 0$, then the matrix of

$$\mathsf{T}: \mathscr{U}_i \to \mathscr{U}_i$$

with respect to this basis is

$$\begin{pmatrix} 0 & 0 & & & 0 \\ 1 & 0 & & & \\ & \ddots & 0 & & \\ & & \ddots & \ddots & \\ 0 & 0 & 0 & 1 & 0 \end{pmatrix}.$$

Assembling these blocks we obtain a matrix of the stated form representing T. □

Remark. In a finite-dimensional vector space $\mathscr{V}$ an ascending chain of subspaces

$$\{0\} = \mathscr{U}_0 \subseteq \mathscr{U}_1 \subseteq \cdots \subseteq \mathscr{U}_n \subseteq \cdots \mathscr{V}$$

must eventually stabilize; that is, there must be an integer m such that

$$\mathscr{U}_i = \mathscr{U}_m \quad \text{if} \quad i \geq m.$$

This is because

$$0 = \dim \mathscr{U}_0 \leq \dim \mathscr{U}_1 \leq \cdots \leq \dim \mathscr{U}_n \leq \cdots \leq \dim \mathscr{V},$$

and since $\mathscr{U}_j \subseteq \mathscr{U}_{j+1}$ $\dim \mathscr{U}_j = \dim \mathscr{U}_{j+1}$ iff $\mathscr{U}_j = \mathscr{U}_{j+1}$.

Similarly, a descending chain of subspaces

$$\mathscr{V} = \mathscr{W}_0 \supseteq \mathscr{W}_1 \supseteq \cdots \supseteq \mathscr{W}_n \supseteq \cdots \supseteq \{0\}$$

must also stabilize at a subspace $\mathscr{W}_l$ such that

$$W_j = W_l \quad \text{if} \quad j \geq l$$

as one easily sees from the inequality

$$\dim \mathscr{V} = \dim \mathscr{W}_0 \geq \dim \mathscr{W}_1 \geq \cdots \geq \dim \mathscr{W}_n \geq \cdots \geq 0.$$

This simple property of *finite-dimensional* vector spaces will play a central role in the construction of the Jordan normal form.

We want to obtain a direct sum decomposition of $\mathscr{V}$ into T-invariant subspaces for an arbitrary linear transformation

$$\mathsf{T}: \mathscr{V} \to \mathscr{V},$$

where $\mathscr{V}$ is finite dimensional. The idea is to *enlarge* the eigenspaces of T. But how? Here is the idea. Suppose that e is an eigenvalue of T. Then

$$\mathscr{V}_e = \operatorname{Ker}(\mathsf{T} - e\mathsf{I}).$$

Let us now look at

$$\mathscr{V}_{e,i} = \operatorname{Ker}(\mathsf{T} - e\mathsf{I})^i, \qquad i = 1, 2, \ldots .$$

Notice that these subspaces are nested one within the other, that is,

$$\mathscr{V}_e = \mathscr{V}_{e,1} \subseteq \mathscr{V}_{e,2} \subseteq \mathscr{V}_{e,3} \subseteq \cdots \subseteq \mathscr{V}.$$

If $\mathscr{V}$ is finite dimensional then this chain of subspaces cannot go on getting bigger forever, for that would lead to an infinite set of linearly independent vectors in $\mathscr{V}$, contradicting the finite dimensionality of $\mathscr{V}$. Therefore there must be an integer n (depending on T, e, etc.) such that

$$\mathscr{V}_{e,j} = \mathscr{V}_{e,n} \quad \text{for all} \quad j \geq n.$$

The vector space $\mathscr{V}_{e,n} = \operatorname{Ker}(\mathsf{T} - e\mathsf{I})^n$ at which the chain

$$\mathscr{V}_e = \mathscr{V}_{e,1} \subseteq \mathscr{V}_{e,2} \cdots \subseteq \mathscr{V}_{e,n} \subseteq \mathscr{V}$$

stops growing is T-invariant, since

$$(\mathsf{T} - e\mathsf{I})^n = \mathsf{T}^n - e\mathsf{T}^{n-1} + \cdots + (-1)^n e^n \mathsf{I}$$

is a polynomial in T and we may therefore apply (17.8). Let us introduce a name for $\mathscr{V}_{e,n}$.

Definition. Let $\mathsf{T}: \mathscr{V} \to \mathscr{V}$ be a linear transformation with eigenvalue e. Let

$$\mathscr{V}_{e,i} = \operatorname{Ker}(\mathsf{T} - e\mathsf{I})^i, \qquad i = 1, 2, \ldots .$$

Let n be an integer such that

$$\mathscr{V}_{e,n} = \mathscr{V}_{e,n+1} = \cdots,$$

that is

$$\mathscr{V}_{e,j} = \mathscr{V}_{e,n} \quad \text{if} \quad j \geq n.$$

The subspace $\mathscr{V}_{e,n}$ is called the *closure* of the eigenspace $\mathscr{V}_e$. The closure of $\mathscr{V}_e$ will be denoted by $\overline{\mathscr{V}}_e$. Note

$$\mathscr{V}_e = \mathscr{V}_{e,1} \subset \mathscr{V}_{e,2} \subset \cdots \subset \mathscr{V}_{e,n} = \overline{\mathscr{V}}_e$$

is a chain of subspaces of $\mathscr{V}$.

Remark. Suppose that somewhere in the chain

$$\mathscr{V}_e = \mathscr{V}_{e,1} \subseteq \mathscr{V}_{e,2} \subseteq \cdots \subseteq \overline{\mathscr{V}}_e,$$

we find $\mathscr{V}_{e,i} = \mathscr{V}_{e,i+1}$. Then, in fact

$$\mathscr{V}_{e,i} = \mathscr{V}_{e,i+1} = \mathscr{V}_{e,i+2} = \cdots = \overline{\mathscr{V}}_e,$$

that is, the first time two of the subspaces $\mathscr{V}_{e,i}$ agree, the chain stabilize; that is, nothing "new" happens after that point.

To see this suppose $\mathscr{V}_{e,i} = \mathscr{V}_{e,i+1}$. By definition $\mathscr{V}_{e,i+2} = \text{Ker}(\mathsf{T} - e\mathsf{I})^{i+2}$. Suppose $\mathbf{A} \in \mathscr{V}_{e,i+2}$. Then

$$(\mathsf{T} - e\mathsf{I})(\mathbf{A}) \in \mathscr{V}_{e,i+1} = \mathscr{V}_{e,i},$$

so

$$(\mathsf{T} - e)^{i+1}(\mathbf{A}) = (\mathsf{T} - e\mathsf{I})^i((\mathsf{T} - e\mathsf{I})(\mathbf{A})) = 0,$$

since $\mathscr{V}_{e,i} = \text{Ker}(\mathsf{T} - e\mathsf{I})^i$. But this says $\mathbf{A} \in \mathscr{V}_{e,i+1}$. Therefore $\mathscr{V}_{e,i+2} = \mathscr{V}_{e,i+1}$. Since the reverse inclusion also holds we must have equality.

The closure of each eigenspace $\mathscr{V}_e$ of a linear transformation T is T-invariant. Therefore $\mathsf{T}(\overline{\mathscr{V}}_e) \subset \overline{\mathscr{V}}_e$. The following proposition tells us about the behavior of the linear transformation $\mathsf{T} - e\mathsf{I} : \overline{\mathscr{V}}_e \to \overline{\mathscr{V}}_e$.

Proposition 17.11. *Let $\mathsf{T}: \mathscr{V} \to \mathscr{V}$ be a linear transformation with eigenvalue e. Let $\overline{\mathscr{V}}_e$ be the closure of the corresponding eigenspace $\mathscr{V}_e$. Then the induced linear transformation*

$$\mathsf{T} - e\mathsf{I}: \overline{\mathscr{V}}_e \to \overline{\mathscr{V}}_e$$

is nilpotent of index k. Furthermore

$$\text{Ker}(\mathsf{T} - e\mathsf{I})^i = \overline{\mathscr{V}}_e, \qquad i \geq k.$$

PROOF. Recall the chain of subspaces

$$\mathscr{V}_{e,i} = \text{Ker}(\mathsf{T} - e\mathsf{I})^i, \qquad i = 1, 2, \ldots .$$

We have $\overline{\mathscr{V}}_e = \mathscr{V}_{e,n}$ where n is, say, the smallest integer such that

$$\mathscr{V}_{e,j} = \mathscr{V}_{e,n} \quad \text{if} \quad j \geq n.$$

Therefore by the definitions

$$(\mathsf{T} - e\mathsf{I})^n = 0: \overline{\mathscr{V}}_e \to \overline{\mathscr{V}}_e,$$

so

$$(\mathsf{T} - e\mathsf{I}): \overline{\mathscr{V}}_e \to \overline{\mathscr{V}}_e$$

is nilpotent of index $k \leq n$. To see that the index is exactly n we reason as follows:

$$(\mathsf{T} - e\mathsf{I})^k = 0: \overline{\mathscr{V}}_e \to \overline{\mathscr{V}}_e$$

implies

$$\operatorname{Ker}(\mathsf{T} - e\mathsf{I})^k = \mathscr{V}_{e,k} \supseteq \overline{\mathscr{V}}_e = \mathscr{V}_{e,n}.$$

But n is the smallest integer such that the chain

$$\mathscr{V}_e = \mathscr{V}_{e,1} \subseteq \cdots \subseteq \mathscr{V}_{e,n} = \overline{\mathscr{V}}_e$$

stabilizes, so $\mathscr{V}_{e,k} \subseteq \mathscr{V}_{e,n} = \overline{\mathscr{V}}_e$. Therefore $\mathscr{V}_{e,k} \subseteq \overline{\mathscr{V}}_e \subseteq \mathscr{V}_{e,k}$ and $\overline{\mathscr{V}}_e = \mathscr{V}_{e,k}$, which says $k = n$. □

We are now in a position to state the main result of this chapter

Theorem 17.12. *Let* $\mathsf{T}: \mathscr{V} \to \mathscr{V}$ *be a linear transformation in the finite-dimensional complex vector space* $\mathscr{V}$. *Let*

$$p(t) = (t - e_1)^{d_1} \cdots (t - e_q)^{d_q}$$

be the characteristic polynomial of T. *Then*:

(1) *The closure of each eigenspace* $\overline{\mathscr{V}}_{e_i}$ *is fully* T*-invariant.*

(2) $\dim \overline{\mathscr{V}}_{e_i} = d_i$.

(3) *The characteristic polynomial of the induced linear transformation*

$$\mathsf{T}: \overline{\mathscr{V}}_{e_i} \to \overline{\mathscr{V}}_{e_i}$$

is

$$(t - e_i)^{d_i}.$$

(4) $\mathscr{V} = \overline{\mathscr{V}}_{e_1} \oplus \cdots \oplus \overline{\mathscr{V}}_{e_q}$.

We will need a number of preliminary steps before undertaking a proof of this result. But before embarking on this undertaking notice that combining (17.12) with the canonical matrix representation for nilpotent transformations of (17.10) will give us a matrix canonical form

$$\begin{pmatrix} \mathbf{M}_1 & & 0 \\ & \ddots & \\ 0 & & \mathbf{M}_q \end{pmatrix},$$

where

$$\mathbf{M}_j = \begin{pmatrix} \mathbf{J}_{1,j} & & 0 \\ & \ddots & \\ 0 & & \mathbf{J}_{r_j,j} \end{pmatrix},$$

and

$$\mathbf{J}_{i,j} = \begin{pmatrix} e_j & & & 0 \\ 1 & & & \\ & \ddots & & \\ 0 & & 1 & e_j \end{pmatrix}, \qquad 1 \geq i \geq r_j; \quad 1 \geq j \geq q.$$

(See (17.14) and the ensuing discussion.) We separate out the crucial technical step in the proof of (17.12) in the following lemma:

Lemma 17.13. *Suppose* $\mathbf{S}\colon \mathscr{V} \to \mathscr{V}$ *a linear transformation in the finite-dimensional vector space* $\mathscr{V}$. *Then there exists an integer k such that:*

$$\left.\begin{aligned} \operatorname{Im} \mathbf{S}^j &= \operatorname{Im} \mathbf{S}^k, \\ \operatorname{Ker} \mathbf{S}^j &= \operatorname{Ker} \mathbf{S}^k, \end{aligned}\right\} \quad \textit{whenever } j \geq k,$$

and, moreover, Ker $\mathbf{S}^k$ *and* Im $\mathbf{S}^k$ *are a complementary pair of* $\mathbf{S}$*-invariant subspaces that is, both are* $\mathbf{S}$*-invariant and*

$$\mathscr{V} = \operatorname{Im} \mathbf{S}^k \oplus \operatorname{Ker} \mathbf{S}^k.$$

Proof. The nested chains of subspaces

$$\{0\} = \operatorname{Ker} \mathbf{S}^0 \subseteq \operatorname{Ker} \mathbf{S}^1 \subseteq \cdots \subseteq \operatorname{Ker} \mathbf{S}^n \subseteq \cdots \subseteq \mathscr{V},$$

$$\mathscr{V} = \operatorname{Im} \mathbf{S}^0 \supseteq \operatorname{Im} \mathbf{S}^1 \supseteq \cdots \supseteq \operatorname{Im} \mathbf{S}^n \supseteq \cdots \supseteq \{0\},$$

must eventually stabilize because $\mathscr{V}$ is finite dimensional. Say

$$\operatorname{Ker} \mathbf{S}^i = \operatorname{Ker} \mathbf{S}^{n'} \quad \text{if} \quad i \geq n',$$

$$\operatorname{Im} \mathbf{S}^i = \operatorname{Im} \mathbf{S}^{n''} \quad \text{if} \quad i \geq n''.$$

If $k = \max\{n', n''\}$, then from the kth term, both sequences are certainly stable. We claim

$$\operatorname{Im} \mathbf{S}^k \cap \operatorname{Ker} \mathbf{S}^k = \{0\}.$$

For if $\mathbf{A} \in \operatorname{Im} \mathbf{S}^k \cap \operatorname{Ker} \mathbf{S}^k$, then there is a vector $\bar{\mathbf{A}} \in \mathscr{V}$ such that

$$\mathbf{A} = \mathbf{S}^k \bar{\mathbf{A}}.$$

Then, of course

$$0 = \mathbf{S}^k \mathbf{A} = \mathbf{S}^{2k} \bar{\mathbf{A}},$$

so $\bar{\mathbf{A}} \in \operatorname{Ker} \mathbf{S}^{2k}$. But by choice of k

$$\operatorname{Ker} \mathbf{S}^k = \operatorname{Ker} \mathbf{S}^{k+1} = \cdots = \operatorname{Ker} \mathbf{S}^{2k},$$

so $\bar{\mathbf{A}} \in \operatorname{Ker} \mathbf{S}^k$. Then

$$\mathbf{A} = \mathbf{S}^k \bar{\mathbf{A}} = 0$$

as claimed. For the linear transformation

$$\mathbf{S}^k\colon \mathscr{V} \to \mathscr{V},$$

we have

$$\dim \mathscr{V} = \dim \operatorname{Im} \mathbf{S}^k + \dim \operatorname{Ker} \mathbf{S}^k.$$

On the other hand, since

$$\operatorname{Im} \mathbf{S}^k \cap \operatorname{Ker} \mathbf{S}^k = \{0\},$$

we have

$$\dim(\text{Im } \mathsf{S}^k + \text{Ker } \mathsf{S}^k) = \dim \text{Im } \mathsf{S}^k + \dim \text{Ker } \mathsf{S}^k$$

also, therefore

$$\dim \mathscr{V} = \dim(\text{Im } \mathsf{S}^k + \text{Ker } \mathsf{S}^k),$$

so $\text{Im } \mathsf{S}^k + \text{Ker } \mathsf{S}^k$ is a subspace of $\mathscr{V}$ with the same dimension as $\mathscr{V}$, and so must equal $\mathscr{V}$. □

PROOF OF 17.12. We begin by applying (17.13) to the linear transformation

$$\mathsf{S}_1 = \mathsf{T} - e_1\mathsf{I} : \mathscr{V} \to \mathscr{V}.$$

We conclude that for a suitable integer k

$$\mathscr{V} = \text{Ker}(\mathsf{T} - e_1\mathsf{I})^k \oplus \text{Im}(\mathsf{T} - e_1\mathsf{I})^k,$$

where by the definition of $\overline{\mathscr{V}}_{e_1}$ we have

$$\overline{\mathscr{V}}_{e_1} = \text{Ker}(\mathsf{T} - e_1\mathsf{I})^k.$$

By (17.7) the subspaces

$$\text{Im}(\mathsf{T} - e_1\mathsf{I})^k, \text{Ker}(\mathsf{T} - e_1\mathsf{I})^k$$

are T-invariant, so by (17.6) T has a matrix representation

$$\mathbf{M} = \begin{pmatrix} \mathbf{A} & 0 \\ 0 & \mathbf{B} \end{pmatrix},$$

where $\mathbf{A}$ is a matrix representation for

$$\mathsf{T}: \overline{\mathscr{V}}_{e_1} \to \overline{\mathscr{V}}_{e_1}$$

and $\mathbf{B}$ is a matrix representation for

$$\mathsf{T}: \text{Im}(\mathsf{T} - e_1\mathsf{I})^k \to \text{Im}(\mathsf{T} - e_1\mathsf{I})^k.$$

From the matrix $\mathbf{M}$ we can compute the characteristic polynomial of T. We get

$$p(t) = p_1(t) \cdot p_2(t),$$

where

$$p_1(t) = \det(\mathbf{A} - t\mathbf{I}),$$

$$p_2(t) = \det(\mathbf{B} - t\mathbf{I}),$$

are the characteristic polynomials of T on $\overline{\mathscr{V}}_{e_1}$ and $\hat{\mathscr{V}}_{e_1} = \text{Im}(\mathsf{T} - e_1\mathsf{I})^k$, respectively. Since $\overline{\mathscr{V}}_{e_1}$ contains the eigenspace $\mathscr{V}_{e_1}$ it follows that

$$p_1(t) = (t - e_1)^{d_1}\hat{p}(t)$$

and $p_2(t)$ cannot contain $(t - e_1)$ as a factor. We claim $\hat{p}(t) = +1$. To this end recall that

$$\mathsf{T} - e_1\mathsf{I}: \overline{\mathscr{V}}_{e_1} \to \overline{\mathscr{V}}_{e_1}$$

is nilpotent of order k, so has a matrix representation in the form

$$\begin{pmatrix} 0 & & 0 \\ * & \ddots & \\ 0 & * & 0 \end{pmatrix}, \qquad * = 0 \text{ or } 1.$$

Therefore $\mathsf{T}: \overline{\mathscr{V}}_{e_1} \to \overline{\mathscr{V}}_{e_1}$ has a matrix representation

$$\begin{pmatrix} e_1 & & 0 \\ * & \ddots & \\ 0 & * & e_1 \end{pmatrix}$$

and therefore

$$p_1(t) = \det\begin{pmatrix} e_1 - t & & 0 \\ * & \ddots & \\ 0 & * & e_1 - t \end{pmatrix} = (e_1 - t)^{\dim \overline{\mathscr{V}}_{e_1}}.$$

Therefore $\dim \overline{\mathscr{V}}_{e_1} = d_1$ and $\hat{p}_1(t) = +1$, because the factorization of a polynomial into linear factors is unique. Therefore

$$\begin{cases} p_1(t) = (t - e_1)^{d_1}, \\ p_2(t) = (t - e_2)^{d_2} \cdots (t - e_q)^{d_q} \end{cases}$$

If we now repeat the preceding argument for the eigenvalue e_2 of the linear transformation

$$\mathsf{T}: \hat{\mathscr{V}}_{e_1} \to \hat{\mathscr{V}}_{e_1}$$

and so on, we arrive at the desired direct sum decomposition. □

Let us put this all together and see what it says about matrix representations of linear transformations. We start with a linear transformation

$$\mathsf{T}: \mathscr{V} \to \mathscr{V}$$

in a finite-dimensional complex vector space $\mathscr{V}$. Let

$$p(t) = (t - e_1)^{d_1} \cdots (t - e_q)^{d_q}$$

be the characteristic polynomial of T, where $e_1, \ldots, e_q$ are the distinct eigenvalues of T. We apply (17.12) to obtain the direct sum decomposition

$$\mathscr{V} = \overline{\mathscr{V}}_{e_1} \oplus \cdots \oplus \overline{\mathscr{V}}_{e_q}$$

of $\mathscr{V}$ into the sum of the closures of the eigenspaces. If we look at (17.6) again we see that we can certainly obtain a block matrix representation

$$\begin{pmatrix} \mathbf{M}_1 & & 0 \\ & \ddots & \\ 0 & & \mathbf{M}_q \end{pmatrix}$$

for T, where $\mathbf{M}_j$ represents

$$\mathsf{T}: \overline{\mathscr{V}}_{e_j} \to \overline{\mathscr{V}}_{e_j}$$

for $j = 1, \ldots, q$. We can use (17.12) to obtain more information about

$$\mathsf{T}: \overline{\mathscr{V}}_{e_j} \to \overline{\mathscr{V}}_{e_j}, \qquad j = 1, \ldots, q,$$

because

$$(\mathsf{T} - e_j \mathsf{I}) = \mathbf{N}_j: \overline{\mathscr{V}}_{e_j} \to \overline{\mathscr{V}}_{e_j}$$

is a nilpotent transformation of index d_j. We apply the knowledge we gained in (17.10) to $\mathbf{N}_1, \ldots, \mathbf{N}_q$, to find a suitable basis for $\overline{\mathscr{V}}_{e_j}$ with respect to which the nilpotent transformation $\mathbf{N}_j$ has a matrix representative

$$\underbrace{\begin{pmatrix} \mathbf{N}_{n_j, j} & & 0 \\ & \ddots & \\ 0 & & \mathbf{N}_{s_j, j} \end{pmatrix}}_{d_j} \Bigg\} d_j, \qquad n_j \geq \cdots \geq s_j \geq 1,$$

and each block $\mathbf{N}_{r_j, j}$ that appears has the form

$$\mathbf{N}_{r_i, j} = \underbrace{\begin{pmatrix} 0 & & & 0 \\ 1 & \ddots & & \\ & \ddots & 0 & \\ 0 & & 1 & 0 \end{pmatrix}}_{r_j} \Bigg\} r_j$$

Since $\mathbf{N}_j = \mathsf{T} - e_j \mathsf{I}$ this means that with respect to the same basis

$$\mathsf{T}: \overline{\mathscr{V}}_{e_j} \to \overline{\mathscr{V}}_{e_j}$$

is represented by the matrix

$$\mathbf{J}_j = \begin{pmatrix} \mathbf{J}_{n_j, j} & & 0 \\ & \ddots & \\ 0 & & \mathbf{J}_{s_j, j} \end{pmatrix},$$

where

$$\mathbf{J}_{r_i, j} = \begin{pmatrix} e_j & & & 0 \\ 1 & \ddots & & \\ & \ddots & e_j & \\ 0 & & 1 & e_j \end{pmatrix}.$$

Altogether this says that T has the matrix representation:

$$\begin{pmatrix} \mathbf{J}_{e_1,1} & & & & & \\ & \mathbf{J}_{r_1,1} & & & \mathbf{0} & \\ & & \mathbf{J}_{e_2,2} & & & \\ & & & \mathbf{J}_{r_2,2} & & \\ & \mathbf{0} & & & \ddots & \\ & & & & & \mathbf{J}_{r_q,q} \end{pmatrix},$$

where

$$\mathbf{J}_{s_i j} = \underbrace{\begin{pmatrix} e_j & & 0 \\ 1 & & \\ & & e_j \\ 0 & 1 & e_j \end{pmatrix}}_{s_i},$$

or

$$\begin{pmatrix} \overbrace{\begin{matrix} e_1 & & & \\ * & e_1 & & \\ & \ddots & \ddots & \\ & & * & e_1 \end{matrix}}^{d_1} & & & \mathbf{0} \\ 0 & \overbrace{\begin{matrix} e_2 & & & \\ * & e_2 & & \\ & \ddots & \ddots & \\ & & * & e_2 \end{matrix}}^{d_2} & & \\ & 0 & \ddots & \\ \mathbf{0} & & & \overbrace{\begin{matrix} \ddots & \\ * & e_q \end{matrix}}^{d_q} \end{pmatrix},$$

where $* = 0$ or 1. The pattern of 0's and 1's being determined as in (17.10) by the nilpotent transformations

$$\mathsf{T} - e_i \mathsf{I} : \overline{\mathscr{V}}_i \to \overline{\mathscr{V}}_i, \qquad i = 1, \ldots, q.$$

Let us put this together as a theorem.

Theorem 17.14. *Let $\mathsf{T} : \mathscr{V} \to \mathscr{V}$ be a linear transformation in the finite-dimensional complex vector space $\mathscr{V}$. Let*

$$p(t) = (t - e_1)^{d_1} \cdots (t - e_q)^{d_q}$$

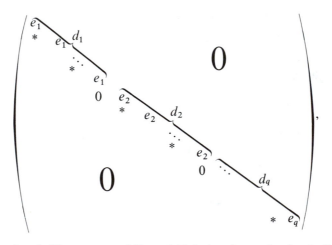

be the characteristic polynomial of T. *Then there exists a basis for* $\mathscr{V}$ *with respect to which the matrix representation of* T *has the block matrix form*

$$\begin{pmatrix} \mathbf{M}_1 & & 0 \\ & \ddots & \\ 0 & & \mathbf{M}_q \end{pmatrix}$$

and

$$\mathbf{M}_j = \underbrace{\begin{pmatrix} \mathbf{J}_{n_j, j} & & 0 \\ & \ddots & \\ 0 & & \mathbf{J}_{s_j, j} \end{pmatrix}}_{d_j} \Bigg\} d_j, \qquad n_j \geq \cdots \geq s_j \geq 1,$$

and

$$\mathbf{J}_{r_j j} = \begin{pmatrix} e_j & & & 0 \\ 1 & \ddots & & \\ & \ddots & e_j & \\ 0 & & 1 & e_j \end{pmatrix}, \qquad j = 1, \ldots, q. \qquad \square$$

This matrix representation is called the *Jordan canonical* form. The matrices

$$\mathbf{J}_{r_j j}: j = 1, \ldots, q, n_j \geq r_j \geq s_j$$

are called the *Jordan blocks.*

Let us look at some examples!

EXAMPLE 6. Let $\mathsf{T}: \mathbb{C}^3 \to \mathbb{C}^3$ be given by

$$\mathsf{T}(x, y, z) = (5x + 4y + 3z, -x - 3z, x - 2y + z).$$

With respect to the standard basis for $\mathbb{C}^3$ the matrix representation of T is

$$\mathbf{A} = \begin{pmatrix} 5 & 4 & 3 \\ -1 & 0 & -3 \\ 1 & -2 & 1 \end{pmatrix}.$$

The characteristic polynomial of T is

$$p(t) = \det(\mathbf{A} - t\mathbf{I}) = \det\begin{pmatrix} 5 - t & 4 & 3 \\ -1 & -t & -3 \\ 1 & -2 & 1 - t \end{pmatrix}$$

$$\begin{aligned} &= (5 - t)[-t(1 - t) - 6] - 4[-1(1 - t) + 3] + 3[2 - (-t)] \\ &= -t^3 + 6t^2 - 32 \\ &= -(t - 4)^2(t + 2), \end{aligned}$$

so there are two eigenvalues

$$\lambda_1 = 4 \text{ of algebraic multiplicity } 2,$$
$$\lambda_2 = -2 \text{ of algebraic multiplicity } 1.$$

Let us find the eigenspace corresponding to the eigenvalue 4. To do so we must solve the linear system

$$\begin{pmatrix} 1 & 4 & 3 \\ -1 & -4 & -3 \\ 1 & -2 & -3 \end{pmatrix}\begin{pmatrix} x \\ y \\ z \end{pmatrix} = \begin{pmatrix} 0 \\ 0 \\ 0 \end{pmatrix}$$

This leads to the equivalent systems

$$\begin{pmatrix} 1 & 4 & 3 \\ 0 & 0 & 0 \\ 1 & -2 & -3 \end{pmatrix}\begin{pmatrix} x \\ y \\ z \end{pmatrix} = \begin{pmatrix} 0 \\ 0 \\ 0 \end{pmatrix},$$

$$\begin{pmatrix} 2 & 2 & 0 \\ 0 & 0 & 0 \\ 1 & -2 & -3 \end{pmatrix}\begin{pmatrix} x \\ y \\ z \end{pmatrix} = \begin{pmatrix} 0 \\ 0 \\ 0 \end{pmatrix},$$

so

$$x = -y = z$$

and the eigenspace is one-dimensional with basis

$$(1 \quad -1 \quad 1) = E_1.$$

To find a basis for the eigenspace corresponding to the eigenvalue -2 we need to solve the linear system

$$\begin{pmatrix} 7 & 4 & 3 \\ -1 & 2 & -3 \\ 1 & -2 & 3 \end{pmatrix}\begin{pmatrix} x \\ y \\ z \end{pmatrix} = \begin{pmatrix} 0 \\ 0 \\ 0 \end{pmatrix},$$

which leads to the solution

$$x = -y = -z.$$

Therefore the eigenspace corresponding to the eigenvalue -2 has dimension 1 with basis $\mathbf{E}_2 = (1, -1, -1)$.

The linear transformation T *cannot* be put into diagonal form because the eigenvalue 4 has algebraic multiplicity 2 but the geometric multiplicity is only 1.

The Jordan normal form of T is

$$\begin{pmatrix} 4 & 0 & 0 \\ * & 4 & 0 \\ 0 & 0 & -2 \end{pmatrix},$$

where $* = 0, 1, *$ cannot be 0 because T is not diagonalizable. Thus we already know

$$\begin{pmatrix} 4 & 0 & 0 \\ 1 & 4 & 0 \\ 0 & 0 & -2 \end{pmatrix}$$

is the Jordan canonical form for T *although we have not yet found the basis with respect to which* T *is represented by this matrix.*

To find the corresponding basis for $\mathscr{V} = \mathbb{C}^3$ we must find a basis for $\overline{\mathscr{V}}_4$. Since we know $\dim \overline{\mathscr{V}}_4 = 2$ and $\mathscr{V}_4 \subset \overline{\mathscr{V}}_4$ we must find a vector $\mathbf{F}_4$ such that

$$(\mathsf{T} - 4\mathsf{I})^2(\mathbf{F}_4) = 0,$$

but

$$(\mathsf{T} - 4\mathsf{I})(\mathbf{F}_4) \neq 0.$$

To find $\mathbf{F}$ notice

$$(\mathsf{T} - 4\mathsf{I})^2 = \begin{pmatrix} 1 & 4 & 3 \\ -1 & -4 & -3 \\ 1 & -2 & -3 \end{pmatrix}^2 = \begin{pmatrix} 0 & -18 & -18 \\ 0 & 18 & 18 \\ 0 & 18 & 18 \end{pmatrix}.$$

So a basis for $\operatorname{Ker}(\mathsf{T} - 4\mathsf{I})^2$ is the solution space of the linear system

$$y + z = 0.$$

One solution is the eigenvector $\mathbf{E}_1 = (1, -1, 1)$ and a *linearly independent* solution is $\mathbf{F} = (0, 1, -1)$. The matrix of T with respect to the ordered basis $(\mathbf{F}_1, \mathbf{F}_2, \mathbf{F}_3)$ is the Jordan normal form

$$\begin{pmatrix} 4 & 0 & 0 \\ 1 & 4 & 0 \\ 0 & 0 & -2 \end{pmatrix},$$

where $\mathbf{F}_1 = \mathbf{F}$, $\mathbf{F}_2 = (\mathsf{T} - 4\mathsf{I})(\mathbf{F}_1)$ and $\mathbf{F}_3 = \mathbf{E}_2$. Another way to obtain the missing vector $\mathbf{F}$ is to reason as follows.

To find $\operatorname{Ker}(\mathsf{T} - 4\mathsf{I})^2$ we can use the fact that $\mathbf{E}_1 \in \operatorname{Ker}(\mathsf{T} - 4\mathsf{I})$ and look for a solution $\mathbf{F}$ to the vector equation

$$(\mathsf{T} - 4\mathsf{I})(\mathbf{F}) = \mathbf{E}_1.$$

We will then have

$$(\mathsf{T} - 4\mathsf{I})^2(\mathbf{F}) = (\mathsf{T} - 4\mathsf{I})(\mathbf{E}_1) = 0.$$

To solve the vector equation

$$(\mathsf{T} - 4\mathsf{I})(\mathbf{F}) = \mathbf{E}_1,$$

we solve the linear system

$$\begin{pmatrix} 1 & 4 & 3 \\ -1 & -4 & -3 \\ 1 & -2 & -3 \end{pmatrix}\begin{pmatrix} x \\ y \\ z \end{pmatrix} = \begin{pmatrix} 1 \\ -1 \\ 1 \end{pmatrix},$$

or what is the same

$$\begin{pmatrix} 1 & 4 & 3 \\ 0 & 0 & 0 \\ 1 & -2 & -3 \end{pmatrix}\begin{pmatrix} x \\ y \\ z \end{pmatrix} = \begin{pmatrix} 1 \\ 0 \\ 1 \end{pmatrix},$$

so

$$\begin{pmatrix} 2 & 2 & 0 \\ 0 & 0 & 0 \\ 1 & -2 & -3 \end{pmatrix}\begin{pmatrix} x \\ y \\ z \end{pmatrix} = \begin{pmatrix} 2 \\ 0 \\ 1 \end{pmatrix},$$

so

$$\begin{aligned} x + y \qquad\quad &= 1, \\ x - 2y - 3z &= 1. \end{aligned}$$

A solution is

$$\begin{aligned} x &= 0, \\ y &= 1, \\ z &= -1, \end{aligned}$$

so $\mathbf{F} = (0, 1, -1)$ is a vector of the type we are looking for.

This second method is the more systematic of the two but does not always apply as the following example serves to illustrate.

EXAMPLE 7. Let $\mathsf{T}\colon \mathbb{C}^3 \to \mathbb{C}^3$ be the linear transformation represented by the matrix

$$\begin{pmatrix} 2 & 2 & -1 \\ -1 & -1 & 1 \\ -1 & -2 & 2 \end{pmatrix}$$

with respect to the standard basis of $\mathbb{C}^3$.

The characteristic polynomial of T is

$$p(t) = \det\begin{pmatrix} 2-t & 2 & -1 \\ -1 & -1-t & 1 \\ -1 & -1 & 2-t \end{pmatrix} = \cdots = -(t-1)^3.$$

Therefore $t = 1$ is an eigenvalue of algebraic multiplicity 3. To determine the corresponding eigenspace we solve the linear system

$$\begin{pmatrix} 1 & 2 & -1 \\ -1 & -2 & 1 \\ -1 & -2 & 1 \end{pmatrix}\begin{pmatrix} x \\ y \\ z \end{pmatrix} = \begin{pmatrix} 0 \\ 0 \\ 0 \end{pmatrix}.$$

This leads to the equation

$$x + 2y - z = 0,$$

which defines a plane in $\mathbb{C}^3$ with a basis $\mathbf{E}_1 = (1, 0, 1)$, $\mathbf{E}_2 = (0, 1, 2)$. Since the eigenspace $\mathscr{V}_1$ is two-dimensional and 1 is the only eigenvalue we have $\overline{\mathscr{V}_1} = \mathbb{C}^3$, and so $\overline{\mathscr{V}_1}$ has dimension 1 over $\mathscr{V}_1$. To find a basis for $\overline{\mathscr{V}_1}$ over $\mathscr{V}_1$ we may choose any vector in $\mathbb{C}^3$ not in $\mathscr{V}_1$, since it will automatically be in $\operatorname{Ker}(\mathsf{T} - \mathsf{I})^2$. (*Note*: Of course $(\mathsf{T} - 1\mathsf{I})^2 = 0$ in this case.)

One such vector is $\mathbf{F}_1 = (1, 0, 0)$. We now follow the prescription of (17.14)

$$(\mathsf{T} - \mathsf{I})(\mathbf{F}_1) = \begin{pmatrix} 1 & 2 & -1 \\ -1 & -2 & 1 \\ -1 & -2 & 1 \end{pmatrix}\begin{pmatrix} 1 \\ 0 \\ 0 \end{pmatrix} = \begin{pmatrix} 1 \\ -1 \\ -1 \end{pmatrix}$$

and $\mathbf{F}_2 = (\mathsf{T} - \mathsf{I})(\mathbf{F}_1) = (1, -1, -1)$ belongs to $\mathscr{V}_1$. If we extend $\{\mathbf{F}_2\}$ to a basis for $\mathscr{V}_1$ by adjoining say $\mathbf{F}_3 = (0, 1, -2)$ we find that the matrix of T with respect to the basis $\{\mathbf{F}_1, \mathbf{F}_2, \mathbf{F}_3\}$ is in the Jordan normal form

$$\begin{pmatrix} 1 & 0 & 0 \\ 1 & 1 & 0 \\ 0 & 0 & 1 \end{pmatrix}.$$

Note that in this example we could not simply proceed as in Example 6, and once we found $\mathscr{V}_1$ look for a vector $\mathbf{F}$ that satisfies

$$(*) \qquad (\mathsf{T} - \mathsf{I})(\mathbf{F}) = \mathbf{E}_1 \quad (\text{or } \mathbf{E}_2).$$

This is because $\mathscr{V}_1 = \mathscr{L}(\mathbf{E}_1, \mathbf{E}_2)$ is two-dimensional. In fact $(*)$ has no solutions whatsoever in this case. What we could do instead is look for a vector $\mathbf{F} \neq 0$ such that

$$(\mathsf{T} - \mathsf{I})(\mathbf{F}) \in \mathscr{L}(\mathbf{E}_1, \mathbf{E}_2).$$

This approach will be illustrated in the next example.

Example 8. As a last example we consider the linear transformation represented by the 4×4 matrix

$$\mathbf{A} = \begin{pmatrix} 7 & 1 & 2 & 2 \\ 1 & 4 & -1 & -1 \\ -2 & 1 & 5 & -1 \\ 1 & 1 & 2 & 8 \end{pmatrix}.$$

We proceed step-by-step as in the two preceding examples. First we compute the characteristic polynomial

$$
\begin{aligned}
p(t) &= \det\begin{pmatrix} 7-t & 1 & 2 & 2 \\ 1 & 4-t & -1 & -1 \\ -2 & 1 & 5-t & -1 \\ 1 & 1 & 2 & 8-t \end{pmatrix} \\
&= (7-t)[\cdots] \\
&= (t-6)^4.
\end{aligned}
$$

So there is exactly one eigenvalue, namely $t = 6$ with algebraic multiplicity 4.

To find a basis for the eigenspace $\mathscr{V}_6$ we must solve the linear system

$$
0 = (\mathbf{A} - 6\mathbf{I})\begin{pmatrix} x \\ y \\ z \\ w \end{pmatrix} = \begin{pmatrix} 1 & 1 & 2 & 2 \\ 1 & -2 & -1 & -1 \\ -2 & 1 & -1 & -1 \\ 1 & 1 & 2 & 2 \end{pmatrix}\begin{pmatrix} x \\ y \\ z \\ w \end{pmatrix}.
$$

Since the first and last row are identical and the sum of the second and third is the negative of the first row we can work with the equivalent system

$$
\begin{aligned}
x + y + 2z + 2w &= 0, \\
x - 2y - z - w &= 0,
\end{aligned}
$$

which easily leads to

$$
\begin{aligned}
y + z + w &= 0, \\
x + z + w &= 0,
\end{aligned}
$$

and the basis

$$
\mathbf{A}_1 = (-1, -1, 1, 0), \qquad \mathbf{A}_2 = (-1, -1, 0, 1)
$$

for the solution space. Therefore

$$
2 = \dim \mathscr{V}_6 < \dim \overline{\mathscr{V}}_6 = 4.
$$

In the next step we need to extend $\mathbf{A}_1, \mathbf{A}_2$ to a basis for

$$
\mathscr{V}_{6,2} = \operatorname{Ker}(\mathbf{A} - 6\mathbf{I})^2.
$$

So we look for vectors $\mathbf{F}$ such that

$$
(\mathbf{A} - 6\mathbf{I})(\mathbf{F}) \in \mathscr{L}(\mathbf{A}_1, \mathbf{A}_2) = \mathscr{V}_{6,1}
$$

That means we are looking for solutions to

$$
\begin{pmatrix} 1 & 1 & 2 & 2 \\ 1 & -2 & -1 & -1 \\ -2 & 1 & -1 & -1 \\ 1 & 1 & 2 & 2 \end{pmatrix}\begin{pmatrix} x \\ y \\ z \\ w \end{pmatrix} = a\mathbf{A}_1 + b\mathbf{A}_2 = \begin{pmatrix} -a & -b \\ -a & -b \\ & a \\ & b \end{pmatrix},
$$

where a and b must first be chosen so that solutions exist. From the first and fourth equation we see $b = -a - b$, so $2b = -a$. If we set $b = 1$ and $a = -2$ we obtain the *consistant* system

$$\begin{aligned} x - 2y - z - w &= 1, \\ -2x + y - z - w &= -2. \end{aligned}$$

(*Note*: Equations 1 and 4 are consequences of these.) From these we obtain

$$y = x - 1 = -(z + w)$$

so a solution is

$$\mathbf{A}_3 = (1, 0, 0, 0).$$

There are in fact no further linear independent solutions to the system

$$(\mathbf{A} - 6\mathbf{I})^2 \begin{pmatrix} x \\ y \\ z \\ w \end{pmatrix} = \begin{pmatrix} 0 \\ 0 \\ 0 \\ 0 \end{pmatrix}$$

because any such solution must be of the form

$$c\mathbf{A}_3 + \mathbf{B}, \qquad \mathbf{B} \in \mathscr{L}(\mathbf{A}_1, \mathbf{A}_2).$$

Finally, we must extend $\{\mathbf{A}_1, \mathbf{A}_2, \mathbf{A}_3\}$ to a basis for

$$\overline{\mathscr{V}}_6 = \mathscr{V}_{6,3} = \operatorname{Ker}(\mathbf{A} - 6\mathbf{I})^3.$$

This is equivalent to finding a solution to the system

$$(\mathbf{A} - 6\mathbf{I}) \begin{pmatrix} x \\ y \\ z \\ w \end{pmatrix} = a_1\mathbf{A}_1 + a_2\mathbf{A}_2 + a_3\mathbf{A}_3.$$

Thus we must solve the system

$$\begin{pmatrix} 1 & 1 & 2 & 2 \\ 1 & -2 & -1 & -1 \\ -2 & 1 & -1 & -1 \\ 1 & 1 & 2 & 2 \end{pmatrix} \begin{pmatrix} x \\ y \\ z \\ w \end{pmatrix} = \begin{pmatrix} -a_1 - a_2 + a_3 \\ -a_1 - a_2 \\ a_1 \\ a_2 \end{pmatrix}.$$

From the first and fourth row we get

$$-a_1 - a_2 + a_3 = a_2$$

and since the sum of the two middle rows is the negative of the first we get

$$-a_1 - a_2 + a_1 = -a_1 - a_2 + a_3.$$

From which we see

$$a_1 = 0, \qquad a_2 = 1, \qquad a_3 = -3$$

leads to a consistent system with solution

$$\mathbf{A}_4 = \begin{pmatrix} 0 \\ 0 \\ 0 \\ 1 \end{pmatrix}.$$

As basis for $\mathbb{C}^4$ we have

$$\mathbf{B}_1 = \mathbf{A}_4,$$

$$\mathbf{B}_2 = (\mathbf{A} - 6\mathbf{I})(\mathbf{B}_1) = \begin{pmatrix} 2 \\ -1 \\ -1 \\ 2 \end{pmatrix},$$

$$\mathbf{B}_3 = (\mathbf{A} - 6\mathbf{I})(\mathbf{B}_2) = \begin{pmatrix} 3 \\ 3 \\ -6 \\ 3 \end{pmatrix},$$

$$\mathbf{B}_4 = \begin{pmatrix} 1 \\ 1 \\ 1 \\ -2 \end{pmatrix},$$

where $\mathbf{B}_4$ is an eigenvector linearly independent from the eigenvector $\mathbf{B}_1$. With respect to this basis the matrix of T is

$$\begin{pmatrix} 6 & 0 & 0 & 0 \\ 1 & 6 & 0 & 0 \\ 0 & 1 & 6 & 0 \\ 0 & 0 & 0 & 6 \end{pmatrix}.$$

The main results of this chapter may be summarized in the following two theorems.

Theorem 17.15. *Let A, $B \in \mathscr{M}_{nn}$. Then A and B represent the same linear transformation $\mathsf{T}: \mathbb{C}^n \to \mathbb{C}^n$ with respect to different bases iff A and B have the same Jordan normal form.* □

Theorem 17.16. *Let* S, T: $\mathscr{V} \to \mathscr{V}$ *be linear endomorphisms of the finite-dimensional complex vector space* $\mathscr{V}$. *Then there exists an isomorphism* P: $\mathscr{V} \to \mathscr{V}$ *such that* $\mathsf{T} = PSP^{-1}$ *iff* S *and* T *have the same Jordan normal form.* □

EXERCISES

1. Give an example of a linear transformation in a finite-dimensional *real* vector space with no nontrivial invariant subspace.

2. Let $\mathsf{T}: \mathbb{C}^2 \to \mathbb{C}^2$ be given the linear transformation given by the matrix

$$\begin{pmatrix} 0 & 0 \\ 0 & 1 \end{pmatrix},$$

with respect to some basis. What are the invariant subspaces of T.

3. Let $\mathsf{T}: \mathscr{V} \to \mathscr{V}$ be a linear transformation with invariant subspaces $\mathscr{U}$, $\mathscr{W}$. Show $\mathscr{U} + \mathscr{W}$ and $\mathscr{U} \cap \mathscr{W}$ are also invariant subspaces of T.

4. Let $\mathsf{D}: \mathscr{P}_n(\mathbb{R}) \to \mathscr{P}_n(\mathbb{R})$ be the differentiation operator. Show $\mathscr{P}_i(\mathbb{R})$, $i = 1, \ldots, n$ are all invariant subspaces of **D**. Does any of them have an invariant complement?

5. Which of the following subspaces of $\mathbb{C}^3$ are complementary to

$$\mathscr{U} = \{(x, y, z) \mid x + y + z = 0\}.$$

(a) $\{(x, y, z) \mid x + 2y + z = 0\}$;
(b) $\{(x, y, z) \mid x - y + z = 0\}$;
(c) $\{(x, y, z) \mid x + 2y + z = 0 \text{ and } x - y + z = 0\}$;
(d) $\mathscr{L}((1, -1, 1))$;
(e) $\mathscr{L}((1, -2, 1))$.

6. Let us define two subspaces of $\mathscr{P}_n(\mathbb{R})$ by

$$\mathscr{S} = \{p \in \mathscr{P}_n(\mathbb{R}) \mid p(x) = p(-x)\},$$

$$\mathscr{A} = \{p \in \mathscr{P}_n(\mathbb{R}) \mid p(x) = -p(-x)\}.$$

Show that $\mathscr{S}$ and $\mathscr{A}$ are complementary subspaces. Are they left invariant by the differentiation operator **D**?

7. Let $\mathscr{V}$ be an inner product space and let $\mathsf{T}: \mathscr{V} \to \mathscr{V}$ be a linear transformation with invariant subspace $\mathscr{W}$. Show that $\mathscr{W}^{\perp}$ is invariant for the adjoint $\mathsf{T}^*: \mathscr{V} \to \mathscr{V}$.

8. For each of the subspaces $\mathscr{W}$ of $\mathbb{C}^3$ in Exercise 5 find a basis for $\mathbb{C}^3$ over $\mathscr{W}$.

9. Verify that each of the following transformations is nilpotent and find their Jordan normal form

$$\mathbf{A} = \begin{pmatrix} 0 & 0 & 0 \\ 1 & 0 & 0 \\ 0 & 1 & 0 \end{pmatrix},$$

$$\mathbf{B} = \begin{pmatrix} 0 & 1 & 2 \\ 0 & 0 & 3 \\ 0 & 0 & 0 \end{pmatrix},$$

$$\mathbf{C} = \begin{pmatrix} 1 & 2 & -1 \\ -1 & -2 & 1 \\ -1 & -2 & 1 \end{pmatrix},$$

$$\mathbf{D} = \begin{pmatrix} 0 & 0 & 0 & 0 & 1 & 0 \\ 0 & 0 & 1 & 1 & 0 & -1 \\ 0 & 0 & 0 & 0 & 0 & 0 \\ 0 & 0 & 0 & 0 & 0 & 0 \\ 0 & 0 & 1 & 0 & 0 & -1 \\ 0 & 0 & 0 & 0 & 0 & 0 \end{pmatrix}.$$

10. Let $\mathsf{T}: \mathbb{C}^2 \to \mathbb{C}^2$ with just one eigenvalue λ. Show that $\mathsf{T} - \lambda\mathsf{I}$ is a nilpotent linear transformation.

11. $\mathsf{N}: \mathscr{V} \to \mathscr{V}$ be a nilpotent transformation. Show that for any linear transformation $\mathsf{T}: \mathscr{V} \to \mathscr{V}$ that $\mathsf{NT} - \mathsf{TN} = \mathsf{S}$ is a nilpotent transformation.

12. Let $\mathsf{N}: \mathscr{V} \to \mathscr{V}$ be a nilpotent transformation in an n-dimensional vector space $\mathscr{V}$. Show that the characteristic polynomial $\mathbf{P}_N(t)$ is $(-1)^n t^n$.

13. Show that the trace of a nilpotent transformation is zero. Is the converse true?

14. Let $\mathsf{T}: \mathbb{C}^3 \to \mathbb{C}^3$ be represented with respect to the standard basis by each of the following matrices in turn. Find the Jordan normal form of T

$$\mathbf{A} = \begin{pmatrix} 5 & 4 & 3 \\ -1 & 0 & -3 \\ 1 & -2 & 1 \end{pmatrix} \qquad \mathbf{D} = \begin{pmatrix} 0 & 0 & 1 \\ 0 & 0 & 0 \\ 1 & 0 & 0 \end{pmatrix},$$

$$\mathbf{B} = \begin{pmatrix} 0 & 1 & 1 \\ 1 & 0 & 1 \\ 1 & 1 & 0 \end{pmatrix}, \qquad \mathbf{E} = \begin{pmatrix} 1 & 0 & 1 \\ 0 & 0 & 0 \\ 0 & 0 & -1 \end{pmatrix},$$

$$\mathbf{C} = \begin{pmatrix} 5 & -6 & -6 \\ -1 & 4 & 2 \\ 3 & -6 & -4 \end{pmatrix}, \qquad \mathbf{F} = \begin{pmatrix} 0 & 1 & 0 \\ 0 & 0 & 1 \\ -1 & 1 & 1 \end{pmatrix}.$$

15. Let $\mathsf{T}: \mathbb{C}^6 \to \mathbb{C}^6$ be represented by the matrix

$$\begin{pmatrix} 1 & 0 & 0 & 0 & 1 & 0 \\ 0 & 1 & 1 & 1 & 0 & -1 \\ 0 & 0 & 1 & 0 & 0 & 0 \\ 0 & 0 & 0 & 1 & 0 & 0 \\ 0 & 0 & 1 & 0 & 1 & -1 \\ 0 & 0 & 0 & 0 & 0 & 1 \end{pmatrix}$$

with respect to the standard basis of $\mathbb{C}^6$. Find the Jordan normal form of T.

16. Let $\mathsf{T}: \mathscr{V} \to \mathscr{V}$ be a linear transformation in a finite-dimensional complex vector space T. Show that $\mathsf{T} = \mathbf{A} + \mathsf{N}$, where $\mathbf{A}: \mathscr{V} \to \mathscr{V}$ is an isomorphism and $\mathsf{N}: \mathscr{V} \to \mathscr{V}$ is nilpotent.

Definition. Let $\mathsf{T}: \mathscr{V} \to \mathscr{V}$ be a linear transformation in the finite-dimensional complex vector space $\mathscr{V}$. For each eigenvalue e_i of T let the integer n_i be defined by $n_i = \dim \overline{\mathscr{V}}_{e_i}$. Let $m(t) = (t - e_1)^{n_1} \cdots (t - e_q)^{n_q}$ where $e_1, \ldots, e_q$ are the distinct eigenvalues of T. The polynomial $m_{\mathsf{T}}(t)$ is called the *minimal polynomial* of T.

17. Show that n_i is the smallest integer such that $\operatorname{Ker}(\mathsf{T} - e_i \mathsf{I})^{n_i} = \overline{\mathscr{V}}_{e_i}$.

18. Show that $(\mathsf{T} - e_i \mathsf{I}): \mathscr{V}_{e_i} \to \mathscr{V}_{e_i}$ is nilpotent of index n_i.

19. Show that n_i is less than or equal to the algebraic multiplicity of the eigenvalue e_i.

20. Show that the linear transformation $m(\mathsf{T}): \mathscr{V} \to \mathscr{V}$ is the zero transformation.

21. (Hamilton–Cayley Theorem). Show that the linear transformation $p(\mathsf{T}): \mathscr{V} \to \mathscr{V}$ is the zero transformation where $p(t)$ is the characteristic polynomial of T. (*Hint*: Use Exercise 19 to show $m(t)$ divides $p(t)$; that is, that there exists a third polynomial $q(t)$ such that $p(t) = q(t) \cdot m(t)$. Then use Exercise 20.)

22. Let $\mathsf{T}: \mathscr{V} \to \mathscr{V}$ be a nonsingular linear transformation in a finite-dimensional complex vector space $\mathscr{V}$. Show that there exists a linear transformation $\mathsf{S}: \mathscr{V} \to \mathscr{V}$ such that $\mathsf{S}^2 = \mathsf{T}$. (*Hint*: Use Jordan normal form.) We call S the square root of T.

23. Find a square root of each of the following

$$\begin{pmatrix} 0 & 1 \\ 1 & 0 \end{pmatrix},$$

$$\begin{pmatrix} 0 & 1 & 1 \\ 1 & 0 & 1 \\ 1 & 1 & 0 \end{pmatrix},$$

$$\begin{pmatrix} 2 & -1 & -1 \\ -1 & 2 & -1 \\ -1 & -1 & 2 \end{pmatrix}.$$

18 Applications to linear differential equations

Differential equations have played an important part in the development of mathematics since the time of Newton. They have also been central to many applications of mathematics to the physical sciences and technology. Often the equations relevant to practical applications are so difficult to solve explicitly that they can only be handled with approximation techniques on large computer systems. In this chapter we will be concerned with a simple form of differential equation, and systems thereof, namely, linear differential equations with constant coefficients. These systems have a great deal in common with systems of linear equations, and we are in a position to apply the hard-won knowledge about Jordan canonical forms to solve such systems. We begin with a short excursion into the subject of differential equations.

EXAMPLE 1. We are looking for all differentiable functions $y\colon \mathbb{R} \to \mathbb{R}$ with the property that the derivative dy/dt of y satisfies the equation

$$\frac{dy}{dt}(t) = r \cdot y(t), \qquad x \in \mathbb{R},$$

where r is some fixed number.

Solution. $y = ce^{rt}$ where c is a constant.

Reason. Fundamental theorem of the calculus.

Definition. Let n be a positive integer, $B \subset \mathbb{R}^{n+2}$ a nonempty subset, and

$$\mathbf{F}\colon B \to \mathbb{R}$$

a function. A function

$$y : J \to \mathbb{R}$$

J an interval (i.e., $J = \{t \in \mathbb{R},\ a \leq t \leq b\}$) is called a *solution to the differential equation*

$$\mathsf{F}\left(t, y, \frac{dy}{dt}, \ldots, \frac{d^n y}{dt^n}\right) = 0$$

iff the following conditions are fulfilled:

(a) y is an n-times differentiable functions;
(b) $\forall x \in J$ the $(n + 2)$-tuple

$$\left(x, y(t), \frac{dy}{dt}(t), \ldots, \frac{d^n y}{dt^n}(t)\right)$$

belongs to B; and

(c) $$\forall x \in J : \mathsf{F}\left(x, y(t), \frac{dy}{dt}(t), \ldots, \frac{d^n y}{dt^n}(t)\right) = 0.$$

A class of differential equations of particular interest are those of the form

$$\frac{d^n y}{dt^n} - f\left(t, y, \ldots, \frac{d^{n-1} y}{dt^{n-1}}\right) = 0,$$

or, equivalently

$$\frac{d^n y}{dt^n} = f\left(t, y, \ldots, \frac{d^{n-1} y}{dt^{n-1}}\right).$$

Such an equation is of the nth *order*.

We can also consider *systems* of differential equations. They are analogous to simultaneous equations.

Definition. A solution to the first-order system

$$\frac{dy_1}{dt} = f_1(t, y_1, \ldots, y_n),$$

$$\frac{dy_2}{dt} = f_2(t, y_1, \ldots, y_n),$$

$$\frac{dy_n}{dt} = f_n(t, y_1, \ldots, y_n),$$

where $f_i : B \to \mathbb{R}$, $i = 1, \ldots, n$, are functions defined on $\varnothing \neq B \subset \mathbb{R}^{n+1}$, consists of an interval $J \subset \mathbb{R}$ and functions

$$y_i : J \to \mathbb{R}, \qquad i = 1, \ldots, n,$$

such that:

(a) $y_1, \ldots, y_n$ are differentiable on J;
(b) $\forall t \in J$ the $(n + 1)$-tuple $(x, y_i(t), \ldots, y_n(t))$ belongs to B; and
(c) $\forall t \in J$ and each $1 \leq i \leq n$

$$\frac{dy_i}{dt}(t) = f_i(t, y_1(t), \ldots, y_n(t)).$$

If each f_i, $1 \leq i \leq n$ is a linear function, that is

$$f_i(t_1, \ldots, t_{n+1}) = \sum_{j=1}^{n+1} a_{ij}t_j, \qquad 1 \leq i \leq n,$$

for $a_{ij} \in \mathbb{R}$, then the above system is called a *linear differential system of the first order*. If, furthermore, $a_{i1} = 0$, $i = 1, \ldots, n$ (i.e., "x does not appear on the right-hand side") then the system is called *homogeneous*.

Remarks. 1. Do not let these definitions scare you! We will only be concerned with linear differential systems of the first order.

2. Note that for linear systems that the functions f_i are defined on all of $\mathbb{R}^{n+1}$.

3. We can write a linear differential system in the matrix form

$$\begin{pmatrix} \frac{dy_1}{dt}(t) \\ \vdots \\ \frac{dy_n}{dt}(t) \end{pmatrix} = bx + \mathbf{A}\begin{pmatrix} y_1(t) \\ \vdots \\ y_n(t) \end{pmatrix}$$

where

$$b = \begin{pmatrix} a_{11} \\ \vdots \\ a_{n1} \end{pmatrix}, \qquad \mathbf{A} = \begin{pmatrix} a_{12} & \cdots & a_{1,n+1} \\ & \ddots & \\ a_{n2} & \cdots & a_{n,n+1} \end{pmatrix}.$$

The system is homogeneous when $b = 0 \in \mathbb{R}^n$.

Suppose that $f: B \to \mathbb{R}$, $B \subset \mathbb{R}^{n+1}$, and that we are looking for a solution $y: J \to \mathbb{R}$ to the nth-order equation

$$(*) \qquad \frac{d^n y}{dt^n}(t) = f\left(t, y, \frac{dy}{dt}, \ldots, \frac{d^{n-1}y}{dt^{n-1}}\right).$$

If we set

$$y_1 = y,$$

$$y_2 = \frac{dy}{dt} = \frac{dy_1}{dt},$$

$$\vdots$$

$$y_n = \frac{d^{n-1}y}{dt^{n-1}} = \frac{dy_{n-1}}{dt}$$

then y is a solution to $(*)$ iff $(y_1, \ldots, y_n)$ is a solution to the first-order system

$$\frac{dy_1}{dt} = y_2,$$

$$\frac{dy_{n-2}}{dt} = y_{n-1},$$

$$\frac{dy_{n-1}}{dt} = y_n,$$

$$\frac{dy_n}{dt} = f(t, y_1, \ldots, y_{n-1}).$$

In this way solving a single differential equation of order n is equivalent to solving a first-order system.

Before we take up some examples let us consider the structure of the set of all solutions to a linear differential system (compare (13.6)).

Proposition 18.1. *Suppose*

$$(*) \qquad \begin{pmatrix} \dfrac{dy_1}{dt} \\ \vdots \\ \dfrac{dy_n}{dt} \end{pmatrix} = bt + \mathbf{A} \cdot \begin{pmatrix} y_1 \\ \vdots \\ y_n \end{pmatrix}$$

$\mathbf{A} \in \mathcal{M}_{nn}$, $b \in \mathbb{R}^n$, is a linear differential system of the first order. Let $\mathscr{V}$ be the set of all solutions on some interval $J = [a, b]$ to the associated homogeneous system

$$(\text{H}*) \qquad \frac{dy}{dt} = Ay, \qquad y = \begin{pmatrix} y_1 \\ \vdots \\ y_n \end{pmatrix}.$$

Then

(a) *$\mathscr{V}$ is a vector subspace of $\mathscr{C}(a, b) \oplus \cdots \oplus \mathscr{C}(a, b)$ (Chapter 4, Example 5).*
(b) *If $z = (z_1, \ldots, z_n)$ is a solution to $(*)$ then every solution to $(*)$ is of the form $z + y$ for some $y \in \mathscr{V}$.*

Proof. (a) First of all $\mathbf{0} \in \mathscr{V}$ because if

$$\theta = (0, \ldots, 0),$$

then

$$\frac{d\theta}{dt} = \begin{pmatrix} 0 \\ 0 \\ \vdots \\ 0 \end{pmatrix} = A \begin{pmatrix} 0 \\ \vdots \\ 0 \end{pmatrix} = A\theta.$$

Next, if $\varphi, \psi \in \mathscr{V}$ and $a, b \in \mathbb{R}$ then using the linearity of differentiation we find

$$\begin{aligned}\frac{d}{dt}(a\varphi + b\psi) &= a\frac{d\varphi}{dt} + b\frac{d\psi}{dt}\\ &= a\mathbf{A}\varphi + b\mathbf{A}\psi\\ &= \mathbf{A}(a\varphi + b\psi),\end{aligned}$$

which says $a\varphi + b\psi$ solves the equation (H$*$) and so $a\varphi + b\psi \in \mathscr{V}$

(b) If z is a fixed solution to ($*$) and w any other solution, then

$$\begin{aligned}\frac{d}{dt}(z - w) &= \frac{dz}{dt} - \frac{dw}{dt}\\ &= bx + \mathbf{A}z - (bx + \mathbf{A}w)\\ &= \mathbf{A}(z - w),\end{aligned}$$

so $z - w$ solves (H$*$) and hence $z - w = y \in \mathscr{V}$. □

Definition. A linear homogeneous differential system of the first order

$$(*) \qquad \frac{dy}{dt} = \mathbf{A}y, \qquad \mathbf{A} \in \mathscr{M}_{nn},$$

is said to be *diagonalizable* if the matrix **A** is diagonalizable.

Suppose that

$$(*) \qquad \frac{dy}{dt} = \mathbf{A}y$$

is a diagonalizable system. Then there exists a diagonal matrix **D** and an invertible matrix $\mathbf{P} \in \mathscr{M}_{nn}$ such that

$$\mathbf{A} = \mathbf{PDP}^{-1}.$$

Thus we can write ($*$) in the form

$$\frac{dy}{dt} = \mathbf{PDP}^{-1}y.$$

If we multiply both sides of this equation by $\mathbf{P}^{-1}$ we obtain the equivalent equation

$$\mathbf{P}^{-1}\frac{dy}{dt} = \mathbf{DP}^{-1}y.$$

Since $\mathbf{P}^{-1}$ is a matrix of constants, i.e., does not depend on the variable t, we can write further

$$\mathbf{P}^{-1}\frac{dy}{dt} = \frac{d}{dt}(\mathbf{P}^{-1}y).$$

Now comes the crucial step. Set

$$w = \begin{pmatrix} w_1 \\ \vdots \\ w_n \end{pmatrix} = \mathbf{P}^{-1}\begin{pmatrix} y_1 \\ \vdots \\ y_n \end{pmatrix},$$

and combine the two preceding equations in

$$\frac{dw}{dt} = \mathbf{D}w.$$

Now **D** being a diagonal matrix, this latter system is just

$$\begin{aligned} \frac{dw_1}{dt} &= d_1 w_1, \\ \vdots \quad & \quad \vdots \\ \frac{dw_1}{dt} &= d_n w_n, \end{aligned}$$

and so has the solution

$$\begin{pmatrix} w_1 \\ \vdots \\ w_n \end{pmatrix} = \begin{pmatrix} c_1 e^{d_1 t} \\ \vdots \\ c_n e^{d_n t} \end{pmatrix}.$$

But since

$$\begin{pmatrix} w_1 \\ \vdots \\ w_n \end{pmatrix} = \mathbf{P}^{-1}\begin{pmatrix} y_1 \\ \vdots \\ y_n \end{pmatrix},$$

we have

$$\begin{pmatrix} y_1 \\ \vdots \\ y_n \end{pmatrix} = \mathbf{P}\begin{pmatrix} w_1 \\ \vdots \\ w_n \end{pmatrix} = \mathbf{P}\begin{pmatrix} c_1 e^{d_1 t} \\ \vdots \\ c_n e^{d_n t} \end{pmatrix}.$$

and thus we have shown:

Theorem 18.2. *Let*

$$(*) \qquad \frac{dy}{dt} = \mathbf{A}y, \qquad \mathbf{A} \in \mathcal{M}_{nn},$$

be a diagonalizable linear system, with

$$\mathbf{A} = \mathbf{PDP}^{-1}$$

for a diagonal matrix **D** *and invertible matrix* **P**. *Then the solutions of* (∗) *are of the form*

$$\begin{pmatrix} y_1 \\ \vdots \\ y_n \end{pmatrix} = \mathbf{P}\begin{pmatrix} c_1 e^{d_1 t} \\ \vdots \\ c_n e^{d_n t} \end{pmatrix},$$

where $\mathbf{D} = \operatorname{diag}\{d_1, \ldots, d_n\}$ *and* $c_1, \ldots, c_n$ *are constants.* □

EXAMPLE 2. Solve the linear differential system

$$\frac{dy_1}{dt} = y_2,$$

$$\frac{dy_2}{dt} = y_1.$$

In matrix form this is the linear differential system

$$\frac{dy}{dt} = \begin{pmatrix} 0 & 1 \\ 1 & 0 \end{pmatrix}\begin{pmatrix} y_1 \\ y_2 \end{pmatrix}.$$

The coefficient matrix

$$\mathbf{A} = \begin{pmatrix} 0 & 1 \\ 1 & 0 \end{pmatrix}$$

is symmetric, and therefore diagonalizable. To diagonalize **A** we first compute the characteristic polynomial

$$p(t) = \det\begin{pmatrix} -t & 1 \\ 1 & -t \end{pmatrix} = (t^2 - 1) = (t - 1)(t + 1).$$

There are two eigenvalues $t = \pm 1$ and corresponding eigenvectors can be easily found, viz.,

$$t = 1: \quad \begin{pmatrix} -1 & 1 \\ 1 & -1 \end{pmatrix}\begin{pmatrix} x \\ y \end{pmatrix} = \begin{pmatrix} 0 \\ 0 \end{pmatrix}. \qquad \text{Sol.: } \mathbf{E}_1 = \begin{pmatrix} 1 \\ 1 \end{pmatrix},$$

$$t = -1: \quad \begin{pmatrix} 1 & 1 \\ 1 & 1 \end{pmatrix}\begin{pmatrix} x \\ y \end{pmatrix} = \begin{pmatrix} 0 \\ 0 \end{pmatrix}. \qquad \text{Sol.: } \mathbf{E}_2 = \begin{pmatrix} 1 \\ -1 \end{pmatrix}.$$

We can therefore write

$$\mathbf{A} = \mathbf{P}\begin{pmatrix} 1 & 0 \\ 0 & -1 \end{pmatrix}\mathbf{P}^{-1},$$

where

$$\mathbf{P} = \begin{pmatrix} 1 & 1 \\ 1 & -1 \end{pmatrix} = (\mathbf{E}_1 \quad \mathbf{E}_2)$$
$$\begin{matrix} \uparrow & \uparrow \\ \mathbf{E}_1 & \mathbf{E}_2 \end{matrix}$$

Now the associated diagonal system is

$$\frac{dw_1}{dt} = w_1,$$

$$\frac{dw_2}{dt} = -w_2,$$

and has as solutions the functions

$$\begin{pmatrix} w_1 \\ w_2 \end{pmatrix} = \begin{pmatrix} c_1 e^t \\ c_2 e^{-t} \end{pmatrix}.$$

Thus the solutions to the equation (∗) are

$$\begin{aligned} \begin{pmatrix} y_1 \\ y_2 \end{pmatrix} = \mathbf{P}\begin{pmatrix} w_1 \\ w_2 \end{pmatrix} &= \begin{pmatrix} 1 & 1 \\ 1 & -1 \end{pmatrix}\begin{pmatrix} c_1 e^t \\ c_2 e^{-t} \end{pmatrix} \\ &= \begin{pmatrix} c_1 e^t + c_2 e^{-t} \\ c_1 e^t - c_2 e^{-t} \end{pmatrix} \quad **. \end{aligned}$$

EXAMPLE 3. Let us solve the differential equation

$$(*) \qquad \frac{d^3 z}{dt^3} + 4\frac{d^2 z}{dt^2} + \frac{dz}{dt} - 6z = 0.$$

First of all we obtain an equivalent first-order system by setting

$$y_1 = z,$$

$$y_2 = \frac{dz}{dt},$$

$$y_3 = \frac{d^2 z}{dt^2}.$$

We then have

$$\frac{dy_1}{dt} = y_2,$$

$$\frac{dy_2}{dt} = y_3,$$

and the original equation now reads

$$\frac{dy_3}{dt} = -4y_3 - y_2 + 6y_1,$$

so we have the system

$$\frac{d}{dt}\begin{pmatrix} y_1 \\ y_2 \\ y_3 \end{pmatrix} = \begin{pmatrix} 0 & 1 & 0 \\ 0 & 0 & 1 \\ 6 & -1 & -4 \end{pmatrix}\begin{pmatrix} y_1 \\ y_2 \\ y_3 \end{pmatrix}.$$

The characteristic polynomial of the coefficient matrix is

$$\begin{aligned} p(\lambda) &= \det\begin{pmatrix} -\lambda & 1 & 0 \\ 0 & -\lambda & 1 \\ 6 & -1 & -4-\lambda \end{pmatrix} \\ &= -\lambda(\lambda(\lambda+4)+1)+6(1) \\ &= -\lambda^3 - 4\lambda^2 - \lambda + 6 \\ &= -(\lambda-1)(\lambda+2)(\lambda+3). \end{aligned}$$

Therefore the coefficient matrix has three distinct eigenvalues and can be diagonalized. The computation of eigenvectors proceeds as usual. We find

$$\lambda_1 = 1: \quad \begin{pmatrix} -1 & 1 & 0 \\ 0 & -1 & 1 \\ 6 & -1 & -5 \end{pmatrix}\begin{pmatrix} x \\ y \\ z \end{pmatrix} = \begin{pmatrix} 0 \\ 0 \\ 0 \end{pmatrix}$$

$$\Rightarrow \quad \mathbf{E}_1 = \begin{pmatrix} 1 \\ 1 \\ 1 \end{pmatrix} \quad \text{is an eigenvector.}$$

$$\lambda_2 = -2: \quad \begin{pmatrix} 2 & 1 & 0 \\ 0 & 2 & 1 \\ 6 & -1 & -2 \end{pmatrix}\begin{pmatrix} x \\ y \\ z \end{pmatrix} = \begin{pmatrix} 0 \\ 0 \\ 0 \end{pmatrix}$$

$$\Rightarrow \quad y = -2x,\ z = -2y \quad \text{so} \quad \mathbf{E}_2 = \begin{pmatrix} 1 \\ -2 \\ 4 \end{pmatrix} \quad \text{is an eigenvector.}$$

$$\lambda_3 = -3: \quad \begin{pmatrix} 3 & 1 & 0 \\ 0 & 3 & 1 \\ 6 & -1 & -1 \end{pmatrix}\begin{pmatrix} x \\ y \\ z \end{pmatrix} = \begin{pmatrix} 0 \\ 0 \\ 0 \end{pmatrix}$$

$$\Rightarrow \quad y = -3x,\ z = -3y \quad \text{so} \quad \mathbf{E}_3 = \begin{pmatrix} 1 \\ -3 \\ 9 \end{pmatrix} \quad \text{is an eigenvector.}$$

The matrix of the basis change is

$$\mathbf{P} = \begin{pmatrix} 1 & 1 & 1 \\ 1 & -2 & -3 \\ 1 & 4 & 9 \end{pmatrix},$$

that is,

$$\begin{pmatrix} 0 & 1 & 0 \\ 0 & 0 & 1 \\ 6 & -1 & -4 \end{pmatrix} = \mathbf{P}\begin{pmatrix} 1 & 0 & 0 \\ 0 & -2 & 0 \\ 0 & 0 & -3 \end{pmatrix}\mathbf{P}^{-1}.$$

Set

$$\begin{pmatrix} w_1 \\ w_2 \\ w_3 \end{pmatrix} = \mathbf{P}^{-1}\begin{pmatrix} y_1 \\ y_2 \\ y_3 \end{pmatrix}.$$

The diagonal system

$$\frac{d}{dt}\begin{pmatrix} w_1 \\ w_2 \\ w_3 \end{pmatrix} = \begin{pmatrix} 1 & 0 & 0 \\ 0 & -2 & 0 \\ 0 & 0 & -3 \end{pmatrix}\begin{pmatrix} w_1 \\ w_2 \\ w_3 \end{pmatrix}$$

has the solutions

$$w_1 = c_1 e^t,$$
$$w_2 = c_2 e^{-2t},$$
$$w_3 = c_3 e^{-3t}.$$

If

$$\begin{pmatrix} w_1 \\ w_2 \\ w_3 \end{pmatrix} = \mathbf{P}^{-1}\begin{pmatrix} y_1 \\ y_2 \\ y_3 \end{pmatrix},$$

then

$$\begin{pmatrix} y_1 \\ y_2 \\ y_3 \end{pmatrix} = \mathbf{P}\begin{pmatrix} w_1 \\ w_2 \\ w_3 \end{pmatrix},$$

and thus

$$y_1 = c_1 e^t + c_2 e^{-2t} + c_3 e^{-3t},$$
$$y_2 = c_1 e^t - 2x_2 e^{-2t} - 3c_3 e^{-3t},$$
$$y_3 = c_1 e^t + 4c_2 t + 9c_3 e^{-3t}.$$

Since $z = y_1$ we obtain that

$$z = c_1 e^t + c_2 e^{-2t} + c_3 e^{-3t}$$

is the general solution to (∗). ∗∗

This example illustrates that the differential equation

$$\frac{d^n z}{dt^n} + a_{n-1}\frac{d^{n-1} z}{dt} + \cdots + a_1 \frac{dz}{dt} + a_0 z = 0$$

is equivalent to the linear system

$$\frac{d}{dt}\begin{pmatrix} y_1 \\ \vdots \\ y_n \end{pmatrix} = \begin{pmatrix} 0 & 1 & 0 & \cdots & 0 \\ 0 & 0 & 1 & \cdots & 0 \\ \vdots & \ddots & & \vdots & \\ 0 & & 0 & & 1 \\ -a_0 & \cdots & a_{n-2} & & -a_{n-1} \end{pmatrix}\begin{pmatrix} y_1 \\ \vdots \\ y_n \end{pmatrix}$$

To apply the preceding solution method we need to know when a matrix of the form

$$\begin{pmatrix} 0 & 1 & \cdots & & 0 \\ & & & & \\ 0 & \cdots & 0 & & 1 \\ -a_0 & \cdots & & & -a_{n-1} \end{pmatrix} = \mathbf{A}$$

is diagonalizable. Matrices of this form are called *companion matrices.*

Proposition 18.3. *The characteristic polynomial of the matrix*

$$\mathbf{B} = \begin{pmatrix} 0 & 1 & & 0 \\ \vdots & \ddots & \ddots & \vdots \\ 0 & & 0 & 1 \\ b_0 & & \cdots & b_{n-1} \end{pmatrix}$$

is

$$p(t) = (-1)^n(t^n - b_{n-1}t^{n-1} - \cdots - b_1 t - b_0).$$

PROOF. For $n = 2$ we have

$$\begin{aligned} p(t) &= \det\begin{pmatrix} -t & 1 \\ b_0 & b_1 - t \end{pmatrix} = -t(b_1 - t) - b_0 \\ &= t^2 - b_1 t - b_0, \end{aligned}$$

so the proposition is true for $n = 2$. Suppose that we knew that the proposition were true for all companion matrices of size $n - 1$ and that $\mathbf{B}$ is a companion matrix of size n. Then using Lagrange expansion with respect to the first column gives

$$p(t) = \det(\mathbf{B} - t\mathbf{I}) = \det\begin{pmatrix} -t & 1 & 0 & & 0 \\ 0 & -t & 1 & 0 & 0 \\ \vdots & \vdots & \vdots & \vdots & \vdots \\ 0 & \cdots & 0 & -t & 1 \\ b_0 & \cdots & \cdots & \cdots & b_{n-1} - t \end{pmatrix}$$

$$= -t \det\begin{pmatrix} -t & 1 & 0 & & 0 \\ 0 & -t & 1 & 0 & 0 \\ \vdots & \vdots & \vdots & \ddots & \vdots \\ b_1 & b_2 & \cdots & & b_{n-1} - t \end{pmatrix} + (-1)^{n+1} b_0 \det\begin{pmatrix} 1 & & 0 \\ & \ddots & \\ 0 & & 1 \end{pmatrix}.$$

Now notice that

$$\det\begin{pmatrix} -t & 1 & 0 & \cdots & 0 \\ 0 & -t & 1 & \cdots & 0 \\ \vdots & \vdots & \vdots & & \vdots \\ 0 & 0 & -t & & 1 \\ b_1 & \cdots\ b_{n-2} & b_{n-1} & & -t \end{pmatrix}$$

is the characteristic polynomial of the companion matrix

$$\begin{pmatrix} 0 & 1 & 0 & \cdots & 0 \\ 0 & 0 & 1 & \cdots & 0 \\ \vdots & \vdots & \ddots & & \vdots \\ 0 & & 0 & & 1 \\ b_1 & & \cdots & & b_{n-1} \end{pmatrix}$$

of size $n - 1$. By our inductive assumption we know its characteristic polynomial, and substituting this and also that the determinant of the identity matrix is 1 in the preceding equation gives

$$\begin{aligned} p(t) &= -t((-1)^{n-1}(t^{n-1} - b_{n-1}t^{n-1} - \cdots - b_1)) + (-1)^{n+1}b_0 \\ &= (-1)^n[t^n - b_{n-1}t^{n-1} - \cdots - b_1 t - b_0], \end{aligned}$$

establishing the proposition by induction. □

Theorem 18.4. *Consider the differential equation*

$$(*) \qquad \frac{d^n z}{dt^n} + a_{n-1}\frac{d^{n-1}z}{dt^{n-1}} + \cdots + a_1\frac{dz}{dt} + a_0 z = 0.$$

If the polynomial

$$p(\lambda) = \lambda^n + a_{n-1}\lambda^{n-1} + \cdots + a_1\lambda + a_0$$

has n distinct real roots, $\lambda_1, \ldots, \lambda_n$, then the general solution of (∗) *is*

$$z = c_1 e^{\lambda_1 t} + \cdots + c_n e^{\lambda_n t}.$$

PROOF. Set

$$\begin{aligned} y_1 &= z, \\ y_2 &= \frac{dz}{dt}, \\ \vdots \quad & \quad \vdots \\ y_n &= \frac{d^n z}{dt^n}. \end{aligned}$$

Then (∗) is equivalent to the linear differential system

$$\frac{d}{dt}\begin{pmatrix} y_1 \\ \vdots \\ y_n \end{pmatrix} = \begin{pmatrix} 0 & 1 & & 0 \\ \vdots & \ddots & & \\ 0 & \cdots & 0 & 1 \\ -a_0 & \cdots & & -a_{n-1} \end{pmatrix}\begin{pmatrix} y_1 \\ \vdots \\ y_n \end{pmatrix} = \mathbf{A}\begin{pmatrix} y_1 \\ \vdots \\ y_n \end{pmatrix}.$$

The characteristic polynomial of the coefficient matrix is (by (18.3))

$$p(\lambda) = \lambda^n + a_{n-1}\lambda^{n-1} + \cdots + a_0.$$

By hypothesis this has n distinct roots $\lambda_1, \ldots, \lambda_n$, so the coefficient matrix $\mathbf{A}$ is diagonalizable, say

$$\mathbf{A} = \mathbf{P}\begin{pmatrix} \lambda_1 & \cdots & 0 \\ \vdots & \ddots & \vdots \\ 0 & \cdots & \lambda_n \end{pmatrix}\mathbf{P}^{-1}.$$

The corresponding diagonal system

$$\frac{d}{dt}\begin{pmatrix} w_1 \\ \vdots \\ w_n \end{pmatrix} = \begin{pmatrix} \lambda_1 & \cdots & \\ \vdots & \ddots & 0 \\ 0 & & \lambda_n \end{pmatrix}\begin{pmatrix} w_1 \\ \vdots \\ w_n \end{pmatrix}$$

has as solution

$$\begin{aligned} w_1 &= c_1 e^{\lambda_1 t}, \\ \vdots \quad & \quad \vdots \\ w_n &= c_n e^{\lambda_n t}, \end{aligned}$$

where $c_1, \ldots, c_n$ are arbitrary constants, and

$$\begin{pmatrix} w_1 \\ \vdots \\ w_n \end{pmatrix} = \mathbf{P}^{-1}\begin{pmatrix} y_1 \\ \vdots \\ y_n \end{pmatrix}.$$

So, in particular

$$z(t) = y_1(t) = p_{11}c_1e^{\lambda_1 t} + \cdots + p_{1n}c_ne^{\lambda_n t}.$$

It only remains to show that none of the entries p_{1j} is zero, so that none of the constants $p_{1j}c_j$ vanish, and thus remain arbitrary. So suppose $p_{1j} = 0$. Then from the equation

$$y_2 = \frac{dy_1}{dt}$$

we get by substitution

$$\begin{aligned} p_{21}c_1e^{\lambda_1 t} + \cdots + p_{2j}c_je^{\lambda_j t} + \cdots + p_{2n}c_ne^{\lambda_n t} \\ = p_{11}c_1e^{\lambda_1 t} + \cdots + p_{2j}c_je^{\lambda_j t} + \cdots + p_{1n}c_ne^{\lambda_n t}, \end{aligned}$$

for all values of the constants $c_1, \ldots, c_n$. In particular, for $c_1 = \cdots = c_{j-1} = c_{j+1} = \cdots = c_n = 0$ and $c_j = 1$. This gives

$$p_{2j}e^{\lambda_j t} = 0$$

so $p_{2j} = 0$. Continuing in this way we obtain in succession $p_{3j} = 0, \ldots, p_{nj} = 0$. But this says that **P** has a column of zeros. So the determinant of **P** is zero, contrary to the fact that **P** is invertible. □

Let us try another example.

EXAMPLE 4. Solve the differential equation

$$\frac{d^3z}{dt^3} - \frac{dz}{dt} = 0.$$

Solution. The associated polynomial is

$$p(\lambda) = \lambda^3 - \lambda = \lambda(\lambda^2 - 1) = \lambda(\lambda - 1)(\lambda + 1).$$

So by (18.4) the general solution we seek is (remember $e^0 = 1$)

$$z = c_1e^{0t} + c_2e^t + c_3e^{-t} = c_1 + c_2e^t + c_3e^{-t}. \quad **$$

Up until now we have only considered linear differential systems

$$\frac{dy}{dt} = \mathbf{A}y,$$

where the coefficient matrix **A** is diagonalizable. We know, of course, that not all matrices are diagonalizable. What happens then? Here is where the Jordan canonical form comes into play! Let us begin by looking at a simple example.

EXAMPLE 5. Consider the differential system

$$\begin{pmatrix} \dfrac{dy_1}{dt} \\ \dfrac{dy_2}{dt} \end{pmatrix} = \begin{pmatrix} a & 0 \\ 1 & a \end{pmatrix} \begin{pmatrix} y_1 \\ y_2 \end{pmatrix},$$

where a is a fixed real number. The characteristic polynomial of the coefficient matrix is

$$p(\lambda) = (a - \lambda)^2$$

so a is a double root, and the coefficient matrix cannot be diagonalized. (Think this through!) On the other hand, we can easily solve this differential system. We have

$$\frac{dy_1}{dt} = ay_1 \quad \Rightarrow \quad y_1 = c_1 e^{at}.$$

If we put this into the second equation we find

$$\frac{dy_2}{dt} = c_1 e^{at} + y_2.$$

Consider $y_2 e^{-at}$, and compute its derivative by the product rule. We find

$$\begin{aligned} \frac{d}{dt}(y_2 e^{-at}) &= -ay_2 e^{-at} + (c_1 e^{at} + ay_2)e^{-at} \\ &= -ay_2 e^{-at} + c_1 + ay_2 e^{-at} \\ &= c_1. \end{aligned}$$

That is, $y_2 e^{-at}$ has a constant derivative. Therefore by the fundamental theorem of the calculus

$$y_2 e^{-at} = c_1 t + c_2,$$

or

$$y_2 = c_1 t e^{-at} + c_2 e^{at},$$

where c_2 is a second arbitrary constant. So the general solution is

$$\begin{aligned} y_1 &= c_1 e^{at}, \\ y_2 &= c_1 t e^{at} + c_2 e^{at}. \quad ** \end{aligned}$$

Here is another example.

EXAMPLE 6. We want to solve the equation

$$\frac{d^2 z}{dt^2} + \frac{2\,dz}{dt} + z = 0.$$

The equivalent linear differential system is

$$\begin{pmatrix} \dfrac{dy_1}{dt} \\ \dfrac{dy_2}{dt} \end{pmatrix} = \begin{pmatrix} 0 & 1 \\ -1 & -2 \end{pmatrix}\begin{pmatrix} y_1 \\ y_2 \end{pmatrix}.$$

The characteristic polynomial of the coefficient matrix $\mathbf{A}$ is

$$p(\lambda) = \lambda^2 + 2\lambda + 1 = (\lambda + 1),^2$$

which has $\lambda = -1$ as a double root. The matrix $\mathbf{A}$ is therefore not diagonalizable. (Because, if it were, the diagonal form would be scalar, and since scalar matrices commute with every matrix the equation $\mathbf{A} = \mathbf{PDP}^{-1}$ would yield $\mathbf{A} = \mathbf{D}$. But $\mathbf{A}$ is not itself diagonal.) On the other hand $\mathbf{A}$ has Jordan canonical form

$$\mathbf{J}(\mathbf{A}) = \begin{pmatrix} -1 & 0 \\ 1 & -1 \end{pmatrix},$$

and we saw how to solve

$$\frac{dw}{dt} = \mathbf{J}(\mathbf{A})w$$

In Example 5. To make use of this information we need the transformation matrix $\mathbf{Q}^{-1}$ such that

$$\mathbf{A} = \mathbf{QJ}(\mathbf{A})\mathbf{Q}^{-1}.$$

That means we must find the Jordan basis. This proceeds as in Chapter 17. Let $\mathsf{T}\colon \mathbb{R}^2 \to \mathbb{R}^2$ be the linear transformation represented by the matrix $\mathbf{A}$ with respect to the standard basis of $\mathbb{R}^2$. Let $\mathbf{F}_1 = (u, v) \in \mathbb{R}^2$ be a vector such that

$$\mathbf{F}_2 = \mathbf{A}\begin{pmatrix} u \\ v \end{pmatrix} + \begin{pmatrix} u \\ v \end{pmatrix} \neq \begin{pmatrix} 0 \\ 0 \end{pmatrix}.$$

Such a vector exists because $\mathbf{A}$ is not diagonalizable. Furthermore, since -1 is the only eigenvalue of $\mathbf{A}$ it follows $\mathbf{F}_2 \in \mathscr{L}(\mathbf{F}_1)$. Therefore, $\{\mathbf{F}_1, \mathbf{F}_2\}$ is a basis for $\mathbb{R}^2$. Moreover

$$\begin{aligned} \mathsf{T}(\mathbf{F}_1) &= \mathbf{F}_1 + \mathbf{F}_2, \\ \mathsf{T}(\mathbf{F}_2) &= (\mathsf{T} + \mathsf{I} - \mathsf{I})(\mathsf{T} + \mathsf{I})(\mathbf{F}_1) \\ &= (\mathsf{T} + \mathsf{I})^2(\mathbf{F}_1) - (\mathsf{T} + \mathsf{I})(\mathbf{F}_1) \\ &= 0 - \mathbf{F}_2, \end{aligned}$$

because $\mathsf{T} + \mathsf{I}$ is nilpotent of index 2. Thus $\mathbf{J}(\mathbf{A})$ is the matrix of T with respect to $\{\mathbf{F}_1, \mathbf{F}_2\}$. A possible choice for $\mathbf{F}_1$ is $(0, 1)$ whence $\mathbf{F}_2 = (1, -1)$. Therefore setting

$$\mathbf{Q} = \begin{pmatrix} 0 & 1 \\ 1 & -1 \end{pmatrix},$$

we have

$$\mathbf{A} = \mathbf{Q}\begin{pmatrix} -1 & 0 \\ 1 & -1 \end{pmatrix}\mathbf{Q}^{-1}.$$

We now proceed as in the diagonalizable case and set

$$\begin{pmatrix} w_1 \\ w_2 \end{pmatrix} = Q^{-1}\begin{pmatrix} y_1 \\ y_2 \end{pmatrix},$$

so

$$\frac{d}{dt}\begin{pmatrix} w_1 \\ w_2 \end{pmatrix} = \mathbf{J}(\mathbf{A})\begin{pmatrix} w_1 \\ w_2 \end{pmatrix} = \begin{pmatrix} -1 & 0 \\ 1 & -1 \end{pmatrix}\begin{pmatrix} w_1 \\ w_2 \end{pmatrix}.$$

Example 5 gives

$$w_1 = c_1 e^{-t},$$

$$w_2 = c_1 t e^{-t} + c_2 e^{-t},$$

and so,

$$\begin{pmatrix} y_1 \\ y_2 \end{pmatrix} = Q\begin{pmatrix} w_1 \\ w_2 \end{pmatrix} = \begin{pmatrix} 0 & 1 \\ 1 & -1 \end{pmatrix}\begin{pmatrix} c_1 e^{-t} \\ c_1 t e^{-t} + c_2 e^{-t} \end{pmatrix}$$

$$\begin{pmatrix} c_1 t e^{-t} + c_2 e^{-t} \\ c_1 e^{-t} - c_1 t e^{-t} - c_2 e^{-t} \end{pmatrix},$$

whence

$$z = c_1 t e^{-t} + c_2 e^{-t},$$

where c_1, c_2 are arbitrary constants. **

This example leads naturally to the differential system

$$\frac{dy}{dt} = \mathbf{J}_n(a)\,\mathbf{Y},$$

where

$$\mathbf{J}_n(a) = \begin{pmatrix} a & & & 0 \\ 1 & \ddots & & \\ & \ddots & \ddots & \\ 0 & & 1 & a \end{pmatrix}$$

is a typical Jordan block. What does the solution to such a system look like? That is easy. We proceed as in Example 5

$$\frac{dy_1}{dt} = ay_1$$

so

$$y_1 = ce^{at},$$

$$\frac{dy_2}{dt} = y_1 + ay_2,$$

so by the product rule (see Example 5 again)

$$\frac{d}{dt}(e^{-at}y_2) = c_1.$$

This says the function $y_2 e^{-at}$ is a constant. Therefore

$$y_2 e^{-at} = c_1 t + c_2,$$

and

$$y_2 = c_1 te^{at} + c_2 e^{at} = (c_1 t + c_2)e^{at}.$$

Next

$$\frac{dy_3}{dt} = y_2 + ay_3$$

so by the product rule

$$\frac{d}{dt}(y_3 e^{-at}) = y_2 e^{-at}$$

$$= c_1 t + c_2.$$

So by the fundamental theorem of the calculus

$$y_3 e^{-at} = \frac{c_1}{2} t_2 + c_2 t + c_3,$$

where c_3 is yet another constant. This says

$$y_3 = \left(\frac{c_1}{2} t^2 + c_2 t + c_3\right)e^{at}.$$

Continuing in this way gives

$$y_k = \left(\frac{c_1}{(k-1)!} t^{k-1} + \frac{c_1}{(k-2)!} t^{k-2} + \cdots + c_{k-1} t + c_k\right) e^{at}$$

for $k = 1, 2, \ldots, n$.

We may summarize this in a theorem.

Theorem 18.5. *The general solution to the linear system*

$$\frac{dy}{dt} = \mathbf{J}_n(a)y$$

is given by

$$y_k = \left(\frac{c_1}{(k-1)!}t^{k-1} + \cdots + c_{k-1}t + c_k\right)e^{at}, \qquad k = 1, \ldots, n,$$

where $c_1, \ldots, c_n$ are arbitrary constants. □

This result can now be applied to nondiagonalizable systems as the following example illustrates.

EXAMPLE 7. Solve the differential system

$$\frac{dy}{dt} = \begin{pmatrix} 2 & 2 & -1 \\ -1 & -1 & 1 \\ -1 & -2 & 2 \end{pmatrix}\begin{pmatrix} y_1 \\ y_2 \\ y_3 \end{pmatrix}.$$

Solution. The characteristic polynomial of the coefficient matrix **A** is

$$p(\lambda) = \det\begin{pmatrix} 2-\lambda & 2 & -1 \\ -1 & -1-\lambda & 1 \\ -1 & -2 & 2-\lambda \end{pmatrix} = \cdots = (\lambda - 1)^3.$$

There is therefore just one eigenvalue 1, of multiplicity 3. The Jordan normal form of **A** is (see Example 7 in Chapter 17)

$$\mathbf{J}(\mathbf{A}) = \begin{pmatrix} 1 & 0 & 0 \\ 1 & 1 & 0 \\ 0 & 0 & 1 \end{pmatrix}$$

and a corresponding Jordan basis for $\mathbb{R}^3$ is

$$\mathbf{F}_1 = (1, 0, 0), \qquad \mathbf{F}_2 = (1, -1, 1), \qquad \mathbf{F}_3 = (0, 1, 2).$$

So

$$\mathbf{A} = \mathbf{Q}\mathbf{J}(\mathbf{A})\mathbf{Q}^{-1},$$

where

$$\mathbf{Q} = \begin{pmatrix} 1 & 1 & 0 \\ 0 & -1 & 1 \\ 0 & 1 & 2 \end{pmatrix}.$$

If we set

$$\begin{pmatrix} w_1 \\ w_2 \\ w_3 \end{pmatrix} = \mathbf{Q}^{-1}\begin{pmatrix} y_1 \\ y_2 \\ y_3 \end{pmatrix},$$

then

$$\frac{dw}{dt} = \mathbf{J}(\mathbf{A})w = \begin{pmatrix} \mathbf{J}_2(1) & 0 \\ 0 & 1 \end{pmatrix} \begin{pmatrix} w_1 \\ w_2 \\ w_3 \end{pmatrix}$$

and working block by block we find

$$\begin{pmatrix} w_1 \\ w_2 \end{pmatrix} = \begin{pmatrix} c_1 e^t \\ (c_1 t + c_2)e^t \end{pmatrix}$$

and

$$w_3 = c_3 e^t,$$

where c_1, c_2, c_3 are arbitrary constants. Thus

$$\begin{pmatrix} y_1 \\ y_2 \\ y_3 \end{pmatrix} = Q \begin{pmatrix} w_1 \\ w_2 \\ w_3 \end{pmatrix} = \cdots = \begin{pmatrix} c_1 e^t + \dfrac{c_1}{2} te^t + c_2 e^t \\ -c_1 e^t - c_2 e^t + c_3 e^t \\ c_1 te^t + c_2 e^t + 2c_3 e^t \end{pmatrix}. \quad **$$

As a last example we consider

EXAMPLE 8. Solve the differential equation

$$\frac{d^3y}{dt^3} - \frac{d^2y}{dt^2} - \frac{dy}{dt} + y = 0.$$

Solution. The coefficient matrix of the associated linear system is

$$\mathbf{A} = \begin{pmatrix} 0 & 1 & 0 \\ 0 & 0 & 1 \\ -1 & 1 & 1 \end{pmatrix}.$$

The characteristic polynomial is (by (18.3))

$$\begin{aligned} p(t) &= (-1)^3(t^3 - t^2 - t + 1) \\ &= -(t-1)^2(t+1), \end{aligned}$$

so $\mathbf{A}$ has two distinct eigenvalues $\lambda_1 = 1$ and $\lambda_2 = -1$ with multiplicities 2 and 1. The Jordan form of $\mathbf{A}$ is therefore

$$\mathbf{J}(\mathbf{A}) = \begin{pmatrix} 1 & 0 & 0 \\ 1 & 1 & 0 \\ 0 & 0 & -1 \end{pmatrix},$$

Following the procedure of Chapter 17 we find the corresponding Jordan basis for $\mathbb{R}^3$ is

$$\mathbf{F}_1 = (0, 1, 2), \qquad \mathbf{F}_2 = (1, 1, 1), \qquad \mathbf{F}_3 = (1, -1, 1).$$

So if we set

$$\mathbf{P} = \begin{pmatrix} 0 & 1 & 1 \\ 1 & 1 & -1 \\ 2 & 1 & 1 \end{pmatrix},$$

then

$$\mathbf{A} = \mathbf{P}\mathbf{J}(\mathbf{A})\mathbf{P}^{-1}.$$

Set

$$\begin{pmatrix} w_1 \\ w_2 \\ w_3 \end{pmatrix} = \mathbf{P}^{-1}\begin{pmatrix} y_1 \\ y_2 \\ y_3 \end{pmatrix},$$

where as usual

$$y_1 = y,$$

$$y_2 = \frac{dy}{dt},$$

$$y_3 = \frac{d^2y}{dt^2}.$$

Our original differential equation is equivalent to the system

$$\frac{d}{dt}\begin{pmatrix} y_1 \\ y_2 \\ y_3 \end{pmatrix} = \mathbf{A}\begin{pmatrix} y_1 \\ y_2 \\ y_3 \end{pmatrix},$$

which in turn is equivalent to

$$\frac{d}{dt}\begin{pmatrix} w_1 \\ w_2 \\ w_3 \end{pmatrix} = \mathbf{J}(\mathbf{A})\begin{pmatrix} w_1 \\ w_2 \\ w_3 \end{pmatrix}.$$

By (18.5) the general solution of the linear system

$$\frac{d}{dt}\begin{pmatrix} w_1 \\ w_2 \\ w_3 \end{pmatrix} = \begin{pmatrix} 1 & 0 & 0 \\ 1 & 1 & 0 \\ 0 & 0 & -1 \end{pmatrix}\begin{pmatrix} w_1 \\ w_2 \\ w_3 \end{pmatrix}$$

is given by (work block by block)

$$\begin{pmatrix} w_1 \\ w_2 \\ w_3 \end{pmatrix} = \begin{pmatrix} c_1 e^t \\ (c_1 t + c_2)e^t \\ c_3 e^t \end{pmatrix}.$$

From the definition of (w_1, w_2, w_3) we thus find

$$\begin{aligned} \begin{pmatrix} y_1 \\ y_2 \\ y_3 \end{pmatrix} &= \mathbf{P} \begin{pmatrix} w_1 \\ w_2 \\ w_3 \end{pmatrix} \\ &= \begin{pmatrix} 0 & 1 & 1 \\ 1 & 1 & -1 \\ 2 & 1 & 1 \end{pmatrix} \begin{pmatrix} c_1 e^t \\ (c_1 t + c_2)e^t \\ c_3 e^t \end{pmatrix} \begin{pmatrix} (c_1 t + c_2)e^t + c_3 e^{-t} \\ c_1 e^t + (c_1 t + c_2)e^t - c_3 e^{-t} \\ 2c_1 + (c_1 t + c_2)e^t + c_3 e^{-t} \end{pmatrix}, \end{aligned}$$

so we conclude

$$y = y_1 = (c_1 t + c_2)e^t + c_3 e^{-t}$$

is the general solution to the original equation $(*)$. $**$

EXERCISES

1. Among all the solutions of the differential system

$$\frac{dy_1}{dt} = -y_2,$$

$$\frac{dy_2}{dt} = -y_1,$$

find a solution that satisfies the *initial conditions*

$$y_1(0) = 2,$$

$$y_2(0) = 0.$$

(*Hint*: First find all solutions. Then determine for which values of the constants the initial conditions are satisfied.)

2. Solve the differential system

$$\frac{dy_1}{dt} = y_2 + y_3,$$

$$\frac{dy_2}{dt} = y_1 + y_3,$$

$$\frac{dy_3}{dt} = y_1 + y_2.$$

Find a solution satisfying

$$y_1(0) = 1,$$
$$y_2(0) = 0,$$
$$y_3(0) = -1.$$

3. Solve each of the following differential equations

$$\frac{d^2y}{dt^2} + 3\frac{dy}{dt} + 2y = 0,$$

$$\frac{d^2y}{dt^2} - 3\frac{dy}{dt} + 2y = 0,$$

$$\frac{d^2y}{dt^2} - y = 0,$$

$$\frac{d^2y}{dt^2} + y = 0.$$

4. Solve each of the following linear systems

(a) $$\frac{d}{dt}\begin{pmatrix} y_1 \\ y_2 \\ y_3 \end{pmatrix} = \begin{pmatrix} 2 & -1 & -1 \\ -1 & 2 & -1 \\ -1 & -1 & 2 \end{pmatrix}\begin{pmatrix} y_1 \\ y_2 \\ y_3 \end{pmatrix}.$$

(b) $$\frac{d}{dt}\begin{pmatrix} y_1 \\ y_2 \\ y_3 \end{pmatrix} = \begin{pmatrix} 1 & 0 & 1 \\ 0 & 1 & 0 \\ -1 & 0 & 1 \end{pmatrix}\begin{pmatrix} y_1 \\ y_2 \\ y_3 \end{pmatrix}.$$

(c) $$\frac{d}{dt}\begin{pmatrix} y_1 \\ y_2 \\ y_3 \end{pmatrix} = \begin{pmatrix} 1 & 1 & 1 \\ 1 & 1 & 1 \\ 1 & 1 & 1 \end{pmatrix}\begin{pmatrix} y_1 \\ y_2 \\ y_3 \end{pmatrix}.$$

5. Solve each of the following differential equations

(a) $$\frac{d^3y}{dt^3} + \frac{d^2y}{dt^2} - \frac{dy}{dt} - y = 0.$$

(b) $$\frac{d^3y}{dt^3} + 3\frac{d^2y}{dt^2} + 3\frac{dy}{dt} + y = 0.$$

(c) $$\frac{d^3y}{dt^3} - 6\frac{d^2y}{dt^2} - 7\frac{dy}{dt} - 6y = 0.$$

In each case determine a solution satisfying the initial conditions

$$y(0) = 1,$$

$$\frac{dy}{dt} = 0,$$

$$\frac{d^2y}{dt^2} = 1.$$

Index

List of Notations

det $\mathbf{C}$	the determinant of the matrix $\mathbf{C}$	182
D	differentiation operator	74
$\Delta(t)$	characteristic polynomial	207
dim	dimension of	48
$\mathbf{E}_i$	$(0, 0, \ldots, 1, 0 \cdots 0) \in \mathbb{R}^n$; standard basis vector	85
$\mathbf{E}_{rs}$	the matrix with a 1 in the (r, s) position and zero otherwise	113
$\mathscr{F}(S)$	the set of all real valued functions defined on S	31
$\mathscr{F}(S, T)$	the elements of $\mathscr{F}(S)$ that vanish on $T \subseteq S$	31
$\mathscr{F}_{\mathbb{C}}(S)$	complex valued functions defined on S	31
$g\|_T$		80
I	the identity transformation	81
$\mathbf{I}$	the identity matrix	118
Im T	the image of T	72
ker T	the kernel of T	72
$\mathscr{L}(\ )$	the linear span of	22, 23
$\mathsf{L}_{k(x)}$		97
L	integration operator	74
$\mathsf{M}_{k(x)}$	multiplication operator by $k(x)$	100
$\mathscr{M}_{mn}$	the set of all $m \times n$ matrices	113
$\mathscr{P}(\mathbb{R})$	the set of all polynomials	28
$\mathscr{P}_n(\mathbb{R})$	the set of all polynomials of degree $\leq n$	27
$\mathbb{R}$	the real numbers	16
$\mathbb{R}^k$	the real Cartesian space	16
R_T		81
γ	the solution space	170
S	the shift operator	87
$\mathsf{T}_{\mathbf{A}}$		275
tr$(\mathbf{A})$	trace of $\mathbf{A}$	248
$\mathsf{T}(\mathscr{U})$		69
T_φ		86
$\mathsf{T}\|_{\mathscr{W}}$	restriction of T to the subspace $\mathscr{W}$	301
$\mathscr{V}_e$	the eigenspace corresponding to eigenvalue e	198
$\mathscr{V}_{xy}$	the xy plane	72
$\mathscr{W}^\perp$	the orthogonal complement of	265
χ_s	the characteristic function of s	37
$\varnothing$	the empty set	94
$\mathscr{V}, \mathscr{W}, \mathscr{U}, \mathscr{L}, \mathscr{T}$	vector spaces	21
$\mathsf{T}, \mathsf{S}, \ldots$	linear transformations	66
iff	if and only if	

(continued)

Martin: The Foundations of Geometry and the Non-Euclidean Plane.
Martin: Transformation Geometry: An Introduction to Symmetry.
Millman/Parker: Geometry: A Metric Approach with Models. Second edition.
Owen: A First Course in the Mathematical Foundations of Thermodynamics.
Palka: An Introduction to Complex Function Theory.
Peressini/Sullivan/Uhl: The Mathematics of Nonlinear Programming.
Prenowitz/Jantosciak: Join Geometries.
Priestly: Calculus: An Historical Approach.
Protter/Morrey: A First Course in Real Analysis. Second edition.
Protter/Morrey: Intermediate Calculus. Second edition.
Ross: Elementary Analysis: The Theory of Calculus.
Samuel: Projective Geometry.
Readings in Mathematics.
Scharlau/Opolka: From Fermat to Minkowski.
Sigler: Algebra.
Silverman/Tate: Rational Points on Elliptic Curves.
Simmonds: A Brief on Tensor Analysis.
Singer/Thorpe: Lecture Notes on Elementary Topology and Geometry.
Smith: Linear Algebra. Second edition.
Smith: Primer of Modern Analysis. Second edition.
Stanton/White: Constructive Combinatorics.
Stillwell: Mathematics and Its History.
Strayer: Linear Programming and Its Applications.
Thorpe: Elementary Topics in Differential Geometry.
Troutman: Variational Calculus with Elementary Convexity.
Wilson: Much Ado About Calculus.